Multigrammatical Framework for Knowledge-Based Digital Economy

Igor A. Sheremet

Multigrammatical Framework for Knowledge-Based Digital Economy

Springer

Igor A. Sheremet
National Computer Corporation
Moscow, Russia

ISBN 978-3-031-13860-7 ISBN 978-3-031-13858-4 (eBook)
https://doi.org/10.1007/978-3-031-13858-4

This Springer imprint is published by the registered company Springer Nature Switzerland AG
The registered company address is: Gewerbestrasse 11, 6330 Cham, Switzerland

Preface

This monograph introduces a new multiset-based conceptual, mathematical, and knowledge engineering framework for planning and scheduling in resource-consuming, resource-producing (industrial), and resource-distributing (economical) sociotechnological systems (*STSs*) in order to enable their smart operation not only in a "business-as-usual" mode, but also in highly volatile and hazardous environments. This framework, named multigrammatical framework (*MGF*), is the result of convergence and deep integration into a unified, flexible, easily applied, and effectively implemented formalism, operating on multisets, of several well-known paradigms from classical operations research and modern knowledge engineering: mathematical programming, optimal scheduling, and constraint programming. The mathematical background of the *MGF*, its algorithmics and applications, implementation issues, as well as its nexus with known models from operations research and theoretical computer science areas are considered. *STSs*' resilience and recovery issues are studied by applying the *MGF* toolkit, and special attention is paid to multigrammatical assessment of resilience of energy infrastructures. *MGF*-represented resource-based games, allowing modeling and simulation of various conflicts in which participants try to eliminate and/or capture one another's resources while their actions consume their own resources, are introduced. Directions of further development of the *MGF* are discussed.

This book is addressed to scholars working in the areas of theoretical and applied computer science, artificial intelligence, systems analysis, operations research, mathematical economy, and critical infrastructure protection, to engineers developing software-intensive solutions for implementation of the knowledge-based digital economy and Industry 4.0, as well as to students, aspirants, and university staff.

Foundational knowledge of set theory, mathematical logic, and routine operations on data bases is needed to read this book. The content of the monograph is presented from simple to complex, in a well-understandable step-by-step manner. Multiple examples and figures are included in order to support the explanation of basic notions, expressions, and algorithms.

Moscow, Russia Igor A. Sheremet

Acknowledgments

The author is thankful to his colleagues, Full Members of the Russian Academy of Sciences (*RAS*) President of the RAS Prof. Gennadiy Krasnikov; Vice-President of the RAS, Chair of the RAS Committee for Systems Analysis (CSA) Prof. Vladislav Panchenko; Prof. Yuriy Gulyaev; and Head of the *RAS* Section for Information Technologies and Automation Prof. Igor Sokolov, as well as to Prof. Vladimir Betelin, Prof. Aleksander Bugaev, Prof. Igor Bychkov, Prof. Yuri Chaplygin, Prof. Mikhail Fedoruk, Prof. Igor Kalyaev, Prof. Andrey Kokoshin, Prof. Arkadiy Kryazhimskiy, Prof. Alexander Kuleshov, Prof. Nikolay Kuznetsov, Prof. Alexander Latyshev, Prof. Yuri Popkov, Prof. Vladislav Pustovoyt, Prof. Alexander Saurov, Prof. Yuri Shokin, Prof. Alexander Sigov, Prof. Victor Soifer, and Prof. Alexander Stempkovskiy, for useful discussions and support. The author is especially thankful to Full Member of *RAS* and Vice-Chair of the *RAS CSA* Prof. Alexei Gvishiani for the promotion of the *MGF* in the international systems analysis community as well as for initiating the author's activities in the area of *MGF* application to resilience issues.

The author is forever deeply grateful to the Nobel Prize winner Full Member of the *RAS* Prof. Zhores Alferov for his attention and support.

Thanks a lot to Full Member of the *RAS* Prof. Evgeny Velikhov for his definite and decisive aid to the author at one of the toughest periods of his scientific career.

Heartfelt gratitude forever to Full Member of the *RAS* Prof. Yuri Borodakiy, who was an active promoter of the author's approach and activities.

The author is thankful to Prof. Mikhail Popov, Dr. Sergey Maev, Dr. Nikolay Makarov, Prof. Nikolay Turko, Prof. Vladimir Zolotarev, Prof. Vladimir Tyminskiy, Prof. Victor Nikolayevskiy, as well as to Dr. Lyudmila Obukhova for their sincere and long-term aid.

An extremely valuable collaboration which began in 2003 and resulted in the first papers dedicated to the multigrammatical framework, which have appeared at 2005, has been with Dr. Vladimir Muravnik, who had formulated a substantial basis of the possibility of application of objective-driven scheduling to a state-controlled sector of the Russian economy in the early 2000s. The author is forever grateful to him for sharing his ideas and proposals.

The author is obliged to his teachers Prof. Vladlen Lebedev, Prof. Victor Zakharov, and Prof. Gheorghiy Stel'mash for basic knowledge in the areas of theoretical computer science, operations research, and the most complicated segments of applied systems analysis.

The author is proud for his pupils to whom he has dedicated the best years of his life and who are his closest aids today—first of all, Dr. Sergey Lebed, Prof. Alexey Markov, and Dr. Evgeny Shaburov as his most diligent and successful followers—and also Dr. Roman Karasev, who has made the pilot software implementation of a practically applied subset of the *MGF* toolkit (Sheremet and Karasev 2018).

For a long time, the author has collaborated with the International Institute for Applied Systems Analysis (*IIASA*), being in 2014–2020 a member of the *IIASA* Scientific Advisory Committee, as well as with the Committee on Data (*CODATA*) of the International Science Council, being since 2018 Co-Chair of the *CODATA* Task Group "Advanced mathematical tools for data-driven applied systems analysis," and in 2020–2022 Chair of the *IIASA-CODATA* Joint Working Group on Open Data, Big Data, and Systems Analysis. This collaboration has enriched the author with an advanced vision of the most topical and complicated problems and mathematical tools of modern systems analysis. Special appreciations go to Prof. Pavel Kabat, Dr. Albert van Jaarsveld, Prof. Donald Saari, Prof. Michael Clegg, Prof. Mary Scholes, Prof. Jim Hall, Prof. Nebojsa Nakicenovic, Dr. Leena Srivastava, Dr. Elena Rovenskaya, Prof. Barend Mons, Dr. Simon Hodson, and Dr. Alena Rybkina for this extremely useful collaboration.

The author had short but remarkable communications with Prof. Noam Chomsky and Prof. Jeffrey Ullman, whose opinions and comments on the proposed multigrammatical framework contributed to the quality of this monograph.

The author extends sincere appreciations to Prof. Fred Roberts for a long-term and creative collaboration and deep understanding of the capabilities of the *MGF*. The author is also thankful to Prof. Javier Rubio-Herrero for his careful edition of the survey of the early results in the area of application of the *MGF* to various problems concerning resilience and recovery of industrial systems.

The author is thankful to Mr. Ronan Nugent and Dr. Alexandru Ciolan for their deep attention to this monograph at the decisive step of its preparation to publication, as well as to Mrs. Daria Saveleva and Mr. Aliaksandr Birukou who were active author's supporters during a long period of processing the book by the Springer Nature team. Thanks a lot to Dr. Phil Watson for his mindful and the most thorough copyediting of this book.

Cordial thanks to my family, and firstly to my mother Maria Andreyevna and my wife Elena, for their lifelong everyday patience and aid.

Contents

Chapter 1
Introduction

Work on this monograph was initiated in 2018, the year of Karl Marx's 200th anniversary. Marx's influence on the evolution of world economical science and its projection on global political processes are well known. Being historically the first thinker who understood in the most general terms the main laws which form the economical basis of capitalism and its political system, Marx subjected them to deep analysis, with results which were tested and, to a certain extent, validated by the future development of mankind and multiple attempts to establish alternatives to capitalist economical and political systems in various countries.

Today's global industry, economy, and policy are radically different from that which was in place while Marx was writing his famous *Das Kapital. Kritik der politischen Ökonomie* (Marx 2018). Currently, mankind exists in the form of a great number of distributed sociotechnological systems (*STSs*), interconnected by an Internet-based global information infrastructure (*GII*), which enables the collaboration of hundreds of billions of computerized technological devices, forming together the Internet of Things (*IoT*), and billions of people, interacting in both physical space and cyberspace (Baller et al. 2016; Bentley and Whitten 2007; Blanchard and Fabrycky 2010). The modern technosphere, created by mankind, provides a lot of new opportunities for improving the quality of life of humans all over the world and achieving ***Sustainable Development Goals*** (*SDGs*), as they were announced by the United Nations (Sustainable Development 2019; Reyers and Selig 2020).

However, there is a global problem to be solved in the nearest future, if people are really to apply effectively all the potential capabilities of the aforementioned technosphere. Namely, it is well known that until now manufacturers produce goods on the basis of a demand forecast, which very often diverges from reality. That is why a significant part of produced material objects remains for a long time in stores and shops, many of them never being acquired by consumers and, finally, being disposed of. It is hard to believe, but some **20% *(one fifth) of*** *the aforementioned* ***produced material objects (some 884 million tons!) are thrown away annually all over the world*** (Shchukin 2021). (*This is typical even for the most flexible market economies with their competing manufacturers. Due to the*

I. A. Sheremet, *Multigrammatical Framework for Knowledge-Based Digital Economy*, https://doi.org/10.1007/978-3-031-13858-4_1

availability of practically the same technologies, manufacturers produce practically identical objects and win competitions for market share by price dumping and/or successful marketing. The loser's production is, as a consequence, useless, and resources spent during these products' manufacture are, in fact, lost by mankind. Moreover, they essentially complicate the persistent global problem of waste management (Giusti 2009; Gollakota et al. 2020). *At the same time, there may be a shortage of some valuable products—first of all, food—in many places and areas where they are extra-desirable.*)

The rising global ***digital economy*** (*DE*) is expected to provide consumers with necessary goods ***on request***. To acquire goods, it would be (and today in a lot of cases already it is) sufficient to use an Internet portal of a manufacturer or a retailer to place an ***order by a query***, specifying what items are required, in what amounts, at what time (not later than or the earliest possible time), and to what locations they must be delivered. After receiving such a query, a management center of an *STS* creates an optimal (rational) schedule of order completion, based on current resources available (including tools of manufacturing), or detects impossibility of completion and replies to the query initiator. A reply to such a query would contain the cost of the order and, if it is impossible to complete the order by a specified time, the earliest possible time at which the order may be completed, what part of the order may be completed by the time specified in the query, or what additional amounts of resources are necessary for order completion by a specified time (if there is a lack of available resources, and a customer is expected to be able to find somewhere missing amounts of resources and to transfer it to an *STS*). After the consumer's payment, a technological process driven by a created schedule of order completion would start, resulting in the manufacture of necessary elements (spare parts) by *3D* printing, or, in some other way, their delivery to the appropriate facilities, the assembly of more complicated items, and so on, until the final objects specified in the initial query would arrive at the designated locations.

Internet trading, in fact, is the simplest implementation of the described process in the case that the aforementioned final objects were manufactured beforehand and stored somewhere. (*Some items may also be manufactured in advance to reduce the time necessary for order completion, but this would be done in a much more precise mode than in contemporary industry and trading.*)

A ***digital economy paradigm***, being the background of the above-described near-future smarter behavior of mankind, based on much more rational production-consumption, simplifies and unifies the representation of all possible *STSs*, which, independent of their functions, may be naturally described in a network-centric framework. Following this paradigm, a human society operates as a set of customers, generating requests (orders) for necessary material (physical) resources (artifacts) and via the *GII* transferring these orders to various trading and/or manufacturing facilities, where these resources may be simply selected (if they are already available at stores and shops) or produced (otherwise); by means of the global transportation infrastructure (*GTI*), joining together all available transportation networks, needed resources are delivered to requests-defined destinations (Gvishiani et al. 2018). Information resources are requested in the same manner, but their delivery to

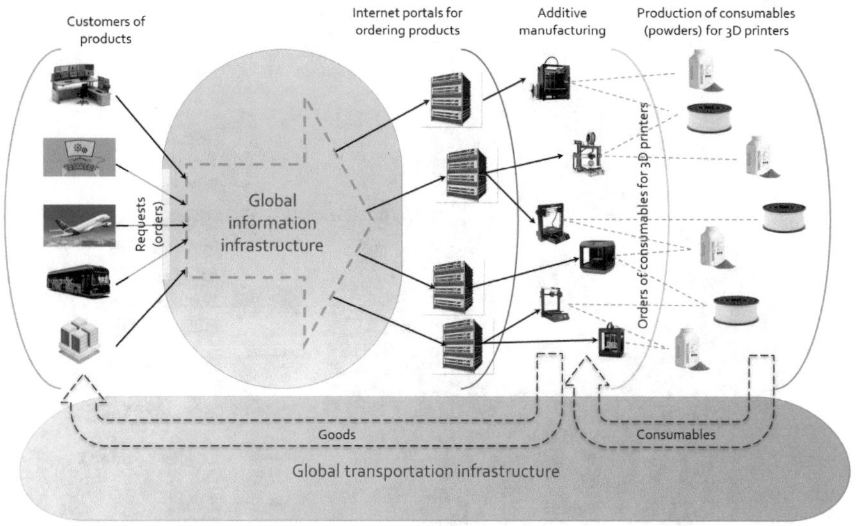

Fig. 1.1 Modern industry, based on additive technologies

customers does not require transportation and is implemented directly by means of the *GII*. Humans may be not only consumers but also participants of the described producing/delivery process in a "*Human-as-a-Device*" (*HaaD*) mode, and, in the general case, every human may play both roles.

Due to total robotization, cyberphysical manufacturing, massively based on additive technologies, and networking by means of the Industrial Internet of Things (*IIoT*), modern industry is much more productive and, what is especially important, ***more flexible***, than even 30 years ago (Colombo et al. 2014; Lee et al. 2015). Such flexibility is a result of the widespread application of *3D* printing as a foundation, which provides the fastest possible transfer from one set of types of produced objects to another by just changing their input *3D* representations (Fig. 1.1).

So the ***Industry 4.0*** concept, integrating all modern innovations in this area and being in fact a conceptual basis of the *DE*, demands a systemic consideration of a whole possible set of scientific and technological problems to be solved in order to implement all advantages which may be eventually available due to the emerging basic capabilities of a new technosphere (Schwab 2015; Wang and Wang 2016).

In fact, the aforementioned set of problems to be solved has only one ***supreme objective***—to create a ***"brain"*** of any *STS* (and, finally, of an ***entire global digital economy***)—which would enable rational interaction and cooperation of devices and humans involved in a process of completion of a flow of orders to this *STS* (Fig. 1.2).

Such a "brain," whose key component will be named below an ***STS scheduler*** (*STSS*), would be able to create schedules of completion of orders, incoming from consumers at unpredicted time moments and determining collections of resources to be produced and/or delivered to specified locations up to specified time moments. For this purpose, an *STSS*, processing an incoming order, would apply also

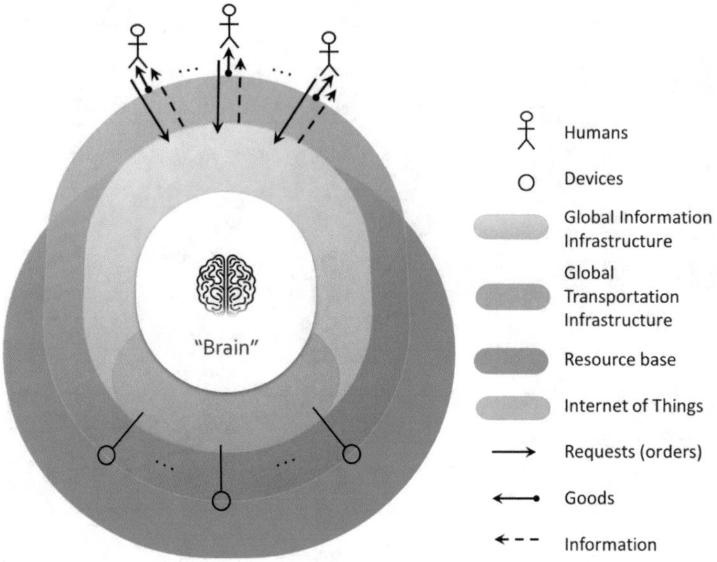

Fig. 1.2 Smart digital economy

knowledge about capabilities of all available devices, their current workload, and also amounts of resources currently available to these devices. So we shall consider below just such ***knowledge-based STSSs***. The knowledge base (*KB*) of any such *STSS* contains all the above-listed knowledge about the current set of devices of the *STS*, named its ***technological base*** (*TB*), as well as about its, in the general case, distributed, current collection of resources, named below its ***resource base*** (*RB*). We shall represent every device as a "black box" with input and output collections of resources, and a time interval needed for this device to produce an output collection given an input collection *(IC)*, if the latter is available from the *RB* (Fig. 1.3a). Here and above, we understand ***production*** in a maximally wide sense, covering, for example, relocation of physical objects (being, in fact, production of new couples <*object, its location*> from current ones by expending some resources for this aim). Hence, a "black box" in the general case is interpreted not only as a ***manufacturing device*** (*MD*) but as ***any activity*** changing a resource base by replacing some part of it by a new one.

By this assumption, the technological base of an *STS* operates asynchronously in such a way that any device may start and execute its operation cycle (*OC*), during which an output collection is produced, at any moment when an *RB* contains the amounts of resources making up its input collection (Fig. 1.3b). The produced output collection enters the *RB* without any delay at the moment when an *OC* ends. Orders completed by an *STS* may be represented as "black boxes" as well, with the only difference being that their output collections are not loaded to an *RB*, but are received by consumers, being the sources of these orders. Also, it is assumed that a collection needed by a consumer becomes available immediately once it appears in a resource

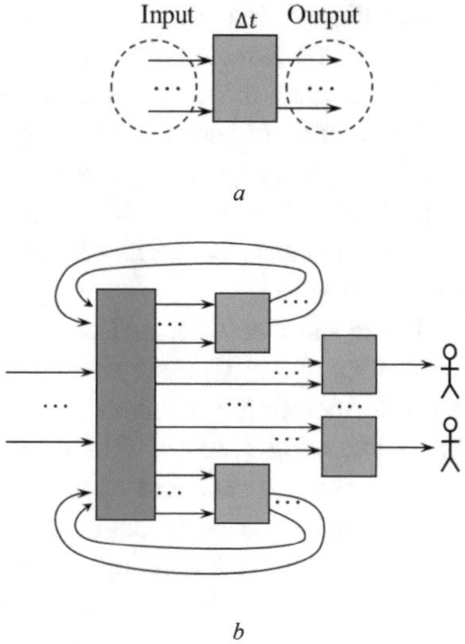

Fig. 1.3 Sociotechnological system: (**a**) device; (**b**) resource base, technological base, and orders

base, so the values of time intervals of such "black boxes" are null. A current ***STS schedule*** is, in fact, a set of schedules of its devices. A ***schedule of*** any ***device***, in turn, is a sequence of couples, each of which includes start and stop moments of its *OC*. A device is not able to begin its next operation cycle until the previous one is finished. In the general case, an *STSS* may correct (modify) already executed schedules when there are no other ways to complete some incoming orders. The activation of an *STSS* and execution of *STS* schedules created (corrected) by it by activating appropriate devices at proper time moments are implemented under the supervision of an ***STS controller*** (*STSC*), which is an *STS*-supervising entity enabling purposeful and coherent operation of an entire *STS* technological base.

This quite simple basic scheme is the background for consideration of a large variety of problems associated with the development of a DE "brain" and studied in this book.

No doubt, any *STS* and, as a whole, global digital economy would be expected to be smart, when it is initially targeted to implement such ambitious concepts as ***Smart City***, ***Smart Nation***, and finally ***Smart Planet***, fully meeting the ***SDGs***.

However, the notion of smartness may be related to either of two entities—to a sociotechnological system (digital economy, as a whole) or, more locally, to its scheduler. We assume that, ***to be smart, an STS would, obligatorily, be supported by a smart scheduler***, so we shall pay primary attention to smart *STSSs*.

We shall call an *STS* scheduler a ***smart*** one if, mainly, due to created schedules, a ***maximal possible number of incoming orders are completed with regard to the current technological and resource bases of the STS*** and, at the same time, it requires the ***minimal possible time for schedule creation*** and, if necessary, ***correction*** (let us recall Marx's "any saving is finally a time saving" (Marx 2018)).

Concerning orders which are not feasible, an *STSS* would create a demand for additional resources and/or devices, which would be included in the *RB* and the *TB* to complete these orders, or assess what delay regarding the term of completion of the order would occur if the order were to be completed without any extensions of the *RB* and the *TB*. The first case is inherent to so-called open STSs, which have the opportunity to request needed resources and devices from some external systems. The second case concerns ***closed STSs***, which do not have the aforementioned opportunity.

In the most complicated case, some ***devices in an STS TB may be capable of producing*** not only passive resources, whose only purpose is to be consumed during an *STS* operation, but also ***other manufacturing devices***, usually named ***tools of manufacturing*** (*ToMs*), able to produce the aforementioned passive resources as well as other *ToMs* and so on. Produced devices, thus, may belong to output collections of producing devices. From the moment they are produced, such devices are included in the technological base of an *STS* and are promptly taken into account by the *STS* scheduler, allowing enhancement of currently executing schedules or schedules associated with the next incoming order. So, in the general case, the technological base of an *STS* may also be changeable, like its resource base. Any produced device or its output may be an objective of a consumer's activity regarding an *STS* and thus may be included in an order.

However, even if smart in stable and somewhat comfortable conditions associated with "business as usual," any considered *STS* may fail under possible ***destructive impact (or a flow of such impacts)*** on either of its two components—technological, which may be affected by various accidental or deliberate events (malfunctions, cyberattacks, etc.), as well as human (*COVID-19* is the most topical example)—or on both of them (natural hazards, international sanctions, geopolitical conflicts, etc.) (Alcaraz and Zeadally 2015; He and Cha 2018; Li et al. 2005; Sheremet 2021a; Scheffer 2009; Scheffer et al. 2009). Any destructive ***impact may eliminate some resources from a resource base or may destroy some devices or reduce their capabilities*** (finally increasing the time period necessary for producing an output collection). In addition to the damage caused by ***initial impacts***, their destructive consequences may be essentially amplified by multiple ***cascade, or chain, effects***, being the result of numerous and very often non-evident interconnections and interdependencies between various remote from each other elements and segments of a modern deeply integrated technosphere formed by critical infrastructures (*CIs*)—*GII*, *GTI*, energy, industrial, banking and finance, etc. (Carreras et al. 2005; Eusqeld et al. 2011; Haimes and Jiang 2001; Kaper et al. 2022; Katay 2010; Kurian and McCarney 2010; Larsen et al. 2017; Liu et al. 2018; Lund et al. 2014; Mazher et al. 2018; Mills et al. 2011, 2012; Nepal and Jamasb 2013; Pederson et al. 2006; Rinaldi et al. 2001; Rehak et al. 2018, 2019; Sheremet 2019c, 2020e, 2021b; Stergiopoulos et al. 2016; Vespignani 2010; Werner 2017). So ***smartness of any STSS in a general case presumes its operation not only in rather stable and predictable but also in extraordinary, hazardous, and highly volatile conditions,*** and this presumption makes the task of creating an *STS* "brain" much more complicated compared to the case of a stable environment. In fact, any *STSS* would be capable of ***contributing to the resilience of a sociotechnological system*** affected by a destructive impact by a ***prompt correction of the current schedule*** and, if necessary and possible, by interaction with external systems that are capable of recovering eliminated (destroyed) resources or participating in completion of those orders interrupted by the impact.

In the most general terms, ***resilience*** is the ability of an *STS* to withstand and recover from disruption (Kaper et al. 2022). We shall say that an *STS* affected by some destructive impact is ***resilient*** to this impact if, as a result of an *STSS* operation, ***no order*** from all orders being processed by this affected *STS* at the moment of impact ***becomes unfeasible***. Otherwise, we shall call the *STS* processing the aforementioned orders ***vulnerable*** to the impact.

It is evident that in order to contribute to the resilience of an *STS*, its scheduler would be capable ***of changing the current STS schedule*** by adapting it to the current capabilities of the affected *STS*, i.e., to the resource base and technological base that remain after the impact. This response may be implemented by a "***rescheduling***" *STS*, which would be concerned with schedules of orders already being processed and, in the case of a flow of distributed impacts, executing this job immediately after any applied impact. In the case of a ***vulnerable closed STS***, its scheduler would solve the task of completion of all or the most valuable orders despite a reduced *RB* and/or *TB*. In the case of a ***vulnerable open STS***, its scheduler would be able to determine a set of resources and devices to be requested from some external friendly systems to recover its *RB* and/or *TB* or to determine some parts of orders which have become unfeasible as a result of an impact, to be completed by such friendly *STSs*.

The described generalized scheme of operation of *STSs*, supported by smart *STSSs*, may serve as the background of an effective implementation of ***large-scale objective-driven projects***, which until now have been associated with such conventional tools of project management as network schedules and various forms of their graphical display (Gantt charts (Klein 1999; Geraldi and Lechter 2012), arrow diagrams, precedence diagrams, etc.). The ideology of such projects in the previous period of life of the technosphere, synchronized with the earliest steps of computer science, was well developed in the most advanced countries of the twentieth century, and their implementation contributed to the greatest space and nuclear programs. In the Soviet Union, one such ideology was called "***program-objective planning***" (Pospelov and Irikov 1976; Raizberg and Lobko 2002), while in the United States, a similar approach was called *PPBS* (***planning, programming, and budgeting system***) (Don Vito 1969) or ***purposeful planning*** (Ackoff 1981), and nowadays, it is known most often as ***output budgeting*** (Robinson 2007). A principal advantage of the application of the described approach to project implementation is that ***smart STSSs*** may create schedules of *STSs* capable of ***producing not only passive (consumed) resources*** but also ***active resources (manufacturing devices)*** *and* ***of operating under flows of destructive impacts***, breaking and shifting initial schedules by eliminating from them some passive and active resources. Another principal advantage of the described approach is that it is ***resource-based***, i.e., ***operating on primary knowledge*** available to *STS* personnel, ***not on some intermediate data*** (like the aforementioned charts and diagrams) obtained as a ***result of preprocessing this***

knowledge, and thus really has the potential to do the job of the *STS* staff, but with much better quality.

However, evidently, the ***smartness of any STS depends on more than the quality of its scheduler***. The latter may be the finest of all possible, but the ***humans*** in this system may operate in such a way that, generating a flow of requests to an *STSS* and obtaining needed assets, they ***never achieve their inherent objectives*** and thus, in fact, they do not assess the *STS* they belong to as a smart one. Thus, there must be introduced and applied a ***wider understanding of smartness***—namely, regarding not only *STS* schedulers but regarding ***entire sociotechnological systems and their operation***. A task now is ***how to apply smart STSSs to make the STSs supported by them smart***. This understanding needs much deeper investigation of various features of modern *STSs* and their rational behavior in the sense of the established criteria of the aforementioned rationality and ways to meet these criteria.

There are a lot of ***new challenges, born from an arising global digital economy and global digitalization***.

A total networked robotization of manufacturing and services (transport and trading being the most demanded) is making useless numerous *HaaD* professions in modern sociotechnological systems—plant and factory workers, car drivers, and shop personnel. This, in turn, generates ***unemployment*** to such an extent that, perhaps, 50% or more of today's working class, who were referred by Marx as ***proletarians***, will become jobless. At the same time, a rather numerous socium of "proletarians of intellectual labor" (*knowledge workers*), or a ***"cybertariat"*** (Huws 2014), is arising—various *IT* personnel, operating and developing various segments of the global information infrastructure as well as of all other critical infrastructures and, in all, a global digital economy. Another socium of "proletarians of intellectual labor," functionally close to the cybertariat, is the ***"cognitariat"***—scholars, engineers, and technicians producing new knowledge about nature, the technosphere, and mankind and applying this knowledge to the development and implementation of new, more effective, and at the same time ***nature-friendly*** (eco-friendly (Bonds and Downey 2012)) and in some cases even ***nature-like*** (Kovalchuk et al. 2019)) sociotechnological solutions.

So the first question to policymakers of the near future would be: ***how to handle such large-scale and unpredictable unemployment***, which, as the *COVID-19* pandemic has demonstrated, may be additionally amplified by a global human-targeting impact? Perhaps an advanced technosphere, supported by a much cheaper and ecologically clean ("green") energy industry (Rodrik 2014), will be so flexible and efficient that it will produce all resources necessary for a decent life for all these jobless people? (Much of this we may see today in the most advanced countries, where unemployment benefit is paid to tens of millions of jobless people. Moreover, a social experiment implemented by the Icelandic government has verified that a 4-day 35-h working week is already possible (Kuznetsov 2021).) If so, a reasonable question arises: has maybe mankind or, at least, some of its numerous sociums at this new stage of development of a global technosphere already came close to ***communism*** as imagined by Marx and whose main law is "***from each according to his ability, to each according to his needs***"? Or has mankind arrived at ***cybersocialism***, of which the first edition was developed by Stafford Beer (1959, 1988) and partly implemented in Chile by Salvador Allende as the Cybersyn/Cybernet project? However, as this does not make the object of this book, we leave the answer to such questions to those readers passionate about economics, history, and politics.

A widely discussed shift "***from ownership to usership***" inspired by the global expansion of *Mobility as a Service (MaaS)* (Baziari 2021), whose background is carsharing, is in fact a remarkable pointer to such stage of the mankind development when the technosphere at its current state makes usership more rational than ownership, and thus property in a form of devices and resources in a natural way evolutionary becomes more and more common than private. Along with such globally addressed and "high-flown" ***abstract*** issues concerning a current step of mankind's history, there are a lot of not less important "down-to-earth" ***concrete*** issues, permanently considered by governments and businesses all over the world.

What socio-political consequences would follow after a mass deployment and putting into operation of the specifically novel elements of a technosphere in some country, region, or corporation? What maximally possible level of the aforementioned human needs may be supported by a specific *STS* given an available resource base? How to optimize a currently operating *STS* or what additional amounts of resources and/or devices would it require to satisfy a given level of needs? What additional pressure on the environment would follow this additional supply? How to achieve both economically and ecologically optimal *STS* operation? Would it be reasonable to establish some regulations limiting the amounts of resources available to individuals and/or sociums, and if so, then how to determine such amounts and to implement the respective limitations? How to choose developers and suppliers for the design, creation, maintenance, and optimization of an *STS*? More generally, how to develop and create an *STS* possessing some predefined properties? And finally, how to invest money in various possible projects to get the most profit and avoid possible risks?

All these issues, whose list may be extended much further, concern a ***higher level of STS control regarding their scheduling***, which, as we have presumed above, may be and would be implemented by *STSSs*. We shall address this level as ***planning*** and presume that the ***smartness of an STS would depend, finally, on the quality of the planning***, which, in turn, is an activity aimed at ***STS goal setting***, while scheduling is an activity aimed at the ***achievement of already established STS goals***. This capability may be implemented by the application of an *STSS* in a ***"what-if"*** mode, so the ***"brain" of a smart STS would provide both planning and scheduling***. Similar to a human brain, whose right hemisphere is responsible for abstract thinking and left for concrete thoughts, an *STS* "brain," as it is evident now, would also consist of two hemispheres (Fig. 1.4), one responsible for planning and the other for scheduling.

Planning ("abstract" thinking) is usually performed by an *STSs* management staff over some more or less long time intervals and associated with the integral parameters of an *STS* operation, particularly the total amounts of resources consumed and/or produced by a system.

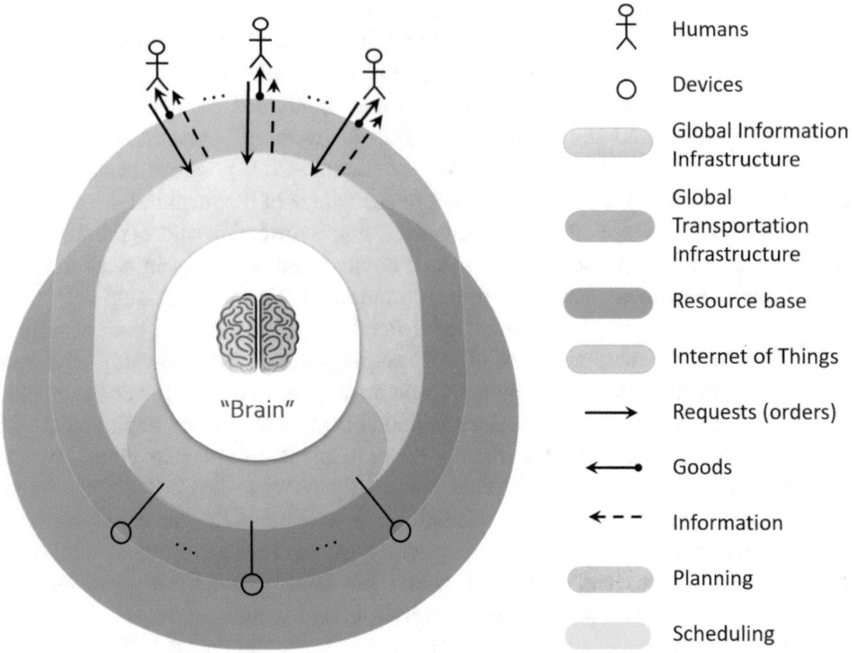

Fig. 1.4 Planning and scheduling in a digital economy

In fact, any operating smart *STSS* provided with a proper knowledge base would be able to implement a ***Government-as-a-Service*** **(*GaaS*)** concept, replacing a lot of low-level office personnel as well as low-skilled policymakers and politicians, who are an initial source of poor decision-making, which practically always and everywhere results in the setting of the wrong goals, i.e., ***the establishment of non-achievable objectives***. If decision-makers operating in areas of policy and economy were able to obtain ***promptly in a "what-if" mode*** results of an ***adequate predictive assessment of the consequences of their declared objectives*** (and, thus, explication of the behavior of controlled *STSs* trying to achieve these objectives), then, obviously, the ***quality of governance would be certain to rise***.

Non-achievable objectives would be identified promptly and would be explicated to decision-makers with an understandable explanation. At the same time, ***achievable goals*** would be provided by provable schedules of *STS* operation, and all that needed to be done during an implementation of these schedules would be their online local correction by *STSSs* according to possible local deviations of the technological and resource bases of the *STSs* and their environment. There would be a ***radical shift in the proportions of efforts spent on objective setting and objective achievement***. If now perhaps no more than the two or three most evident and thus most often non-achievable objectives are considered by decision-makers, and practically all their time is spent on current *STS* control, trying to make achievable a wrongly determined goal, then by application of adequate *STSSs*, this situation may radically change, thus providing choice and formulation of much more rational objectives due to the opportunity for an adequate assessment of a full set of variants.

Here, it would be appropriate to recall that the development of a highly interconnected global information infrastructure and a *GII*-based technosphere have made extremely important the problem of ***assessment of consequences of possible destructive impacts on elements and segments of the global economy***, and this problem concerns not so much accidental (nature-triggered), but mostly deliberate impacts. In this case, one who wants to solve a disagreement (conflict) with his counterpart by application of force must, at minimum, ***compare the resources which he would consume*** for the preparation and implementation of impacts (diplomatic, economical, military, etc.), ***with the benefits he would obtain*** as a result of these impacts: from the simplest case, when there would be no response from the object of an impact, to more and more sophisticated cases, corresponding to ***branching chains of mutual impacts***, each depending on the results of previous ones. The high interconnectivity of the modern technosphere as well as of the global economy makes this problem extremely topical.

On the other hand, if some destructive action were to achieve its target, and some objects or segments of a technosphere become disabled, the staff of the affected side must promptly react to this situation by developing an ***optimal plan of STS recovery*** and its return to normal operation. Today, processes which take place in an industry, economy, or polity are so dynamic that necessary decisions must be made in hard real time, and all possible chains of actions which may be performed by conflicting sides must be simulated with a high degree of confidence for an assessment of all possible results of a confrontation. In the best case, this simulation would be executed **a priori**, in a predictive mode, and exact, unambiguous, easily interpreted results would be available to a decision-making staff well before some real extreme situations would occur. This implies the ***hardest "what-if" application of STSSs***, covering issues usually considered in such advanced area of operations research (*OpR*) as ***game theory***.

Let us note that since a computer defeated the human world chess champion, it has become clear that there is no need to apply human intelligence to the solution of well-defined combinatorial problems (including those which have their place in economical and political spheres). Much more efficient is to implement such human-computer interaction where a human's role would be reduced to a ***formulation of objectives***, done by decision-makers, while a computer's (namely, *STSS's*) work would be search and selection in a full space of possible strategies permitting the achievement of an established objective. ***One who would possess and apply a smart STSS, given a minimal time for creation of a comprehensive strategy for achievement of a formulated goal, would be a winner in all possible competitions of the future.***

Let us note also that, following the concept of knowledge-based scheduling, we do not follow the conventional for an *AI* understanding of a *KB* as a store of ***heuristic expert-originated implications***, which would be extracted from experts by knowledge engineers and then would be applied by a knowledge interpretation engine (expert system, *ExS*) to solve the problem in question (Kendal and Green 2007). The efficiency of any decision support system implementing such an approach is crucially dependent on the expert's skills, experience, and subjectively understood ways of solving specific problems. So, as this varies from one to another expert or even expert community, knowledge engineers responsible for *KB* maintenance may obtain different logics of *STS* scheduling and different results in practical situations.

The approach described in this monograph is based on ***fundamentally different techniques of extraction of knowledge*** to be applied to various tasks expected to be solved by *STSSs*. Let us recall that the conceptual background of the aforementioned approach to the design of smart *STSSs* is understanding an *STSS's* knowledge base as a set of "black boxes" implicitly interconnected via a resource base, each representing some device consuming some input collection of resources and producing another collection during some adequate time interval (let us recall once more that we understand "producing" in a wider sense than "manufacturing" that permits correct consideration of a much wider set of systems and their activities). The role of knowledge engineers in this approach is limited to checking that ***all devices and resources physically forming part of the STS are represented in the KB by their "digital twins"*** (*DiTs*). (It is presumed that knowledge about any device or resource item at some moment entering a *TB* or an *RB* is immediately included in the *KB* and any change of their actual status leads to a respective change in the *KB*.) There is no place for any ambiguity or subjectivity in such a *KB*, and any schedule created by an *STSS* enables the completion of a consumer's order, which, in turn, strictly defines a collection of resources to be produced, as well as terms of production and, eventually, amounts of resources which may be joined by a customer to the *STS* resource base for the completion of this order (in the simplest case, such resource may be money). Similarly, ***there is no place for any heuristics*** in the process of schedule creation by an *STSS*, and all reduction of a redundant search during this process would be done by cutting off branches provably generating deadlocks and not leading to a solution. In fact, an explication of the aforementioned resource interconnections between devices involved in order completion in the form of an *STS* schedule is a core intellectual function of an *STSS*. By such interpretation of a knowledge base, an order, and a solution search, the approach described here is conceptually close to the ideology of such segments of operations research as scheduling theory (first of all, scheduling of resource-constrained projects (Klein 1999; Trojet et al. 2011; Abdolshah 2014)), but covers ***essentially wider area of STSs and associated problems***.

All that is said above is, clearly, sufficient to affirm that ***development of a theoretical (conceptual, mathematical, and knowledge engineering) framework for the design and application of knowledge-based schedulers, ensuring smartness and resilience of advanced sociotechnological systems and, as a whole, the future global digital economy, is one of the most important, valuable, and ambitious tasks of modern computer science.***

At present, ***there exist two different approaches which may be applied to the creation of schedulers of smart STSs of any nature.***

The first, originating from the ***operations research*** area, is based on the application of various optimization models developed in the mathematical programming, game theory, scheduling theory, and other *OpR* segments (Cavdaregla et al. 2013; Conway et al. 2003; Dabrowski and Hunt 2011; Dabrowski et al. 2011; D'Andrea and Dullerud 2003; Dullerud and Paganini 2005; Gass and Assad 2005; Harrington 2014; Hemmecke et al. 2009; Hermann 2006; Hespanha et al. 2007; Hillier and Lieberman 2014; Karlot 2005; Lasdon 2013; Leung 2004; Papayoanou 2010; Taha 2016). The main advantage of this approach is the ***provability of solutions of formulated problems***: an algorithmics of any such model enables search for a strictly optimal solution of a problem, which is why there is no necessity for decision-makers to verify and/or validate results obtained from a software/hardware engine implementing this algorithmics. Another advantage of this approach is the application of multiply improved and refined algorithms which provably reduce redundant search in a solution space without any losses of possible solutions. However, to obtain such optimal solutions, it is necessary to formulate a considered ***practical*** problem in such a way that it would be correctly represented as a specific ***OpR theoretical*** problem. As is well known from long-term experience, ***many real practical problems are difficult to embed in abstract theoretical OpR models***, thus requiring various extensions and modifications of classic theoretical models.

For this reason, an alternative approach, originating from the ***artificial intelligence*** (*AI*) area, was proposed, and multiple general-purpose or somewhat specialized knowledge representation models (*KRMs*) were developed that allow the search for solutions by logical inference, implementing the application of currently available knowledge to input data (Akerkar and Sajja 2010; Apt 2003; Bratko 2012; Cross 2017; Frunkwirth and Abdennadher 2003; Kendal and Green 2007; Pannu 2015; Sheremet 1994, 2013; Wallace 2002; Yeoh and Yokoo 2012). This approach, whose mathematical background is, finally, mathematical logic, is free of the *OpR* initial limitations, and for this reason, it is indeed more flexible and potentially more widely applicable. However, the aforementioned flexibility is connected with great implementational difficulties, arising from the so-called combinatorial explosion, which is an implication of the necessity of a total search of the solution space (Abraham et al. 2010; Gondran and Minoux 2008; Roberts 2009). To avoid such difficulties, various heuristic-based problem-oriented models and inference engines associated with them were developed (Belardinelli and Argento 2017; Klavins et al. 2006; Marriott and Stucky 2003; Mesbahi and Egerstedt 2010; Olfati-Saber et al. 2007; Pena et al. 2019; Sycara 1998; Waldrop 2018), and multiple successful application attempts are known (Rzevski and Skobelev 2014; Sainter et al. 2000; Salamon 2011; Schatten et al. 2015; Shoham and Leyton-Brown 2009; Wooldridge 2009). However, heuristically obtained, i.e., ***not provable in a strict mathematical sense***, results of application of such models and engines require additional validation and postprocessing by decision-makers, thus sometimes reducing the value of this approach.

This monograph introduces a new direction in modern computer science—***multiset-based knowledge engineering for planning and scheduling in sociotechnological systems***—in order to enable their smart operation in highly volatile and hazardous environments (Sheremet 2010, 2011a, 2016a, 2018, 2019a, b, 2021a; Sheremet and Karasev 2018). A family of syntactically, semantically, and pragmatically interconnected knowledge representation models named ***multiset grammars*** (multigrammars, *MGs*) is joined to form the ***multigrammatical framework (MGF)***, which is described in this book. This is the result of convergence and deep integration into a unified, flexible, easily applied, and effectively implemented formalism (treating multisets similar to the way that classic string-operating grammars (Chomsky 2005) treat strings) of ***several well-known paradigms from classical operations research and modern knowledge engineering***: mathematical programming, game theory, optimal scheduling, logic programming, and constraint programming. The main objective of the aforementioned convergence and integration was to create a ***general-purpose KRM initially oriented to a simple and natural representation of STSs, orders, and impacts***, and implemented by an inference engine, providing, as in classic models of *OpR*, ***provable solutions*** not requiring regular a posteriori postprocessing by decision-makers.

A multiset (*MS*) is a generalization of the classical notion of a set, which is defined as a collection of distinguishable items (elements). Unlike a set, an *MS* may contain indistinguishable (countable or measured) elements, thus providing an opportunity for processing numeric and symbolic information by one and the same unified multiset-centered toolkit (Blizard 1989; Hickman 1980; Lake 1976; Meyer and McRobbie 1982; Petrovskiy 2002, 2003, 2018; Red'ko et al. 2015; Singh et al. 2007). In fact, programming by multiset transformation was introduced in Banatre and Le Metoyer (1993), while a notion of a constraint multiset grammar as a tool for description and recognition of visual objects with complex structure appeared in Marriott (1994, 1996) and Marriott and Meyer (1997). ***The author's approach to the definition, development, and application of multiset grammars is driven by the above-described scope and needs of smart and resilient STSs,*** which, interconnecting and interacting with one another, would form in the near future a ***smart and resilient digital economy***.

The content of this monograph is as follows. Classes of sociotechnological systems—resource-consuming, resource-producing (industrial), and resource-distributing (economical)—and associated problems to be solved are introduced in Chap. 2. A mathematical theory of recursive multisets whose main contents are syntax, semantics, and formal mathematical properties of various classes of *MGs* is considered in Chap. 3, which fully represents the main feature of *MGs* as a new non-conventional model of computing—***goal- and knowledge-driven computation by ubiquitous multiset generation***. *MGs'* enhanced algorithmics, enabling development of *MG*-centered *STS* schedulers, implemented in conventional and highly parallel (multi-agent-based) computing environments and effectively operating on very large knowledge bases (*KBs*), is considered in Chap. 4. (Such initial attention to algorithmic issues of the *MGF* is explained by their key role in the whole proposed approach: it is evident that lack of techniques providing essential reduction of the computational complexity of the application of a current *KB* to an incoming data flow would make meaningless all attempts at development and useful dissemination of *MGF*-based software and hardware.) Chapter 5 is dedicated to multigrammatical modeling of resource-consuming *STSs*. Industrial *STSs* are considered in Chap. 6, while economical *STSs* along with the multigrammatical models of some valuable problems of an economical combinatorics are addressed in Chap. 7. Resilience and recoverability issues, including cascade effects, in application to industrial *STSs* and key issues of the assessment of resilience of energy infrastructures are considered in Chap. 8. Resource-based games *(RBGs)* as a flexible and general tool for modeling various conflicts and cooperation of, primarily, economical *STSs* are introduced in Chap. 9. *RBGs* may be applied efficiently at any real situation usually shifting upon a time in both directions along the axis “cooperation-competition-confrontation-conflict.” Multigrammatical representations of classic problems of operations research (optimization on networks and flows, integer linear programming, multiobjective Boolean programming, etc.) are described thoroughly in Chap. 10, illustrating the *MGF's* capabilities to provide solutions of various already known combinatorial problems. Interconnections of the *MGF* with well-known models of computation and knowledge engineering tools (systems of vector addition/subtraction, Petri nets, string-operating grammars) are investigated in Chap. 11. The rest of the book is Chap. 12. A generalized scheme of an *MGF*-centered smart *STS* and its operation in a hazardous environment under supervision of a knowledge-based *STS* controller is introduced there, as well as practically oriented issues of accumulation and maintenance of very large *MGF* knowledge bases. Also, directions of future development of the multigrammatical framework (including its hardware implementation on highly parallel and unconventional computational bases) are discussed. All chapters dedicated to the description of possible applications of the *MGF* should be perceived by an interested reader as no more than a primary illustration of its capabilities.

Chapter 2
Sociotechnological Systems and Associated Problems

To consider the main problems associated with the development of smart and resilient sociotechnological systems, we shall begin from some primary ontology based on some key features of all possible *STSs*. In this chapter, we shall introduce *three basic classes of STSs* regarding operations they execute on available resources: resource-consuming, resource-producing, and resource-distributing. Any real *STS*, in fact, may combine some or all basic features of these three classes, including subsystems of any classes. (Let us note here that any human, in the general case, may be present in at least two segments of any *STS*—resource-consuming and resource-producing—being a goods consumer as an element of the first one and a goods producer as an element of the second (Sheremet 2019c).) However, to formulate and formalize tasks to be solved regarding real *STSs*, it would be convenient and useful to do this work sequentially step by step in accordance with an increasing complexity of systems.

2.1 Classes of Sociotechnological Systems

Any sociotechnological system, being the join of social (*humans*) and technological (*devices*) components (segments), is a collection of three components (Fig. 2.1):

1. Organizational structure
2. Technological base
3. Resource base

*The **organizational*** (or organizational staff) ***structure*** of an *STS* represents the set of its subordinated positions filled by real persons carrying out their functions associated with their areas of responsibility.

*The **technological base*** of an *STS* includes various technological equipment (*TE*) and devices assigned to the aforementioned positions—and, thus, to real persons—and enabling their efficient performance. In the general case, a *TB* may include

I. A. Sheremet, *Multigrammatical Framework for Knowledge-Based Digital Economy*, https://doi.org/10.1007/978-3-031-13858-4_2

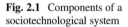

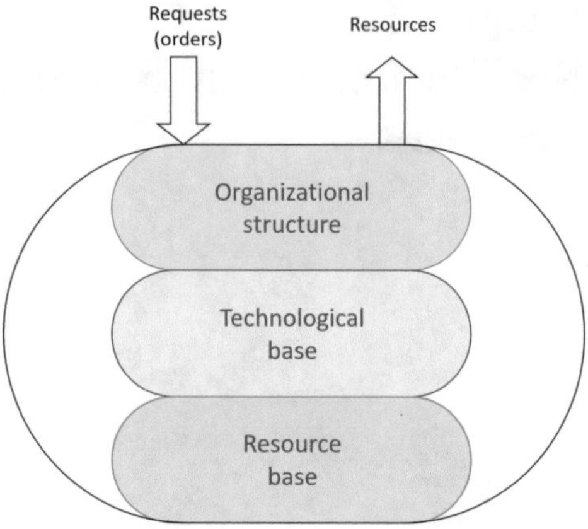

Fig. 2.1 Components of a sociotechnological system

various manufacturing devices and facilities producing some physical (material) objects (including other *MDs* and systems). This segment of a *TB* will be called a ***manufacturing technological base*** *(MTB).*

The ***resource base*** of an *STS* is nothing but various resources consumed by humans, devices, and facilities and necessary to all of them for operation.

From the most general point of view, technological devices and facilities capable of manufacturing various objects may be considered as some kind of resources, usually called ***active***, thus distinguishing them from ***passive*** resources, which are only consumed during a manufacturing process and delivered to consumers. By this generalization, a technological base may be considered as a part of a resource base. In what follows in this book, in some cases, it will be convenient to use this interpretation of a *TB*.

Let us note also that in the general case, an *STS* may be not only some government structure, industrial corporation, bank, or company but any urban agglomeration, living house, hotel, or mobile object like an aircraft or ship. The only difference is that in the latter case, positions may reflect not the formally defined functionality of persons being on duty (e.g., an aircraft crew), but more or less temporary places in a social segment of an *STS* (passenger, habitant, client, etc.). Such *STSs* will be called *life-supporting*.

Elements of an *STS* may be located at points close to one another, and such *STSs* will be called ***local*** (*LSTS*), or they may be at some considerable distance from one another, and such *STSs* will be called ***distributed*** (*DSTS*).

Let us also note that, as everywhere in systems analysis, there may be two variants of *STSs* regarding their connectivity with the environment—closed and open (von Bertalanffy 1988; Vernadsky 1998; Whitten and Bentley 2006). A ***closed*** *STS* is isolated from its environment and thus has no opportunity to get (receive) resources from external *STSs*, so any "refreshment" of its *RB* and *TB* may be a result of only its own activity. An ***open*** *STS*, on the contrary, is capable of receiving active and/or passive resources from external systems on an ad hoc request or on some regular basis.

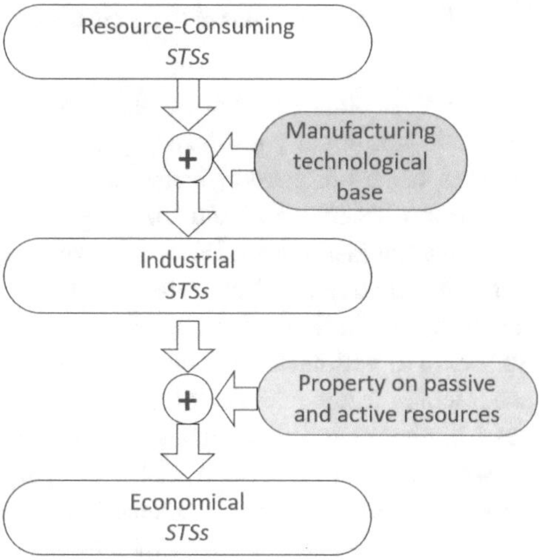

Fig. 2.2 Nexus between classes of sociotechnological systems

As was mentioned above, all possible *STSs* may be divided into three primary classes by their capabilities regarding available resources (Fig. 2.2):

1. Resource-consuming
2. Resource-producing or industrial
3. Resource-distributing or economical

A ***resource-consuming system*** (*RCS*) may only consume resources which allow its operation and has no capability to produce new resources. A ***resource-producing (industrial) system*** (*IS*) has a ***manufacturing technological base*** capable of producing (manufacturing) new resources—both passive (consumables) and active (devices and, among them, tools of manufacturing). However, no one device and no one object produced by an *IS* are assigned to any person or socium. Such assignment or reassignment is out of the *IS's* mission, but is a key feature of ***resource-distributing, or economical, systems*** (*ESs*). A basic category introduced and commonly applied regarding *ESs* is a ***property on resources*** (both passive and active). It is assumed that any owner (holder), which may be a person or a socium, may possess any amount of any resource, while any unit of any resource (passive or active) belongs to only one owner, who has the right to transfer it to other owners. A general case of common ownership of $N > 1$ holders on some amount of resources may be considered as ownership by any holder of some mutually agreed material or virtual unit representing a $1/N$ share of this amount.

So, in fact, economical *STSs* are the most general class of sociotechnological systems. Industrial *STSs* correspond to a subclass of economical *STSs* with only one owner of all resources of an *STS*. Resource-consuming *STSs*, in turn, correspond to a subclass of industrial *STSs* with an empty manufacturing technological base. Also, there may exist *ESs* without any *MTB*; thus, all that is done by any such system is consumption of resources and their motion between owners.

Let us consider the introduced classes of *STSs* in more detail.

2.2 Resource-Consuming Sociotechnological Systems

The simplest class of *STSs* is ***resource-consuming systems***, which, as was said above, do not directly produce any resources, and all they do is to consume some amounts of resources which are necessary for their operation. The most typical examples of *RCSs* are armed forces, police, emergency services, and other similar governmental institutions. From the other side, there are millions of pure office *STSs* with invisible output all over the world, and a lot of such systems as a whole do fit the above definition of an *RCS*, as well as millions of life-supporting *STSs*, consisting of various living houses, inhabited by humans who only consume various goods and nutrition necessary for their life. The technological base of any *STS* of this class in the majority of cases is a collection of computers, smartphones, and networking hardware, which enable carrying out of routine office functions by *STS* personnel or everyday communication among citizenry (passengers, clients).

A common feature of governmental-like *RCSs* is their hierarchical organizational structure, representing nested composition of subsystems (divisions, subdivisions, etc.) down to positions of employees and servicemen, as well as subordination of such positions. All such subsystems (positions) are equipped with various devices in accordance with supply standards established by the *RCS* management (and/or management of higher levels of subordination). Similar standards establish amounts of various resources consumed by devices and *RCS* personnel filling *RCS* positions. Resource consumption, fixed by supply standards, is with respect to some basic period of time (day, month, year, etc.). In fact, practically, the same hierarchical structures, representing sets of more or less temporary positions, may have life-supporting *RCSs* (without assigning positions in such hierarchies any authority rights and powers) (Sheremet 2019c). In this case, supply standards concern some usual amounts of resources consumed by humans filling various stable (as "citizen") and/or temporary ("passenger") positions.

The following are some examples of everyday tasks that are usually solved by an *RCS's* management and associated with ***planning of an RCS operation***:

1. What *RB* would be sufficient for an *RCS* operation given the considered time interval and established supply standards?
2. What money would be spent for the acquisition of amounts of resources necessary for system operation during some fixed time period, given prices of acquired resources?
3. For what maximal time interval would an *RCS* be able to operate, given the available resource base and/or money?
4. If a current *RB* is non-sufficient for an *RCS* operation during the given time interval, what amounts of resources would be acquired and what money would be spent for this objective?
5. What amounts of various resources would be acquired, given alternative market proposals of possible suppliers, to minimize the money spent when there exist alternative ways to supply the *RCS* as well as alternative ways to use the acquired resources, etc.?

However, the most complicated task to be solved by an *RCS's* management is ***to assess (usually in a "what-if" regime) the reasonability of possible transformations of the RCS's structure and supply standards***, i.e., what real consequences would arise in the case of the implementation of such transformations? In the hardest case, the task is to ***construct a new RCS*** given the commercially available proposals as well as the capabilities of industrial partners which may be involved in a solution of this task. In contrast with all the above-listed tasks, which relate to the area of ***analysis*** of large-scale *RCSs*, the last case is nothing but a multi-dimensional and multiobjective task of ***synthesis*** of a new such system. One more, perhaps the most difficult, task in this class of problems is ***systems reverse engineering***, which in essence is to determine a possible structure of an *RCS* given a possible (more or less wide) set of its structures, technological bases, and supply standards, as well as a fully or partly known collection of resources consumed by this system during some fixed period of time.

Let us note that all resources consumed by an *RCS* would be produced (manufactured) by some industrial systems possessing adequate capabilities, and, in fact, ***resource-consuming STSs are the main source of requests (orders) to industrial systems***. Moreover, to return to the set of problems associated with the creation of smart *STSs* announced in the Introduction of this book, it is obvious that *RCSs* are a key segment of mankind whose total supply by necessary goods determines in a quantitative sense the notion of "needs" in Marx's own words, "***from each according to his ability, to each according to his needs***." Comparing the aforementioned total needs regarding any specific *STS* with the total real capabilities of its industrial segment, explicated as its schedule, the governing staff of this *STS* would be able to implement ***smart planning*** by a search for solutions balancing both totals. The result of such planning may be a schedule of the *STS* matching the initial or a partially reduced set of needs, which would be fixed as an objective of the *STS*. Having in mind these remarks, let us consider in more detail key issues concerning industrial systems, which are, in many dimensions, a basic class of *STSs*.

2.3 Resource-Producing (Industrial) Sociotechnological Systems and Smart Industry

Resource-producing (industrial) sociotechnological systems are the result of an extension of *RCSs* by a manufacturing technological base, including various tools of manufacturing, producing various resources—passive (which are only consumed) as well as active (again tools of manufacturing, capable of producing other resources, which may be also *ToMs*, etc.).

The main feature of modern industry is its fast transformation from "*usual*" ("*old-fashioned*") industrial systems to much more flexible and efficient ***Industry 4.0***, massively based on industrial robotics, cyberphysical (additive) manufacturing, smart logistics, computer networking, and alternative ("*green*") power engineering.

Altogether, they integrate a new generation of *ToMs* together with new logistical and energy infrastructures into a highly interconnected global technosphere, rapidly transforming into the "*Internet of Everything*" (including along with *IoT* and *IIoT* also the "*Internet of Humans*" and "*Internet of Services*") as the technological base of the global digital economy. This ***new industrial paradigm may radically boost the efficiency of supply chains and may also reduce redundant manufacturing operations and resource consumption taking place in modern industry, commerce, and the economy as a whole***.

Until now, a manufacturer produced goods on the basis of ***demand forecast***, which very often (even in a "*business-as-usual*" mode, not to speak about various force majeure situations like the *COVID-19* pandemic) is not precise. That is why a significant proportion of produced material objects is left for a long time in stores and shops, and many of them are never acquired by consumers. (It is typical even of the most flexible *market economies* with their competing manufacturers producing practically the same objects and winning competition for market share by price dumping and/or successful marketing; the loser's production is, as a consequence, out of usage, and all resources consumed during the manufacturing of these products are, in fact, lost by mankind.) Not by the chance, the most expensive production, such as of aircraft and helicopters, elite yachts and cars, etc., is performed on demand and after the consumer has paid a large part of the entire cost (or even all of it) necessary for the compensation of all expenses which would occur during the manufacturing cycle. At the same time, there may be a lack of some valuable goods at many places and areas where they are extra-desirable.

An *Industry 4.0* may provide consumers with necessary goods ***on request***. To acquire goods, one would access the Internet portal of their holder (producer or trader) to place an order containing information about what items are required, in what amounts, at what time (not later than, at the earliest possible time, or exactly at some time moment), and to what locations they would be delivered (Figs. 1.1 and 2.3).

(*Internet trading, in fact, is the simplest implementation of the described process in application to the aforementioned objects, which were manufactured beforehand and stored somewhere. Of course, some items in the described process may also be manufactured in advance to reduce the time necessary for order completion, but this is done in a much more "precise" mode than in present industry and commerce.*)

After receiving such a request, the supervising control entity of an industrial system, called below an ***IS controller*** (***ISC***), activates the "***brain***" of the *IS*, named an **"*IS scheduler*"** (***ISS***), which creates an optimal (or at least rational) schedule for the completion of this order, based on current available resources (including the current *RB* and the current workload of *ToMs*), or detects the impossibility of completion and returns a rejection reply to the *ISC* (and the latter to the request initiator), or creates a set of proposed variants of completion by reduction of the order (in the case of a ***closed*** *IS*) or by extension of the *RB* and/or *TB* by means of some friendly external systems (for an ***open*** *IS*). In a positive case, a technological process, driven by a created schedule of order completion, which may be represented as some kind of Gantt chart or precedence diagram (Klein 1999), would start,

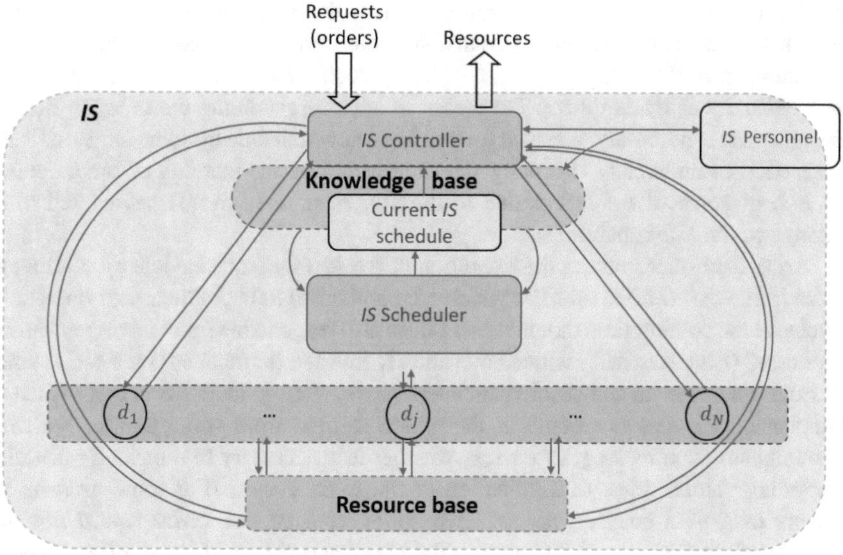

Fig. 2.3 A general scheme of operation of a smart knowledge-based industrial system

resulting in the manufacturing of necessary consumables (spare parts) by *3D* printing or, in some other way, their delivery to facilities, assembly of more complicated items, and so on, until the final objects specified in the initial request arrive occur at the designated points.

Flows of orders (requests) are received by an *ISC* from multiple customers (resource consumers) belonging to various external *STSs* (both *RCSs* and *ISs*) at unpredictable moments and in unpredictable sequence. An *ISS* permits online creation of schedules, ensuring completion of order, by the application of the following objects:

1. A current ***knowledge base***, which accumulates unified formal representations of ***operational capabilities of all manufacturing devices*** in the *IS's* manufacturing technological base (***digital twins of MDs***)
2. ***Data about the available resource base*** of the *IS*, which, as was said above in Sect. 2.1, contains passive resources (materials, spare parts, microchips, fuels, electrical energy, etc.), so all such data is a collection of ***digital twins of real resource items***
3. The *IS's* ***current schedule***, ensuring completion of all previous orders and reflecting current and future workload of all devices in the *MTB*

Along with orders, the *IS* input flow may contain messages about changes to the *IS's MTB*: some *MDs* may be added to the *MTB* during its operation as a result of a cooperation of the *IS* with friendly external *STSs*; some, on the contrary, may be eliminated from it by some destructive impacts (malfunction, natural disasters, deliberate impacts of hostile *STSs*, etc.). Just the same changes may appear in the

resource base of the *IS*—some resources may be added to the *RB* by friendly external *STSs*, some become unavailable, some are simply consumed during a preceding manufacturing process, etc. Every such change is processed by the ***IS controller*** by activating the *ISS* to assess whether rescheduling the *IS* regarding its current state is necessary; a reason for making a new schedule by some large or small correction of an already executing schedule would be the inability of timely completion of some of the orders due to the aforementioned events having led to a change of the *IS's* capabilities.

An *IS* controller initiates the execution of the *IS's* current schedule by sending a "*start message*" (*Sm*) to all *MDs* which are preplanned to begin their next operation cycles at the coming time moment. An *ISC* is also responsible for receiving a "*finish message*" (*Fm*) from *MDs* which have already finished (terminated) their *OCs* at the current time moment and sending an "*alert message*" (*Am*) to an *ISS* in cases when a preplanned *Fm* has not come at the expected time from some device. An *ISS* accumulates such messages to assess whether it is necessary to remake a schedule regarding shifted times of termination of operation cycles. If at some moment it occurs that, as a consequence of such shifts, ***at least one order would not be completed on time***, the *ISS* reports to the *ISC* that a ***threat of breaking a current schedule*** has appeared, and further actions would be performed by *IS* management personnel, who may apply the *ISS* in a "what-if" mode to make a decision on rescheduling combined with communication with consumers of which orders may be affected and with friendly external systems, which may have capabilities to support this *IS*.

All knowledge and data about the current state of the *MTB* and the *RB* are maintained in an up-to-date state by the *IS* controller, whose *IS* knowledge base includes digital twins of all *MDs* in the *MTB* and of all objects in the *RB*. In case of their unavailability, as well as in case of the aforementioned delays to *OC* termination, the *ISC* would activate the *ISS*, initiating *IS* rescheduling. Such ***permanent online rescheduling*** allows timely order completion despite all possible impacts eliminating elements of the *RB* or the *MTB* or decreasing the capabilities of some *MDs*. This property is inherent in an *IS* due to its knowledge base and associated knowledge interpretation engine (*KIE*), which is a key element of an *ISS* and whose application excludes the necessity of permanent remaking of *IS* software and by this reason provides prompt reaction to all fluctuations of the *MTB* and *RB*. Due to such flexibility, the considered *IS* may be called ***smart,*** thus being the basis of a *Smart Industry*, *Smart Cities*, and *Smart Nations*.

As may be seen from this short description, from our initial considerations, we ***are referring such two basic features of industrial systems as smartness and resilience, in close interconnection and interdependency***.

The aforementioned Industrial Internet of Things infrastructure provides simple, flexible, and reliable implementation of the described smart (knowledge-based) industrial systems (Fig. 2.4).

Flexibility is achieved due to the development and application of some general-purpose interfaces and network protocols for connection to *STS* devices and facilities

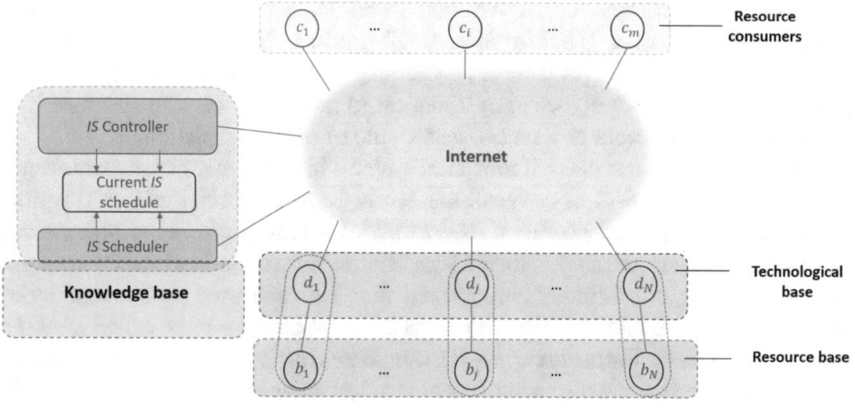

Fig. 2.4 Operation of a smart (knowledge-based) industrial system on an (Industrial) Internet of Things infrastructure

via *IIoT*. Thus, due to a knowledge-based and networked *ISS*, such industrial systems provide smart (rational, if not optimal) behavior in any conditions.

Concerning *IS* personnel, there may be three key groups of specialists:

1. Management staff performing *IS* control
2. *IT* personnel operating software and hardware all over the *IS*
3. Knowledge engineers maintaining the *IS* knowledge base, which in the case of a large-scale *MTB* may contain thousands or even millions of digital twins of manufacturing devices in the *MTB* and items in the *RB*

Due to knowledge-based scheduling, there would be a minimal number of highly skilled and thus highly paid software engineers necessary for a permanent remaking of the *IS's* software. They would be replaced by a smaller number of lower-paid knowledge engineers and helpdesk operators.

Evidently, an *IS's* personnel being humans are also resource consumers, and this obstacle would be taken into account, especially in any cases concerning *IS* resilience issues.

It may be predicted that the extra-large volumes of the knowledge bases of smart *ISs*, being a consequence of the number of *MDs* in *ISs*, and their central position in these *ISs*, make urgent the development of advanced toolkits allowing monitoring, validation, and verification of such *KBs* (being, in fact, ***Big Knowledge***), as is already done regarding ***Big Data*** via implementation of the ***FAIR Data*** concept and introducing ***Data Stewardship*** as a specific kind of *IT* activity (Colombo et al. 2014; Franks 2014; Lee et al. 2015; Roberts 2016; Wang and Wang 2016; Wilkinson et al. 2016).

Note once more (as was first said in Sect. 2.1) that in the general case, ***new active resources (manufacturing devices) may be produced inside a described process by other such devices already in operation***. For example, it may be quite usual for one *3D* printer to print another one and so on until there exist the technological capability and resources necessary for the required work. In fact, today, such a service is already available, as a "*Factory in a Day*" (Factory 2013), which allows rapid deployment and operation of facilities possessing wide industrial capabilities. Active resources which are an objective of a request to an *IS* may also be specified by a query, or they may occur as a part of a schedule of order completion.

As seen, the simplest class of industrial systems includes *ISs*, which manufacture only passive resources. Let us recall that ***we refer to as passive*** not only various consumables but also all ***produced devices which do not perform manufacturing***, i.e., which have no capability to produce new resources (so various devices and systems carrying out different complicated functions and used by humans during their lifetime are considered below to be passive). Such *ISs* will be called *ISs MPR* (***Industrial Systems Manufacturing Passive Resources***). *ISs* which may manufacture not only passive but also active resources, i.e., ***tools of manufacturing***, will be called *ISs MPAR* (***Industrial Systems Manufacturing Passive and Active Resources***), and their *MTBs* will be called ***self-producing***. Let us note that, as was mentioned above, transportation vehicles "produce" objects with new locations and, hence, in such an interpretation may be considered to be active resources.

Any of the industrial systems introduced above may be ***open*** or ***closed***. An ***open IS***, similar to the abovementioned open *STSs*, has opportunities to interact with some external *ISs* capable of delivering to it all or some of the necessary passive and/or active resources, thus allowing recovery or upgrade of the resource base of such an *STS*. In the alternative case, i.e., if an *IS* is ***closed***, there are no other choices than to try to complete not all, but some part, of an order.

Also, any *IS* may be ***local*** or ***distributed***. All manufacturing devices of a ***local IS*** operate at some one compact place (territory); thus, no time is lost for the relocation of consumed and produced resources between *IS* elements. On the contrary, elements of a ***distributed IS*** are located at considerable distances from one another, so transportation of the aforementioned resources between *IS* elements is necessary, and, hence, additional time and resources are required for resource relocation.

Most typical ***tasks usually solved by IS management*** and thus obligatory to be solved by *IS* schedulers are ***assessment of order feasibility*** and ***creating schedules permitting order completion***.

The first task associated with the assessment of order feasibility is to determine ***whether the resources available are sufficient for order completion***. If these amounts of resources are not sufficient, there are two additional tasks to be solved:

1. What amounts of resources must additionally be acquired, if the order is to become feasible?
2. What part of an order and/or at what time may it be completed without additional resources?

The first of these additional tasks is relevant for ***open ISs***. In the alternative case, i.e., if an *IS* is ***closed***, it is necessary to determine what part of an order may be completed given the available resources.

Creating schedules permitting order completion usually covers the following activities:

1. An implementation of competitions (tenders) among possible participants in order completion to find the most rational way (ways) of completion
2. A rational distribution of work and resources among *IS* units (down to separate manufacturing devices) involved in order completion by the results of a tender
3. Creation of a schedule of order completion by a set of selected participants (in the form of a Gantt chart or in some other comprehensive representation used by the *SIS* for control of facilities and manufacturing devices assigned to relevant segments of the whole process of completion)

The first two steps are executed by the ***IS staff*** according to relevant antimonopoly regulations, while the third one may be implemented ***directly by the IS scheduler*** without any human intervention.

Now for the first time, we shall mention ***the basic mathematical toolkit*** which is necessary not only for the aforementioned tasks regarding industrial systems but also for solution of all such tasks concerning all described classes of *STSs* and announced in the previous part of this chapter. In this section, we shall make the first step to such a unified and flexible toolkit, specific to industrial systems described in the Introduction as ***"black box" representations of STSs*** (Fig. 1.3). Following this representation, every manufacturing device will be represented below as a "black box" with input and output collections of resources and a time interval consumed by an *MD* to produce an output collection from an input one, if the latter is available from an *RB*. One and the same name (identifier) of an *MD* may be present in different "black box" representations, differing by *IC*s; such "black boxes" represent *MDs* capable of processing various *ICs*. However, to represent a specific way of operation of an *MD* associated with a respective *IC*, some identifying information is concatenated to the name of an *MD*, forming the unique name of a respective "black box." If some manufacturing device is produced by another *MD* (in the case of *IS MPR*), its name is present in the output collection of the latter, just like the name of any passive resource or device which is not a manufacturing one.

The whole set of *MDs* of an *IS* operates asynchronously, so any *MD* may start and execute its operation cycle, manufacturing an output collection, at any moment when the *RB* contains amounts of resources making up its input collection. The produced output collection enters the *RB* without any delay at the moment when the *OC* ends, so a needed collection becomes available to a consumer immediately it appears in the resource base; thus, values of time intervals of such "black boxes" are null. A current *IS* schedule is a set of some $n > 0$ schedules each ensuring completion of some one order. From the other side, any current *IS* schedule is a composition of $N > 0$ particular schedules of manufacturing devices of an *IS*. Any such particular schedule, in turn, is a sequence of couples, including start and stop moments of *OCs* of this *MD*. A manufacturing device is not able to begin the next operation cycle until the previous one is finished. In the general case, an *ISS* may correct (modify) already executing schedules, when there are no another ways to complete some incoming orders.

Also an ISS may apply data about locations of all elements of its MTB and RB, information about the transportation infrastructure used by the IS during its

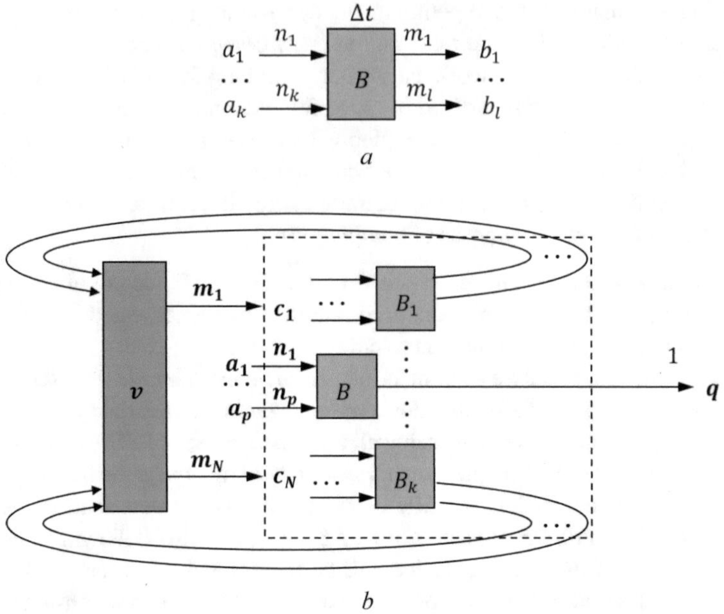

Fig. 2.5 A "black box" explication of a local industrial system: (**a**) a manufacturing device; (**b**) an industrial system

operation, and also data about current positions of transportation vehicles (man-driven as well as unmanned), which may be used, if necessary, for the relocation of consumed and produced resources—both passive and active. As was already mentioned in the Introduction to this book, from a general point of view, any unit (item) of an arbitrary passive resource may be considered as a physical object with its location, while a transportation vehicle is some kind of active resource, "manufacturing" a transported object with its new location from its current state and consuming some other resources (fuel, electrical power, etc.*) necessary for transportation.*

Let us consider the simplest case of a ***closed local industrial system manufacturing passive resources***. Namely, we shall explicate a manufacturing technological base of such an *IS MPR*, represented as a set of manufacturing devices, in the following way. Each *MD* will be represented as a "black box" *B* with *k* inputs and *l* outputs (Fig. 2.5a). We shall denote the *i*-th input by a_i, which is the name of a resource, and denote by n_i the amount (volume, quantity) of elementary units (or units of measurement) of this resource. This means n_1 units of resource a_1, ..., n_k units of resource a_k as well as *n* units of time Δt are required by an *MD B* for creating (producing, manufacturing) m_1 units of resource b_1, ..., m_l units of resource b_l. The resource base of an *IS* is represented by a collection of resources $\boldsymbol{v}$, containing $\boldsymbol{m_1}$ units of resource $\boldsymbol{c_1}$, ..., $\boldsymbol{m_N}$ units of resource $\boldsymbol{c_N}$. Created resources enter the resource base and may be extracted from the *RB* by any manufacturing device that may start its operation cycle at any moment when the *RB* contains all resources defined by its inputs. Let us recall that an *IS* completing any order operates on the presumption that as soon as all necessary resources are available, the next operation cycle of an *MD* may begin (of course, if two or more *MDs* need the same resources, there may be various ways in which the *IS* can operate, differing by the activation sequence of competing devices). By reason of the locality of the *IS*, resource exchanges between the *RB* and devices are assumed to be immediate. A process of *IS* operation is driven by a flow of input orders (Fig. 2.5b). Each such order defines a collection of resources, including $\boldsymbol{n_1}$ units of resource $\boldsymbol{a_1}$, ..., $\boldsymbol{n_p}$ units of resource $\boldsymbol{a_p}$, which are to be manufactured by the system. As may be seen from Fig. 2.5b, an input order is also represented as the same "black box," with inputs corresponding to an order-defined resource collection, and a one-element output collection $\boldsymbol{q}$, where $\boldsymbol{q}$ is an order identifier. Let us underline that ***there are no firm (rigid) links between devices for resource exchange***; instead of such links, a more general and flexible scheme is implemented, allowing the aforementioned ***exchange* via *a resource base***.

Clearly, every operation cycle of any manufacturing device *B* is executed during some determined time interval. Due to a unified representation, this information is simply implanted in every "black box" as a specific resource Δt, whose amount *n* is the duration of the aforementioned interval measured by the application of a fixed unit Δt (second, minute, etc.). The only substantial difference between the resource Δt and other resources is that Δt is not fully additive, because all manufacturing devices in an *IS* may operate in parallel; time is additive only regarding one such device, i.e., to produce *l* items, each created during *n* time units Δt, an *MD* would operate for $l \cdot n$ time units Δt. But if there are two or more such devices, they would produce the aforementioned *l* items during time $k \cdot \Delta t < l \cdot n \cdot \Delta t$ due to the parallel operation of the manufacturing devices and rational distribution of jobs among them. Time may also feature in an order in such a way that, along with the resource amounts required by the customer, it may contain a time interval $\boldsymbol{n} \cdot \Delta t$ as a precise duration of order completion or upper bound ("no more than $\boldsymbol{n} \cdot \Delta t$") of this duration. Along with these two cases, there may be a third one, where $\boldsymbol{n} \cdot \Delta t$ is declared by the customer to be the minimum of all possible values corresponding to various possible ways of completing the order. Also, a customer may include in an order amounts of some resources that must be available when the order is completed, so there may be spent no more than a declared amount of every listed resource, or even a minimum of all possible values. These restrictions may be established by a customer as well as by an *IS* or as a result of their mutual agreement. As may be seen from this short description, an order declared by a customer may be represented as a couple $<$ *Obj, Rs* $>$, where *Obj* is a list of resources that are the objective of the request, while *Rs* is a list of restrictions on amounts of resources and time that are available for achievement of the objective. Let us emphasize once more that time restrictions may be various; they may define that the order must be completed:

1. In the minimal timeframe
2. In no longer than some fixed value $\boldsymbol{n} \cdot \Delta t$
3. Precisely in a fixed timeframe $\boldsymbol{n} \cdot \Delta t$

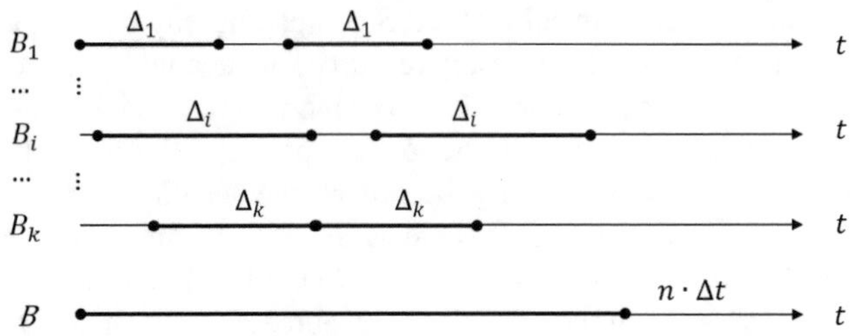

Fig. 2.6 An *IS* schedule

4. In a combination of 1 and 2, i.e., the minimal timeframe possible, being no longer than $\boldsymbol{n} \cdot \Delta t$

As is known from Sect. 2.3, to achieve an objective represented by a couple < *Obj, s* >, a schedule must be created, including interconnected operation cycles, to be executed by manufacturing devices ("black boxes") in an *IS*. Every such *OC* begins with extraction from a resource base of the necessary amounts of resources and finishes with inclusion of the created objects in an *RB*. A result of an *ISS* operation, i.e., a schedule, may be represented in a form similar to a Gantt chart (mentioned in the Introduction and Sect. 2.3 in association with *ISSs* and now explicated at Fig. 2.6), which contains *K* parallel axes, each corresponding to its own manufacturing device ("black box") B_i. A time interval $\left[t_{i_j},\ t_{i_j} + \Delta t_i\right]$, marked by a start point t_{i_j} and a finish point being a moment preceding $t_{i_j} + \Delta t_i$ on the *i*-th axis, represents the aforementioned operation cycle of B_i. All such operations are synchronized by a unitary discrete time scale (with the step of discretization being the value of the variable Δt, i.e., second, minute, etc.) for all *k* time axes, and an operation cycle of B_i may start only inside a time interval when the resource base contains all amounts of resources defined by B_i's inputs, i.e., as soon as the *RB* contains such amounts. The only difference between the representation of an *IS* schedule explicated at Fig. 2.6 and a Gantt diagram is that in the last one, the start point of any operation cycle (named *work*) is connected by arrows with final points of works whose results are necessary for its beginning. Because in the proposed representation of *IS* resource exchange between *MDs* is implemented via a resource base, and every operation cycle starts immediately once a necessary resource set is available to an *MD*, such arrows are redundant.

Let us recall that in the general case, various manufacturing devices may compete for resources; so starting one of them causes a delay of the others. By this fundamental feature, the creation of any IS (and in the general case of any STS) schedule permitting completion of an order requires a massive search in a solution space, and various algorithms, implemented by various ISSs (STSSs), may differ in the computational complexity of such a search. In general, the creation of such a schedule by an ISS (STSS) is a sophisticated combinatorial problem, and the development of algorithms providing an efficient search for solutions of this problem

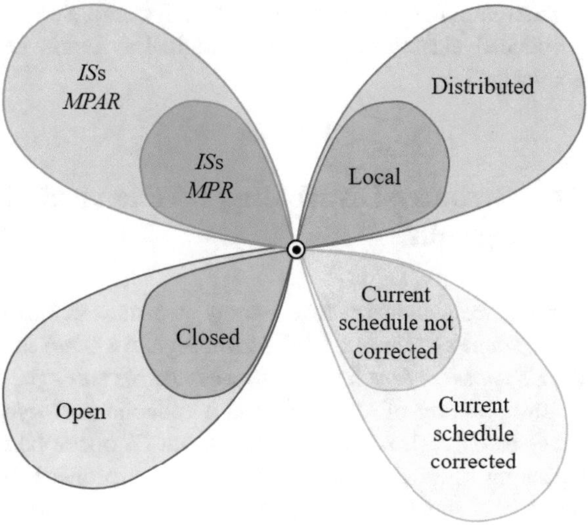

Fig. 2.7 Ontology of industrial systems

is one of the most important tasks that must be carried out to create smart and resilient sociotechnological systems in the future digital economy.

Let us consider now more general cases concerning a ***local IS application and operation***. "The closest" generalization is an *IS* which ***completes a flow of orders*** in such a way that each next order may arrive at the *IS* during processing of previous orders. The simplest but the worst discipline of such flow handling is sequential, i.e., when an *IS* begins to execute operations to complete the next order only after the previous order is already completed. However, to achieve maximal productivity, an *ISS* would provide a maximal parallelization of sequentially incoming orders: operations ensuring completion of every incoming order would begin as soon as possible, in parallel with operations already executing to complete earlier orders. Thus, an *ISS* operates upon a flow of orders processed in a parallel mode. There may be two approaches to the creation of such *IS* schedules. ***The first*** is based on a presumption that schedules ensuring completion of the previous orders are ***not corrected*** during the creation of a schedule permitting completion of the next-arriving order. ***The second*** approach is based on the ***possibility of correction of already executing schedules*** when the creation of the next one takes place. Of course, any possible correction would not cause a shift of the timing of any previous order beyond its deadline.

The ontology of industrial systems in the form of a "flower" with four "petals," representing the introduced ontology, is represented in Fig. 2.7. As can be seen, there are $2^4 = 16$ such classes.

Let us note that all resources produced by an *IS* are not assigned to anybody, and thus, these resources belong to all humans in this *STS* or, on the contrary, do not belong to anybody. In reality, any resource item, since it is produced, is somebody's property, and operation of any *STS*, ultimately, is a ***permanent redistribution of property*** among persons (sociums) in this *STS* (and, possibly, persons and sociums in external *STSs* interacting with this system). So let us move to the third of the introduced classes of sociotechnological systems—resource-distributing, or economical, *STSs*.

2.4 Resource-Distributing (Economical) Sociotechnological Systems

Turning from industry to economy, we may say that in the general case, ***every economical system is a sociotechnological system in which humans (subjects of the ES) possess (are holders, or owners, of) some property.***

The property of any subject is a collection of *objects* (items of passive and/or active resources). As was said above, an *ES* operation is a permanent redistribution (speaking more generally, a motion) of a property between holders, and, if the property of some of these holders includes active resources, then the aforementioned *ES* operation includes also manufacturing new resources (active as well as passive), which after their creation become somebody's property and in this mode are used in the following steps of an *ES* operation. In other words, ***economy begins where there appears ownership of resources***, which are used for the consumption, exchange, and manufacturing of new resources. *ESs* differ from *ISs* by one essential feature: their passive resources, being property of the *ES's* subjects, may be, along with various material and virtual objects, such assets as ***money and stock*** (various security papers including financial derivatives), ***created and used specially for exchange operations***.

The simplest *ESs* are called below ***"exchange economical systems"*** (*EESs*). Their main feature is that no new objects are produced by them during their life cycle, and all that is done by the *EES's* subjects is the mutual exchange of objects which are their property. In the general case, an *EES* operates in such a way that objects belonging to some subjects are eliminated from their property at some moments and are immediately added to the property of other subjects. Examples of *EESs* are stock exchanges, shops, currency exchanges, etc. All such systems do not possess any tools of manufacturing, so all they do is a permanent redistribution of objects being, becoming, and leaving their subjects' property. The most important thing is that the ***total set of objects circulating among EES subjects is extended only as a result of the action of external systems***. ***If an EES is closed*** *(isolated)*, then the aforementioned ***set may only stay constant or reduce*** by reason that some objects are lost during the system's life cycle.

As in the case of the above-considered "*IS of the future*," an operation of any *EES* may be organized on request, and processing such requests would be performed by some control engine whose core intellectual component will be called, similar to the case of *ISs*, an "*EES scheduler*" (*EESS*). The latter processes information about the needs of *EES* subjects, represented by their requests, which specify types and amounts (quantities) of necessary objects, as well as types and amounts (quantities) of objects, being the property of the request initiator, which he agrees to give away from his property for this acquisition. An *EES* scheduler in hard real time creates exchange chains leading to the satisfaction of all current requests (if at least one such chain is possible); otherwise, it constructs replies containing information about lack of some elements (resources and/or transportation links), necessary for creating exchange chains. ***This kind of scheduling is very close to what brokers do at stock exchanges.***

It is essential that every exchange chain in an *EES* is implemented immediately, i.e., it is postulated that a new distribution of property created by an exchange chain replaces the initial one promptly at the moment when the currently processed set of requests come to the *EES* scheduler. (We assume that the duration of the time interval spent by an *EES* scheduler for constructing an exchange chain satisfying a set of requests which have come to the *EES* at some moment is close to zero, so for exchange participants, this interval is not noticeable.)

The next subclass of *ESs*, called ***"lending economical systems"*** (*LESs*), differs from *EESs* in how it treats a lack of the necessary amounts of some resources in some subjects' ownership and their desire to acquire these resources, as the basis for a property redistribution. Namely, in any *LES,* one subject *L* ("***lender***") may credit another subject *B* ("***borrower***" or "***debtor***") by giving to *B* some part of *L*'s property on a mutual agreement that *B* will return it at a mutually agreed time moment, and also at the same or another moment, *B* will transfer to *L* some "***payment***," i.e., a mutually agreed part of his property, eliminating it from his ownership.

As seen, in such systems, the lender "produces" or "acquires" some parts of his property by applying an aforementioned mutual agreement for a removal of this property from the borrower. However, as in the previous case of *EESs*, the total set of objects circulating among subjects of any such system does not increase, because ***no new objects are produced by LESs***; ***all that is done is a redistribution of amounts of objects among subjects***. Such redistribution, unlike in *EESs*, continues over time, so different exchange chains or their parts may proceed in parallel, and, in the general case, one and the same subject at any moment may participate in different exchange chains.

Similar to the case of *EESs*, exchange chains are created by an *LES* scheduler (*LESS*).

As may be seen, *EESs* and *LESs* have one basic feature in common with *RCSs*: ***they do not produce new resources***. The difference is that ***humans in RCSs are passive*** in the sense they only consume their resources, while ***humans in EESs and LESs are active*** in the sense they may acquire additional resources by exchanges (however, without any manufacturing).

A more advanced subclass of economical systems includes *ESs* whose subjects along with passive resources possess also active resources (tools of manufacturing) capable of producing various passive resources. As we have seen, the technological (industrial) background of these economical systems is *ISs MPR*, so we shall call these *ESs* ***"economical systems operating passive resources"*** (*ESs OPR*). Schedulers of such systems would be an extension of their *ISs MPR* prototypes, planning

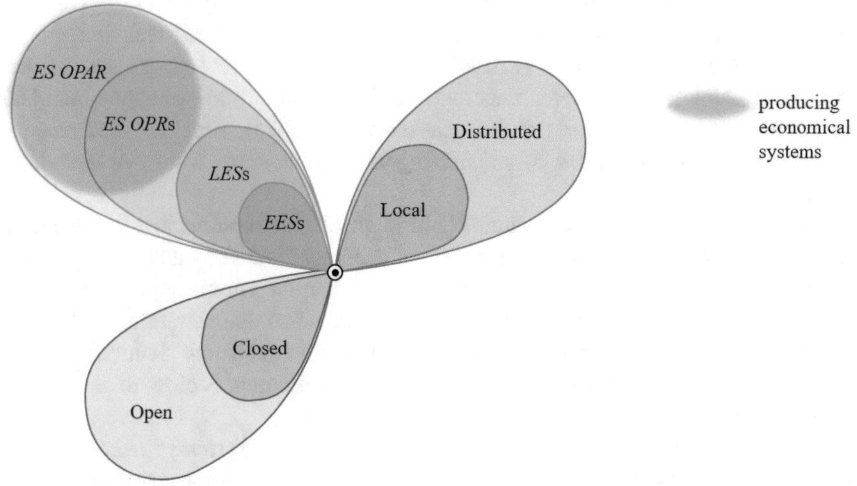

Fig. 2.8 Ontology of economical systems

not only manufacturing processes but also intersubject exchanges necessary for the production of new objects, which in this case are only passive resources.

A further generalization of *ESs OPR* is economical systems whose subjects, along with passive resources, possess also active resources, which are capable of producing other active resources. Evidently, the technological (industrial) background of such *ESs* is *ISs MPAR*, so we shall call these *ESs* ***"economical systems operating passive and active resources"*** (*ESs OPAR*). As in *ESs OPR*, every active resource, being the property of some subject, may process (apply) passive resources which are also the property of this subject. So schedulers of such economical systems would implement the most sophisticated algorithmics, enabling request execution (order completion) by creating supply chains including not only intersubject exchanges but also manufacturing active resources, which when put into action may produce new resources, which, in turn, may be passive or active.

We shall join *ESs OPR* and *ESs OPAR* into a general subclass of *ESs* and call it "***producing economical systems***" (*PESs*).

As may be seen, there exists an evident hierarchy of introduced classes of *ESs*. Namely, *ESs OPR* are a subclass of *ESs OPAR* not producing active resources. In turn, *LESs* are a special case of *PESs* with an empty producing capability. Finally, *EESs* are a special case of *LESs* with an empty lending capability.

Like industrial systems, economical systems may be open or closed, local or distributed.

The ontology of economical systems in the form of a "flower" with three "petals," representing the introduced ontology, is represented in Fig. 2.8. As can be seen, there are $4 \cdot 2 \cdot 2 = 16$ such classes.

In the real world, practically all economical systems are mixed, in the sense that every produced resource item may be put into an exchange or an action, and any item obtained by an exchange may be consumed or applied during manufacturing.

Describing in short the declared four basic subclasses of economical systems, we deliberately did not mention such fundamental categories of economical theory and practice as the ***cost*** of asset production by manufacturers and the ***price*** of their acquisition by consumers. Both play a key role in consideration of Marx's surplus-value (or "added-value") and capital accumulation categories as well as in his "***goods-money-goods***" and "***money-goods-money***" cycles (Marx 2018).

However, price and cost still fall within the described scheme of *ES* operation, because they are covered by the more general categories "property" and "resources," which allow a correct consideration both on verbal and on strict mathematical levels. Let us pay, however, some attention to money and its place in the introduced simple ontology of economical systems.

Every active resource in any *PES* is characterized, along with functional parameters including types and quantities of input and output objects as well as time of manufacturing, also by economical parameters, such as money, which is a passive resource, representing the cost of one application of active resources for manufacturing output objects. This money, being the property of *ES* subjects, is transferred by exchange operations between these subjects, thus creating a money circulation process inside an economical system. As may be seen, this process is synchronized with manufacturing and other exchanges and prolonged by a flow of requests arriving from *ES* subjects and/or external systems—all such requests specifying the subject's requirements and initiating redistribution of property among *ES* subjects.

Note also that future economical systems based on an advanced Industry 4.0 at the same time would be and, in fact, are totally based on electronic commerce and banking. Money, being "the blood of an economy," is already rapidly transforming from cash to a virtual form. This transformation rather naturally has led to a cybercurrency (Vandezande 2018), which, in turn, gave birth to such still-exotic versions of money as cryptocurrency (Houben 2015) and associated technologies (Dannen 2017). In fact, there is no technological difficulty in introducing some new cryptocurrency; nowadays, it is rather more difficult to ensure its more or less wide application and circulation.

Being massively based on digitized industry, commerce and banking, the economy also becomes digital. However, a technological form of property does not change the described general algorithmics of ***ES scheduling*** or the problems considered by economy decision-makers and their staffs in routine everyday management, including an assessment of *ESs'* capability to execute incoming requests and to win various competitions in a volatile market environment. Similar problems are considered by the aforementioned economy decision-makers during strategic ***planning***, including assessment of feasibility of investments as well as assessment of *ES* vulnerability/resilience to various real or possible destructive impacts.

Now after all considerations, it is rather evident that from a technological point of view, the list of problems associated with *ESs* is practically the same as regarding

ISs, with the only difference that every object, which in the case of *ISs* has no owner, from the moment of its appearance in a resource base would become somebody's property and in this capacity could be referred to by any subject of the *ES*. Also, it is clear that the ongoing request-driven creation/correction of any *ES* schedules permitting property manufacturing, redistribution, and consumption would be done by an ***economical system scheduler*** (***ESS***). The latter is the most universal and effective engine responsible for the rational behavior of any *STS* of this class and, thus, of any *RCSs* and *ISs* that are particular cases of *ESs*. So from this place, it would be reasonable, if necessary, not to split our consideration into all possible cases of schedulers, each regarding its own class of *STSs*, but to refer to the general notion of ***STS scheduler***, discussed for the first time in the Introduction to this book and providing all actions addressed above to *ESSs*, as the most representative of all such engines.

However, it is clear that there is a high-priority problem to be considered if one is really planning to develop any kind of *STS* scheduler as well as to allow stable and efficient operation of any *STS* ***in any possible conditions***, not only in more or less comfortable and stable conditions characterized as "business as usual." This problem, not once mentioned in the Introduction and in the previous sections of this Chap. 2, flows out from the fact that any *STS* operates in a volatile and hazardous environment which is a permanent source of threats to the system's normal operation. If an *STS* is affected by an accidental or deliberate destructive impact (or even a batch of such impacts) eliminating some segments of this system (including humans, devices, and facilities, as well as the passive resources consumed by them), all the above-considered tasks would be solved by an *STS* scheduler, given that it remained part of the *STS* after the impact. In a successful case, an affected system would be capable of completing all orders which are already being executed. Otherwise, some orders would be eliminated or shifted out of the schedule, and, most importantly, some actions would be performed to recover the system and to return it to a normal state (Cavdaregla et al. 2013). If the affected *STS* is open, there would be possible participation of some external *STSs* (one or more) in these activities, so it would be necessary to assess whether it would be possible for these *STSs* to recover an affected sociotechnological system or to take some part of the orders not yet completed and try to complete them instead of the affected system. Such tasks relate to the area of *STS* resilience and recoverability.

Due to the great importance of this issue, we shall dedicate to it the special Sect. 2.5.

2.5 Resilience and Recovery of Sociotechnological Systems

By digitalization, robotization, and networking, sociotechnological systems of any of the considered classes become more and more "technological," interconnected, and, thus, vulnerable to even local malfunctions of devices in these systems, as well as to the loss of some relatively small parts (quantities) of the resources necessary for

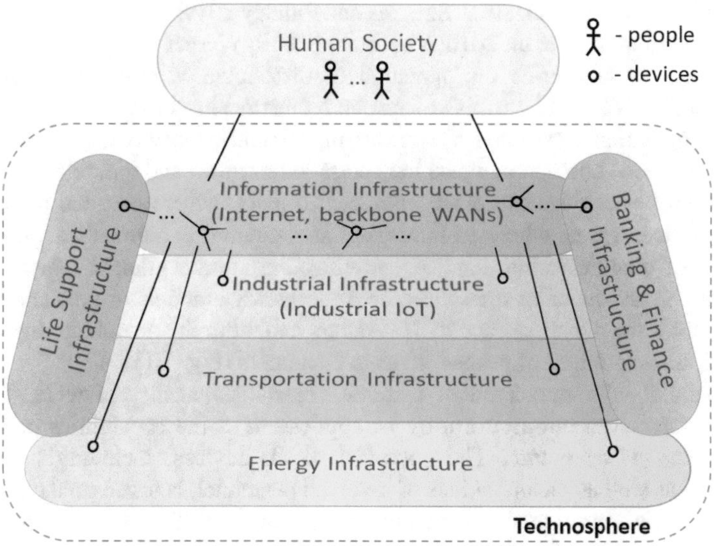

Fig. 2.9 Basic critical infrastructures enabling operation and interaction of *STSs* in a global digital economy

their operation. Every such malfunction or loss, by multiple chain (or cascade) effects, may lead to serious negative consequences, which may occur far from the place (area) of the initial accident (Dobson et al. 2007; Rinaldi et al. 2001; Rehak et al. 2018). On the other hand, as is shown by the *COVID-19* pandemic, all *STSs* may become no less vulnerable as a result of impact on their human segments. We have already considered resilience issues briefly in the previous sections, but now, paying attention to the exclusive importance of this area, we shall do it in a more integrated sense.

A digital economy paradigm simplifies and unifies representation of all possible sociotechnological systems, which, independent of their functions, may be naturally described in a network-centric framework. Let us repeat once more that according to the *Industry 4.0* concept (Baller et al. 2016; Colombo et al. 2014; Lee et al. 2015; Schwab 2015; Wang and Wang 2016), which we have addressed already in the Introduction, a human society operates as a set of customers, generating requests for necessary resources via a common Internet-based information infrastructure connecting a set of various devices and facilities, which produce and deliver the aforementioned resources. These interconnected devices and facilities together form industrial, transportation, energy, life support, and banking and finance infrastructures (Fig. 2.9), all of them named *critical (CIs)* for their critically important place in the current technosphere (Alcaraz and Zeadally 2015; Eusqeld et al. 2011; Gvishiani et al. 2018; Haimes and Jiang 2001; He and Cha 2018; Katay 2010; Larsen et al. 2017; Li et al. 2005; Liu et al. 2018; Lund et al. 2014; Mazher et al. 2018; Nepal and

Jamasb 2013; Rehak et al. 2019; Sharkey and Pinkley 2019; Sheremet 2019c, 2020e, 2021b; Stergiopoulos et al. 2016; Vespignani 2010; Werner 2017).

Let us now redescribe the operation of distributed *STSs* of all classes on a background of *CIs*. The process of resource creation in the general case includes multiple operations, executed by various manufacturing devices, located more or less far from one another, and also transportation of items and their elements (spare parts) from places where they are produced to places where they are used. In the most advanced cases, where an industrial infrastructure implements additive technologies of manufacturing, e.g., *3D* printing, the aforementioned logistics is the simplest one and includes transportation of necessary amounts of powders or other consumables from their storage to *3D* printers and, after the manufacturing cycle is finished, transportation of printed items to customers (Fig. 1.1).

In parallel with relocation of material objects, payments for work done are performed in an electronic form by an application of the capabilities of banking and finance infrastructure. The operation of all devices, including networking hardware, as well as the operations of involved personnel, is based on the consumption of electric energy produced by power plants and transferred by electrical grids, forming an energy infrastructure, which is often considered as a whole with fuel storage, pipelines, and terminal fuel stations. Relocation of products and persons is performed by various vehicles upon railroads, highways, and water and air routes, and all these vehicles utilize fuels and/or electric energy (the latter in a case of electric transport). A life-support infrastructure includes technological objects providing people with residences, food, water supply, and other services (including emergency services, healthcare, etc.) necessary for their existence and functionality. The described process of a *DSTS's* operation is highly decentralized, and interactions between devices are organized in a peer-to-peer mode via information infrastructure.

A more complicated case associated with destructive impacts addresses the aforementioned ***cascade (or chain) effects***. Such effects emerge in the propagation and multiplication of failures (damage) without additional impacts from external sources, but only by reason of existing internal interconnections between subsystems (elements) of an affected *STS*; the destruction (malfunction) of such an interconnection causes an immediate or a delayed destruction (malfunction) of its neighboring (in a space and/or a functional sense) elements. For example, a failure of an electrical power plant causes a failure of an electricity-driven railway system as well as of fuel stations that allow refueling of automotive transport, even if there is no direct impact on transportation vehicles; as a consequence, all ground transport loses the capability to relocate resources and products necessary for industrial facilities that produce goods and food for humans. These industrial facilities also fail, and so on.

An objective of all considered *STSS* applications is the creation of schedules permitting the completion of all orders arriving at the *STS* despite any destructive impacts and chain effects. If this is possible, the *STS* is called ***resilient*** to an impact.

However, in some circumstances, there may also be a negative result when some orders cannot be completed under current resource and time restrictions. In this case, an affected *STS* is called ***vulnerable*** to an impact.

So one more task of an *STSS* application is to provide an assessment of what amounts of resources and devices would need to be produced (acquired) and delivered to (included in) the *RB* and the *TB* of an ***affected STS*** to complete already processed orders. Furthermore, it is extremely important to know what time and resources are necessary to recover the technological capabilities and resource base of an affected *STS*, because resources which must be added after a destructive impact must in turn be produced (manufactured). Thus, impacts generate an additional flow of orders, allowing recovery of an affected *STS* through ***that part of it that had remained in a normal state*** and/or its recovery by some ***external*** sociotechnological systems. In fact, there may be ***three variants of participation of an external STS*** in the mitigation of consequences of a destructive impact on an affected *IS*:

1. Recovery of a *TB* and an *RB* reduced by an impact
2. Support of an affected *IS* by carrying out some of its functions as long as is necessary or possible
3. Both of the above (i.e., recovery and support in parallel)

A ***closed*** affected *STS* operates upon a resource base and by a technological base that cannot be recovered using an outside environment (or external system); the latter generates only destructive impacts. An ***open*** affected *STS*, by contrast, has capabilities for achieving *RB* and *TB* recovery by means of external systems. Nevertheless, this case may be reduced to the previous one by considering a recovery as a "constructive" impact, which "affects" an *STS* by adding some resources to an *RB* or a *TB* instead of eliminating them. Thus, we may unify rescheduling initiated by *RB* and *TB* recovery through the already considered case of a destructive impact.

However, in the general case, there may be situations where neither the *STS*'s own capabilities nor the external system's possible aid can complete an order. That is why another necessary task of an *STSS* application, concerning *STSs* in both affected and normal states, is an assessment of ***what part of an order may be completed given the available resources***. This function is extremely useful if there is no opportunity for recovery of an *RB* and a *TB*, i.e., a destroyed part of the passive resources and technological capabilities of an *IS*.

All that we have said above concerns local *STSs*. Let us consider now the general case—a ***distributed*** sociotechnological system. Formalizing this case, we shall follow the "*Occam's razor*" ideology ("entities should not be multiplied beyond necessity") and use previous constructions and entities, extending them by a minimal number of additional items. Namely, we shall represent a *DSTS* in the same way as a local *STS*, adding their locations to every resource collection containing units of some resource (note that such a technique was mentioned in the Introduction and in Sect. 2.3). We shall do the same with inputs and outputs of "black boxes," adding locations to names of resources. In this way, we implant all necessary geospatial information into our representations of the technological and resource bases of local sociotechnological systems, thus providing a unified form of representation of the *TB* and *RB* of any distributed *STS*. Similarly, geospatial information may be added to orders, which also consist of the aforementioned resource collections. This will be the only syntactic extension for a located resource representation—instead of primary names of resources, we shall use the construction (a,z), where z is the name of a point (place, area) where resource a is located and "(", ",", and ")" are delimiters. ***By this minor change, we expand the introduced intermediate representation of local sociotechnological systems and their resource bases and orders to distributed STSs.*** From this moment, an *RB*, explicated in Fig. 2.5, will contain $\boldsymbol{m_1}$ units of resource $\boldsymbol{c_1}$ located at point $\boldsymbol{z_1}$, ... , $\boldsymbol{m_N}$ units of resource $\boldsymbol{c_N}$ located at point $\boldsymbol{z_N}$. Similarly, a manufacturing device ("black box") allows the creation of m_1 units of resource b_1 located at point z_1, ... , m_l units of resource b_l located at point z_l, if there are n_1 units of resource a_1 located at point y_1, ... , n_m units of resource a_m located at point y_m, during a time interval of n time units Δt. And, finally, an order defines $\boldsymbol{n_1}$ units of resource $\boldsymbol{a_1}$ located at point $\boldsymbol{y_1}, \ldots, \boldsymbol{n_q}$ units of resource $\boldsymbol{a_q}$ located at point $\boldsymbol{y_q}$ (Sheremet 2018).

The presence of locations in all components of a described intermediate formalization of *DSTSs* provides a natural, flexible, and simple representation of logistical capabilities of such systems, including both transportation and storage logistics. A storage located at point z may be modeled by its presence in the resource base collections, containing $\boldsymbol{m'}_1$ units of resource $\boldsymbol{c'}_1$ located at z, ..., $\boldsymbol{m'}_N$ units of resource $\boldsymbol{c'}_N$ located also at z. Transportation capabilities, relocating one unit of a resource a from a point y to a point z, may be simply modeled by a "black box" with an input (a, y) with the number of units of this resource being 1, output (a,z), and inputs representing the amounts of resources necessary for this operation (among these resources may be an engaged transportation tool, i.e., cargo ship, etc., amounts of fuel necessary for relocation of this tool and the transported resource from y to z, and some others) (Sheremet 2019c). As may be seen, any such approach to modeling transportation operations unifies all objects participating in a relocation—active (transporting) as well as passive (transported). A destructive impact may be represented in the described framework as the set of locations affected by it, so the result of an impact may be modeled simply by eliminating all resource collections whose locations are in the aforementioned set. If the destruction is partial, an impact may be represented as in the local *STS* case, i.e., in the form of a list of eliminated collections.

As may be seen from the above description, a unified representation of distributed and local sociotechnological systems makes redundant the development of a special scheduler for *DSTSs,* if one has already been developed for local *STSs*. On the other hand, the same techniques may be applied to the representation of ownership of resources: it would be sufficient to add (concatenate) to a construction (a, z) one more component x being the name of some subject of an economical *DSTS*. So application of objects of the form $(a : x, z)$, where ":" is one more delimiter, means that subject x is the holder (owner) of objects of type a located at place z. (The same notation will be used below in Chaps. 7 and 9, dedicated, respectively, to economical systems and resource-based games.) In this way, we unify *RCSs*, *ISs*, and *ESs*, regarding the engine which will be developed and applied for their scheduling.

Thus, there is one general problem for solution, i.e., ***creating a unitary STS scheduler, ensuring smart behavior of STSs by the application of unitary algorithmics to any class of STSs (resource-consuming, industrial, economical), any***

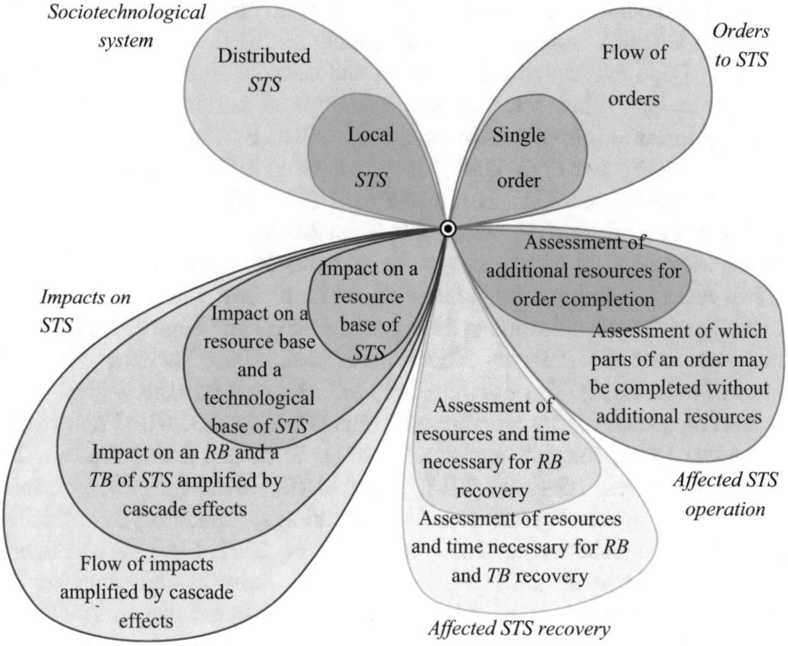

Fig. 2.10 Ontology of *STSS* tasks

resource bases and technological bases, various types of orders, flows of orders, types of impacts on STSs' resources (RBs as well as TBs), and flows of such impacts.

The ontology of tasks to be considered for the creation of the aforementioned unitary *STSS* and mutual interconnections of these tasks is presented in Fig. 2.10 in the usual for this chapter form of a "flower," whose "petals" correspond to the directions of a described ontology of situations in which sociotechnological systems may operate, and thus an *STS* scheduler would ensure provably smart behavior of an *STS*.

As everywhere above, each "petal" extends from a simple to a more complicated case; thus, every tuple of five different values (more or less bulky) defines a specific subclass of *STSs* and, thus, a complex of problems to be solved. This solution now may be organized in the most rational direction—from simple to complex. As may be seen, there are $2{\cdot}2{\cdot}4{\cdot}2{\cdot}2 = 64$ classes of problems to be solved. All problems described in this and the previous sections, in some more or less general formulation, were considered in many earlier works in the operations research area and various segments of theoretical computer science. Perhaps the closest to the set of tasks presented in Fig. 2.10 are the models and algorithms developed in scheduling theory (Abdolshah 2014; Conway et al. 2003; Hermann 2006; Hillier and Lieberman 2014; Klein 1999; Leung 2004). Various asynchronous systems operating upon limited resources are modeled by different classes of Petri nets (David and Alla 2005; Hespanha et al. 2007; Hopcroft et al. 2001; Meduna 2014).

Models of optimal resource distribution and transportation are thoroughly considered in the operations research area by means of various matrix, graph, and network optimization models (Abraham et al. 2010; Bast et al. 2009; Dabrowski and Hunt 2011; Fiedler et al. 2006; Gass and Assad 2005; Gondran and Minoux 2008; Gutin and Punnen 2006; Hemmecke et al. 2009; Hillier and Lieberman 2014; Karlot 2005; Lasdon 2013; Roberts 2009; Taha 2016).

Along with classical approaches to the formulation and solution of the above-described tasks, some knowledge-based models are already known that also have potential for efficient application to these tasks (Akerkar and Sajja 2010; Cross 2017; Kendal and Green 2007; Pannu 2015; Sainter et al. 2000; Sheremet 1994, 2013; Yeoh and Yokoo 2012). The most advanced *AI* technologies in this context are logic programming and constraint programming (Apt 2003; Bratko 2012; Frunkwirth and Abdennadher 2003; Marriott and Stucky 2003; Siekmann 2014; Wallace 2002), deductive databases (Darwen 2012; Date 2012; Siekmann 2014; Sheremet 2020a, b), and, in recent years especially, multi-agent systems (Belardinelli and Argento 2017; Jadbabiae et al. 2003; Klavins et al. 2006; Mesbahi and Egerstedt 2010; Olfati-Saber et al. 2007; Pena et al. 2019; Rzevski and Skobelev 2014; Salamon 2011; Schatten et al. 2015; Shoham and Leyton-Brown 2009; Sycara 1998; Van der Hoog 2018; Waldrop 2018; Wooldridge 2009). Capabilities and limitations of the listed approaches regarding problems important for *STSS* development were considered from the most general positions in Gvishiani et al. (2018).

As may be seen from the introduced ontologies, explicated in Figs. 2.7, 2.8, and 2.10, the research space of the considered area is extra-large, and it is impossible to cover all this space in one book. So the main purpose of this monograph is to ***initiate broad consideration of the proposed multigrammatical framework as a theoretical and implementational background for the development of theoretical and practically oriented tools enabling smart behavior of sociotechnological systems in both regular and extreme situations***.

Chapter 3
Syntax, Semantics, and Mathematical Properties of Multiset Grammars

This chapter contains a broad introduction to a *mathematical theory of recursive multisets* and a description of various classes of multiset grammars, which all together form the multigrammatical framework and serve as a mathematical background for representation and solution of all tasks introduced in Chap. 2. As was already mentioned in the Introduction, the main feature of *MGs* as a non-conventional model of computing is *ubiquitous goal-driven and knowledge-driven computation, implemented by multiset generation.*

Basic notions of this theory—*multiset, filter, multiset grammar, filtering multiset grammar* (*FMG*)—are introduced in Sect. 3.1 sequentially and in such a way that their main features, useful for application to *STSs*, are illustrated by simple examples associated with *STS* components and clarifying the practical sense of these notions. *MGs* as well as *FMGs* generate and in the case of *FMGs* select new multisets, which represent resource collections and have no connection with the time scale. By this limitation, the applicability of these classes of multigrammars is limited on tasks with the character of *planning*, which commonly operates on integral (aggregated) collections of resources consumed and/or produced by various actors of *STSs* during some considered time interval.

To represent and solve tasks from the *STS scheduling* area, it is necessary to develop and apply some minimal extension of *MGs* and *FMGs* with time marks implanted in the generated multisets, representing *STS* resource bases varying upon a time scale as a result of application of manufacturing devices. This extension, named *temporal multiset grammars* (*TMGs*), is introduced and formally defined in Sect. 3.2. The technique of *TMG* representation of manufacturing devices and, as a whole, of manufacturing technological bases is considered in detail, and filtering *TMGs* (*FTMGs*) are described as a tool of representation of *STSs* and orders to such systems. The most interesting subclass of temporal *MGs* is so-called self-generating TMGs (*SG TMGs*), which allow direct representation of *STSs* capable of producing tools of manufacturing (such *STSs* may be industrial systems manufacturing passive and active resources as well as economical systems operating, as well, passive and active resources), which, in turn, after their production and inclusion in an *MTB* may

I. A. Sheremet, *Multigrammatical Framework for Knowledge-Based Digital Economy*, https://doi.org/10.1007/978-3-031-13858-4_3

be applied for the same purpose, and so on. There are no basic limitations on the recursion depth in *SG TMGs*.

All classes of multiset grammars presented in Sects. 3.1 and 3.2 have one common feature: they allow representation of known *STSs* and solution of all tasks concerning such *STSs*. Such tasks usually concern *systems assessment* (or *analysis*). If it is necessary to formulate and solve some inverse task, whose purpose is search in some large set of *STSs* and selection from this set of one or several systems possessing some predefined features, i.e., in fact, to design a new system (such tasks fall under *systems synthesis*), then all introduced classes of multigrammars support only one solution method: *synthesis-by-analysis*, i.e., sequential generation and assessment of all possible variants, which in the most practically interesting cases is simply unimplementable. To make a tool capable of operating not only one *STS* but sets of such systems, a technique called *metagrammatical extension of multiset grammars* is proposed and considered in Sect. 3.3. A multiset metagrammar due to the presence of variables in its constructions and determination of domains of these variables by filters provides a natural and compact description of a set of multigrammars each representing some one specific system. A filter comprises conditions determining the multisets (i.e., resource collections) which are selected from those generated by these multigrammars (i.e., produced by the respective *STS*); thus we, in fact, formulate a task of systems synthesis. The syntax and semantics of multiset metagrammars (*MMGs*), including temporal and self-generating temporal *MMGs*, described in this section, may become a basis for effective application of the multigrammatical framework to a lot of complicated problems from the areas of strategic planning, economical combinatorics, and systems reverse engineering, which until now have been difficult to formalize by means of conventional matrix-vector-based mathematical tools.

The novelty of the multigrammatical framework makes it reasonable to carefully study the *formal mathematical properties of various classes of multiset grammars* and their subclasses. This is done in Sect. 3.4.

To make the *MGF* more suitable for representation of some of the most frequently considered practical tasks from both planning and scheduling, some *descriptional extensions* are joined to the basic syntax and semantics of *MGs* and their variations. These extensions are discussed in Sect. 3.5.

Let us begin our consideration of the *MGF* with its basic notions and constructions.

3.1 Multisets, Multiset Grammars, and Filtering Multiset Grammars

3.1.1 Basic Notions and Definitions

Classical set theory is based on the concept of a set as an unordered collection of elements different ("distinguishable") from one another. The basis of the theory of multisets is the notion of a *multiset*, which is understood as a collection of so-called multiobjects composed of indistinguishable elements (*objects*) (Blizard 1989; Hickman 1980; Lake 1976; Meyer and McRobbie 1982; Petrovskiy 2002, 2003, 2018; Red'ko et al. 2015; Singh et al. 2007). A *multiobject* containing n identical objects a is denoted as $n \cdot a$, where n is called the *multiplicity* of the object a (here and below " · " is a composing symbol). (We shall use everywhere small symbols "a" with or without lower indices as well as small "b", "c", etc. to denote objects; multisets will be denoted by small "v" or "w" with or without indices.)

A record

$$v = \{n_1 \cdot a_1,\ \ldots,\ n_m \cdot a_m\} \tag{3.1}$$

means that the multiset v contains n_1 objects a_1, …, n_m objects a_m. The set $\beta(\text{-}v) = \{a_1, \ldots, a_m\}$ is called the *basis* of the multiset v. We shall use one and the same symbol " ∈ " to denote that an object or a multiobject is a member of (belongs to) a multiset v, so $a_i \in v$ and $n_i \cdot a_i \in v$ mean that object a_i belongs to *MS* v as well as multiobject $n_i \cdot a_i$ belongs to this *MS*. From a substantial point of view, a set $\{a_1, \ldots, a_m\}$ and a multiset $\{1 \cdot a_1, \ldots, 1 \cdot a_m\}$ are equivalent, as are an object a and a multiobject $1 \cdot a$. If the multiplicity of an object a in v is zero, this is equivalent to the absence of a in the multiset v, which is written, as usual, $a \notin v$. The empty multiset and the empty set are denoted by $\{\varnothing\}$ (in both cases, the statement "no objects enter a set (multiset)" is denoted, so it is not necessary to introduce additional denotation for empty multiset regarding an empty set).

We shall say that a multiset

$$v = \{n_1 \cdot a_1,\ \ldots,\ n_\mathrm{m} \cdot a_m\} \tag{3.2}$$

is *included* in a multiset

$$v' = \{n'_1 \cdot a'_1,\ \ldots,\ n'_{m'} \cdot a'_{m'}\} \tag{3.3}$$

which is denoted $v \subseteq v'$, if

$$(\forall n \cdot a \in v)\ (\exists n' \cdot a \in v')\ \ n \leq n', \tag{3.4}$$

and in this case, v is called a *submultiset* of v'. Further, v is *strictly included* in v', which is denoted $v \subset v'$, if $v \subseteq v'$, and $v \neq v'$; in this case, v is called a *strict submultiset* of v'. If $v \subseteq v'$ ($v \subset v'$), we shall also say that v' *includes (strictly includes)* v. If necessary, we shall write $v' \supseteq v$ and $v' \supset v$, which is equivalent to the previous notation.

Example 3.1

The multiset $v = \{3 \cdot a, 4 \cdot b, 1 \cdot c\}$ is strictly included in the multiset $v' = \{4 \cdot a, 4 \cdot b, 1 \cdot c, 1 \cdot d\}$.

In this book we shall use five basic operations on multisets: addition, subtraction, multiplication by integer number, join, and intersection.

The result of *addition* of multisets v and v' (denoted below by the bold symbol " **+** ") is the multiset

$$v + v' = \bigcup_{\substack{a \in \beta(v) \cup \beta(v') \\ n \cdot a \in v \\ n' \cdot a \in v'}} \{(n + n') \cdot a\}. \tag{3.5}$$

The result of *subtraction* of v' from v (denoted by the bold symbol " **−** ") is defined as follows:

$$v - v' = \bigcup_{\substack{a \in \beta(v) \cap \beta(v') \\ n \cdot a \in v \\ n' \cdot a \in v' \\ n \geq n'}} \{(n - n') \cdot a\}. \tag{3.6}$$

We use the aforementioned equivalence of $a \notin v$ and $0 \cdot a \in v$ in (3.5) and (3.6) to shorten the definitions.

The result of *multiplication* of a multiset v by an integer number n (denoted by $v * n$) is the multiset

$$v * \mathrm{n} = \{(\mathrm{n} \times \mathrm{n}_1) \cdot \mathrm{a}_1, \ \ldots, \ (\mathrm{n} \times \mathrm{n}_m) \cdot \mathrm{a}_m\}, \tag{3.7}$$

i.e., the multiplicity n_i of every object a_i is multiplied by n (here " $\times$ " denotes usual multiplication of integer numbers). This operation is commutative, so writing $n * v$ is also correct. Obviously, $v * 0 = 0 * v = \{\varnothing\}$ for every multiset v.

Multisets v and v' may be *joined* (denoted $v \cup v'$) and *intersected* (denoted $v \cap v'$):

$$v \cup v' = \bigcup_{\substack{a \in \beta(v) \cup \beta(v') \\ n \cdot a \in v \\ n' \cdot a \in v'}} \{max\,(n, \ n') \cdot a\}, \tag{3.8}$$

$$v \cap v' = \bigcup_{\substack{a \in \beta(v) \cup \beta(v') \\ n \cdot a \in v \\ n' \cdot a \in v'}} \{min\,(n, \ n') \cdot a\}. \tag{3.9}$$

Statement (3.8) defines that the join of v and v' contains all objects which belong to at least one of these multisets, and the multiplicity of each such object is the maximum of the multiplicities of the corresponding objects from v and v' (if the object belongs to only one of these multisets, and its multiplicity is n, then $max(-n, 0) = n$). Statement (3.9) is similar, but the intersection of v and v' contains only those objects which belong to both multisets, and the multiplicity of each such object is the minimum of the two multiplicities.

Example 3.2
Let $v = \{3 \cdot a, 2 \cdot b, 1 \cdot d\}$, $v' = \{1 \cdot a, 2 \cdot c, 2 \cdot d\}$. Then

$$v + v' = \{4 \cdot a,\ 2 \cdot b,\ 2 \cdot c,\ 3 \cdot d\},$$
$$v - v' = \{2 \cdot a,\ 2 \cdot b\},$$
$$3 * \mathrm{v}' = \{9 \cdot \mathrm{a},\ 6 \cdot \mathrm{b},\ 3 \cdot \mathrm{d}\},$$
$$v \cup v' = \{3 \cdot a,\ 2 \cdot b,\ 2 \cdot c,\ 2 \cdot d\},$$
$$v \cap v' = \{1 \cdot a,\ 1 \cdot d\}.$$

All the defined operations are known from the well-known sources (Petrovskiy 2002, 2003; Singh et al. 2007). At the same time, the filtering operations introduced in (Sheremet 2010, 2011a) and defined below operate on *sets of multisets (SMSs)*, creating subsets of these sets by selection of those multisets which satisfy some conditions from the operands of these operations.

3.1.2 Filters

Filters are a tool which allows selection of subsets of *SMSs*.

Filters consist of atomic constructions called *conditions*. There are boundary conditions (*BCs*) and optimizing conditions (*OpCs*). The first, in turn, may be elementary and chain.

An *elementary boundary condition (EBC)*, also called a *restriction*, may have the form $a\rho n$, $n\rho a$, or $a\rho a'$, where a and a' are objects, n is an integer number, and $\rho \in \{=, \geq, >, <, \leq\}$ is the symbol of a relation. The semantics of *EBCs* is as follows.

Let V be a set of multisets. A multiset $v \in V$ *satisfies* an *EBC* $\bar{n}\rho a$ if $n \cdot a \in v$ and $\bar{n}\rho n$ is true. Similarly, an *MS* $v \in V$ satisfies an *EBC* $a\rho\bar{n}$ if $n\rho\bar{n}$ is true. And, finally, an *MS* $v \in V$ satisfies an *EBC* $a\rho a'$, if $\bar{n} \cdot a \in v, \overline{n'} \cdot a' \in v$ and $\bar{n}\rho\overline{n'}$ is true.

Example 3.3
Let $V = \{v_1, v_2, v_3\}$, where

$$v_1 = \{3 \cdot a, 2 \cdot b, 4 \cdot c\},$$
$$v_2 = \{1 \cdot a, 3 \cdot b\},$$
$$v_3 = \{1 \cdot b, 5 \cdot c\}.$$

Then the result of the application of the *EBC* $a < 2$ to the set V is the set $\{v_2, v_3\}$ (note that v_3 is included in this set because, as said above, the absence of an object in a multiset is equivalent to the zero multiplicity of this object). If the *EBC* is $b > 3$, then the result is the empty set; if $c = 5$, then the result is $\{v_3\}$; if $b > c$, then the result is $\{v_2\}$.

An *optimizing condition* has the form $a = opt$, where a is an object and $opt \in \{min, max\}$. A multiset $v \in V$ satisfies an optimizing condition $a = min$, where $n \cdot a \in v$, if every multiset $v' \in V - \{v\}$ satisfies the *EBC* $a \geq n$. Such a condition is also called "minimizing." Similarly, an *MS* $v \in V$ satisfies an optimizing condition $a = max$, where $n \cdot a \in v$, if every multiset $v' \in V - \{v\}$ satisfies the *EBC* $a \leq n$. Such a condition also called "maximizing."

As above, the case $a \notin v$ ($a' \notin v'$) is equivalent to $0 \cdot a \in v$ ($0 \cdot a' \in v'$).

Example 3.4
Let V be the same as in the previous example and the optimizing condition be $a = max$. Then the result of its application to V is the set $\{v_1\}$. If the optimizing condition $b = min$, then the result is $\{v_3\}$.

A *filter F* is a set of both boundary and optimizing conditions and may be represented as the join of two *subfilters* named boundary and optimizing and denoted, respectively, $F_{\leq}$ and F_{opt}:

$$F = F_{\leq} \cup F_{opt}, \tag{3.10}$$

where $F_{\leq}$ is a set of *EBCs* and F_{opt} is a set of *OpCs*.

The result of filtration of an *MS V* by a filter F is denoted $V \downarrow F$ and is defined as follows:

$$V \downarrow F = (V \downarrow F_{\leq}) \downarrow F_{opt}, \tag{3.11}$$

i.e., V is filtered by a boundary subfilter, and the obtained result is filtrated by an optimizing subfilter.

The complete top-down formal definition of the semantics of boundary subfilters corresponding to the above verbal description is as follows:

$$v \downarrow F_{\leq} = \bigcap_{i=1}^{m} (V \downarrow bc_i) = \bigcap_{i=1}^{m} \left[\bigcup_{v \in V} (\{v\} \downarrow bc_i) \right] \tag{3.12}$$

$$\{v\} \downarrow \{a\theta n\} = \begin{cases} \{v\}, \text{ if } l \cdot a \in v \& \ l\theta n \vee a \notin v \& 0\theta n \\ \{\varnothing\} \text{ otherwise,} \end{cases} \tag{3.13}$$

$$\{v\} \downarrow \{a\theta a'\} = \begin{cases} \{v\}, \text{ if } l \cdot a \in v \& l' \cdot a' \in v \& l\theta l' \vee \\ a \notin v \& l' \cdot a' \in v \& 0\theta l' \vee \\ \ l \cdot a \in v \& a' \notin v \& l\theta 0 \vee \\ a \notin v \& a' \notin v \& 0\theta 0 \\ \{\varnothing\} \text{ otherwise.} \end{cases} \tag{3.14}$$

The semantics of optimizing subfilters is defined similarly:

$$v \downarrow F_{opt} = \bigcap_{j=1}^{k} \left(V \downarrow opt_j\right) \tag{3.15}$$

$$v \downarrow \{a = min\} = \{ v \mid l \cdot a \in v \& (\forall l' \cdot a' \in v) l \le l' \vee a \notin v\}, \tag{3.16}$$

$$v \downarrow \{a = max\} = \{ v \mid l \cdot a \in v \& (\forall l' \cdot a' \in v) l' \le l\}. \tag{3.17}$$

As can be seen, the result of an application of a filter $F_{\le}$ to a set of multisets V is a subset of V containing multisets satisfying all conditions in $F_{\le}$. Relations (3.13) and (3.14) define the results of filtering a one-element set $\{v\}$ by a one-element filter $\{a\theta n\}$ and $\{a\theta a'\}$, respectively. These relations are quite evident; the only singularity concerns objects which do not belong to a filtered multiset, which corresponds to zero multiplicities of such objects. The result of an optimizing filtration (3.15)–(3.17) also fully corresponds to the usual sense of such a filtration: those multisets are selected which contain multiobjects $l \cdot a$ with the minimal (maximal) multiplicity l in comparison with all other multisets containing multiobjects $l' \cdot a$ (as above, some heterogeneity is produced by cases where an object a does not belong to a multiset which is an element of set V).

Example 3.5

Let the *SMS V* be the same as in the two previous examples, and the filter

$$F = \{a \ge 1, b \le 3, c = min\}.$$

Then

$$V \downarrow F = (V \downarrow F_{\le}) \downarrow F_{opt} = (V \downarrow \{a \ge 1, \ b \le 3\})$$
$$\downarrow \{c = min\} = \left(\{v_1, \ v_2\} \bigcap \{v_1, \ v_2, \ v_3\}\right) \downarrow \{c = min\} = \{v_1, v_2\}$$
$$\downarrow \{c = min\} = \{v_2\},$$

because the multiplicity of the object c is zero ($c \notin v_2$).

Table 3.1 Boundary conditions

CBC	$V \downarrow CBC$
$2 \leq a \leq 4$	$\{v_2\}$
$1 \leq c \leq b \leq 3$	$\{v_1\}$
$a = b < 5$	$\{\emptyset\}$

Note that due to the commutativity of the operations of set-theoretical join and intersection, filtration inside subfilters may be executed in an arbitrary order.

There may be more sophisticated boundary subfilters containing not only elementary but also so-called chain boundary conditions (*CBCs*). Each *CBC* is constructed from *EBCs* by writing them sequentially:

$$e_1 \rho_1 e_2 \rho_2 \ldots e_i \rho_i e_{i+1} \ldots e_m \rho_m e_{m+1}, \tag{3.18}$$

where $e_1, \ldots, e_{m+1}$ are objects or non-negative integers, while $\rho_1, \ldots, \rho_m$ are symbols of relations: $\rho_i \in \{=, \geq, >, <, \leq\}$.

The semantics of *CBCs* is defined as follows. It is affirmed that (3.18) represents a boundary filter

$$\{e_1 \rho_1 e_2,\ e_2 \rho_2 e_3,\ \ldots,\ e_i \rho_i e_{i+1},\ \ldots,\ e_m \rho_m e_{m+1}\}, \tag{3.19}$$

and a multiset $v \in V$ satisfies *CBC* (3.18) if and only if it satisfies (3.19).

Example 3.6
Let $V = \{v_1, v_2\}$, where

$$v_1 = \{5 \cdot a,\ 3 \cdot b,\ 1 \cdot c\},$$
$$v_2 = \{2 \cdot a,\ 4 \cdot c,\ 3 \cdot d\},$$

and the chain boundary conditions are $2 \leq a \leq 4$, $1 \leq c \leq b \leq 3$, and $a = b < 5$.

Table 3.1 contains the results of an application of the listed boundary conditions to *V*.

Moreover, boundary subfilters may be sets of logical expressions consisting of boundary conditions and logical operations (&, $\vee$ and $\neg$) composed, if necessary, with brackets to any level of complexity. A similar structure may have optimizing subfilters. By this feature, the language of filters becomes close to the *SQL*-like query languages (Darwen 2012; Date 2012) providing access to relational data bases.

Let us note that a set of all multisets with the same basis is partly ordered by the relation of strict inclusion (Sheremet 2010, 2011a). If we consider some set of multisets *V*, then it may contain a multiset *v* which is not a strict submultiset of any other multiset $v' \in V$. The set of all such multisets will be denoted **max***V*:

$$\mathbf{max}\, V = \{ v \in V \mid v \in V \& (\nexists v' \in V)\, v \subset v' \}. \tag{3.20}$$

Similarly, there may be a multiset $v \in V$ such that there is no multiset $v' \in V$ which is a strict submultiset of v. The set of all such multisets will be denoted **min** V:

$$\mathbf{min}\, V = \{ v \in V \mid v \in V \& (\nexists v' \in V)\, v' \subset v \}. \tag{3.21}$$

Both **max** and **min** operations on multisets may be very useful in systems analysis and operations research in the *MGF*, and some such applications will be considered below in this book. (In papers on relational algebras and partially ordered sets, such operations are usually denoted by **sup** and **inf,** and their results are called a least upper bound and a greatest lower bound, respectively (Schmidt 2010; Simovici and Djeraba 2008).)

3.1.3 Multiset Grammars and Filtering Multiset Grammars

Multisets may be used for generating other multisets by means of *multiset grammars* ("multigrammars"). By analogy with classical grammars operating on strings of symbols (Chomsky 2005), we shall define a multiset grammar as a couple

$$S = < v_0, R >, \tag{3.22}$$

where v_0 is a multiset called a *kernel*, while R, called a *scheme,* is a finite set of so-called rules which are used for generation of new multisets from multisets already generated. A *rule* has the form

$$v \rightarrow v', \tag{3.23}$$

where v and v', called the left and the right parts of the rule, respectively, are multisets, and $v \neq \{\varnothing\}$.

The semantics of a rule is defined as follows. Let $\overline{v}$ be a multiset. We shall say that a rule (3.23) is applicable to $\overline{v}$ if

$$v \subseteq \overline{v}, \tag{3.24}$$

and shall define the result of an application as the multiset

$$\overline{v}' = \overline{v} - v + v', \tag{3.25}$$

i.e., if $\overline{v}$ includes v, then v is replaced by v'. This operation is called a *generation step*.

A generation step generating an *MS* $\overline{v}'$ from an *MS* $\overline{v}$ by an application of a rule $r \in R$ is denoted by

$$\overline{v} \overset{r}{\Rightarrow} \overline{v}', \tag{3.26}$$

while the fact that an *MS* $\overline{v}'$ is generated from an *MS* $\overline{v}$ by any (possibly empty) sequence of generation steps, called a *generation chain*, will be written as

$$\overline{v} \overset{R}{\Rightarrow} \overline{v}', \tag{3.27}$$

or, if the *MG* need not be named, then, as in classical string-operating grammars (Hopcroft et al. 2001; Meduna 2014),

$$\overline{v} \overset{*}{\Rightarrow} \overline{v}'. \tag{3.28}$$

If the generation chain is non-empty, instead of the $\overset{*}{\Rightarrow}$ symbol, $\overset{+}{\Rightarrow}$ is used.

The *set of multisets generated by an MG* $S = \ < v_0, R>$ is denoted V_s and is formally defined as follows:

$$V_s = \{v | v_{0\Rightarrow}\}. \tag{3.29}$$

By A_s we shall designate the set of all objects occurring in rules in R.

An *MS* v is called a *terminal multiset (TMS)* if there is no rule $r \in R$ that may be applied to v. The *set of terminal multisets (STMS)* will be denoted $\overline{V}_s$. Obviously, $\overline{V}_s \subseteq V_s$.

All introduced notions and constructions are illustrated by Fig. 3.1.

Example 3.7

Let us consider a multiset grammar $S = \ < v_0, R>$, where $v_0 = \{2 \cdot a, 3 \cdot b\}$ and $R = \{r_1, r_2\}$, where

$$r_1 : \{1 \cdot a,\ 2 \cdot b\} \rightarrow \{3 \cdot b\},$$
$$r_2 : \{2 \cdot a,\ 1 \cdot b\} \rightarrow \{2 \cdot b\}.$$

According to (3.22)–(3.29),

$$\{2 \cdot a,\ 3 \cdot b\} \overset{r_1}{\Rightarrow} \{1 \cdot a,\ 4 \cdot b\} \overset{r_1}{\Rightarrow} \{5 \cdot b\},$$

$$\{2 \cdot a,\ 3 \cdot b\} \overset{r_2}{\Rightarrow} \{4 \cdot b\},$$

$$\{2 \cdot a,\ 3 \cdot b\} \overset{+}{\Rightarrow} \{5 \cdot b\},$$

$$\{2 \cdot a,\ 3 \cdot b\} \overset{*}{\Rightarrow} \{5 \cdot b\},$$

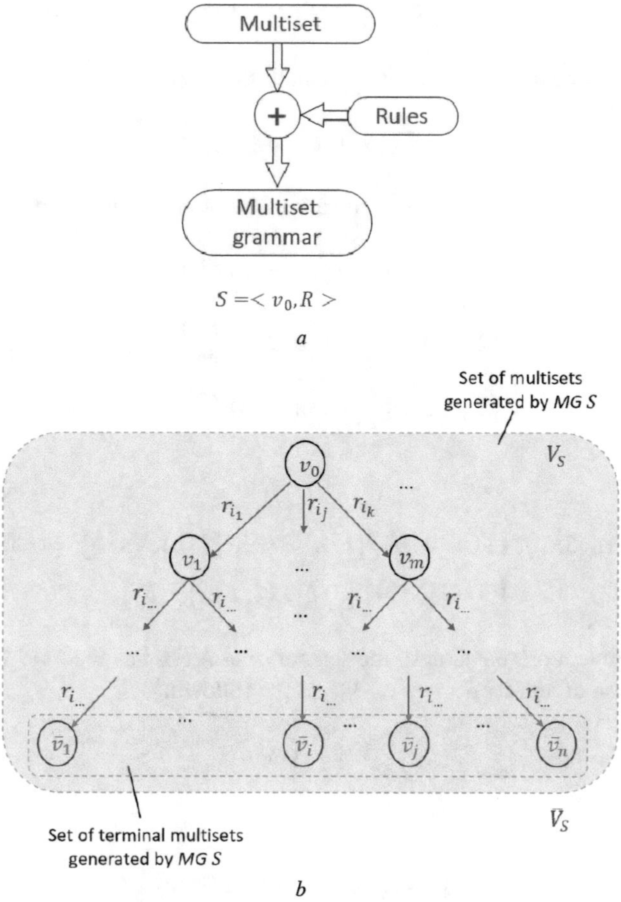

Fig. 3.1 Multiset grammar: (**a**) components; (**b**) set of multisets and set of terminal multisets generated by an *MG*

$$\{2 \cdot a, 3 \cdot b\} \overset{R}{\Rightarrow} \{5 \cdot b\},$$

$$\{2 \cdot a, 3 \cdot b\} \overset{+}{\Rightarrow} \{4 \cdot b\},$$

$$\{2 \cdot a, 3 \cdot b\} \overset{*}{\Rightarrow} \{4 \cdot b\},$$

$$\{2 \cdot a, 3 \cdot b\} \overset{R}{\Rightarrow} \{4 \cdot b\},$$

$$V_S = \{\{2 \cdot a, \ 3 \cdot b\}, \{1 \cdot a, \ 4 \cdot b\}, \{4 \cdot b\}, \{5 \cdot b\}\},$$

$$\bar{V}_S = \{\{4\cdot b\}, \{5\cdot b\}\}.$$

If add to R one more rule r_3 with an empty right part

$$\{1\cdot a, 1\cdot b\} \to \{\varnothing\},$$

then there will be three additional generation chains including this new rule:

$$\{2\cdot a, 3\cdot b\} \overset{r_3}{\Rightarrow} \{1\cdot a, 2\cdot b\} \overset{r_1}{\Rightarrow} \{3\cdot b\},$$

$$\{2\cdot a, 3\cdot b\} \overset{r_3}{\Rightarrow} \{1\cdot a, 2\cdot b\} \overset{r_3}{\Rightarrow} \{1\cdot b\},$$

$$\{2\cdot a, 3\cdot b\} \overset{r_1}{\Rightarrow} \{1\cdot a, 4\cdot b\} \overset{r_3}{\Rightarrow} \{3\cdot b\},$$

leading to

$$V_S = \{\{2\cdot a,\ 3\cdot b\}, \{1\cdot a,\ 4\cdot b\}, \{1\cdot a,\ 2\cdot b\}, \{1\cdot b\}, \{3\cdot b\}, \{4\cdot b\}, \{5\cdot b\}\},$$
$$\bar{V}_S = \{\{1\cdot b\}, \{3\cdot b\}, \{4\cdot b\}, \{5\cdot b\}\}.$$

The iterative representation of the semantics of *MGs*, i.e., the *SMS* V_s generated by application of an *MG* $S = \ < v_0, R>$, is the following:

$$V_{(0)} = \{v_0\}, \tag{3.30}$$

$$V_{(i+1)} = V_{(i)} \cup \left(\bigcup_{\bar{v}\in V_{(i)}} \bigcup_{r\in R} \pi\,(\bar{v},\ r) \right), \tag{3.31}$$

$$V_S = V_{(\infty)}, \tag{3.32}$$

where

$$\pi\,(\bar{v},\, v \to v') = \begin{cases} \{\bar{v} - v + v'\}, \text{ if } v\ \bar{v}, \\ \{\varnothing\} \text{ otherwise.} \end{cases} \tag{3.33}$$

As can be seen, (3.33) implements an application of a rule $v \to v'$ as defined by (3.24) and (3.25). The *SMS* V_s is a *fixed point* of the described process, i.e., $V_s = V_{(i)}$ with $i \to \infty$. If for some finite $V_{(i)} = V_{(i+1)}$, then $V_S = V_{(i)}$, and V_s is also finite. For a terminal multiset $\bar{v}$, the following are true:

$$\pi\,(\overline{v},\, r) = \{\varnothing\} \tag{3.34}$$

for all rules $r \in R$, i.e., no multiset may be generated from a *TMS*, and

$$\bar{V}_S = \{\, \overline{v} \mid \overline{v} \in V_S \& (\forall r \in R)\, \pi\,(\,\overline{v},\, r) = \{\varnothing\}\,\}. \tag{3.35}$$

A multigrammar $-S_{\overline{v}} = <\overline{v}, -R>$ is called a *mirror* (*or dual*) *MG* of a multigrammar $S = \ < v_0, R>$, if

$$-R = \{v' \rightarrow v \mid v \rightarrow v' \in R\}, \tag{3.36}$$

and $\overline{v} \in \bar{V}_S$. As can be seen, a mirror multigrammar generates multisets in the reverse order—from one of the terminal multisets generated by the initial *MG S* to its kernel v_0. According to this definition, there are $|\bar{V}_S|$ mirror *MGs* of an *MG S*. They all have the same scheme$-R$ and different kernels $\overline{v} \in \bar{V}_S$. However, in the general case, $\bar{V}_{-S} \neq \{v_0\}$. The only fact which is valid in this context is that $v_0 \in V_{-S}$.

Example 3.8
Let $S = \ < v_0, R>$, where $v_0 = \{2 \cdot a, 5 \cdot b\}$ and $R = \{r_1, r_2\}$, and

$$r_1 : \{1 \cdot a,\, 3 \cdot b\} \rightarrow \{1 \cdot b\},$$
$$r_2 : \{2 \cdot a,\, 2 \cdot b\} \rightarrow \{2 \cdot c\}.$$

As can be seen,

$$\{2 \cdot a,\, 5 \cdot b\} \overset{r_1}{\Rightarrow} \{1 \cdot a,\, 3 \cdot b\},$$
$$\{2 \cdot a,\, 5 \cdot b\} \overset{r_2}{\Rightarrow} \{3 \cdot b,\, 2 \cdot c\}.$$

The multigrammar $-S_v = \ < v, - R>$, where $v = \{1 \cdot b\}$, $-R = \{-r_1, -r_2\}$, and

$$-r_1 : \{1 \cdot b\} \rightarrow \{1 \cdot a,\, 3 \cdot b\},$$
$$-r_2 : \{2 \cdot c\} \rightarrow \{2 \cdot a,\, 2 \cdot b\},$$

is the mirror *MG* of the *MG S*. We have

$$\{1 \cdot b\} \overset{-r_1}{\Rightarrow} \{1 \cdot a,\, 3 \cdot b\} \overset{-r_1}{\Rightarrow} \{2 \cdot a,\, 5 \cdot b\} = v_o \overset{-r_1}{\Rightarrow} \{3 \cdot a,\, 7 \cdot b\} \overset{-r_1}{\Rightarrow} \{4 \cdot a,\, 9 \cdot b\}$$
$$\times \overset{-r_1}{\Rightarrow} \ldots.$$

By analogy with the aforementioned string-operating grammars (Chomsky 2005; Hopcroft et al. 2001; Meduna 2014), multiset grammars may be classified by their left part. Namely, if the left part of any rule of an *MG* is a multiset $\{1 \cdot a\}$, where

a may be any object $a \in A_S$, then the *MG* is called *context-free* (CF). Otherwise, i.e., if there exists at least one rule $v \to v'$ where v is not of the form $\{1 \cdot a\}$, the *MG* is called *general* (*GMG*). To simplify the following text of the monograph, we shall use the notion "multiset grammar(s)" as a *synonym* of "general multiset grammar (s)," thus omitting the attribute "general."

Example 3.9
Consider the context-free multigrammar

$$S = < \{1 \cdot a,\ 2 \cdot b\}, \{r_1, r_2, r_3, r_4\} >,$$

where

$$r_1 :: \{1 \cdot a\} \to \{3 \cdot b, 2 \cdot c\},$$
$$r_2 : \{1 \cdot a\} \to \{1 \cdot b, 3 \cdot c\},$$
$$r_3 : \{1 \cdot b\} \to \{2 \cdot c\},$$
$$r_4 : \{1 \cdot b\} \to \{1 \cdot c\}.$$

Logics of generation of the set $\overline{V}_S$ is represented by Table 3.2, where every line corresponds to one generation step in such a way that the first cell of this line is number of the multiset, to which the rule, which number is in the third cell, is applied. The aforementioned multiset is in the second cell of the line while the number of resulting *MS* in the fourth cell. In lines representing terminal multisets, the third and the fourth cells are empty.

According to the represented generation chains,

$$V_s = \{x_0, x_1, \ \ldots, x_{29}\},$$

$$\overline{V}_s = \bigcup_{i=14}^{29} \{x_i\} = \{\{4 \cdot c\}, \{5 \cdot c\}, \{7 \cdot c\}, \{8 \cdot c\}, \{10 \cdot c\}, \{11 \cdot c\}, \{13 \cdot c\}\}.$$

In the applications considered below, we shall use a simplified way of writing rules in the schemes of CF *MGs*:

$$a \to n_1 \cdot a_1, \ldots, n_m \cdot a_m, \tag{3.37}$$

where the left part, called the *header*, is used instead of a multiset $\{1 \cdot a\}$ and the right part, called the *body*, is a multiset $\{n_1 \cdot a_1, \ldots, n_m \cdot a_m\}$ without the brackets. The set $\overline{A}_s \subseteq A_s$ contains the so-called terminal objects (*TOs*), which are present (if at all) only in the bodies of the rules. The set $A_S^N = A_s - \overline{A}_s$ contains the so-called non-terminal objects (*NTOs*), each being the header of at least one rule. The set of rules with the same header a is denoted R_a; evidently, $R_a \subseteq R$.

Table 3.2 Generation of the set $\overline{V}_S$

Number of the multiset	Multiset	Number of the applied rule	Number of the resulting multiset
0	$\{1 \cdot a, 1 \cdot b\}$	1	1
		2	2
		3	3
1	$\{3 \cdot b, 1 \cdot c\}$	4	4
		5	5
2	$\{2 \cdot b, 3 \cdot c\}$	4	6
		5	7
3	$\{2 \cdot b, 2 \cdot c\}$	4	8
		5	9
4	$\{2 \cdot b, 3 \cdot c\}$	4	10
		5	11
5	$\{2 \cdot b, 5 \cdot c\}$	4	12
		5	13
6	$\{1 \cdot b, 4 \cdot c\}$	4	14
		5	15
7	$\{1 \cdot b, 7 \cdot c\}$	4	16
		5	17
8	$\{1 \cdot b, 3 \cdot c\}$	4	18
		5	19
9	$\{1 \cdot b, 6 \cdot c\}$	4	20
		5	21
10	$\{1 \cdot b, 4 \cdot c\}$	4	22
		5	23
		4	24
11	$\{1 \cdot b, 7 \cdot c\}$	5	25
		4	26
12	$\{1 \cdot b, 6 \cdot c\}$	5	27
		4	28
13	$\{1 \cdot b, 9 \cdot c\}$	5	29
14	$\{5 \cdot c\}$		
15	$\{8 \cdot c\}$		
16	$\{8 \cdot c\}$		
17	$\{11 \cdot c\}$		
18	$\{4 \cdot c\}$		
19	$\{7 \cdot c\}$		
20	$\{7 \cdot c\}$		
21	$\{10 \cdot c\}$		
22	$\{5 \cdot c\}$		
23	$\{8 \cdot c\}$		
24	$\{8 \cdot c\}$		
25	$\{11 \cdot c\}$		

(continued)

Table 3.2 (continued)

Number of the multiset	Multiset	Number of the applied rule	Number of the resulting multiset
26	$\{7 \cdot c\}$		
27	$\{10 \cdot c\}$		
28	$\{10 \cdot c\}$		
29	$\{13 \cdot c\}$		

Rules like (3.37) will be called *unitary rules* (*URs*); similarly, any *MG* whose scheme is a set of *URs* will be called a *unitary multiset grammar* (*UMG*).

The most natural interpretation of (3.37) is as follows: a system (device) a consists of n_1 subsystems (subdevices, spare parts) a_1, …, n_m subsystems (subdevices, spare parts) a_m. This interpretation, called *structural*, is illustrated by the following example.

Example 3.10
Let us consider a computer, consisting of four processors, a memory, a monitor, and a keyboard. A processor, in turn, consists of 16 cores and a memory of 64 1-Gbit modules. This device may be represented by the *UMG* $S = \langle v_0, R \rangle$, where $v_0 = \{1 \cdot (computer)\}$, $R = \{r_1, r_2, r_3\}$, and rules are as follows:

$$r_1 : (\text{computer}) \rightarrow 4 \cdot (\text{processor}), 1 \cdot (\text{memory}), 1 \cdot (\text{monitor}), 1 \cdot (\text{keyboard});$$

$$r_2 : (\text{processor}) \rightarrow 16 \cdot (\text{core});$$

$$r_3 : (\text{memory}) \rightarrow 64 \cdot (1\text{gbit}).$$

If industrial systems are considered, then the multiobjects representing the resources necessary for manufacturing (assembling) an object a from n_1 objects a_1, …, n_m objects a_m may be written as the body of a corresponding *UR:*

$$a \rightarrow n_1 \cdot a_1, \ldots, n_m \cdot a_m, n'_1 \cdot a'_1, \ldots, n'_l \cdot a'_l, \tag{3.38}$$

where the aforementioned amounts of resources (electrical energy, fuel, money, etc.) are n'_1 units of resource $a'_1, \ldots, n'_l$ units of resource a'_l. Henceforth, to distinguish the notion "object," used in the multiset theory, and the notion "object," used in the considered applications, where it means some manufactured device or system as well as a unit of a resource consumed during manufacturing, we shall use below the notion "object-resource" (*OR*), covering all entities occurring in *URs*. (Note that *ORs* like $a'_1, \ldots, a'_l$ from (3.38) may be also headers of some *URs*, which means they are produced by some devices.) Let us illustrate this interpretation, which we call *technological*, by the following example, where durations of time intervals necessary to manufacture devices for producing *ORs* are represented in the form $(\Delta t : x)$, where Δt is a time measurement unit (here it will be "minute," denoted *mnt*), while x is an identifier of an *MD*. The same form of representation of time intervals will be applied everywhere below in this book.

Example 3.11
Let us consider the computer from the previous Example 3.10 and unitary rules describing its manufacturing:

$$r_1' : \ (\text{computer}) \rightarrow 4 \cdot (\text{processor}), 1 \cdot (\text{memory}), 1 \cdot (\text{monitor}), 1 \cdot (\text{keyboard}),$$
$$5 \cdot (\text{mnt : asm} - \text{comp}), 100 \cdot (\text{wh}), 300 \cdot (\text{usd});$$
$$r_2' : \ (\text{processor}) \rightarrow 16 \cdot (\text{core}), 3 \cdot (\text{mnt : asm} - \text{proc}), 30 \cdot (\text{wh}), 500 \cdot (\text{usd});$$
$$r_3' : \ (\text{memory}) \rightarrow 64 \cdot (1\text{gbit}), 1 \cdot (\text{mnt : asm} - \text{mem}), 20 \cdot (\text{wh}), 100 \cdot (\text{usd}).$$

As can be seen, the *UR* r_1' defines that it is necessary to spend 5 min of operation of the assembly line (*AL*) manufacturing computers from their components, 100 Wh h of electrical energy, as well as 300 *dollars* to produce one computer. Similarly, the *UR* r_2' defines that it is necessary to spend 3 min of operation of the *AL* manufacturing processors from the processor cores, 30 Wh h of electrical energy, as well as 500 *dollars* to produce one processor. Finally, the *UR* r_3' defines that it is necessary to spend 1 min of operation of the *AL* manufacturing computer memory from 64 one-gigabit memory modules, 20 Wh h of electrical energy, as well as 100 *dollars* to produce one memory.

Let us now define the syntax and the semantics of *filtering multiset grammars* (*FMGs*).

An *FMG* $S = \ < v_0, R, F>$, where v_0 and R have the same sense as in *MGs* and F is a filter, defines a set of terminal multisets in the following way:

$$\overline{V_s} = \overline{V}_{s'} \downarrow F, \tag{3.39}$$

where $S' = \ < v_0, R>$ is called the *core multigrammar of the FMG S*. As can be seen, the *STMS* generated by a core *MG* S' is filtered by F, and the resulting set of terminal multisets is defined to be generated by the *FMG S* (Fig. 3.2).

Example 3.12
Let *FMG* $S = \ < \{1 \cdot a\}, R, F>$, where the scheme $R = \{r_1, r_2, r_3, r_4\}$ and

$$r_1 : \ \{1 \cdot a\} \rightarrow \{3 \cdot b, 2 \cdot c\},$$
$$r_2 : \ \{1 \cdot a\} \rightarrow \{2 \cdot b, 5 \cdot c\},$$
$$r_3 : \ \{1 \cdot b\} \rightarrow \{2 \cdot c, 1 \cdot d\},$$
$$r_4 : \ \{1 \cdot b\} \rightarrow \{1 \cdot c, 3 \cdot d\},$$

and the filter $F = \{c \leq 8, d > 3, c = max, d = min\}$, i.e.

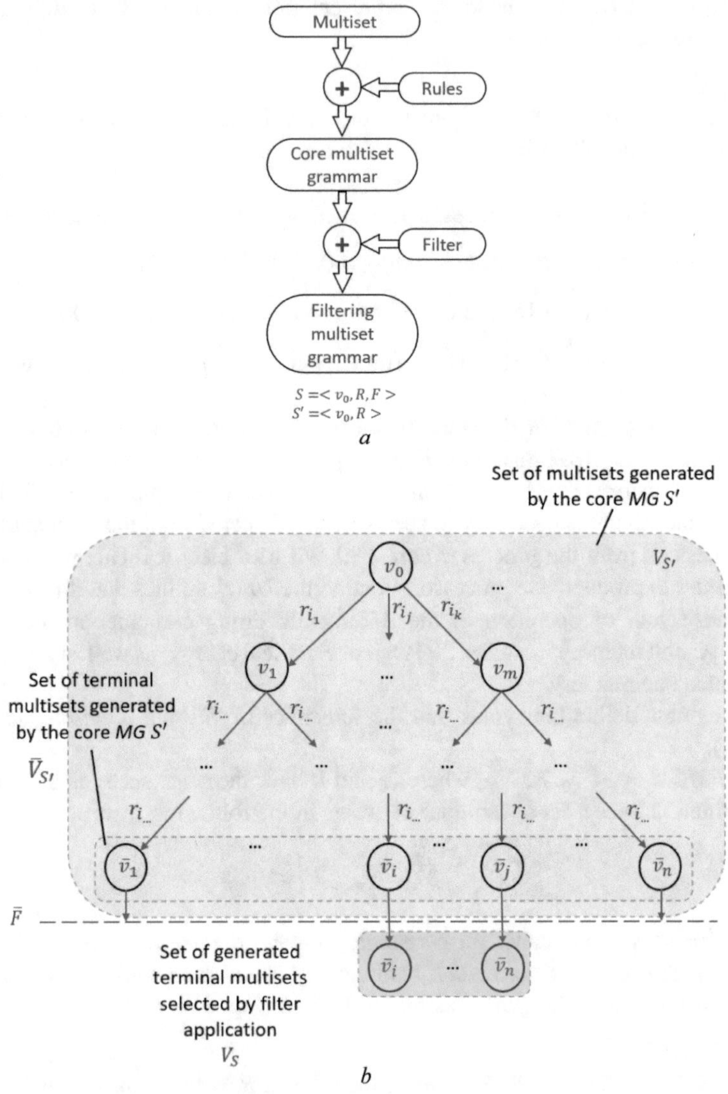

Fig. 3.2 Filtering multiset grammar: (**a**) components; (**b**) set of multisets and set of terminal multisets generated by an *FMG*

$$F_{\leq} = \{c \leq 8,\ d > 3\}, F_{opt} = \{c = max\ ,\ d = min\ \}.$$

Then $S' = \ < \{1 \cdot a\}, R>$, and logics of generation of the set $\overline{V}_{s'}$ is represented by Table 3.3, similar to Table 3.2.

Table 3.3 Generation of the set $\overline{V}_{s'}$

Number of the multiset	Multiset	Number of the applied rule	Number of the resulting multiset
0	$\{1 \cdot a\}$	1	1
		2	2
1	$\{3 \cdot b, 2 \cdot c\}$	3	3
		4	4
2	$\{2 \cdot b, 5 \cdot c\}$	3	5
		4	6
3	$\{2 \cdot b, 4 \cdot c, 1 \cdot d\}$	3	7
		4	8
4	$\{2 \cdot b, 3 \cdot c, 3 \cdot d\}$	3	9
		4	10
5	$\{1 \cdot b, 7 \cdot c, 1 \cdot d\}$	3	11
		4	12
6	$\{1 \cdot b, 6 \cdot c, 3 \cdot d\}$	3	13
		4	14
7	$\{1 \cdot b, 6 \cdot c, 2 \cdot d\}$	3	15
		4	16
8	$\{1 \cdot b, 5 \cdot c, 4 \cdot d\}$	3	17
		4	18
9	$\{1 \cdot b, 5 \cdot c, 4 \cdot d\}$	3	19
		4	20
10	$\{1 \cdot b, 4 \cdot c, 6 \cdot d\}$	3	21
		4	22
11	$\{9 \cdot c, 2 \cdot d\}$		
12	$\{8 \cdot c, 2 \cdot d\}$		
13	$\{8 \cdot c, 4 \cdot d\}$		
14	$\{7 \cdot c, 6 \cdot d\}$		
15	$\{8 \cdot c, 3 \cdot d\}$		
16	$\{7 \cdot c, 5 \cdot d\}$		
17	$\{7 \cdot c, 5 \cdot d\}$		
19	$\{7 \cdot c, 5 \cdot d\}$		
19	$\{7 \cdot c, 5 \cdot d\}$		
20	$\{6 \cdot c, 7 \cdot d\}$		
21	$\{8 \cdot c, 7 \cdot d\}$		
22	$\{5 \cdot c, 9 \cdot d\}$		

As now may be seen,

$$V_{s'} = \{x_0, x_1, \ldots, x_{22}\},$$

$$\overline{V}_{s'} = \bigcup_{i=11}^{22} \{x_i\}$$

$$= \left\{ \begin{array}{l} \{5\cdot c,\ 2\cdot d\}, \{6\cdot c,\ 7\cdot d\}, \{7\cdot c,\ 5\cdot d\}, \{7\cdot c,\ 6\cdot d\}, \{8\cdot c,\ 2\cdot d\}, \\ \{8\cdot c,\ 3\cdot d\}, \{8\cdot c,\ 4\cdot d\}, \{8\cdot c,\ 7\cdot d\}, \{9\cdot c,\ 2\cdot d\} \end{array} \right\},$$

$$\overline{V_s} = \overline{V}_{s'} \downarrow F = \left(\overline{V}_{s'} \downarrow \{c \leq 8,\ d > 3\}\right) \downarrow \{c = max\,, d = min\,\} =$$

$$\{\{5\cdot c,\ 2\cdot\ \ d\}, \{6\cdot c,\ 7\cdot d\}, \{7\cdot c,\ 5\cdot d\}, \{7\cdot c,\ 6\cdot d\}\}$$
$$\downarrow \{c = max\,, d = min\,\} =$$

$$\{\{5\cdot c,\ 2\cdot d\}, \{6\cdot c,\ 7\cdot d\}, \{7\cdot c,\ 5\cdot d\}, \{7\cdot c,\ 6\cdot d\}\} \downarrow \{c = max\,\} \cap$$

$$\{\{5\cdot c,\ 2\cdot d\}, \{6\cdot c,\ 7\cdot d\}, \{7\cdot c,\ 5\cdot d\}, \{7\cdot c,\ 6\cdot d\}\} \downarrow \{d = min\,\} =$$

$$\{\{7\cdot c,\ 5\cdot d\}, \{7\cdot c,\ 6\cdot d\}\} \cap \{\{5\cdot c,\ 2\cdot d\}\} = \{\varnothing\}.$$

If $F_{opt} = \{c = min\,, d = min\,\}$,then

$$\overline{V_s} = \{\{5\cdot c,\ 2\cdot d\}, \{6\cdot c,\ 7\cdot d\}, \{7\cdot c,\ 5\cdot d\}, \{7\cdot c,\ 6\cdot d\}\} \downarrow \{c = min\,\} \cap$$

$$\{\{5\cdot c,\ 2\cdot d\}, \{6\cdot c,\ 7\cdot d\}, \{7\cdot c,\ 5\cdot d\}, \{7\cdot c,\ 6\cdot d\}\} \downarrow \{d = min\,\} =$$

$$\{\{5\cdot c,\ 2\cdot d\}\} \cap \{\{5\cdot c,\ 2\cdot d\}\} = \{\{5\cdot c,\ 2\cdot d\}\}.$$

A *filtering unitary multiset grammar* (*FUMG*) is defined similarly, i.e., as a triple $S = \ < v_0, R, F>$ where the scheme R is a set of unitary rules. The mathematical semantics of *FUMGs* is defined by the same relation (3.39); the only difference is that $S' = \ < v_0, R>$ is the *core unitary multigrammar of the FUMG S*.

Let us note that the introduced knowledge representation does not include time as a basic category necessary for description of time-dependent processes of *STS* operation. A time-incorporating generalization of multiset grammars suitable for solution of such tasks is considered in the next section.

3.2 Temporal Multiset Grammars

3.2.1 Temporal Multisets, Temporal Rules, and Manufacturing Devices

Temporal multiset grammars operate on *temporal multisets* (abbreviated below as *MST*). It is postulated that every *MST* contains a multiobject $n \cdot t$, where t is a special fixed object "time", while n is a number defining a point of a discrete time scale when "the rest" of v, i.e., $v - \{n \cdot t\}$ called the *content of the MST*, is active. A multiobject $n \cdot t$ is called a "time marker" (*TM*) *of an MST*. This technique provides an easy representation of the so-called dynamic multisets, whose content depends on time according to some predefined relations.

A *dynamic multiset* (*DMS*) is a function whose set of values on a discrete time interval $[1, n]$ is a set of multisets $\{v_1 + \{1 \cdot t\}, \ldots, v_n + \{n \cdot t\}\}$. This set may be defined in analytical form as

$$\boldsymbol{v}(n) = \{f_1(n) \cdot a_1, \ \ldots, f_m(n) \cdot a_m, n \cdot t\}, \tag{3.40}$$

where functions $f_1, \ldots, f_m$ on a variable n are, in turn, analytically defined and a value $\boldsymbol{v}(\overline{n})$, where $\overline{n} \in [0, \infty]$ is a discrete time moment, is an *MST*

$$\{f_1(\overline{n}) \cdot a_1, \ \ldots, f_m(\overline{n}) \cdot a_m, \overline{n} \cdot t\}. \tag{3.41}$$

In the case $n = 0$,

$$\boldsymbol{v}(0) = \{f_1(0) \cdot a_1, \ \ldots, f_m(0) \cdot a_m\}, \tag{3.42}$$

so all multisets considered in the previous sections are special cases of *MSTs* corresponding to the initial moment of the time scale ($\overline{n} = 0$).

Temporal multisets are a basic construction in the following definition of temporal multiset grammars.

A *temporal multiset grammar* $S = \ < v_0, R>$ differs from any multigrammar considered earlier in (Sheremet 2010, 2011a, 2019a) and in Sect. 3.1 by two principal features:

1. The kernel v_0 is a temporal multiset representing the resource base of some industrial system at an initial moment of its operation (not necessarily the aforementioned $\overline{n} = 0$).
2. The scheme R representing the manufacturing technological base of this *IS* is a set of *temporal rules* (*TRs*)—couples $\langle r; v \to v' \rangle$, each representing a *manufacturing device* named r, whose inputs are defined by a multiset v and outputs by a multiset v'.

A key feature of any *TMG* is that every temporal rule $\langle r; v \to v' \rangle$ contains the time interval necessary for the *MD* r to produce the resource collection v'. This is specified by the inclusion in the multiset v' of a multiobject $\Delta n \cdot \Delta t$, where Δt is a specific object representing a time measurement unit, so $\Delta n \cdot \Delta t$ is a representation of the duration of the time interval necessary for the *MD* r to produce the output collection of object-resources $v' - \{\Delta n \cdot \Delta t\}$ after the input collection of object-resources v has become available to the *MD* r. (This representation of time intervals was introduced in Sect. 3.1.3 starting with Example 3.11.) A temporal rule may be represented by a "black box," depicted in Fig. 3.3a.

This unified representation of a "time resource" provides a natural and flexible description of *ISs* of any degree of complexity. (*There may be an alternative form of MD representation when a multiobject* $\Delta n \cdot \Delta t$ *appears in not the right but the left part of a TR, thus interpreting this MO as a consumed collection of ORs. Both ways are of equivalent representative power, but here we shall use a "right-part" representation, which follows not a "consuming" but a "generating" role of a*

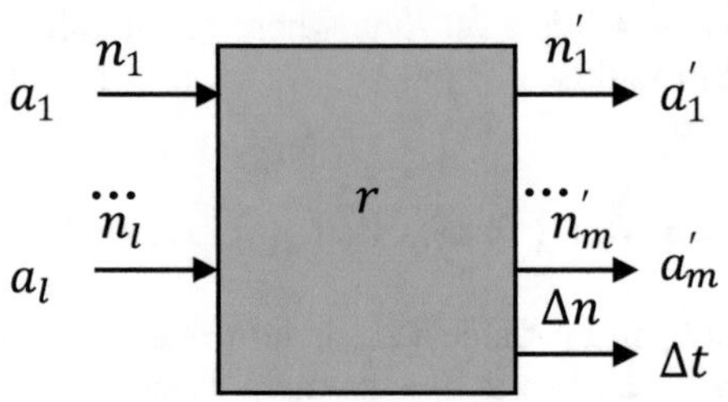

a

$$< r; \{n_1 \cdot a_1, \ldots, n_l \cdot a_l\} \to \{n_1' \cdot a_1', \ldots, n_m' \cdot a_m', \Delta n \cdot \Delta t\} >$$

b

Fig. 3.3 Representation of a manufacturing device used in temporal multiset grammars: (**a**) "black box" *r*; (**b**) rule *r*

"time resource" that appoints a time moment marking the start moment of an MD operation cycle, when an OR collection $\{n_1' \cdot a_1, \ldots, n_m' \cdot a_m'\}$ *will be loaded into the resource base.*)

The result of an application of the *TR* $\langle r; v \to v' \rangle$ to an *MST* $\overline{v}$, where $\Delta n \cdot \Delta t \in v'$ and $n \cdot t \in \overline{v}$, is the temporal multiset

$$(\overline{v} - \{n \cdot t\}) - v + v' + \{(n + \Delta n) \cdot t\}, \tag{3.43}$$

which fully represents the real operation of an *IS*: an input collection of resources v at a time moment n is extracted from a resource base, and the result $v' - \{\Delta n \cdot \Delta t\}$ of operation of an *MD* r appears as an output collection in the *RB* at a time moment $n + \Delta n$. Of course, this action is possible if all object-resources necessary to the *MD* are present in the resource base at time moment n (Fig. 3.4), i.e., $v \subseteq \overline{v}$. If $\Delta n = 0$, then the *TR* is the same as the rules presented in the non-temporal case, and thismeans that the result of any such *MD* operation appears in the *RB* immediately.

Let us illustrate a nexus of the "black box" and *TMG*-based techniques of representation of manufacturing technological bases by the following example.

Example 3.13
Let us consider the manufacturing technological base of some cyberphysical manufacturing facility which is an industrial system manufacturing passive resources, whose elements are depicted in Fig. 3.5.

The considered *MTB* includes some number of robotized assembly lines. The first one, named "asm-line-1," producing cars from their components in such a way that one car is assembled from four wheels, one body, and one engine, and this operation consumes 300 kWh h of electrical power. In 30 min of work of this assembly line

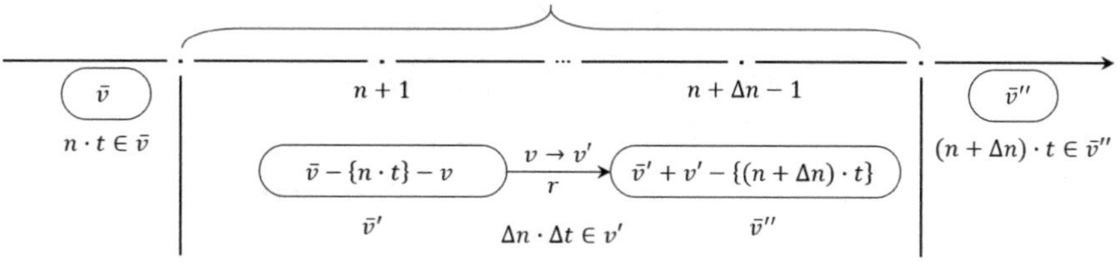

Fig. 3.4 Application of a manufacturing device (temporal rule) r to a resource base (temporal multiset) at a time moment n, when $v \subseteq \bar{v}$

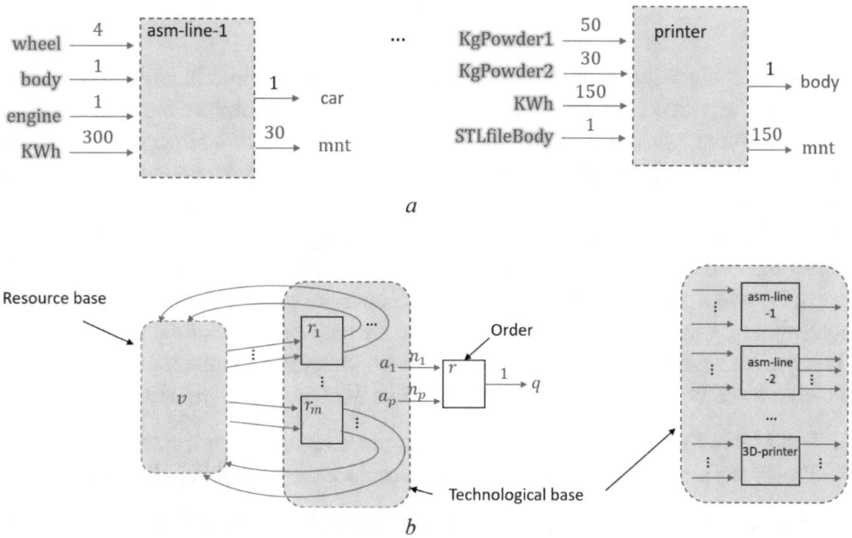

Fig. 3.5 Example of "black box" representation of manufacturing technological bases of *ISs MPR*: (**a**) "black boxes" representing *MDs*; (**b**) "black boxes" representing an *MTB*

after the start of its operation cycle, provided it has the aforementioned car components at its input, a car is manufactured and delivered to the resource base of the *IS*. There are other assembly lines named "asm-line-2", ..., "asm-line-*k*" producing various components from their subcomponents and ready spare parts. However, this facility is equipped with a *3D*-printer producing car bodies. One body is printed from 50 kg of some powder 1 and 30 kg of another powder 2, applying for this purpose an *STL*-file of a body containing information about its layer structure and consuming 150 kWh h of electrical power. This work is done by this *3D*-printer in 150 min.

The scheme of a *TMG* representing this *MTB* contains the following temporal rules:

$$< asm - line - 1;$$

$$\{4 \cdot wheel,\ 1 \cdot body,\ 1 \cdot engine,\ 300 \cdot kWh\}$$

$$\{1 \cdot car,\ 30 \cdot mnt\ \} >,$$

...

$$< printer;$$

$$\{50 \cdot kgPowder1,\ 30 \cdot kgPowder2,\ 150 \cdot kWh,\ 1 \cdot STLfileBody\}$$

$$\{1 \cdot body,\ 150 \cdot mnt\ \} > .$$

The most important thing here is that manufacturing devices in an *IS* may operate in parallel: any two *MDs* which do not require for their operation one and the same object-resource presented in the resource base may start their operation cycles simultaneously and manufacture their output collections independently. So, in the general case there may be many different variants of activation of *MDs* at any moment. Each such sequence of activations, being an implementation of an *IS* schedule, leads to its own resource base and its own time moment when the *IS* stops. As was already said in Sect. 2.3, any *IS* operates under the supervision of an *IS* controller as a set of an interconnected resource base and *N* manufacturing devices. At any time moment the *ISC* may send *"start" messages* to manufacturing devices which are ready to operate. Readiness of some *MD* to operate means:

1. The *MD* is free, i.e., it is not operating at the moment when it receives the *Sm*.
2. The *IS* resource base contains all resources necessary for this *MD* operation.

In this way, an *IS schedule* may be represented as a set of couples $\langle n, R(n)\rangle$ ordered by the size of the first component, where n is a time moment and $R(n) \subseteq R$ is a subset of the manufacturing technological base, whose elements (manufacturing devices) will be activated immediately upon receiving an *Sm* from the *ISC*. It is supposed that collections of object-resources manufactured by *MDs* finishing their operation cycles at a moment n are at the same moment loaded into an *RB*, and from this moment, all these object-resources are available to all manufacturing devices.

According to this description, operation of a manufacturing device is performed as a sequence of non-intersecting *operation cycles*, which are time intervals established by the multiplicities of the object Δt in the temporal rules. Operation of an *IS* as a whole may be represented by a Gantt-like chart (Fig. 3.6), in which every axis, as in Fig. 2.6, corresponds to some one *MD* (and, thus, to the *TR* that represents), and operation cycles are marked by bold segments (this representation differs from "classical" Gantt charts by an absence of pointers/arrows, which in the classical case connect subordinated segments; this is a consequence of the implicit interaction of *MDs* via a resource base).

Operation of an industrial system as a whole is driven by a flow of orders initiated by customers at unpredictable time moments. Every order presumes, in the simplest

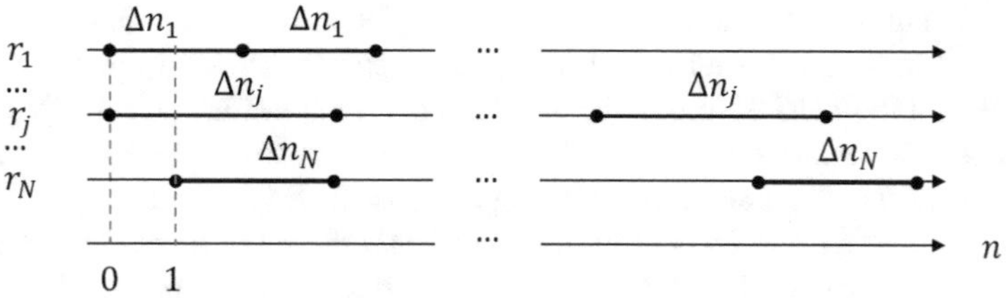

Fig. 3.6 A Gantt-like chart representing operation of an industrial system

case, a collection of resources which would be manufactured by the *IS* for the customer and also a time moment $\mathbf{n}$ when the order would be completed. The value $\mathbf{n}$ may be defined as follows:

1. $\mathbf{n} \leq n$, i.e., this means the order must be completed not later than the moment n.
2. $\mathbf{n} = n$, i.e., the order must be completed precisely at the moment n.
3. $n \leq \mathbf{n} \leq n'$, i.e., inside a time interval $[n, n']$.
4. $\mathbf{n} = min$, i.e., at the earliest moment possible.
5. $\mathbf{n} \leq n$ & $\mathbf{n} = min$, i.e., at the earliest moment possible, not later than n.
6. $n \leq \mathbf{n} \leq n'$ & $\mathbf{n} = min$, i.e., at the earliest moment possible, being inside a time interval $[n, n']$.

(In a general case, any disjunctive-conjunctive form of the basic conditions $n \leq \mathbf{n} \leq n'$, $\mathbf{n} > n$, $\mathbf{n} \leq n$, and $\mathbf{n} = opt$, where $opt \in \{min, max\}$, may be used.)

Also in the general case, there may be some restrictions (limitations) on the amounts of object-resources available to an *IS* for order completion. These limitations may be established by the *IS* (in fact, by some announced regulations developed by the system's management and implemented during *IS* operation), by the customer initiating the order, or a combination of both. In fact, the above-considered time limitations are a special case of the possible set of restrictions on available resources assigned to every incoming order.

Let us move on to the mathematical semantics of *TMGs* formally determining the above verbal description of the process of *IS* operation.

3.2.2 *Mathematical Semantics of* TMGs

Let $S = < v_0, R>$ be a temporal multiset grammar whose kernel v_0 is a temporal multiset and R is a scheme, which is a set of temporal rules $\langle r; v \rightarrow v' \rangle$, such that the *MS* v'contains a multiobject $\Delta n \cdot \Delta t$. The *MST* v_0 represents an *IS* resource base at an initial moment n_0, so $n_0 \cdot t \in v_0$, and R represents the *IS* manufacturing technological base, as described above.

By analogy with general multiset grammars, a mathematical semantics of *TMGs* may be developed in the form of a set of relations describing the sequential creation of a set of temporal multisets V_S. This set is created in full accordance with the verbal description introduced above and reflecting key features of our "black box" representation of industrial systems. Following this approach, the set V_S generated by an application of the *TMG S* from the substantial point of view may be defined as the set of all possible resource bases which may be created from the initial *RB* v_0 by application of the manufacturing technological base *R*. However, *the main objective of application of such a TMG is to obtain not only the set of resource bases which may be created by the MTB of the represented IS but, to a much higher degree, a set of schedules enabling the creation of the aforementioned set of RBs*. In this way, the mathematical semantics of *TMGs* will be formally defined in such a way that the *result of any TMG application is a set of schedules, each enabling creation of the relevant resource base*. Another difference between the mathematical semantics of *TMGs* and of *MGs* will be that in the case of *TMGs*, there are *terminal temporal multisets* (*TMST*) in the set $\overline{V_S}$ that correspond to states of the *IS* in which no manufacturing device may be applied to the current resource base by reason of lack of necessary resources. In the case of *TMGs*, *all MSTs* $v \in V_S$ *are essential*, and some of them, *not only terminal ones*, may be selected by use of the aforementioned predefined relevant criteria to obtain schedules whose execution leads to the required resource bases. This obstacle will be highlighted once more when we consider filtering temporal multiset grammars.

Let us formally define V_S as a result of application of a *TMG S*, using a verbal description of *IS* operation and taking into account all that has been said above about the necessity of inclusion of *IS* schedules in the set V_S (and thus, naturally, in the set $\overline{V_S}$).

A key notion of the definition of the mathematical semantics of *TMGs* proposed and considered below is the *current state of an IS*, represented by a temporal multiset

$$v = v_A + v_R + \{n \cdot t\}, \tag{3.44}$$

which is a sum of three *MSs*:

1. $v_A = \{m_1 \cdot a_1, \ldots, m_M \cdot a_M\}$, which is the resource base of the *IS* at a time moment $n \geq n_0$, where we recall that n_0 is the initial moment of *IS* operation
2. $v_R = \{1 \cdot r_1\tau_1, \ldots, 1 \cdot r_N\tau_N\}$, which is the current schedule of the *IS* at the same time moment (here r_j is the unique name of a manufacturing device and τ_j is its individual schedule (*ISh*), while *N* is the number of *MDs* in the *MTB*)
3. $\{n \cdot t\}$, whose only element is a time marker defining the aforementioned current moment

Below we shall denote by $\bar{R}$ the set $\{r_1, \ldots, r_N\}$ of names of the manufacturing devices in the manufacturing technological base of the *IS*.

Let us consider now the multiset v_R, which is a key element of all further developments, in more detail.

A string $r_j\tau_j$, which is the *composite name of an object* (such names will be called *composites* for short, and they will be considered in more detail in Sect. 3.5.3), consists of a prefix r_j, which is the unique name of a temporal rule (which, recall, represents a manufacturing device, and thus a rule inherits the name of an *MD*), and a suffix τ_j, which is a string $[n_1^j, l_1^j] \ldots [n_{k_j}^j, l_{k_j}^j]$, where "[", "," and "]" are dividers, not parts of r_j, n_1^j, and l_l^j. (The alphabet of symbols used in composites will be denoted *U*.) A suffix τ_j as a whole is a string representation of an individual schedule of a manufacturing device r_j, where each n_i^j is a time moment when this device starts its operation cycle and l_i^j is a time moment when this device finishes this *OC*, becomes free, and may begin its next *OC*. Here for every $\in [1, k_j]$

$$l_i^j - n_i^j = \Delta n_j \tag{3.45}$$

and also

$$n_{i+1}^j \geq l_i^j, \tag{3.46}$$

where Δn_j is the duration of an *OC* of an *MD* r_j, established by the representation of this manufacturing device by the temporal rule

$$< r_j; v_j \ \rightarrow v_j' >, \tag{3.47}$$

such that $\Delta n_j \cdot \Delta t \in v_j'$. Restriction (3.46) excludes intersection of two *OCs* of one and the same *MD*, ensuring a sequential execution of jobs by any such device. As may be seen, if $n_{i+1}^j = l_i^j$, then the next *OC* begins immediately after a previous one ends.

So, as can be seen, a suffix τ_j is basic information for an *IS* scheduler, and an *IS* schedule $\boldsymbol{Sch}(v_R)$, defined above as set of couples $\langle n, R(n) \rangle$ ordered by the size of the first component, may be easily obtained from a multiset v_R:

$$\boldsymbol{Sch}(v_R) = \bigcup_{n \in \boldsymbol{N}(v_R)} \left\{ < n, \bigcup_{r[n_1,\ l_1] \ldots [n_k,\ l_k] \in v_R n \in \{n_1\ ,\ \ldots,\ n_k\}} \{r\} > \right\}, \tag{3.48}$$

where

$$\boldsymbol{N}(v_R) = \bigcup_{r[n_1,\ l_1] \ldots [n_k,\ l_k] \in v_R} \{n_1\ ,\ \ldots,\ n_k\} \tag{3.49}$$

is the set of time moments when at least one *MD* r is activated.

Example 3.14

Let the set of individual schedules of the manufacturing devices of some *IS* be

$$v_R = \{1 \cdot r_1[12.00,12.01][12.05,12.06][12.06,12.07],$$
$$1 \cdot r_2[12.00,12.03][12.03,12.06][12.08,12.11],$$
$$1 \cdot r_3[12.01,12.03][12.05,12.07],$$
$$1 \cdot r_4[12.04,12.08][12.10,12.14][12.14,12.18],$$
$$1 \cdot r_5[12.00,12.05]\},$$

which means the *MD* r_1, whose operation cycle has a duration of 1 unit (let it be a minute, denoted below for short *mnt*), will execute three operation cycles beginning at 12.00, 12.05, and 12.06 (note that the third such *OC* begins at the same moment 12.06 as the second finishes); the *MD* r_2, whose operation cycle has a duration of 3 min, will execute three operation cycles beginning at 12.00, 12.03, and 12.08 (the second such *OC* begins at the same moment 12.03 as the first finishes); the *MD* r_3, whose operation cycle has a duration of 2 min, will execute two operation cycles beginning at 12.01 and 12.05; the *MD* r_4, whose operation cycle has a duration of 4 min, will execute three operation cycles beginning at 12.04, 12.10, and 12.14 (the third such *OC* begins at the same moment 12.14 as the second finishes); and, finally, the *MD* r_5, whose operation cycle has a duration of 5 min, will execute one operation cycle beginning at 12.00. (Here and below for simplicity, we use a common form of time representation instead of integer values as multiplicities, i.e., 12.00 instead of 1200.)

According to (3.48) and (3.49), the *IS* schedule to be applied by the *IS* manager will be as follows:

$$\boldsymbol{Sch}(v_R) = \{ < 12.00, \{r_1, r_2, r_5\} >,$$
$$< 12.01, \{r_3\} >,$$
$$< 12.03, \{r_2\} >,$$
$$< 12.04, \{r_4\} >,$$
$$< 12.05, \{r_1, r_3\} >,$$
$$< 12.06, \{r_1\} >,$$
$$< 12.08, \{r_2\} >,$$
$$< 12.10, \{r_4\} >,$$
$$< 12.14, \{r_4\} > \},$$

and thus $R(12.00) = \{r_1, r_2, r_5\}, \ldots, R(12.14) = \{r_4\}$.

Henceforth we shall consider V_S to be the set of all possible states of an industrial system which may be reached from the initial state, which is a temporal multiset $v_A^0 + v_R^0 + \{n_0 \cdot t\}$, where $v_A^0 + \{n_0 \cdot t\} = v_0$ is the initial resource base of the *IS* and v_R^0 is the initial state of its manufacturing technological base:

$$v_R^0 = \{1 \cdot r_1, \ \ldots, \ 1 \cdot r_N\}, \tag{3.50}$$

which means the schedules of all *MDs* at the initial moment are empty, and all manufacturing devices are not operating.

Before consideration of a *formal mathematical definition of the application of TMGs*, enabling generation of a set V_S, beginning from the initial state of an *IS* represented by an *MST* (3.50), let us give some necessary remarks on some principal ideas lying in the background of this definition.

First of all, as was already mentioned, *manufacturing devices whose operation cycles finish at a moment n are free after this moment*. The results of finished *OCs* (collections of resources produced by *MDs* during these *OCs*) are added ("loaded") to a resource base at a moment *n*. In this case, temporal rules corresponding to such manufacturing devices which have become free at moment *n* may be applied to the current *MST* from exactly this moment. (As may all other *TRs* representing manufacturing devices which have become free earlier and were not assigned until moment *n*)

Second, we shall assume that an *IS* operates in such a way that *any MD is activated as soon as possible*, i.e., at a moment when it is free and the *RB* contains the resources necessary for its operation. So no delays while a free *MD* with resources sufficient for an operation stays in an inactive state are permitted. Such activation of *MDs* will be called "immediate," and by default this, namely, background of *ISs*' schedules will be referred everywhere below. However, in a general case, there may be implemented such scheduling, which allows that any free *MD* may not be activated at a time moment when an *RB* already contains resources sufficient for this *MD* activation. Such "concessions" may, in fact, provide resources, which would be used by this *MD*, to other *MDs*; that's why there might be created *IS* schedules, which would satisfy predefined conditions, while no one schedule, based on exclusively immediate activation of *MDs*, would not satisfy these conditions. We shall consider firstly a basic case of "immediate-activation-based" (*IAB*) scheduling and relevant to it mathematical semantics of *TMGs*, while some very local correction of this semantics, representing a general case, covering also delayed activation of *MDs*, will be described at the bottom of Sect. 3.2.2.

Third, it will be assumed that during time intervals, when no *MD* begins or finishes its operation cycle, the resource base and the manufacturing technological base of an *IS* are unchangeable. In other words, *"inner" operations of "black boxes" during the aforementioned intervals are "invisible" to an IS scheduler.*

Evidently, any set $R(n) \subseteq R$ satisfies the condition

$$v(n) = \sum_{\langle r; v \to v' \rangle \in R(n)} v \subseteq v_A, \tag{3.51}$$

which allows activation of all *MDs* the sum of whose input collections are a submultiset of the *MS* v_A, i.e., the resource base at moment n. Due to (3.51), the set $R^*(n)$ of all possible variants of activation of *MDs* at moment n will be as follows:

$$R^*(n) = \mathbf{max}\ \{\ R(n) \mid v(n) \subseteq v_A\}, \tag{3.52}$$

which ensures *MD* activation at the earliest possible moments: any $R(n)$ in the set $R^*(n)$ ensures the activation of a maximal set of *MDs* in the sense that adding to this set any one more *MD* would break condition (3.51). Of course, in the general case, there may be more than one variant of *MD* activation, so $|R^*(n)| \geq 1$.

Example 3.15
At some moment n, let the multiset $v_A = \{6 \cdot a, 4 \cdot b, 3 \cdot c\}$, and the set R contains the following temporal rules:

$$< r_1; \{3 \cdot a,\ 2 \cdot b\} \to \{1 \cdot c,\ 3 \cdot mnt\} >,$$

$$< r_2; \{4 \cdot a,\ 3 \cdot c\} \to \{2 \cdot b,\ 2 \cdot mnt\} >,$$

$$< r_3; \{2 \cdot a,\ 4 \cdot b\} \to \{4 \cdot c,\ 5 \cdot mnt\} >,$$

$$< r_4; \{3 \cdot b,\ 2 \cdot c\} \to \{1 \cdot a,\ 1 \cdot mnt\} > .$$

According to (3.51) and (3.52)

$$R^*(n) = \mathbf{max}\ \{\ \{r_1\}, \{r_2\}, \{r_3\}, \{r_4\}, \{r_1,\ r_3\}\} = \{\{r_1\}, \{r_4\}, \{r_2,\ r_3\}\}.$$

As can be seen, there are three different ways of applying the considered *MTB* to the current *RB* at the current moment n:

Activation of the *MD* r_1

Activation of the *MD* r_4

Activation of the *MDs* r_2, r_3

Namely, activating the *MD* r_1 excludes the opportunity to activate any other *MD*. The same is true regarding the *MD* r_4. However, the *MDs* r_2, r_3 may begin their operation cycles simultaneously because the sum of their input collections is a submultiset of v_A, i.e., this *RB* is sufficient to provide the necessary resources for both input collections.

Now we may consider the logic of generation of the set of temporal multisets defined by a temporal multiset grammar, i.e., in fact, a formal definition of the mathematical semantics of *TMGs*. The basis of this logic is the function $\boldsymbol{\pi}$ (a bold symbol), which is a generalization of the function π defined by (3.31) regarding the mathematical semantics of *MGs*. The function $\boldsymbol{\pi}$ has only one argument, which is an *MST* $v_A + v_R + \{n \cdot t\}$. Namely,

$$\boldsymbol{\pi}\,(v_A + v_R \ + \{n \cdot t\}) = \{v_A \ - v + v_R \ - \{1 \cdot r\tau\} + \{1 \cdot r\tau[n, \ l]\} + \{n \cdot t\} \mid r\tau \in v_R \ \& \langle r; v \rightarrow v' + \{\Delta n \cdot \Delta t\}\rangle \in R \& v \subseteq v_A \ \& \boldsymbol{T}(\tau) \leq n \& l = n + \Delta n\}. \tag{3.53}$$

In this definition, an auxiliary function $\boldsymbol{T}$ ("Tail") is used, defined as follows:

$$\boldsymbol{T}([n_1, \ l_1] \ldots [n_k, \ l_k]) = l_k, \tag{3.54}$$

where $k \geq 1$. As can be seen, the function $\boldsymbol{T}$ selects the last element of this list, which is the moment when the device finishes its last operation cycle.

So the result of an application of the function $\boldsymbol{\pi}$ is a set of *MSTs*, each corresponding to an activation of some one *MD*. An *MD* r may be activated at a moment n if the *MS* v which is the input collection of object-resources of this *MD* is a submultiset of the resource base v_A and also r is free (i.e., the last operation cycle of r has finished, which is detected by the inequality $\boldsymbol{T}(\tau) \leq n$).

Let us note that any application of the function $\boldsymbol{\pi}$ does not change the time marker of the processed *MST* so at the same time moment n along with an *MD* r other *MDs* in the same element of the set $R^*(n)$ may be activated. To represent this opportunity, we shall introduce a function $\boldsymbol{\pi}^*$ defined as follows:

$$\boldsymbol{\pi}^*\,(v_A + v_R + \{n \cdot t\}) = \begin{cases} \bigcup\limits_{w \in \boldsymbol{\pi}(v_A + v_R + \{n \cdot t\})} \boldsymbol{\pi}^*(w), \\ \text{if } \boldsymbol{\pi}\,(v_A + v_R + \{n \cdot t\}) \neq \{\varnothing\}, \\ \{v_A + v_R + \{n \cdot t\}\} \cup \boldsymbol{J}\,(v_A + v_R + \{n \cdot t\}) \ \text{otherwise.} \end{cases} \tag{3.55}$$

As can be seen, recursive application of the function $\boldsymbol{\pi}^*$ continues, while there is at least one *MD* which may be activated, and thus at least one possible *IS* schedule may be prolonged. In the case no *MD* may be activated during implementation of a chain of recursive applications of the functions $\boldsymbol{\pi}^*$ and $\boldsymbol{\pi}$ except *MDs* already present in the submultiset v_R of the input *MST* $v_A + v_R + \{n \cdot t\}$, this *MST* is added to the already accumulated set of *MST*s generated as a result of previous steps of application of the considered *TMG*, and a function $\boldsymbol{J}$ ("Jump of time") is applied to this *MST*. Before we consider this function, let us illustrate the essence of the functions $\boldsymbol{\pi}$ and $\boldsymbol{\pi}^*$ by a simple example.

Example 3.16
Let the initial moment of an *IS*'s operation be $n_0 = 12.00$ (the time scale begins from the time moment 12.00, and, as in Example 3.14, the time measurement unit is a minute), its resource base at this moment be $v_A^0 = \{5 \cdot a, 3 \cdot b, 1 \cdot c\}$, and its manufacturing technological base, containing two manufacturing devices r_1 and r_2,be represented by two temporal rules

$$< r_1; \{2 \cdot a,\ 1 \cdot b\} \to \{1 \cdot c,\ 1 \cdot mnt\} >,$$
$$< r_2; \{3 \cdot a,\ 1 \cdot c\} \to \{1 \cdot b,\ 2 \cdot mnt\} >,$$

so, according to (3.50), $v_R^0 = \{1 \cdot r_1,\ 1 \cdot r_2\}$.

Due to (3.53)–(3.55),

$$\pi\left(v_A^0 + v_R^0 + \{n_0 \cdot t\}\right) =$$
$$\{\{3 \cdot a,\ 2 \cdot b,\ 1 \cdot c,\ 1 \cdot r_1[12.00{,}12.01],\ 1 \cdot r_2,\ 12.00 \cdot t\},$$
$$\{2 \cdot a,\ 4 \cdot b,\ 1 \cdot r_1,\ 1 \cdot r_2[12.00{,}12.02],\ 12.00 \cdot t\}\}.$$

As can be seen, the result of applying the function π is a two-element set, in which the first element represents the result of application of *TR* r_1 (the individual schedule of the *MD* r_1 becomes [12.00,12.01], i.e., during this time interval, this *MD* will manufacture its output collection $\{1 \cdot c\}$, and after this time, the resource base of the *IS* will be $\{3 \cdot a, 2 \cdot b, 1 \cdot c\}$, while the individual schedule of *MD* r_2 will remain empty), and the second element represents the result of application of *TR* r_2 (the individual schedule of the *MD* r_2 becomes [12.00,12.02], i.e., during this time interval, this *MD* will manufacture its output collection $\{1 \cdot b\}$, and after this time, the resource base of the *IS* will be $\{2 \cdot a, 4 \cdot b\}$, while the individual schedule of the *MD* r_1 will remain empty).

Let now consider the result of applying the function π^*, which, due to (3.55), will be as follows:

$$\pi^*\left(v_A + v_R + \{n \cdot t\}\right) =$$
$$\pi^*\left(\{3 \cdot a,\ 2 \cdot b,\ 1 \cdot c,\ 1 \cdot r_1[12.00{,}12.01],\ 1 \cdot r_2,\ 12.00 \cdot t\}\right)$$
$$\cup \pi^*\left(\{2 \cdot a,\ 3 \cdot b,\ 1 \cdot r_1,\ 1 \cdot r_2[12.00{,}12.02],\ 12.00 \cdot t\}\right) =$$
$$\pi^*\left(\{2 \cdot b,\ 1 \cdot r_1[12.00{,}12.01],\ 1 \cdot r_2[12.00{,}12.02],\ 12.00 \cdot t\}\right)$$
$$\cup \pi^*\left(\{2 \cdot b,\ 1 \cdot r_1[12.00{,}12.01],\ 1 \cdot r_2[12.00{,}12.02],\ 12.00 \cdot t\}\right) =$$
$$\pi^*\left(\{2 \cdot b,\ 1 \cdot r_1[12.00{,}12.01],\ 1 \cdot r_2[12.00{,}12.02],\ 12.00 \cdot t\}\right) =$$
$$J\left(\{2 \cdot b,\ 1 \cdot r_1[12.00{,}12.01],\ 1 \cdot r_2[12.00{,}12.02],\ 12.00 \cdot t\}\right).$$

Application of the function π^* results in one application of the function J, whose argument corresponds to the *RB* containing the multiobject $2 \cdot b$, and two individual schedules of the manufacturing devices in the *MTB*, namely, [12.00,12.01] for the *MD* r_1 and [12.00,12.02] for the *MD* r_2. This means that

$$R^*(12.00) = \mathbf{max}\ \{\ \{r_1\}, \{r_2\}, \{r_1, r_2\}\}, = \{\{r_1,\ r_2\}\}.$$

So both *MDs* r_1 and r_2 may begin their operation cycles at the moment $n_0 = 12.00$ simultaneously.

Let us consider now the function ***J***, which performs a join on a set of generated temporal multisets with time markers in the interval $[n, \boldsymbol{\mu}(v_R, n)]$, when the state of the *IS* is unchangeable, i.e., during this interval, no *MD* finishes its operation cycle. So a *jump of time is done*, and the process of schedule creation repeats from the moment $\boldsymbol{\mu}(v_R, n)$. The value $\boldsymbol{\mu}(v_R, n)$ is determined by the following obvious relation:

$$\boldsymbol{\mu}(v_R, n) = min \bigcup_{r[n_1, l_1] \ldots [n_k, l_k] \in v_R} \{\ m \mid m \in \{l_1\ , \ldots, l_k\} \& m > n\ \} \quad (3.56)$$

At this moment $\boldsymbol{\mu}(v_R, n)$, all *MDs* whose *OCs* finish load their produced collections into the resource base. To represent "loading," we shall use the special function ***L*** ("Load") with two arguments: the first is the manufacturing technological base v_R at the considered time moment, and the second is $\boldsymbol{\mu}(v_R, n)$. The result of application of ***L*** to these arguments is a multiset representing the collection of resources loaded into the resource base at moment $\boldsymbol{\mu}(v_R, n)$:

$$\boldsymbol{L}(v_R, n) = \sum_{\substack{r[n_1, l_1] \ldots [n_k, l_k] \in v_R \\ \langle r; v \to v' + \{\Delta n \cdot \Delta t\}\rangle \in R \\ \boldsymbol{\mu}(v_R, n) \in \{l_1\ , \ldots,\ l_k\}}} v', \quad (3.57)$$

i.e., multisets v' in this sum are the results of operation cycles of corresponding *MDs* finishing at the moment $\boldsymbol{\mu}(v_R, n)$.

Let us note that the existence of a value $\boldsymbol{\mu}(v_R, n)$ is equivalent to an opportunity to prolong the current *IS* schedule. If such a value does not exist, this means that the current generated *MST* is terminal and may be included in the set of *TMSTs*.

Now we may introduce a formal definition of the function ***J*** which is the final element in the constructed definition of the mathematical semantics of *TMGs* and is as follows:

$$\boldsymbol{J}\,(v_A + v_R + \{n \cdot t\}) = \\ = \begin{cases} \boldsymbol{V}\,(v_A + v_R + \{n \cdot t\}) \cup \boldsymbol{\pi}^*(v_A + v_R + \boldsymbol{L}(v_R, n) + \{\boldsymbol{\mu}(v_R, n) \cdot t\}\,) \\ \quad \text{if} \exists \boldsymbol{\mu}(v_R, n), \\ \{\varnothing\} \text{ otherwise}, \end{cases} \tag{3.58}$$

where

$$\boldsymbol{V}\,(v_A + v_R + \{n \cdot t\}) = \bigcup_{i=n}^{\boldsymbol{\mu}(v_R,\, n)-1} \{v_A + v_R + \{i \cdot t\}\} \tag{3.59}$$

is the function of one argument which is the current state of the *IS*, and the value of this function is the aforementioned set of *MSTs* with time markers in the interval $[n, \boldsymbol{\mu}(v_R, n) - 1]$, when the state of the *IS* is unchangeable, i.e., during this interval, no *MD* finishes its operation cycle and loads its result into the *RB*. The set of *MSTs* representing the set of unchangeable states of the *IS*, which at least includes the input *MST* of the function $\boldsymbol{V}$ with the time marker $n \cdot t$, is joined to the accumulated set of generated *MSTs*. The last is joined also with the result of the application of the function $\boldsymbol{\pi}^*$ to the argument $v_A + v_R + \boldsymbol{L}(v_R, n) + \{\boldsymbol{\mu}(v_R, n) \cdot t\}$, which is the *IS* state at the moment $\boldsymbol{\mu}(n)$. If such a moment does not exist, i.e., all operation cycles have already finished, and the current *RB* is not sufficient to allow activation of at least one *MD*, then the result of applying the function $\boldsymbol{J}$ to the *MST* $v_A + v_R + \{n \cdot t\}$ will be the empty set $\{\varnothing\}$. So a temporal multiset w such that $\boldsymbol{J}\,(w) = \{\varnothing\}$ is terminal: no *MST* may be generated from w.

As can be seen from this description, *all possible branches corresponding to all possible IS schedules are generated without loss, until all MSTs have been obtained.*

Example 3.17

As in Example 3.16, let the initial moment of *IS* operation be $n_0 = 12.00$, its resource base at this moment be $v_A^0 = \{5 \cdot a,\, 3 \cdot b,\, 1 \cdot c\}$, and its manufacturing technological base, containing two manufacturing devices r_1 and r_2,be represented by two temporal rules

$$< r_1; \{2 \cdot a,\, 1 \cdot b\} \rightarrow \{1 \cdot c,\, 1 \cdot mnt\} >,$$

$$< r_2; \{3 \cdot a,\, 1 \cdot c\} \rightarrow \{1 \cdot b,\, 2 \cdot mnt\} >,$$

and, according to (3.50), $v_R^0 = \{1 \cdot r_1,\, 1 \cdot r_2\}$. As a result of Example 3.16, we have $\boldsymbol{J}\,(\{2 \cdot b, 1 \cdot r_1[12.00,12.01], 1 \cdot r_2[12.00,12.02], 12.00 \cdot t\})$. Let us consider the application of this function. According to (3.53)–(3.60),

$$\boldsymbol{J}\left(\{2\cdot b,\ 1\cdot r_1[12.00,12.01],\ 1\cdot r_2[12.00,12.02],\ 12.00\cdot t\}\right)$$

$$=\boldsymbol{C}\left(\{2\cdot b, 1\cdot r_1[12.00,12.01], 1\cdot r_2[12.00,12.02], 12.00\cdot t\}\right)\cup$$

$$\cup\boldsymbol{\pi}^*\left(\begin{array}{l}\{2\cdot b,\ 1\cdot r_1[12.00,12.01],\ 1\cdot r_2[12.00,12.02]\}+\\ \boldsymbol{L}\left(\begin{array}{l}\{1\cdot r_1[12.00,12.01],\ 1\cdot r_2[12.00,12.02]\}+\\ \boldsymbol{\mu}(\{1\cdot r_1[12.00,12.01],\ 1\cdot r_2[12.00,12.02]\},\ 12.00)\end{array}\right)\\ +\boldsymbol{\mu}(\{1\cdot r_1[12.00,12.01],\ 1\cdot r_2[12.00,12.02]\},\ 12.00)\cdot t\end{array}\right),$$

$$\boldsymbol{\mu}(\{1\cdot r_1[12.00,12.01],\ 1\cdot r_2[12.00,12.02]\},\ 12.00)=12.01,$$

$$\boldsymbol{C}\left(\{2\cdot b,\ 1\cdot r_1[12.00,12.01],\ 1\cdot r_2[12.00,12.02],\ 12.00\cdot t\}\right)=$$
$$\{2\cdot b,\ 1\cdot r_1[12.00,12.01],\ 1\cdot r_2[12.00,12.02],\ 12.00\cdot t\},$$

$$\boldsymbol{L}(\{1\cdot r_1[12.00,12.01],\ 1\cdot r_2[12.00,12.02]\},\ 12.01)=\{1\cdot c\},$$

and

$$\boldsymbol{J}\left(\{2\cdot b,\ 1\cdot r_1[12.00,12.01],\ 1\cdot r_2[12.00,12.02],\ 12.00\cdot t\}\right)=$$
$$\{\{2\cdot b,\ 1\cdot r_1[12.00,12.01],\ 1\cdot r_2[12.00,12.02],\ 12.00\cdot t\}\}$$
$$\cup\boldsymbol{\pi}^*\left(\{2\cdot b,\ 1\cdot c,\ 1\cdot r_1[12.00,12.01],\ 1\cdot r_2[12.00,12.02],\ 12.01\cdot t\}\right).$$

The result of applying $\boldsymbol{\pi}^*$ to the temporal multiset

$$\{2\cdot b,\ 1\cdot c,\ 1\cdot r_1[12.00,12.01],\ 1\cdot r_2[12.00,12.02],\ 12.01\cdot t\}$$

will be the empty set, and, finally,

$$\boldsymbol{J}\left(\{2\cdot b,\ 1\cdot r_1[12.00,12.01],\ 1\cdot r_2[12.00,12.02],\ 12.00\cdot t\}\right)=$$
$$\{\{2\cdot b,\ 1\cdot r_1[12.00,12.01],\ 1\cdot r_2[12.00,12.02],\ 12.00\cdot t\},$$

$$\{2 \cdot b,\ 1 \cdot c,\ 1 \cdot r_1[12.00{,}12.01],\ 1 \cdot r_2[12.00{,}12.02],\ 12.01 \cdot t\}\}.$$

As can be seen, the result of applying the function $\boldsymbol{J}$ to the input *MSTs* includes two *MSTs* joined to the set of generated *MSTs* accumulated before the function $\boldsymbol{J}$ application. As can be seen, the second of these *MSTs*, namely,

$$\{2 \cdot b,\ 1 \cdot c,\ 1 \cdot r_1[12.00{,}12.01],\ 1 \cdot r_2[12.00{,}12.02],\ 12.01 \cdot t\},$$

is terminal, because no *MST* may be generated from this temporal multiset.

Having introduced all the basic relations constructions, we may now *formally define the mathematical semantics of temporal multiset grammars.*

Statement 3.1 Let $S = \ < v_0, R>$ be a temporal multiset grammar, where v_0 is a temporal multiset such that $n_0 \cdot t \in v_0$. Then the set of all temporal multisets generated by application of this *TMG* is

$$V_S = \boldsymbol{\pi}^* \left(v_0 + v_R^0\right), \tag{3.60}$$

and the set of terminal *MSTs* generated by application of this *TMG* is

$$\bar{V_S} = \{w \mid w \in V_S \& \boldsymbol{J}\,(w) = \{\varnothing\}\}. \tag{3.61}$$

We shall not give any proof of this statement, on the supposition that the above detailed description of the functions $\boldsymbol{\pi}^*$ and $\boldsymbol{J}$, which is the basis of any possible proof, is sufficient.

As may be seen, from the substantial point of view, the logic defining the generation of a set $\boldsymbol{V_S}$, serving as the mathematical semantics of the temporal multiset grammars, allows generation of all possible schedules of the industrial system represented by a TMG $\boldsymbol{S}$, beginning from its resource base and manufacturing technological base at an initial moment $\boldsymbol{n_0}$.

All the said above refers the basic case of *IAB scheduling*. As we have promised at the beginning of this section, now we shall consider a general case presuming *not only IAB but also delayed activation of manufacturing devices*. To cover this kind of activation, it is sufficient to modify slightly the definition of the function $\boldsymbol{\pi}$ regarding (3.53). Namely, we shall redefine this function as

$$\boldsymbol{\pi}\,(v_A + v_R\ + \{n \cdot t\}) =$$

$$\{\{v_A\ - v + v_R\ - \{1 \cdot r\tau\} + \{1 \cdot r\tau[n,\ \ l]\} + \{n \cdot t\}\} \cup$$

$$\{v_A\ + v_R\ - \{1 \cdot r\tau\} + \{1 \cdot r\tau/n,\ m/\} + \{n \cdot t\}\} \quad \Bigg|$$

$$r\tau \in v_R\ \& \langle r; v \rightarrow v' + \{\Delta n \cdot \Delta t\}\rangle \in R \& v \subseteq v_A\ \&$$

$$T(\tau) \leq n\&l = n + \Delta n\&m = n + 1\}. \tag{3.62}$$

As seen, application of the function π to a current *MST* results in appearance of two-element set, which the first element is the same as in (3.53), while the second represents non-activation of the free *MD r* with sufficient input resource collection $v \subseteq v_A$ by replacement of the multiobject $1 \cdot r\tau$ by the *MO* $1 \cdot r\tau/n, m/$, where $m = n + 1$. Here and everywhere below, a string$/n, m/$, where symbol "/" is a divider, represents the time interval beginning at the time moment n and finishing at the next time moment m. During this interval, the aforementioned *MD r* remains inactive besides having both preconditions of its activation satisfied. By reason of this inactivity, the resource base of the *IS* remains the same, thus modelling "concession" of the *MDr* and enabling an opportunity for activation of some other *MDs* at the next generation step corresponding to the next time moment n'.

Since redefinition of the function π by the equality (3.62), the individual schedule of any *MD* acquires a form of a string $x_1 \ldots x_i \ldots x_k$, where x_i is either $[n, l]$ or $/n, m/$, the first case corresponding to the operation cycle of this *MD* while the second to its interval of inactivity being an implication of delayed activation. As seen, according to (3.55)–(3.61) and (3.62), any *MD* may stay inactive for any time period. However, this does not make the set V_S certainly infinite, because presence of substrings $/n, m/$ in individual schedules in no way implies result of test whether $\boldsymbol{J}(w) = \{\varnothing\}$ which positive output terminates generation branches. Remind that the time jump regarding a considered time moment is possible if there exists at least one *MD* which would finish its operation cycle. By reason that no one inactive *MD* finishes its *OC*, a result of the aforementioned test does not depend on such *MDs*.

All the rest of the equalities introduced above regarding *IAB* activation remain without any corrections.

Now let us consider a natural and practically oriented generalization of *TMGs*—filtering temporal multiset grammars (*FTMGs*)—and their application to *IS* scheduling.

3.2.3 Filtering Temporal Multiset Grammars

By analogy with multiset grammars and filtering *MGs*, we shall introduce filtering *TMGs* by adding to the couple $<v_0, R>$ in the definition of a *TMG* a third component, called, as everywhere above, a filter, and by a very local correction of the *TMGs'* mathematical semantics.

Namely, a *filtering temporal multiset grammar* will be a triple $S = \; < v_0, R, F>$, where v_0 and R are the same as in *TMGs* and F is a filter, which is a set of conditions allowing selection of the elements of the set $V_{S'}$ satisfying F. Here $S' = \; < v_0, R>$ is called the *core TMG of FTMG S*. So

$$V_S = V_{S'} \downarrow F, \tag{3.63}$$

where, as everywhere above, $\downarrow$is the symbol of filtration of the set $V_{S'}$ by filter F. As may be seen, (3.63) differs from the definition (3.39), where in the left part of the equality, $\overline{V_S}$ appears while in the right part $\overline{V_{S'}} \downarrow F$ appears, so that what is filtered by F is the set of *all MSTs* $V_{S'}$ generated by application of the core *TMG* of *FTMG S*, not the set of *only terminal MSTs* as in the case of *FMG*. It is clear why it is necessary to introduce a generalized mathematical semantics of *FTMGs* different than the mathematical semantics of the *FMGs*: *Associating a filter with an order, we assume that the order may be completed by an IS in any state, not only in terminal states, where the IS stops*.

A graphical representation of the sense of the definition of *FTMGs* is shown in Fig. 3.7.

Given these basic definitions, let us consider *FTMG's* mathematical semantics, developed on the foundation of the recursive functions $\boldsymbol{\pi}^*$ and $\boldsymbol{J}$ introduced in Sect. 3.2.2. We shall define *FTMG's* mathematical semantics, supposing that object t may be used in filters like all other objects representing ordinary additive resources. This unification is the simplest and requires the minimal changes in the definition (3.51)–(3.61).

Let us repeat that, unlike *FMGs*, where only terminal multisets are filtered, a filtering *TMG* is applied in such a way that every newly generated *MST* is tested, and such filtration is done precisely at the step when this *MST* is created.

So, let $S = \langle v_0, R, F \rangle$ be a filtering temporal multiset grammar. Let us redefine functions $\boldsymbol{\pi}^*$, $\boldsymbol{J}$, and $\boldsymbol{V}$, making them two-argument in such a way that the first argument is the same as in the initial definitions (3.55), (3.58), and (3.59), respectively, while the second is a boundary subfilter $F_{\leq}$ of the filter F:

$$\boldsymbol{\pi}^* (v_A + v_R + \{n \cdot t\}, F_{\leq}) =$$

$$= \begin{cases} \bigcup\limits_{w \in \boldsymbol{\pi}(v_A + v_R + \{n \cdot t\})} \boldsymbol{\pi}^*(w, F_{\leq}), \\ \quad \text{if } \boldsymbol{\pi}(v_A + v_R + \{n \cdot t\}) \neq \{\varnothing\}, \\ (\{v_A + v_R + \{n \cdot t\}\} \downarrow F_{\leq}) \cup \boldsymbol{J}(v_A + v_R + \{n \cdot t\}, F_{\leq}) \text{ otherwise,} \end{cases} \tag{3.64}$$

and

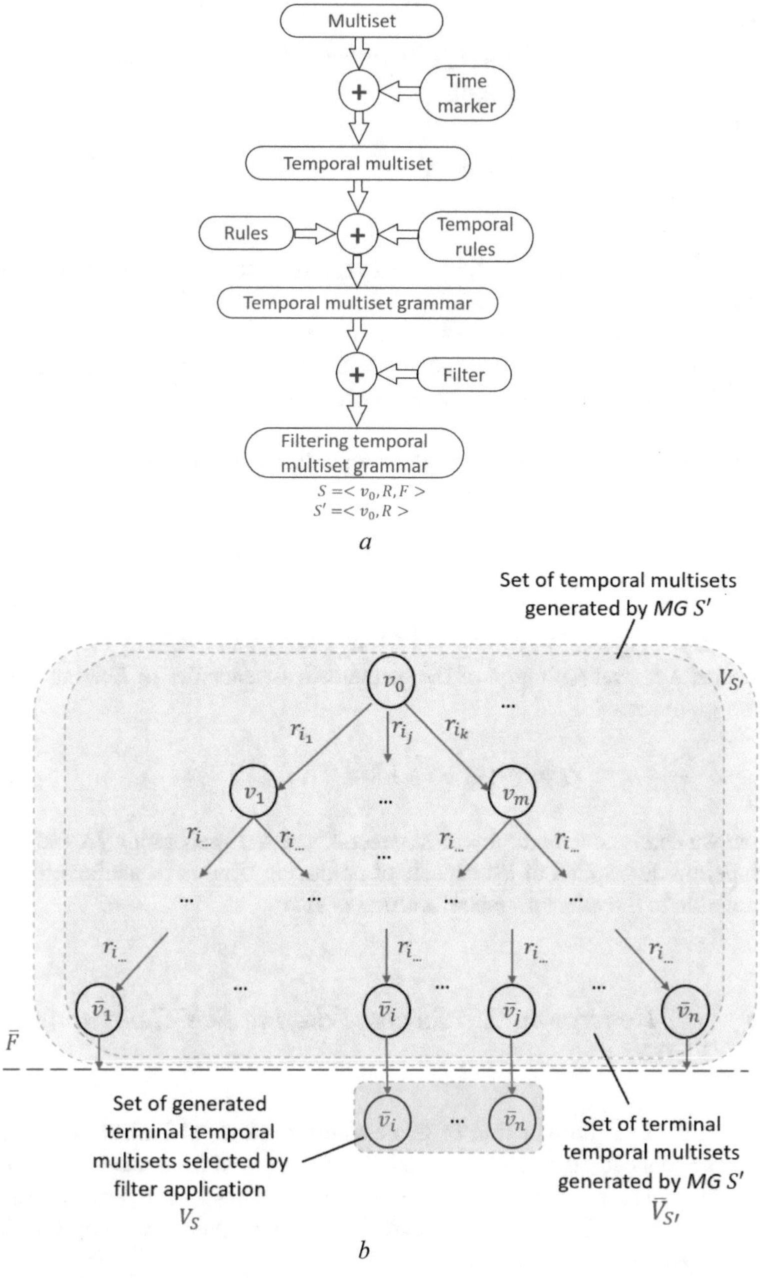

Fig. 3.7 Filtering temporal multiset grammars: (**a**) components; (**b**) set of multisets and set of terminal multisets generated by an *FTMG*

$$\boldsymbol{J}\,(v_A + v_R + \{n \cdot t\}, F_{\leq}) = \\ = \begin{cases} \boldsymbol{V}\,(v_A + v_R + \{n \cdot t\}, F_{\leq}) \bigcup \boldsymbol{\pi}^*(v_A + v_R + \boldsymbol{L}(v_R, n) + \{\boldsymbol{\mu}(v_R, n) \cdot t\}, F_{\leq}), \\ \qquad \text{if} \exists \boldsymbol{\mu}(v_R, n), \\ \\ \qquad \{\varnothing\} \text{ otherwise.} \end{cases} \tag{3.65}$$

As can be seen, due to (3.64), any *MST* represents the state of an *IS* such that no *MD* may be applied to the *RB*, and for this reason, this *MST* is included in $V_{S'}$ and must be tested by a boundary subfilter $F_{\leq}$. The result of $\{v_A + v_R + \{n \cdot t\}\} \downarrow F_{\leq}$ may be $\{\varnothing\}$, and in this case, the empty set will be joined to the accumulated set of generated *MSTs* V_S. Otherwise a non-empty one-element set $\{v_A + v_R + \{n \cdot t\}\}$ will be joined to V_S.

A similar local correction will be done regarding the function $\boldsymbol{V}$:

$$\boldsymbol{V}\,(v_A + v_R + \{n \cdot t\},\, F_{\leq}) = \left(\bigcup_{i=n}^{\boldsymbol{\mu}(v_R,\ n)-1} \{v_A + v_R + \{i \cdot t\}\} \right) \downarrow F_{\leq}. \tag{3.66}$$

Thus, replacing (3.55) by (3.64), (3.58) by (3.65), and (3.59) by (3.66), we have constructed a formal definition of the mathematical semantics of filtering temporal multiset grammars:

$$V_S = \boldsymbol{\pi}^*\left(v_0 + v_R^0 + \{n_0 \cdot t\},\, F_{\leq}\right) \downarrow F_{opt}. \tag{3.67}$$

Now we shall move on to the most general case, self-generating *TMGs/FTMGs*, which permit description of *ISs* capable of producing "tools of manufacturing," i.e., they are able to manufacture manufacturing devices.

3.2.4 *Self-Generating* TMGs *and Filtering Self-Generating* TMGs

As was said in the introduction to this chapter, a principal limitation of multiset grammars and metagrammars is their inability to represent *ISs* manufacturing *ToMs*. This disadvantage is the main reason for the development of *self-generating TMGs/FTMGs*, whose capabilities are fully sufficient for the representation of any kind of industrial system.

The main difference between *TMGs* and *SG TMGs* is that the scheme *R* of a self-generating *TMG* $S = \,< v_0, R>$ may contain rules $\langle r; v \rightarrow v' \rangle \in R$ such that v', in turn, may contain multiobjects like $n' \cdot r'$, where n' is an integer number and r' is the *name of a temporal rule* (i.e., *the name of a manufacturing device represented by this TR*).

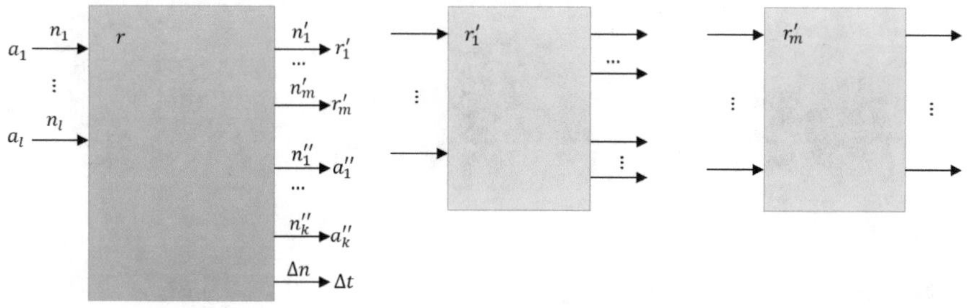

a

$$\langle r; \{n_1 \cdot a_1, \ldots, n_l \cdot a_l\} \rightarrow \{n_1' \cdot r_1', \ldots, n_m' \cdot r_m', n_1'' \cdot a_1'', \ldots, n_k'' \cdot a_k'', \Delta n \cdot \Delta t\}\rangle$$

b

Fig. 3.8 Representation of manufacturing devices modeled by self-generating multiset grammars: (**a**) "black box"; (**b**) a temporal rule r

This means that, after extracting a collection of resources v from a resource base, the *IS*, during Δn time units, where $\Delta n \cdot \Delta t \in v'$, will produce a collection of resources v' containing n' devices (of type) r' and put (load) this collection into the manufacturing technological base. It is most important that, from moment n, when the aforementioned n' devices r' are joined to the *MTB*, they may be applied like any other manufacturing devices already in the *MTB* of the *IS*. In other words, *appearance of a multiobject* $n' \cdot r'$ *as a result of an operation cycle of some MD represents addition of* n' *MDs* r' *to the manufacturing technological base of the industrial system.*

To return to a "black box" representation of industrial systems, an *IS MPAR* may be represented as depicted in Fig. 3.8, where r is a manufacturing device which during one operationcycle produces, along with a collection of passive object resources $a_1'', \ldots, a_k''$, also a collection of manufacturing devices $r_1', \ldots, r_m'$ in amounts $n_1', \ldots, n_m'$, respectively. Each of these manufacturing devices may, in turn, after its inclusion in the *IS MTB* produce not only passive *ORs* but also *MDs* or produce only passive *ORs* or only *MDs*.

Let us illustrate the nexus between the "black box" and *SG TMG*-based representations of *MTBs* of such industrial systems.

Example 3.18
Let us consider the manufacturing technological base of some cyberphysical manufacturing facility, being an industrial system producing passive and active resources, whose elements are depicted in Fig. 3.9.

The considered *MTB* includes some number of robotized assembly lines. The first one, similar to that mentioned in Example 3.13 and named "asm-line," assembles cars from their components in such a way that one car is assembled from four wheels, one body, and one engine, and this operation consumes 300 kWh h of

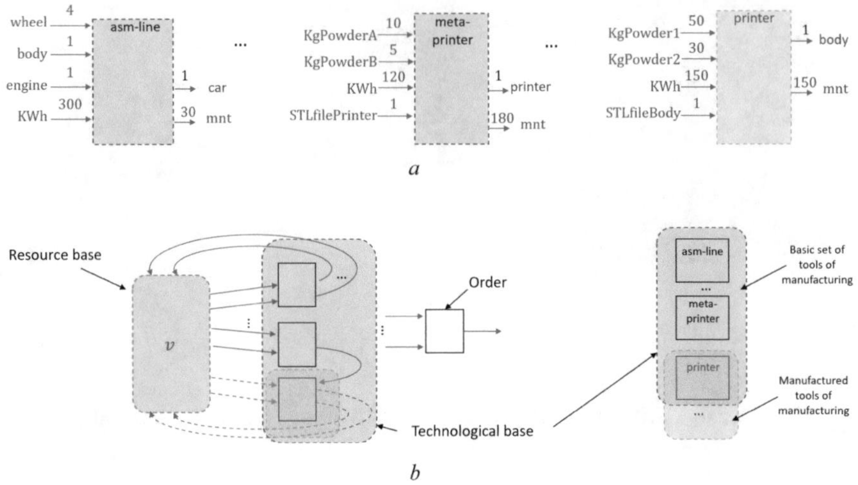

Fig. 3.9 Example of a "black box" representation of the manufacturing technological base of an *IS MPAR*: (**a**) "black boxes" representing *MDs*; (**b**) "black boxes" representing a self-producing *MTB*

electrical power. In 30 min of work of this assembly line after the start of its operation cycle, provided with the aforementioned car components as its input, a car is manufactured and delivered to the resource base of the *IS*. There are other assembly lines assembling various components from their subcomponents and ready spare parts. A new feature of this facility is that it is equipped with a so-called metaprinter capable of producing *3D*-printers, producing, in turn, car bodies. The metaprinter consumes 10 kg of some powder A and 5 kg of another powder B, applying for this purpose the *STL*-file of a printer containing information about its layer structure and consuming 120 kWh h of electrical power. The printer produced after 180 min is capable, in turn, of producing one body from 50 kg of some powder 1 and 30 kg of another powder 2, applying for this purpose the *STL*-file of a body containing information about its layer structure and consuming 150 kWh h of electrical power. This work is done by this *3D*-printer in 150 min.

The scheme of a *TMG* representing this *MTB* contains the following temporal rules:

$$< asm - line;$$

$$\{4 \cdot wheel,\ 1 \cdot body,\ 1 \cdot engine,\ 300 \cdot kWh\} \to$$

$$\{1 \cdot car,\ 30 \cdot mnt\ \} >,$$

$$\cdots$$

$$< metaprinter;$$

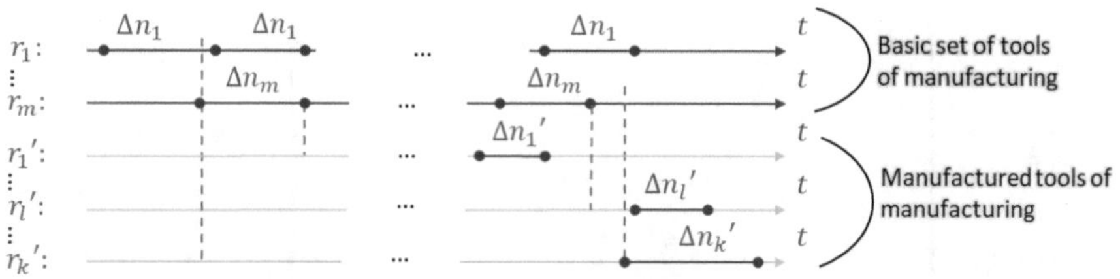

Fig. 3.10 An expandable Gantt-like chart of the operation of an industrial system manufacturing tools of manufacturing

$$\{10 \cdot kgPowderA,\ 5 \cdot kgPowderB,\ 120 \cdot kWh,\ 1 \cdot STLfilePrinter\} \to$$
$$\{1 \cdot printer,\ 180 \cdot mnt\ \} >$$
$$\dots$$
$$< printer;$$
$$\{50 \cdot kgPowder1,\ 30 \cdot kgPowder2,\ 150 \cdot kWh,\ 1 \cdot STLfileBody\} \to$$
$$\{1 \cdot body,\ 150 \cdot mnt\ \} > .$$

Similarly, a *Gantt-like chart* of an *IS* after the introduction of temporal rules with names of rules in their right parts becomes "expandable," i.e., after finishing an operation cycle of any such manufacturing device, there would appear $N = n''_1 + \ldots + n''_k$ new axes, providing explication of the operation cycles of the N new manufactured devices (*MdDs*), as depicted in Fig. 3.10.

So, as can be seen, the *naming of temporal rules and inclusion of their names in the right parts of temporal rules create a descriptional background for a representation of ISs manufacturing tools.*

The *most complicated step* in the development of a mathematical semantics of *SG TMGs* is *to assign a unique name to every new MdD*, while all these manufactured devices are named identically as indistinguishable passive object resources. To avoid this difficulty, we shall use *composite names*, including, along with the name of the type of a manufactured device r', also three additional components:

The name of the manufacturing device r which produced r'

The time moment n at which the corresponding operation cycle of r finished

The serial number i of this *MdD* among n' *MdDs* produced during this *OC*

In this way we cover even cases where *MD* produces $n' > 1$ *MdDs* at a single moment.

Due to the generality of the above-described mathematical semantics of *TMGs*, a formal definition of a mathematical semantics of self-generated *TMGs* may be constructed by a single very local change. Namely, the function $\boldsymbol{L}$, introduced by (3.57), is now redefined as follows:

$$L(v_R,n)=\left(\sum_{\substack{r[n_1,l_1]\ldots[n_k,l_k]\in v_R\\ \langle r;v\to v'+\{\Delta n\cdot\Delta t\}\rangle\in R\\ \boldsymbol{\mu}(v_R,n)\in\{l_1\;,\ldots,l_k\}\\ l'\cdot a'\in v'\\ a'\in A_s}}\{l'\cdot a'\}\right)$$

$$+\left(\sum_{\substack{r[n_1,l_1]\ldots[n_k,l_k]\in v_R\\ \langle r;v\to v'+\{\Delta n\cdot\Delta t\}\rangle\in R\\ \boldsymbol{\mu}(v_R,\mathrm{n})\in\{l_1\;,\ldots,l_k\}\\ l'\cdot r'\in v'\\ r'\in\bar{R}}}\sum_{i=1}^{l'}\{1\cdot r'<r,n,i>\}\right). \tag{3.68}$$

As can be seen, the redefined function $\boldsymbol{L}$ allows the creation of a multiset which is the sum of the results of all operation cycles which have finished at moment n. But, unlike (3.57), there are two summands in this sum. The first concerns passive object resources and is practically the same as in (3.57): the whole difference is that the sum is accumulated from one-element multisets $\{l' \cdot a'\}$, where $a' \in A_S$ is the name of a passive *OR*. The second summand in (3.68) concerns active resources (i.e., manufacturing devices) and is the result of summing one-element multisets of the form $\{1 \cdot r' < r, n, i>\}$, where r' is the name of a type of device produced by the manufacturing device r. (Recall that the set of names of *MDs* is denoted by $\bar{R}$.) So every one of the l' such devices ($l' \cdot r' \in v'$) will have a unique identifier $r' < r, n, i>$, which means this object is the i-th of l' such objects, all l' being included in the manufacturing technological base of the *IS* at moment n. Here "<", ">", and "," are dividers, used for an unambiguous interpretation of such objects with composite names. This construction fully corresponds to the above description of the logic of composition of names of manufactured devices. All schedules of the *MdDs* at moment n, when they are loaded into the *MTB*, are empty.

According to (3.68), after an $\boldsymbol{L}(v_R, n)$ application, the manufacturing technological base of the *IS* will increase by

$$N = \sum_{\substack{r[n_1, l_1] \ldots [n_k, l_k] \in v_R \\ \langle r; v \rightarrow v' + \{\Delta n \cdot \Delta t\}\rangle \in R \\ \mu(v_R, n) \in \{l_1 \ ,\ldots, l_k\} \\ l' \cdot r' \in v' \\ r' \in \bar{R}}} l' \quad (3.69)$$

elements, each representing a device whose manufacturing was finished by an *MD* r at moment n. All these *MdDs* enter the manufacturing technological base of the industrial system at this moment and may be used (must taken into account) in the following steps of *MST* generation (i.e., *IS* schedule creation). The same number of axes will be added to the relevant Gantt-like chart.

However, the correction of the function $\boldsymbol{L}$ is not the only action necessary for an adequate transformation of the mathematical semantics of *TMGs* into the more general case of *SG TMGs*. It is evident that all newly produced manufacturing devices being added to the manufacturing technological base cannot be activated, because the definition (3.53) of function $\boldsymbol{\pi}$, responsible for *MD* activation, concerns only those manufacturing devices which are present in the initial *MTB*.

So we shall correct the definition of $\boldsymbol{\pi}$ in an appropriate way. Henceforth function $\boldsymbol{\pi}$ will be defined as follows (recall that U is the alphabet of symbols used in names of manufacturing devices):

$$\begin{gathered}\boldsymbol{\pi}\ (v_A + v_R \ + \{n \cdot t\}) = \\ \{v_A \ - v + v_R \ - \{1 \cdot r\tau\} + \{1 \cdot r\tau[n, \ \ l]\} + \{n \cdot t\} \mid \\ rp\tau \in v_R \ \ \&p \in \{\nabla\} \cup P\& \\ \langle r; v \rightarrow v' + \{\Delta n \cdot \Delta t\}\rangle \in R\&v \subseteq v_A \ \ \&\boldsymbol{T}(\tau) \leq n\&l = n + \Delta n\}\end{gathered} \quad (3.70)$$

where p may be the empty string ∇ (in this case, r is an *MD* present in the initial manufacturing technological base) or a string $p' \in P$, where $P = \{<\} \cdot U^* \cdot \{>\}$ (in this case, r is an *MdD* produced by this *MTB*, and p' is the production chain which was involved for this purpose; "<" and ">" are, as above, dividers).

As now may be seen, names of manufactured devices have the following structure:

$$r < r^{(k)} < r^{(k-1)} \ldots r'' < r', n_1, i_1 >, n_2, i_2 >, \ldots >, n_{k-1}, i_{k-1} >, n_k, i_k >. \quad (3.71)$$

Due to the logic of composing (or, rather, structuring) the names of produced manufacturing devices, any such name means that the *MD* of type r was produced by the device $r^{(k)}$ at moment n_k and has number i_k; the device $r^{(k)}$ was produced by the device $r^{(k-1)}$ at moment n_{k-1} and has number i_{k-1};...; the device r'' was produced by the device r' at moment n_1 and has number i_1; the device r' belongs to the initial manufacturing technological base.

The *proposed nested structure of MdD names allows storage of the full history of manufacturing of all elements of the manufacturing technological base*, and this is done by some kind of "blockchain" (Dannen 2017; Zheng et al. 2018). However, the manufacturing capabilities of any *MdD* are defined by the prefix *r* of this device, and this prefix identifies a temporal rule whose body describes the inputs and outputs of this *MdD*.

Example 3.19

Let the initial moment of *IS* operation be $n_0 = 12.00$, its resource base at this moment be $v_A^0 = \{5 \cdot a,\ 3 \cdot b,\ 1 \cdot c\}$, and its manufacturing technological base contain two manufacturing devices r_1 and r_2, represented by two temporal rules

$$< r_1; \{2 \cdot a,\ 1 \cdot b\} \rightarrow \{1 \cdot c,\ 1 \cdot mnt\} >,$$

$$< r_2; \{3 \cdot a,\ 1 \cdot c\} \rightarrow \{1 \cdot b,\ 2 \cdot mnt\} >.$$

Along with these two temporal rules, the *MTB* contains one more *TR*

$$< r_3; \{2 \cdot b,\ 1 \cdot c\} \rightarrow \{1 \cdot r_4,\ 4 \cdot mnt\} >,$$

representing an *MD* capable of producing in 4 min one *MD* r_4, represented, in turn, by the *TR*

$$< r_4; \{1 \cdot b\} \rightarrow \{7 \cdot a,\ 2 \cdot mnt\} >,$$

and the initial state of this *MTB* is $v_R^0 = \{1 \cdot r_1,\ 1 \cdot r_2,\ 1 \cdot r_3\}$. (Note that $r_4 \notin v_R^0$, and this means that there is no *MD* r_4 in the *MTB* at the initial moment $n_0 = 12.00$.) According to the mathematical semantics of *SG TMGs*, the set of terminal *MSTs* generated by this *SG TMG* contains the following *MST:*

$$\left\{ \begin{array}{l} 7 \cdot a, \\ 1 \cdot r_1[12.00{,}12.01],\ 1 \cdot r_2[12.00{,}12.02], \\ 1 \cdot r_3[12.01{,}12.05],\ 1 \cdot r_4 < r_3,\ 12.05{,}1 > [12.05{,}12.07], \\ 12.07 \cdot t \end{array} \right\}.$$

As may be seen, at the moment 12.00, two *MDs* r_1 and r_2 are activated, extracting from the *RB* their input collections $\{2 \cdot a, 1 \cdot b\}$ and $\{3 \cdot a, 1 \cdot c\}$, respectively, so the *RB* will become $\{2 \cdot b\}$. At the moment 12.01, the *MD* r_1 will finish its operation cycle and load its output collection $\{1 \cdot c\}$ into the *RB*, which hence will become $\{2 \cdot b, 1 \cdot c\}$. This *RB* will be sufficient for the beginning of the *OC* of the *MD* r_3, which, extracting from the *RB* its input collection $\{2 \cdot b, 1 \cdot c\}$, will make the resource base empty. At the moment 12.02, the *MD* r_2 will finish its operation cycle and load its output collection $\{1 \cdot b\}$ into the *RB*, which at this moment will become $\{1 \cdot b\}$. At the moments 12.03 and 12.04, no *MD* will finish its *OC*, and no *MD* will be activated, because this *RB* is not sufficient for the activation of any manufacturing device. But at the moment 12.05, the operation cycle of the *MD* r_3 will finish, and the *MD* r_4 with the unique identifier $<r_3, 12.05,1>$ (produced by the *MD* r_3 at 12.05, being the first and only device produced by this *MD* at this moment) is activated, extracting the collection $\{1 \cdot b\}$ from the *RB*, which at this moment becomes empty. It will remain empty until the moment 12.07, when the *MD* $r_4 < r_3, 12.05,1>$ will finish its operation cycle and load into the resource base the collection $\{7 \cdot a\}$. No *MD* may be activated with this *RB*, and thus at the moment 12.07, the *IS* stops, and the result of its operation will be the resource base $\{7 \cdot a\}$ and the manufacturing technological base $\{r_1, r_2, r_3, r_4 < r_3, 12.05,1>\}$.

Another schedule of this *IS* may be created by the following generation chain. At the moment 12.00, two *MDs* r_1 and r_3 are activated (the *TRs* r_1 and r_3 are applied), extracting from the *RB* their input collections $\{2 \cdot a, 1 \cdot b\}$ and $\{2 \cdot b, 1 \cdot c\}$, respectively; thus the *RB* becomes $\{3 \cdot a\}$. At the moment 12.01, the *MD* r_1 finishes its operation cycle and loads its output collection $\{1 \cdot c\}$ into the *RB* which, hence, becomes $\{3 \cdot a, 1 \cdot c\}$. This collection is sufficient for application of the temporal rule r_2, and after it is applied, the *RB* becomes empty. At the moment 12.02, no one *OC* is finished, but at the moment 12.03, the operation cycle of the *MD* r_2 ends, and as a result the *RB* becomes $\{1 \cdot b\}$, but this collection of resources is not sufficient for activating of free *MD* r_1, as well as for reactivating the *MD* r_2. At the moment 12.04, the *OC* of the *MD* r_4 finishes, and thus the multiobject $1 \cdot r_4 < r_3, 12.04,1>$ is joined to the generated multiset. Since this moment, the technological base of the considered *IS* contains initially available *MDs* r_1, r_2, r_3 as well as the manufactured *MD* $r_4 < r_3, 12.04,1 >$. Because the resource base at this moment is $\{1 \cdot b\}$, the *MD* r_4 is activated (the *TR* r_4 is applied), and, hence, the *RB* becomes empty. It remains empty until 12.06, when the operation cycle of the *MD* $r_4 < r_3, 12.04,1>$ finishes, and the *RB* as a result of this *OC* becomes $\{7 \cdot a\}$. No one *MD* may be activated by this *RB*, and thus the terminal *MST* generated by the described alternative generation chain is

$$\left\{\begin{array}{l} 7 \cdot a, \\ 1 \cdot r_1[12.00,12.01],\ 1 \cdot r_2[12.01,12.03], \\ 1 \cdot r_3[12.00,12.054],\ 1 \cdot r_4 < r_3,\ 12.04,1 > [12.04,12.06], \\ 12.06 \cdot t \end{array}\right\}.$$

As seen, this schedule of the considered *IS* enables even its quicker operation and manufacturing of the new *MD*.

As can be seen, a recursive representation of multiset generation according to the mathematical semantics of *SG TMGs* provides a precise description of industrial production and supply chains of any complexity. From the knowledge engineering point of view, this process is nothing but *metainference* (Kendal and Green 2007; Siekmann 2014), whose result is *a monotonic extension of a set of temporal rules, applied for the generation of multisets*. From the practical point of view, the defined generation is a multiply repeated manufacturing of tools of manufacturing, and *due*

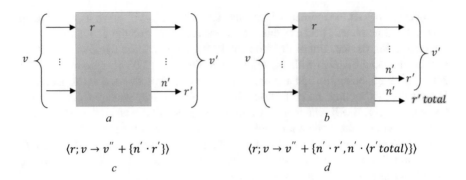

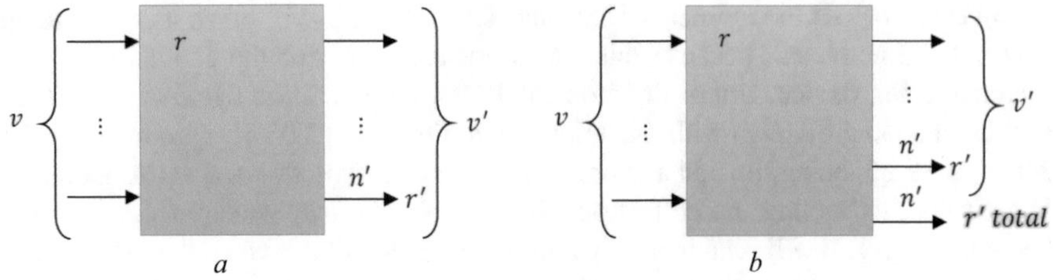

$$\langle r; v \to v'' + \{n' \cdot r'\}\rangle$$

c

$$\langle r; v \to v'' + \{n' \cdot r', n' \cdot \langle r' total\rangle\}\rangle$$

d

Fig. 3.11 Extension of a "black box" by additional information necessary for counting the number of produced devices of different types: (**a**) an initial "black box"; (**b**) an extended "black box"; (**c**) an initial rule; (**d**) an extended rule

to the descriptional capabilities of the multigrammatical framework, there are no descriptional limits on the representation of such chains, which may expand while there are sufficient amounts of passive resources for production of new *MDs*, and the capabilities of the aforementioned tools permit such operations.

Let us now introduce *filtering self-generating* (*FSG*) *TMGs*, which are a simple modification of *SG TMGs*, similarly to what was done in Sect. 3.2.3 with *TMGs*. Following this approach, an *FSG TMG* will be a triple $S = \ < v_0, R, F>$, where $S' = \ < v_0, R>$ is the *core SG TMG of the FSG TMG S* and *F* is a filter. Using (3.63)–(3.67) as a foundation, it is quite clear that the only difference between *SG TMGs* and *FSG TMGs* is the definition of the functions $\boldsymbol{L}$ and $\boldsymbol{\pi}$. Therefore, the mathematical semantics of *FSG TMGs* is the set of relations (3.64)–(3.66) and (3.68)–(3.71). We assume that the names of temporal rules do not appear in filters. As will be clarified below, this restriction does not limit the expressive power of the query language used for order representation and for planning *IS* operations.

To finish the description of *FSG TMGs*, let us comment on some nuances concerning possible restrictions on the number of manufacturing devices, which may be critically necessary for planning *IS* operations, as well as on various time parameters of the created schedules—for example, integral time, when some *MDs* may be in an operational and/or non-operational state, as it usually called in the theory of scheduling. Both amounts may be computed rather simply by including in the right parts of rules multiobjects representing increments of the total number of *MDs* of any type as a result of their production (1) and increments of durations of time intervals when the manufacturing device is active (2).

To implement *the first* of these ideas, it is sufficient to join to a temporal rule, representing a "black box" which is capable of producing n' devices r', one additional multiobject $n' \cdot \langle r' total\rangle$ (Fig. 3.11). Thus $\langle r; v \to v'\rangle$ becomes

$$\langle r; v \rightarrow v' + \{n' \cdot \langle r' total \rangle\} \rangle, \tag{3.72}$$

where $\langle r' total \rangle$ is an *OR* denoting the total number of manufacturing devices r' entering the *RB* at the current generation step.

As is clear, every application of the extended rule adds to the resource base a multiset $\{n' \cdot \langle r' total \rangle\}$, so after M such applications, there will be $M \cdot n'$ objects $\langle r' total \rangle$. In this way, to establish necessary restrictions on the total number of *MDs* r' produced by an *IS* (in the general case of *MDs* r' produced by an *IS*, there may be more than one *MD* producing r', so their capabilities must be summed), it is sufficient to include in a filter F_q an incoming order boundary condition $\langle r' total \rangle \theta m$ and/or an optimizing condition $\langle r' total \rangle = opt$. Such conditions will select precisely such multisets $v \in V_S$ which represent an *RB* with the appropriate number of produced manufacturing devices of type r'.

To implement *the second* of the aforementioned ideas, namely, concerning the *duration of a time interval when a manufacturing device is active,* we may use similar techniques. For this purpose, a temporal rule $\langle r; v \rightarrow v' \rangle$, such that $\Delta n \cdot \Delta t \in v'$, may be replaced by an extended *TR*

$$\langle r; v \rightarrow v' + \{\Delta n \cdot (\Delta total)\} \rangle. \tag{3.73}$$

As is clear, any application of this rule adds a multiplicity Δn of object $\langle \Delta total \rangle$ to the already accumulated multiplicity of this object, which, as can be seen, is the current duration of operation of device r. By including a condition $\langle \Delta total \rangle \theta l$ and/or $\langle \Delta total \rangle = opt$ in a filter F_q, one can select those multisets $v \in V_S$, i.e., *IS* schedules, that correspond to applications of an *MD* r whose total duration of operation satisfies this condition. For example, if it is necessary to minimize the total time of inactivity of an *MD* r, it is sufficient to use the condition $\langle \Delta total \rangle = max$.

Let us note that temporal multisets may be operated on *not only by temporal multigrammars* but also by other *MGs* whose schemes do not include temporal rules at all. Just this approach, based on application of *filtering multiset grammars* for representation of loans, permitting implementation of lending activities in the class of lending economical systems, will be described in Sect. 8.2. *TMG* application allows generation of new temporal multisets with new time markers obtained from current time markers and durations of time intervals presented in temporal rules. At the same time, temporal multisets generated by application of *FMGs* inherit time markers occurring in the rules (for this purpose, the mathematical semantics of *FMGs* will be locally corrected).

Now, having introduced the basic syntax and mathematical semantics of classes of multiset grammars, we may move on to their enhancements and a deeper investigation. Consideration of the *MG's, FMG's*, *FTMG's*, and *FSG TMG's* metagrammatical extensions and mathematical properties of the listed classes of *MGs* is the main content of Sects. 3.3 and 3.4, respectively.

3.3 Metagrammatical Extensions of Multiset Grammars

Speaking in terms of mathematical logic and knowledge engineering, all the introduced classes of *MGs* operate on only *"ground rules" interconnecting "ground facts" represented by multisets*. This feature is a consequence of the background of *MGs*, which, as was said above, are string-operating grammars. However, all modern knowledge representation models from the logic programming and the constraint programming areas (Apt 2003; Bratko 2012; Frunkwirth and Abdennadher 2003; Marriott and Stucky 2003; Siekmann 2014; Sheremet 1994, 2013, 2020b; Wallace 2002) are based on "abstract rules," each representing a more or less general (universal) "law" (or "regularity") by the inclusion of variables in their bodies. Any "abstract rule" represents a more or less large set of "ground rules," each formed by a substitution of constants instead of corresponding variables. The application of "abstract rules," allowing their "concretization" in order to achieve a solution of a specific problem, is done by an inference engine, whose operating logic permits creation of the necessary chain of implications leading to a solution.

To extend the capabilities of the introduced classes of multiset grammars up to "abstract rules," the so-called multiset metagrammars (for short, *multimetagrammars*) are introduced here; they are a generalization of the unitary *MMGs* introduced in (Sheremet 2010, 2011a). In developing the necessary tools for the multigrammatical framework, we have tried to follow an "Occam's razor" principle: *do not multiply the number of basic entities without necessity*. Following this principle and trying to make the developed knowledge representation maximally understandable and similar to that already used in knowledge engineering, we have taken filtering multiset grammars as our foundation and have transformed their basic construction—a rule—by including in it the so-called multiplicity-variables (*MVs*). To define sets of values ("domains") of these variables, relevant chain boundary conditions of the form $n \leq \gamma \leq n'$ are included in the *FMG* filters; every such definition means that values of variable γ may be integer numbers from n to n'. (As everywhere above, ≥ 0.) The scope of a variable declaration occurring in a filter is the entire scheme of the *MMG*.

Let us consider formal definitions of these verbally introduced notions.

3.3.1 Multiset Metagrammars

A *multiset metagrammar S* is a triple $<v_0, R, F>$, where v_0 and F are, as above, a kernel and a filter correspondingly, while R is a scheme containing rules and so-called metarules (*MRs*).

A metarule has the form

$$\{\mu_1 \cdot a_1, \ \ldots, \mu_m \cdot a_m\} \rightarrow \{\mu_1{}' \cdot a_1{}', \ \ldots, \mu_n{}' \cdot a_n{}'\}, \quad (3.74)$$

where μ_i is a positive integer number, as everywhere above, or a variable $\gamma \in \Gamma$, where Γ is the universe of variables. If μ_i is a variable, then it is called a *multiplicity-variable (MV)*. As we have seen, a rule is a special case of a metarule in which multiplicities $\mu_1, \ldots, \mu_m$ are constants, which in the case of *MMGs* are named *multiplicity-constants* (*MCs*).

The filter F of an *MMG* $S = \ < v_0, R, F>$ is a set of conditions, which may be boundary and/or optimizing, as in *FMGs*. At the same time, as was already said, F includes chain boundary conditions of form

$$n \leq \gamma \leq n', \tag{3.75}$$

which contain variables not objects; there is one and only one *CBC* (3.75) for every variable γ occurring in unitary metarules in a scheme R. This *CBC* is called a *variable declaration* and has a quite evident semantics, verbally described above. If $n = 0$, a shortened version of (3.75) may be used, namely,$\gamma \leq n'$.

If F includes a subfilter

$$F_\Gamma = \{n_1 \leq \gamma_1 \leq n_1', \ \ldots, n_l \leq \gamma_l \leq n_l'\} \tag{3.76}$$

containing all variable declarations, then every tuple $\langle \overline{n_1}, \ \ldots, \overline{n_l} \rangle$ such that $\overline{n_1} \in [n_1, n_1'], \ldots, \overline{n_l} \in [n_l, n_l']$ allows the creation of one filtering multigrammar by substitution of $\overline{n_1}, \ldots, \overline{n_l}$ into all metarules in the scheme R instead of multiplicity-variables. Rules in R are transferred to a new scheme denoted

$$R \circ \langle \overline{n_1}, \ \ldots, \overline{n_l} \rangle \tag{3.77}$$

without any transformations. Every such *FMG* generates a set of terminal multisets by an application of the filter $\mathcal{F} = F - F_\Gamma$, which contains all the “usual” (without variables) conditions (both boundary and optimizing).

As may be seen from this short informal description, an *MMG* $S = \langle a_0, R, F \rangle$ is a simple unified tool for a compact representation of a *set of FMGs*, which is denoted S^*: for practically important problems, the number of such *FMGs*, i.e., the cardinality of S^*, may reach extremely large numbers. It is easy to see that this number is not greater than the product of the cardinalities of all the variable domains:

$$|S^*| \leq \prod_{j=1}^{l} \left(n_j' - n_j + 1\right) \tag{3.78}$$

(if there are no identical *FMGs* after variable value substitutions, then (3.79) becomes equality).

Let us now give a formal definition of the *mathematical semantics of multiset metagrammars*.

An *MMG* $S = \langle v_0, R, F \rangle$ defines a set of terminal multisets $\overline{V_s}$ in the following way:

$$\overline{V_s} = \left(\bigcup_{\bar{s} \in S^*} \overline{V_{\bar{s}}} \right) \downarrow \mathcal{F}, \tag{3.79}$$

$$S^* = \bigcup_{\bar{n}_1 \in [n_1,\ n'_1]} \cdots \bigcup_{\bar{n}_l \in [n_l,\ n'_l]} \{\langle v_0,\ R \circ \langle \overline{n_1},\ \ldots,\ \overline{n_l} \rangle \rangle\} \tag{3.80}$$

$$R \circ \langle \overline{n_1},\ \ldots, \overline{n_l} \rangle = \{ r \circ \langle \overline{n_1},\ \ldots,\ \overline{n_l} \rangle \mid r \in R \}, \tag{3.81}$$

$$\mathcal{F} = F - F_\Gamma, \tag{3.82}$$

$$F_\Gamma = \bigcup_{j=1}^{l} \left\{ n_j \le \gamma_j \le n'_j \right\}, \tag{3.83}$$

and, if r is

$$\{\mu_1 \cdot a_1,\ \ldots, \mu_m \cdot a_m\} \to \{\mu_1{}' \cdot a_1{}',\ \ldots, \mu_n{}' \cdot a_n{}'\}, \tag{3.84}$$

then $r \circ \langle \overline{n_1},\ \ldots, \overline{n_l} \rangle$ is the rule

$$\begin{aligned} &\{(\mu_1 \circ \langle \overline{n_1},\ \ldots,\ \overline{n_l} \rangle) \cdot a_1,\ \ldots, (\mu_m \circ \langle \overline{n_1},\ \ldots,\ \overline{n_l} \rangle) \cdot a_m\} \to \\ &\{(\mu_1{}' \circ \langle \overline{n_1},\ \ldots,\ \overline{n_l} \rangle) \cdot a_1{}',\ \ldots, (\mu_n{}' \circ \langle \overline{n_1},\ \ldots,\ \overline{n_l} \rangle) \cdot a_n{}'\}, \end{aligned} \tag{3.85}$$

where

$$\mu_i \circ \langle \overline{n_1},\ \ldots, \overline{n_l} \rangle = \begin{cases} \mu_i, \text{ if } \mu_i \in [1,\ \infty], \\ \overline{n_i}, \text{ if } \mu_i \in \Gamma. \end{cases} \tag{3.86}$$

All multiplicities $\mu_i{}' \circ \langle \overline{n_1},\ \ldots, \overline{n_l} \rangle$, $i = 1, \ldots, n$, occurring in the right parts of *MRs* are determined similarly to (3.86). As can be seen, according to (3.85) and (3.86), all multiplicity-variables of an *MR* r (3.84) are replaced by their corresponding values from a tuple $\langle \overline{n_1},\ \ldots, \overline{n_l} \rangle$, while all multiplicity-constants remain unchanged. Evidently, if all $\mu_1, \ldots, \mu_m, \mu'_1, \ldots, \mu'_n$ are constants, i.e., the metarule is a rule, it remains unchanged.

Let us repeat that the *scope of multiplicity-variables is the whole scheme R*, i.e., if there are $m > 1$ occurrences of one and the same variable γ in the different metarules, they are all replaced by one and the same value from an applied tuple $\langle \overline{n_1},\ \ldots, \overline{n_l} \rangle$. This singularity makes the *MMG* semantics (and by this, finally, their pragmatics, i.e., techniques of application) radically different from conventional rule-based knowledge representation models (Akerkar and Sajja 2010; Cross 2017; Kendal and Green 2007; Pannu 2015; Sainter et al. 2000; Sheremet 1994, 2013; Yeoh and

Yokoo 2012), whose common feature is that the scope of any variable is the rule where it occurs.

Let us note also that the presence of filters in multiset metagrammars, by analogy with *FMGs*, would lead us to the notion of a "filtering MMG." However, the notion "metagrammar" presumes the existence of, at least, a filter F_Γ in every *MMG*, so the prefix "filtering" is redundant here, and it will not be used below. But an *MMG* whose filter contains at least one optimizing condition will be called an *optimizing MMG* (Sheremet and Zhukov 2016). The same prefix will be used when we speak about *FMGs* whose filters contain at least one optimizing condition. Such *FMGs* will be called, similarly, *optimizing FMGs*.

From what is said above in this section, the nature of the "multiset metagrammar" notion is obvious: as a "metalanguage" is used for description of other languages, a "multiset metagrammar" is a multigrammar used for description of a set of other MGs by means of metarules, multiplicity-variables, and boundary conditions defining MV's domains.

Example 3.20
Consider the *MMG* $S = \langle v_0, R, F \rangle$, where $v_0 = \{1 \cdot a\}$, R contains three metarules:

$$r_1: \quad \{1 \cdot a\} \to \{2 \cdot b, \gamma_1 \cdot c\},$$
$$r_2: \quad \{1 \cdot b, 1 \cdot c\} \to \{\gamma_2 \cdot d\},$$
$$r_3: \quad \{1 \cdot b, \gamma_2 \cdot c\} \to \{\gamma_2 \cdot d\},$$

while the filter F contains the following conditions:

$$c \leq 1,$$
$$d = max,$$
$$1 \leq \gamma_1 \leq 2,$$
$$0 \leq \gamma_2 \leq 1,$$

the last two being the declarations of the variables γ_1 and γ_2.

According to (3.79)–(3.86), the scheme S defines four *FMGs*:

$$S_1 = \langle v_0, R \;\square\langle 1,\ 0\rangle\rangle$$
$$S_2 = \langle v_0, R \;\square\langle 1,\ 1\rangle\rangle,$$
$$S_3 = \langle v_0, R \;\square\langle 2,\ 0\rangle\rangle,$$
$$S_4 = \langle v_0, R \;\square\langle 2,\ 1\rangle\rangle,$$

and $F = \{c \leq 1, d = max\}$.

As can be seen,

$$R_1 = R \;\square\langle 1, 0\rangle = \{\langle\{1\cdot \mathrm{a}\}\to\{2\cdot \mathrm{b},\ 1\cdot \mathrm{c}\}\rangle,$$
$$\langle\{1\cdot b,\ 1\cdot c\}\to\{\varnothing\}\rangle,$$
$$\langle\{1\cdot b\}\to\{\varnothing\}\rangle\},$$
$$\overline{v}_1 = \{\{\varnothing\}, \{1\cdot c\}\},$$
$$R_2 = R \;\square\langle 1, 1\rangle = \{\langle\{1\cdot \mathrm{a}\}\to\{2\cdot \mathrm{b},\ 1\cdot \mathrm{c}\}\rangle,$$
$$\langle\{1\cdot b,\ \ 1\cdot c\}\to\{1\cdot d\}\rangle\}\rangle,$$

(R_2 includes two, not three rules, because $r_2\ \square\ \langle 1,1\rangle = r_3\ \square\ \langle 1,1\rangle$),

$$\overline{v}_2 = \{\{1\cdot b\}, \{1\cdot d\}\},$$
$$R_3 = R \;\square\langle 2, 0\rangle = \{\langle\{1\cdot \mathrm{a}\}\to\{2\cdot \mathrm{b},\ 2\cdot \mathrm{c}\}\rangle,$$
$$\langle\{1\cdot b,\ 1\cdot c\}\to\{\varnothing\}\rangle,$$
$$\langle\{1\cdot b\}\to\{\varnothing\}\rangle\},$$
$$\overline{v}_3 = \{\{\varnothing\}, \{1\cdot c\}, \{2\cdot c\}\},$$
$$R_4 = R \;\square\langle 2, 1\rangle = \{\langle\{1\cdot \mathrm{a}\}\to\{2\cdot \mathrm{b},\ 2\cdot \mathrm{c}\}\rangle,$$
$$\langle\{1\cdot b,\ \ 1\cdot c\}\to\{1\cdot d\}\rangle\},$$

(R_4, like R_2, includes two, not three rules, because $r_2\ \square\ \langle 2,1\rangle = r_3\ \square\ \langle 2,1\rangle$),

$$\overline{v}_4 = \{\{2\cdot d\}\}.$$

According to (3.78),

$$\overline{V}_S = \left(\overline{V}_1 \cup \overline{V}_2 \cup \overline{V}_3 \cup \overline{V}_4\right) \downarrow \{c \le 1, d = max\}$$
$$= (\{\{\varnothing\},\ \{1\cdot c\},\ \{1\cdot b\},\ \{1\cdot d\},\ \{2\cdot c\},\ \{2\cdot d\}\} \downarrow \{c \le 1\}) \downarrow \{d = max\}$$
$$= \{\{\varnothing\}, \{1\cdot c\}, \{1\cdot b\}, \{1\cdot d\}, \{2\cdot d\}\} \downarrow \{d = max\}$$
$$= \{\{2\cdot d\}\}.$$

(Note that all multisets not containing the object c are included in the result of filtering by the boundary condition $c \le 1$ because of the zero multiplicity of this object).

As we have seen, *MMG* filters contain boundary conditions relating objects and variables, but optimizing conditions relating only objects. That is why from both theoretical and practical points of view, it is reasonable to extend the mentioned filters by *optimizing conditions relating variables* and having a form

$$\gamma = opt. \tag{3.87}$$

This form defines optimality of generated *TMSs* through values of multiplicity-variables used during generation of their terminal multisets.

The meaning of an optimizing condition (3.87) is quite clear: select those *TMSs* which are generated by use of that value of the variable γ which is optimal (minimal, maximal) among all other *TMSs* generated by use of all other values from the domain of variable γ. By this verbal description, *we extend the notion of TMS optimality from multiplicity-constants to multiplicity-variables* occurring in the metarules applied during *TMS* generation.

A formal definition of the meaning of an optimizing condition (3.87), verbally described above, is as follows.

Let us introduce l auxiliary terminal objects $\overline{\gamma_1}, \ldots, \overline{\gamma_l}$ corresponding to variables $\gamma_1, \ldots, \gamma_l$ occurring in an *MMG* $S = \langle v_0, R, F \rangle$, i.e., in metarules and boundary conditions defining domains of multiplicity-variables. After that let us replace the kernel v_0 by a multiset $\{1 \cdot a'\}$,where a' is an auxiliary object, and join to the scheme R one metarule

$$\{1 \cdot a'\} \to v_0' \tag{3.88}$$

where

$$v_0' = v_0 + \{\gamma_1 \cdot \overline{\gamma_1}, \ \ldots, \gamma_l \cdot \overline{\gamma_l}\}. \tag{3.89}$$

After that let us replace all optimizing conditions of the form $\gamma = opt$ in the filter F by $\overline{\gamma} = opt$, thus converting them to a "canonical" object-containing form. The obtained scheme and filter will be denoted R' and F', respectively.

As can now be seen, the *MMG* $S' = \langle \{1 \cdot a'\}, R', F' \rangle$ generates terminal multisets of the form

$$\{n_{i_1} \cdot a_{i_1}, \ \ldots, n_{i_k} \cdot a_{i_k}, \overline{n_{j_1}} \cdot \overline{\gamma_1}, \ \ldots, \overline{n_{j_l}} \cdot \overline{\gamma_l}\}, \tag{3.90}$$

which are selected to $\overline{V_{S'}}$ if and only if the *TMS*

$$\{n_{i_1} \cdot a_{i_1}, \ \ldots, n_{i_k} \cdot a_{i_k}\} \tag{3.91}$$

satisfies all conditions in F concerning terminal objects occurring in the scheme R as well as the *TMS*

$$\{\overline{n_{j_1}} \cdot \overline{\gamma_1}, \ \ldots, \overline{n_{j_l}} \cdot \overline{\gamma_l}\} \tag{3.92}$$

satisfying all optimizing conditions of the form $\overline{\gamma_i} = opt_i$.

It is not difficult to define $\overline{V_s}$ by subtracting multisets of the form (3.92) from all *TMSs* $v \in \overline{V_{\overline{s}}}$, but from a practical point of view, it is much more useful to consider

$\overline{V_{S'}}$ not $\overline{V_s}$ to be the result of application of S. It's clear that all *TMSs* $v \in \overline{V_{S'}}$ contain values of variables $\gamma_1, \ldots, \gamma_l$ as multiplicity-constants of terminal objects $\overline{\gamma_1}, \ldots, \overline{\gamma_l}$, computation of which is usually the main purpose of application of the aforementioned multimetagrammar.

Example 3.21

Consider the *MMG* $S = \langle v_0, R, F \rangle$ from the previous example with one addition: filter F includes one more optimizing condition $\gamma_1 = max$. According to (3.88) and (3.89), the additional metarule

$$\{1 \cdot a'\} \rightarrow \{1 \cdot a, \gamma_1 \cdot \overline{\gamma}_1, \gamma_2 \cdot \overline{\gamma}_2\}$$

is added to the scheme R.

As may be seen, the *MMG* S', like the *MMG* S from the previous example, includes

$$S_1 = \langle v_0, R \;\square \langle 1,\; 0 \rangle \rangle,$$

$$S_2 = \langle v_0, R \;\square \langle 1,\; 1 \rangle \rangle,$$

$$S_3 = \langle v_0, R \;\square \langle 2,\; 0 \rangle \rangle,$$

$$S_4 = \langle v_0, R \;\square \langle 2,\; 1 \rangle \rangle,$$

and $F = \{c \leq 1, \;\; d = max, \;\; \overline{\gamma}_1 = max\}$.

The generation chains lead to the following terminal multisets:

$$\overline{V}_1 = \{\{1 \cdot \overline{\gamma}_1\}, \{1 \cdot c, \; 1 \cdot \overline{\gamma}_1\}\},$$

$$\overline{V}_2 = \{\{1 \cdot b, \; 1 \cdot \overline{\gamma}_1, \; 1 \cdot \overline{\gamma}_2\}, \;\; \{1 \cdot d, \; 1 \cdot \overline{\gamma}_1, \; 1 \cdot \overline{\gamma}_2\}\},$$

$$\overline{V}_3 = \{\{2 \cdot \overline{\gamma}_1\}, \{1 \cdot c, \; 2 \cdot \overline{\gamma}_1\}, \{2 \cdot c, \; 2 \cdot \overline{\gamma}_1\}\},$$

$$\overline{V}_4 = \{\{2 \cdot d, \; 2 \cdot \overline{\gamma}_1, \; 1 \cdot \overline{\gamma}_2\}\}.$$

Thus

$$\begin{aligned}\overline{V}_S = (\{\{1 \cdot \overline{\gamma}_1\}, \; \{1 \cdot c, \;\; 1 \cdot \overline{\gamma}_1\}, \; \{1 \cdot b, \;\; 1 \cdot \overline{\gamma}_1, \;\; 1 \cdot \overline{\gamma}_2\}, \; \{1 \cdot d, \;\; 1 \cdot \overline{\gamma}_1, \;\; 1 \cdot \overline{\gamma}_2\},\\ \{2 \cdot \overline{\gamma}_1\}, \{1 \cdot c, \; 2 \cdot \overline{\gamma}_1\}, \{2 \cdot c, \; 2 \cdot \overline{\gamma}_1\}, \{2 \cdot d, \; 2 \cdot \overline{\gamma}_1, \; 1 \cdot \overline{\gamma}_2\}\}\\ \downarrow \{c \leq 1\}) \downarrow \{d = max, \overline{\gamma}_1 = max\}\\ = \{2 \cdot d, \; 2 \cdot \overline{\gamma}_1, \; 1 \cdot \overline{\gamma}_2\}\},\end{aligned}$$

and the solution of the problem represented by the *MMG* S is the one-element set $\{\{2 \cdot d\}\}$ associated with the maximal value $\gamma_1 = 2$ and the non-optimized value $\gamma_2 = 1$.

As can be seen, it is very useful in practice to consider $\overline{V_{S'}}$ to be the result of application of the *MMG* $S = \langle v_0, R, F\rangle$ (even in cases where the scheme *R* does not contain optimizing conditions of the form $\gamma = opt$). The only inconvenience, which an attentive reader might have noticed in the last example, is the absence of multiobjects of the form $n \cdot \bar{\gamma}$, where $n = 0$, in generated *TMSs*. This inconvenience, however, may be eliminated simply by an imperative inclusion of these *TMS* multiobjects of the form $0 \cdot \bar{\gamma}$.

3.3.2 Unitary Multiset Metagrammars

It is not difficult to project *MMGs* onto unitary multiset metagrammars. Namely, the definition of a *UMMG* is very similar to (3.74)–(3.92).

A *unitary multiset metagrammar S* is a triple $<v_0, R, F>$, where v_0 and *F* are, as above, a kernel and a filter, respectively, while *R* is a scheme containing *unitary rules* and *unitary metarules* (*UMRs*).

A *unitary metarule* has the form

$$a \to \mu_1 \cdot a_1, \ldots, \mu_m \cdot a_m, \tag{3.93}$$

where the μ_i have the same sense as in *MMGs*, and, similarly to *MMGs*, a unitary rule is a special case of a unitary metarule whose multiplicities $\mu_1, \ldots, \mu_m$ are constants. As in (3.37), the object *a* is called the *header*, while the unordered (in the same sense as in the *UR* definition) list $\mu_1 \cdot a_1, \ldots, \mu_m \cdot a_m$ is called the *body of the UMR*.

The *filter F* of a *UMMG* $S = < v, R, F>$ is a set of conditions, which may be boundary and/or optimizing, as in *UMGs*. Variable declarations have the same form as in *MMGs*:

$$n \le \gamma \le n'. \tag{3.94}$$

Due to the fact that *UMGs* are a syntactically simplified version of context-free multigrammars, there is no necessity for any additional definition about the mathematical semantics of unitary multiset metagrammars. So the aforementioned definition is the same as that of *MMGs*.

Example 3.22
Consider the *UMMG* $S = \langle\{1 \cdot a\}, R, F\rangle$, where the scheme *R* contains the following three unitary metarules:

$$r_1 : a \to 3 \cdot b, \gamma_1 \cdot c,$$

$$r_2 : b \to 1 \cdot c, \gamma_2 \cdot d,$$
$$r_3 : b \to \gamma_1 \cdot c, \gamma_2 \cdot d,$$

while the filter F contains the following conditions:

$$2 \leq c \leq 4,$$
$$1 \leq \gamma_1 \leq 2,$$
$$0 \leq \gamma_2 \leq 1,$$
$$d = max\,.$$

According to (3.79)–(3.93), S defines four *UMGs*:

$$S_1 = \langle \{1 \cdot a\}, R \circ \langle 1,\ 0 \rangle \rangle,$$
$$S_2 = \langle \{1 \cdot a\}, R \circ \langle 1,\ 1 \rangle \rangle,$$
$$S_3 = \langle \{1 \cdot a\}, R \circ \langle 2,\ 0 \rangle \rangle,$$
$$S_4 = \langle \{1 \cdot a\}, R \circ \langle 2,\ 1 \rangle \rangle,$$

which correspond to $4 = 2 \times 2$ sets of the two-valued domains of the variables γ_1 and γ_2.

As can be seen,

$$R_1 = R \circ \langle 1, 0 \rangle = \{\langle a \to 3 \cdot b,\ 1 \cdot c \rangle, \langle b \to 1 \cdot c \rangle\},$$
$$\overline{V}_{\bar{s}_1} = \{\{4 \cdot c\}\},$$
$$R_2 = R \circ \langle 1, 1 \rangle = \{\langle a \to 3 \cdot b,\ 1 \cdot c \rangle, \langle b \to 1 \cdot c,\ 1 \cdot d \rangle\},$$
$$\overline{V}_{\bar{s}_2} = \{\{4 \cdot c,\ 3 \cdot d\}\},$$
$$R_3 = R \circ \langle 2, 0 \rangle = \{\langle a \to 3 \cdot b,\ 2 \cdot c \rangle, \langle b \to 1 \cdot c \rangle, \langle b \to 2 \cdot c \rangle\},$$
$$\overline{V}_{\bar{s}_3} = \{\{5 \cdot c\}, \{8 \cdot c\}\},$$
$$R_4 = R \circ \langle 2, 1 \rangle = \{\langle a \to 3 \cdot b,\ 2 \cdot c \rangle, \langle b \to 1 \cdot c,\ 1 \cdot d \rangle, \langle b \to 2 \cdot c,\ 1 \cdot d \rangle\},$$
$$\overline{V}_{\bar{s}_4} = \{\{5 \cdot c,\ 3 \cdot d\}, \{8 \cdot c,\ 3 \cdot d\}\}.$$

After that,

$$\overline{V_s} = \left(\overline{V}_{\bar{s}_1} \cup \overline{V}_{\bar{s}_2} \cup \overline{V}_{\bar{s}_3} \cup \overline{V}_{\bar{s}_4}\right) \downarrow \{2 \leq c \leq 4, d = max\}$$
$$= (\{\{4 \cdot c\},\ \{4 \cdot c,\ 3 \cdot d\},\ \{5 \cdot c\},\ \{8 \cdot c\},\ \{5 \cdot c,\ 3 \cdot d\},\ \{8 \cdot c,\ 3 \cdot d\}\} \downarrow \{2 \leq c \leq d\})$$
$$\downarrow \{d = max\} = \{\{4 \cdot c\}, \{4 \cdot c,\ 3 \cdot d\}\} \downarrow \{d = max\} = \{\{4 \cdot c,\ 3 \cdot d\}\}.$$

Optimizing conditions with variables in *UMMGs* are applied and interpreted as in *MMGs*, which is illustrated by the following example.

Example 2.23
Consider the *UMMG* $S = \langle\{1 \cdot a\}, R, F\rangle$, where the scheme R contains the same unitary metarules as in the previous example, while the filter F contains all conditions from this example and one more optimizing condition $\gamma_1 = max$. Then

$$\overline{R} = R \cup \{r_0\},$$

where

$$r_0 = \langle \bar{a} \to 1 \cdot a,\ \gamma_1 \cdot \overline{\gamma_1},\ \gamma_2 \cdot \overline{\gamma_2}\rangle,$$

and $\overline{S} = \big\langle \{1 \cdot \bar{a}\},\ \overline{R},\ \overline{F}\big\rangle$, where

$$\bar{F} = F \cup \{\overline{\gamma_1} = max\}.$$

According to (3.79)–(3.93), the *UMMG* $\overline{S}$ defines the following four *UMMGs*:

$$\overline{S_1} = \big\langle \{1 \cdot \bar{a}\},\ \overline{R} \circ \langle 1,\ 0\rangle \big\rangle,$$
$$\overline{S_2} = \big\langle \{1 \cdot \bar{a}\},\ \overline{R} \circ \langle 1,\ 1\rangle \big\rangle,$$
$$\overline{S_3} = \big\langle \{1 \cdot \bar{a}\},\ \overline{R} \circ \langle 2,\ 0\rangle \big\rangle,$$
$$\overline{S_4} = \big\langle \{1 \cdot \bar{a}\},\ \overline{R} \circ \langle 2,\ 1\rangle \big\rangle,$$

which, as in the previous example, correspond to four sets of values of the variables γ_1 and γ_2. Then, as above,

$$\overline{R_1} = \overline{R} \circ \langle 1, 0\rangle = \{\langle \bar{a} \to 1 \cdot a,\ 1 \cdot \overline{\gamma_1}\rangle, \langle a \to 3 \cdot b,\ 1 \cdot c\rangle, \langle b \to 1 \cdot c\rangle\},$$
$$\overline{V}_{\bar{s}_1} = \{\{4 \cdot c,\ 1 \cdot \overline{\gamma_1}\}\},$$
$$\overline{R_2} = \overline{R} \circ \langle 1, 1\rangle = \{\langle \bar{a} \to 1 \cdot a,\ 1 \cdot \overline{\gamma_1},\ 1 \cdot \overline{\gamma_2}\rangle, \langle a \to 3 \cdot b,\ 1 \cdot c\rangle, \langle b \to 1 \cdot c,\ 1 \cdot d\rangle\},$$
$$\overline{V}_{\bar{s}_2} = \{\{4 \cdot c,\ 3 \cdot d,\ 1 \cdot \overline{\gamma_1},\ 1 \cdot \overline{\gamma_2}\}\},$$
$$\overline{R_3} = \overline{R} \circ \langle 2, 0\rangle = \{\langle \bar{a} \to 1 \cdot a,\ 2 \cdot \overline{\gamma_1}\rangle, \langle a \to 3 \cdot b,\ 2 \cdot c\rangle, \langle b \to 1 \cdot c\rangle, \langle b \to 2 \cdot c\rangle\},$$
$$\overline{V}_{\bar{s}_3} = \{\{5 \cdot c,\ 2 \cdot \overline{\gamma_1}\}, \{8 \cdot c,\ 2 \cdot \overline{\gamma_1}\}\},$$
$$\overline{R_4} = \overline{R} \circ \langle 2, 1\rangle =$$
$$= \{\langle \bar{a} \to 1 \cdot a,\ 2 \cdot \overline{\gamma_1},\ 1 \cdot \overline{\gamma_2}\rangle, \langle a \to 3 \cdot b,\ 2 \cdot c\rangle, \langle b \to 1 \cdot c,\ 1 \cdot d\rangle, \langle b \to 2 \cdot c,\ 1 \cdot d\rangle\},$$
$$\overline{V}_{\bar{s}_4} = \{\{5 \cdot c,\ 3 \cdot d,\ 2 \cdot \overline{\gamma_1},\ 1 \cdot \overline{\gamma_2}\}, \{8 \cdot c,\ 3 \cdot d,\ 2 \cdot \overline{\gamma_1},\ 1 \cdot \overline{\gamma_2}\}\}.$$

After that,

$$\overline{V_{\bar{s}}} = \left(\overline{V_{\bar{s}_1}} \cup \overline{V_{\bar{s}_2}} \cup \overline{V_{\bar{s}_3}} \cup \overline{V_{\bar{s}_4}}\right) \downarrow \{2 \leq c \leq 4,\ d = max\,,\ \overline{\gamma_1} = max\ \}$$
$$= \{\{4 \cdot c,\ 1 \cdot \overline{\gamma_1}\},\ \{4 \cdot c,\ 3 \cdot d,\ 1 \cdot \overline{\gamma_1},\ 1 \cdot \overline{\gamma_2}\}\} \downarrow \{d = max\,,\ \overline{\gamma_1} = max\}$$
$$= \{\{4 \cdot c,\ 3 \cdot d,\ 1 \cdot \overline{\gamma}_1,\ 1 \cdot \overline{\gamma}_2\}\},$$

i.e., the set of terminal multisets generated by the *UMMG S* contains only the *TMS* $\{4 \cdot c, 3 \cdot d, 1 \cdot \overline{\gamma}_1, 1 \cdot \overline{\gamma}_2\}$, which corresponds to the variable values $\gamma_1 = 1$ and $\gamma_2 = 1$.

3.3.3 Temporal Multiset Metagrammars and Self-Generating Temporal Multiset Metagrammars

It is evident that, as in the cases considered above, to introduce temporal multiset metagrammars (*TMMGs*), it is sufficient to take filtering *TMGs* as a foundation, to define the notion of a temporal metarule (*TMR*) by inclusion of *TR* multiplicity-variables, and, finally, to extend filters by boundary conditions declaring the domains of these *MVs* as well as by optimizing conditions selecting the temporal multisets with optimal (minimal, maximal) values of *MVs* occurring in these conditions.

We shall define a *temporal multiset metagrammar* as a triple $S = \langle v_0, R, F \rangle$, where R is a set of temporal rules and temporal metarules, each *TMR* being of the form

$$\{\mu_1 \cdot a_1,\ \ldots,\ \mu_m \cdot a_m\} \rightarrow \{\mu'_1 \cdot a'_1,\ \ldots,\ \mu'_n \cdot a'_n,\ \mu'_{n+1} \cdot \Delta t\}. \tag{3.95}$$

where, as in (3.74), μ_i and μ_j' may be positive integer numbers or multiplicity-variables. By this definition, a *TR* is a special case of a *TMR* where all μ_i and μ_j' are positive integer numbers.

A filter F of a *TMMG* $S = \langle v_0, R, F \rangle$ is a set of boundary and optimizing conditions, and for every variable γ occurring in at least one *TMR*, there exists one and only one chain boundary condition

$$n \leq \gamma \leq n', \tag{3.96}$$

defining the domain of the *MV* γ as an interval $[\,n, n']$.

As in the case of *MMGs*, a subfilter

$$F_\Gamma = \{n_1 \leq \gamma_1 \leq n_1',\ \ldots,\ n_l \leq \gamma_l \leq n_l'\} \tag{3.97}$$

of a filter F defines that every tuple $\langle \overline{n_1},\ \ldots,\ \overline{n_l} \rangle$ such that $\overline{n_1} \in [n_1, n_1'], \ldots, \overline{n_l} \in [n_l, n_l']$ permits the creation of one *TMG* by substitution of the values $\overline{n_1}, \ldots, \overline{n_l}$ instead of the corresponding multiplicity-variables in all *TMRs* in the scheme R. Temporal rules in R are included in the created scheme $R \circ \langle \overline{n_1},\ \ldots,\ \overline{n_l} \rangle$ without any transformations. So a *TMMG* $S = \langle v_0, R, F \rangle$ represents a set of *TMGs* S^*. The cardinality of this set may be estimated by the same inequality (3.78).

The mathematical semantics of a *TMMG* $S = \langle v_0, R, F \rangle$ is defined by the same set of Eqs. (3.79)–(3.94), because, evidently, a metagrammatical extension of any class of multiset grammars does not transform its basic mathematical semantics. As in the *MMGs*, the scope of all *MVs* in a filter F of the *TMMG* S is the whole scheme R.

Optimizing conditions with variables may also be used in filters of temporal multiset metagrammars, and there is no difference between *MMGs* and *TMMGs* concerning this feature.

The same techniques may be used for a definition of self-generating temporal multiset metagrammars (*SG TMMGs*).

Now we shall analyze sequentially the formal mathematical properties of the introduced and above-described classes of multiset grammars and multiset metagrammars.

3.4 Mathematical Properties of Multiset Grammars and Multiset Metagrammars

Below in this section, multiset grammars as a family of knowledge representation models and a logical formalism will be investigated from the point of view of their mathematical properties, providing both knowledge engineers and scholars with proper tools for the analysis of created knowledge bases and detection of possible anomalies and malfunctions.

3.4.1 Interconnections Between Classes of Multiset Grammars

First of all, it would be very useful to introduce some reasonable order inside the family of multiset grammars described in the above sections and forming together *the multigrammatical framework*. Such an order would provide a methodological basis for studying mathematical properties of various classes of the aforementioned family, which in the most compact and integrated form is presented in Table 3.4. As can be seen, this family includes $3 \times 4 = 12$ classes.

The simplest class is the *unitary multiset grammars*, which are a special case of *MGs*. As there exist *MGs* whose schemes include rules whose left parts are multisets containing $n > 1$ multiobjects, the class of *UMGs* is a proper subclass of the class of *MGs*:

Table 3.4 Classes of multiset grammars and metagrammars

UMGs	*MGs*	*TMGs*	*SG TMGs*
FUMGs	*FMGs*	*FTMGs*	*FSG TMGs*
UMMGs	*MMGs*	*TMMGs*	*SG TMMGs*

$$UMGs \subset MGs. \tag{3.98}$$

Temporal multiset grammars, evidently, are a special case of *self-generating temporal multiset grammars*. It is clear that any *SG TMG* whose scheme *R* does not contain rules $\langle r; v \to v' \rangle$ with multiobjects $l' \cdot r' \in v'$ is a *TMG*. In turn, any *TMG* whose scheme *R* does not contain rules $\langle r; v \to v' \rangle$ with multiobjects $\Delta n \cdot \Delta t \in v'$ is an *MG*. From this, *TMGs* are a special case of *SG TMGs*, and *MGs* are a special case of *TMGs*, so we have the chain of strict inclusions:

$$MGs \subset TMGs \subset SG\ TMGs. \tag{3.99}$$

Filtering *UMGs*, *MGs*, *TMGs*, and *SG TMGs* are a generalization of the corresponding classes considered above, created by the addition of filters to the basic constructions. So every class of *UMGs*, *MGs*, *TMGs*, and *SG TMGs* is a special case (namely, having an empty filter) of its generalization:

$$UMGs \subset \text{FUMGs}, \tag{3.100}$$

$$MGs \subset FMGs, \tag{3.101}$$

$$TMGs \subset FTMGs, \tag{3.102}$$

$$SG\ TMGs \subset FSG\ TMGs. \tag{3.103}$$

For the same reasons that primary ("filterless") classes are proper subclasses of one another, a *similar chain of inclusions holds regarding "filtering" classes*:

$$FUMGs \subset FMGs \subset FTMGs \subset FSG\ TMGs. \tag{3.104}$$

Finally, similar to the relations (3.100)–(3.103) and similar to the chain (3.104),

$$FUMGs \subset \text{UMMGs}, \tag{3.105}$$

$$FMGs \subset MMGs, \tag{3.106}$$

$$FTMGs \subset TMMGs, \tag{3.107}$$

$$FSG\ TMGs \subset SG\ TMMGs, \tag{3.108}$$

$$UMMGs \subset \ MMGs \subset \ TMMGs \subset SG\ TMMGs. \tag{3.109}$$

This is implies the presence of proper versions of metarules (*UMRs*, *MRs*, *TMRs*) in all the introduced metagrammatical extensions, as well as the presence of variable-containing chain boundary conditions used for variable domain declarations.

So, as may be seen, the most powerful class of the introduced *MG* family are *self-generating temporal multiset metagrammars*. The descriptional capabilities of various classes will be studied in more detail in Chaps. 5–9, which are dedicated to applications of the multigrammatical framework.

Any class of the *MG* family possesses its own specific mathematical properties. We shall not study all possible features of all these classes in this book but shall limit our consideration here to only *MGs*, *FMGs*, *UMGs*, and *FUMGs*, because this is critically necessary for the *development and description of the enhanced (improved, optimized) algorithmics of the multigrammatical framework*. However, the techniques proposed for solution of mathematical problems arising while studying the aforementioned properties are very general and flexible; an interested reader may effectively apply them to all other classes on his own.

Some common properties are peculiar to all listed classes of multiset grammars. Such properties are *cyclicity*, *finitarity*, and *usefulness*.

Speaking informally, any multiset grammar is *cyclic* if at least one non-empty generation chain begins from a multiset which is a submultiset of an *MS* created by this chain. An *MG* is *infinitary* if it defines an infinite set of multisets. Otherwise it is *acyclic* and *finitary*. Thus we shall consider criteria for recognition of cyclicity and finitarity of *MGs* belonging to all classes of the *MG* family. Also we shall consider the notion of an *MG's* usefulness and criteria for its recognition, understanding that any *MG* of any class is *useful* if it generates at least one multiset different from its kernel.

Let us begin a consideration of the aforementioned properties from the basic class of the introduced family.

3.4.2 Properties of Multiset Grammars

An *MG* $S = \langle v_0, R \rangle$ will be called *cyclic* if there exists a generation chain

$$v_0 \overset{*}{\Rightarrow} v \overset{+}{\Rightarrow} v' \tag{3.110}$$

such that $v \subseteq v'$. Taking into account that

$$v \subseteq v' \equiv v' = v + \Delta v, \tag{3.111}$$

where $\Delta v \supseteq \{\varnothing\}$, we may see that in this case an infinite generation chain

$$v_0 \overset{*}{\Rightarrow} v \overset{+}{\Rightarrow} v + \Delta v \overset{+}{\Rightarrow} (v + \Delta v) + \Delta v \overset{+}{\Rightarrow} \ldots \overset{+}{\Rightarrow} v + n * \Delta v \overset{+}{\Rightarrow} \ldots . \quad (3.112)$$

is possible. In other words, once executed, such a non-empty chain may repeat infinitely.

Example 3.24
Let $S = \langle v_0, R \rangle$, where $v_0 = \{1 \cdot a, 2 \cdot b\}$, and $R = \{r_1, r_2, r_3\}$, where

$$r_1 \text{ is } \{1 \cdot a\} \rightarrow \{2 \cdot b\},$$
$$r_2 \text{ is } \{2 \cdot b\} \rightarrow \{2 \cdot a\},$$
$$r_3 \text{ is } \{1 \cdot b\} \rightarrow \{1 \cdot c\}.$$

As can be seen, there exists the generation chain

$$\{1 \cdot a,\ 2 \cdot b\} \overset{r_1}{\Rightarrow} \{4 \cdot b\} \overset{r_2}{\Rightarrow} \{2 \cdot a,\ 2 \cdot b\} \overset{r_1}{\Rightarrow} \{1 \cdot a,\ 4 \cdot b\} \overset{r_2}{\Rightarrow} \{6 \cdot b\} \overset{r_1}{\Rightarrow} \ldots,$$

which contains the *MS* v_0 as a submultiset in every *MS* obtained by application of the rule r_2.

Recognition of an MG's cyclicity may be performed in parallel with generation of a set of multisets according to the *MG's* mathematical semantics. To implement this, it is sufficient to extend (3.30)–(3.32) by a check of condition (3.110). For that purpose, we shall redefine (3.30)–(3.32) in such a way that not multisets but couples $\langle v, W \rangle$ are generated, where W is the set of *MSs* which were generated by the execution of all previous steps leading to the creation of the *MS* v, i.e., if

$$v_0 \overset{r_{i_1}}{\Rightarrow} v_1 \overset{r_{i_2}}{\Rightarrow} \ldots \overset{r_{i_j}}{\Rightarrow} v_j \Rightarrow \ldots \overset{r_{i_N}}{\Rightarrow} v \quad (3.113)$$

then the aforementioned couple would be $\langle v, W \rangle$, where

$$W = \{v_0,\ v_1,\ \ldots,\ v_j,\ \ldots,\ v_m\}. \quad (3.114)$$

So the growth of the set W provides an opportunity to check the condition (3.110), and thus (3.30)–(3.33) may be replaced by the following definition, where, as above, $S = \langle v_0, R \rangle$:

$$V_{(0)} = \{\langle v_0,\ \{\varnothing\}\rangle\}, \quad (3.115)$$

$$V_{(i+1)} = V_{(i)} \cup \left(\bigcup_{\langle \overline{v},\ W\rangle \in V_{(i)}} \bigcup_{r \in R} \{\pi'(\langle \overline{v},\ W\rangle,\ r)\} \right), \quad (3.116)$$

$$V_S = V_{(N)}, \tag{3.117}$$

where N is such that

$$V_{(N+1)} = V_{(N)}, \tag{3.118}$$

and

$$\pi'(\langle \overline{v},\ W\rangle, v \to v') = \begin{cases} \varphi(\overline{v} - v + v',\ W \cup \{\overline{v}\}), & \text{if } v \subseteq \overline{v}, \\ \overline{v} \text{ otherwise}, \end{cases} \tag{3.119}$$

$$\varphi(\overline{v},\ W) = \begin{cases} \overline{v} + \{1 \cdot \Delta\}, & \text{if } (\exists \overline{v}' \in W)\ \overline{v}' \subseteq \overline{v}, \\ \langle \overline{v},\ W\rangle \text{ otherwise}. \end{cases} \tag{3.120}$$

The result of an application of an *MG S*, if it is *acyclic*, is a set of couples whose first elements would be generated if the definition (3.30)–(3.33) was used, starting from the obvious kernel. If *S* is *cyclic*, then V_S will include, along with the "usual" multisets, also "marked" *MSs* $\overline{v} + \{1 \cdot \Delta\}$, where Δ is an auxiliary object, and the *MO* $1 \cdot \Delta$ (a "cycle detector") denotes that this *MS* is a "kernel of cycle generation" in full accordance with (3.116). If such an *MS* $\overline{v}$ occurs, then no cyclic chains starting from it are created, and the infinite set of multisets which would be generated, beginning from $\overline{v}$, is marked by the multiobject $1 \cdot \Delta$. By these techniques, we assess also the *finiteness* of an *SMS* V_S: all possible "sources of infiniteness" are "caught" and present in $V_{(i)}$ in the form of *marked MSs*. If so, then there exists a number N such that $V_{(N+1)} = V_{(N)}$, and after this value N, there will be no any increment to $V_{(N+1)}$; hence, $V_{(N+1)}$ is a finite set of multisets and marked multisets defined by a cyclic multigrammar *S*, as is affirmed by (3.115)–(3.120). Such "finitarization" of infinite *SMSs* is very useful not only for recognition of an *MG's* finitarity but, if necessary, for various operations on such *SMSs*.

Let us consider now definitions (3.115)–(3.120) in detail, comparing it with (3.30)–(3.33). As can be seen, (3.115) differs from (3.30) by the presence of a couple $\langle v_0, \{\varnothing\}\rangle$ instead of an *MS* v_0; the empty set as the second component of this couple means that no multisets are generated at this initial moment. The relation (3.116) is very similar to (3.31), but instead of an *MS* $\overline{v}$, a couple $\langle \overline{v}, W\rangle$ is used, and instead of the function π, the function π' is used, of which the first argument is this couple. The function π' is similar to π in the sense of applying a rule $v \to v'$ to a multiset $\overline{v}$. If the rule is not applicable to this *MS*, the result is the same $\overline{v}$ without any changes. However, if the rule is applicable, the auxiliary function φ of two arguments is used, the first being the result of an application $\overline{v} - v + v'$ in the sense of (3.31) and the second the join of sets *W* and $\{\overline{v}\}$, which means the *MS* $\overline{v}$ is included in the set of predecessors of $\overline{v} - v + v'$ in the generation chain beginning from the *MS* v_0 and leading to the *MS* $\overline{v} - v + v'$. The definition (3.120) of the function φ contains two alternatives, the first of which corresponds to the case when $\overline{v}$ and some $\overline{v}' \in W$

correspond to the criteria of *MG* cyclicity, i.e., $\overline{v}' \subseteq \overline{v}$, so $\overline{v}'$ is a predecessor of $\overline{v}$ in a generation chain. So in this case, the result of $\varphi(\overline{v}, W)$ is nothing but a marked multiset $\overline{v} + \{1 \cdot \Delta\}$, terminating further generation. Otherwise the result is a couple $\langle \overline{v}, W \rangle$, which is included in $V_{(i+1)}$.

Let us now study the issue of an *MG's finitarity*. As was defined verbally in the introductory part of this Sect. 3.4, a multiset grammar is *infinitary* if the set of multisets generated by application of this *MG* is infinite.

It is evident that any acyclic *MG* is finitary. However, the inverse statement is not true.

Example 3.25
Let $S = \langle v_0, R \rangle$, where $v = \{1 \cdot a\}$, and $R = \{r_1, r_2\}$, where

$$r_1 \text{ is } \{1 \cdot a\} \rightarrow \{1 \cdot a\},$$
$$r_2 \text{ is } \{1 \cdot a\} \rightarrow \{1 \cdot b\}.$$

As can be seen, although this *MG* is cyclic, $V_S = \{\{1 \cdot a\}, \{1 \cdot b\}\}$, which is finite.

A criterion for deciding an *MG's* finitarity is formulated as follows.

Statement 3.2 A cyclic *MG* $S = \langle v_0, R \rangle$ is infinitary if and only if there exists a generation chain (3.110) such that $v \subset v'$.

Proof If there exists a generation chain $v_0 \overset{*}{\Rightarrow} v \overset{+}{\Rightarrow} v'$ such that $v \subset v'$, then in (3.112) $n > 1$, so the number of multisets generated by prolongation of this chain will only increase, and there is no upper bound for this process, so the *SMS* V_S is infinite. If we suppose that *S* is cyclic and there is no generation chain meeting the aforementioned condition, then the only possibility is that in all cyclic chains like (3.110) $v \subseteq v'$, but not $v \subset v'$, so $v = v'$. If so, then the number of generated *MSs* does not increase, and hence the *MG* is finite.

An algorithmic implementation of the logic of this criterion is obvious. It is sufficient to transform the definition of the function φ from the two-alternative to the three-alternative form:

$$\varphi(\overline{v}, W) = \begin{cases} \overline{v} + \{1 \cdot \Delta\}, & \text{if } (\exists \overline{v}' \in W)\ \overline{v}' \subset \overline{v}, \\ \{\varnothing\}, & \text{if } (\exists \overline{v}' \in W)\ \overline{v}' = \overline{v}, \\ \langle \overline{v}, W \rangle & \text{otherwise.} \end{cases} \tag{3.121}$$

As can be seen, the second alternative joins the empty set to the set V_S when an *MS* $\overline{v}$ appears in a generation chain more than once. As $\overline{v}$ has already been included in V_S at one of the previous generation steps, V_S will not change in this case.

A more careful investigation of cyclicity and finitarity leads us to the notion of *terminal finitarity*. An *MG* $S = \langle v_0, R \rangle$ will be called *terminal finitary* if the set of terminal multisets generated by application of this *MG* is finite:

$$|\overline{V}_S| < \infty. \tag{3.122}$$

It is clear what the difference is between *finitarity* and *terminal finitarity*. An infinite set V_S generated by application of an infinitary *MG S* may include a finite subset $\overline{V}_S$. This possibility may be illustrated by the following example.

Example 3.26
Let $S = \langle v_0, R \rangle$, where $v = \{1 \cdot a\}$, $R = \{r_1, r_2, r_3\}$, and

$$r_1: \quad \{1 \cdot a\} \rightarrow \{2 \cdot b\},$$
$$r_2: \quad \{1 \cdot a\} \rightarrow \{1 \cdot c\},$$
$$r_3: \quad \{1 \cdot b\} \rightarrow \{1 \cdot c,\ 1 \cdot b\}.$$

As can be seen,

$$V_S = \{\{1 \cdot c\}\} \cup \left(\bigcup_{i=0}^{\infty} \{2 \cdot b,\ i \cdot c\} \right),$$

so

$$\overline{V}_S = \{\{1 \cdot c\}\}$$

is a finite set, while $V_S - \overline{V}_S$ is an infinite set in which every i-th element is a non-terminal multiset which may be used for the generation of the i + 1-th element.

However, it is not difficult to obtain a criterion to decide *terminal infinitarity* of a multiset grammar. We shall make one more local correction to the definition (3.120) of the function φ, making it, like (3.121), three-alternative, but with the second alternative different from (3.121):

$$\varphi(\overline{v},\ W) = \begin{cases} \overline{v} + \{1 \cdot \Delta\}, & \text{if } (\exists \overline{v}' \in W)\ \overline{v}' \subset \overline{v}, \\ \overline{v} + \{1 \cdot \Delta'\}, & \text{if } (\exists \overline{v}' \in W)\ \overline{v}' = \overline{v}, \\ \langle \overline{v},\ W \rangle & \text{otherwise}, \end{cases} \tag{3.123}$$

where Δ' is an auxiliary object, distinguishable from Δ, used for marking multisets which are generated from themselves, and thus non-terminal, but which are not "starting points" for infinite *SMSs*.

Statement 3.3 A cyclic *MG* $S = \langle v_0, R \rangle$ is *terminal infinitary* if the set

$$\overline{V}_S' = \{\overline{v} \mid \overline{v} \in V_S \& V_S \cap \{\overline{v} \ + \{1 \cdot \Delta,\ \ 1 \cdot \Delta'\}\} \neq \{\varnothing\}\} \tag{3.124}$$

is non-empty.

As can be seen, the non-empty set $\overline{V}_S'$ is nothing but $V_S - \overline{V}_S$.

The set of multisets generated by application of an *MG* $S = \langle v_0, R\rangle$ in accordance with (3.115)–(3.119) and (3.123) will be called the *finitarized SMS of the MG S* (abbreviated as *FSMS*) and denoted $\overline{V}_S^*$.

Let us now pay some attention to an *MG's usefulness*. Formally, an *MG* $S = \langle v_0, R\rangle$ is *useful* if

$$\overline{V}_S - \{v_0\} \neq \{\varnothing\}, \tag{3.125}$$

i.e., if the *MG* generates at least one terminal multiset which differs from the kernel. It is clear that if v_0 is terminal, then $\overline{V}_S = \{v_0\}$, and there is no use for such a grammar *S*. For this reason, such an *MG* is called *useless*.

A criterion for deciding an *MG's* usefulness is formulated as follows.

Statement 3.4 An *MG* $S = \langle v_0, R\rangle$ is useful if and only if

$$\overline{V}'_S \neq \{v_0\}. \tag{3.126}$$

Thus to decide usefulness of an *MG S*, it is sufficient to create the set $\overline{V}'_S$ and to compare it with $\{v_0\}$. Inequality means that there exists at least one *TMS* different from v_0, so the *MG* is really useful. Otherwise *S* is useless.

3.4.3 Properties of Filtering Multiset Grammars

We shall call an *FMG* $S = \langle v_0, R, F\rangle$ *cyclic* if the core *MG* $S' = \langle v_0, R\rangle$ of this *FMG* is cyclic. Thus techniques *for deciding cyclicity* of an *FMG* would be the same as in the case of *MGs*.

Because a filter is applied only to terminal multisets, there is no sense introducing a notion of *finitarity of an FMG*. This notion applies to the core *MG* of the *FMG*, and the criterion for deciding its finitarity is the same as above. The same applies to *terminal finitarity* of an *FMG*.

A filtering multiset grammar $S = \langle v_0, R, F\rangle$ will be called *terminal finitary* if

$$\left|\overline{V}_{S'} \downarrow F\right| < \infty. \tag{3.127}$$

It is clear that if $\overline{V}_{S'}$ is finite, the result of its filtration will be also finite. So let us consider the general case, when $\overline{V}_{S'}$ is infinite, while the result of its filtration by *F* would be or would not be finite.

Consider $\overline{V'}_{S'}$—the finitarized *SMS* of the core *MG* S' of an *FMG S*. Let us suppose for simplicity that the filter *F* contains only elementary boundary conditions and optimizing conditions. It is evident that only *EBCs* $a \leq n$, $a = n$, and *OpCs* $a = min$ may cut off infinite subsets of a set $\overline{V}_{S'}$. Let us create a subfilter $\overline{F} \subseteq F$ containing only such "cutting" conditions and denote by $A_{\overline{F}}$ the set of so-called restricted objects selected by the subfilter $\overline{F}$. Also let us transform the definition of the function φ as follows:

$$\varphi(\overline{v},\ W)=\begin{cases}\overline{v}+\{1\cdot\Delta\}, & \text{if } (\exists\overline{v}'\in W)\ \overline{v}'\subset\overline{v}\&\beta(\overline{v}-\overline{v}')\cap A_{\overline{F}}=\{\varnothing\}\\ \overline{v}+\{1\cdot\Delta'\}, & \text{if } (\exists\overline{v}'\in W)\ \overline{v}'=\overline{v},\\ \langle\overline{v},\ W\rangle & \text{otherwise.}\end{cases} \tag{3.128}$$

As can be seen, the only difference of (3.128) from (3.123) is that the first alternative contains one additional condition, which provides a check whether there exists at least one restricted object, in the increment $\overline{v}-\overline{v}'$. Obviously, existence of at least one such object a allows all multisets generated from $\overline{v}'$ to be cut off, including multiobjects $m\cdot a$ with $m>n$, where the condition $a\le n$ or $a=n$ is in $\overline{F}$. The same cut off would be done in the case of all multisets generated from $\overline{v}'$ and including multiobjects $m\cdot a$ with $m>\overline{n}$, where $\overline{F}$ includes an optimizing condition $a=min$ and $\overline{n}$ is a multiplicity of an object a in the first generated terminal multiset satisfying this condition. So if no restricted object $a\in A_{\overline{F}}$ is in the aforementioned increment (i.e., if the intersection of the basis of the multiset $\overline{v}-v'$ and $A_{\overline{F}}$ is the empty set), then the *MS* $\overline{v}$ is the root of a tree with an infinite set of nodes, which are generated multiobjects. Thus the result of $\varphi(\overline{v},\ W)$ in this case is a finitarized representation of this set. This representation is, evidently, the multiset $\overline{v}+\{1\cdot\Delta\}$.

If

$$\beta(\overline{v}-\overline{v}')\cap A_{\overline{F}}\neq\{\varnothing\}, \tag{3.129}$$

then the aforementioned tree is finite, and thus the third alternative of (3.128) is applied: this means only that the couple $\langle\overline{v},\ W\rangle$ is included in the resulting set $\varphi(v_0,\ \{\varnothing\})$ used for detection of *FMG* finitarity—not for creation of $\overline{V}_S$.

So a decision criterion for *FMG* terminal finitarity is formally stated as follows.

Statement 3.5 An *FMG* $S=\langle v_0,R,F\rangle$ is terminal infinitary if it satisfies (3.127) with the function φ defined by (3.128).

Concerning an *FMG's* usefulness, which is defined as the existence of at least one generated *TMS* different from the kernel v_0, it is evident that a filter F being applied to the set $\overline{V}_{S'}$ does not change criterion (3.126): it is sufficient to apply this criterion to the core *MG* of the considered *FMG*.

Statement 3.6 An *FMG* $S=\langle v_0,R,F\rangle$ is useless if and only if the core *MG* $S'=\langle v_0,$ $R\rangle$ of this *FMG* is useless.

Let us now consider mathematical properties of unitary multiset grammars and filtering *UMGs*.

3.4.4 Properties of Unitary Multiset Grammars and Filtering UMGs

Because *UMGs* and *FUMGs* are, in essence, syntactically simplified representations of CF *MGs* and CF *FMGs*, respectively, all that was said in Sects. 3.4.2 and 3.4.3 about cyclicity, finitarity, and usefulness of *MGs* and *FMGs* also applies regarding all their special cases, i.e., *UMGs* and *FUMGs* among others. However, there are two additional properties of *UMGs* which we shall consider below in this section: *integrity (connectivity)* and *variativity.*

A *UMG* $S = < v_0, R>$ will be called *integrated* or *connected* (the notions are equivalent) if for every object $a \in A_S$ which does not occur in the kernel v_0,there exists at least one unitary rule whose body contains a multiobject $n \cdot a$. Otherwise the *UMG* is called *disintegrated*, or *disconnected.*

Example 3.27
Let $UMGS = < \{1 \cdot a\}, R>$, where the scheme *R* contains the following unitary rules:

$$r_1 : a \to 2 \cdot b, 3 \cdot c,$$

$$r_2 : b \to 3 \cdot c, 1 \cdot b,$$

$$r_3 : c \to 1 \cdot d.$$

This *UMG* is integrated, because for the non-terminal object *b*, there exist two *UR*s (r_1 and r_2), containing *b* in their bodies, while for *c* there is one such *UR* (r_1). If we join to *R* one more *UR*

$$r_4 : e \to 3 \cdot b, 4 \cdot c,$$

then the created *UMG* will become disintegrated, because there are no *UR*s in *R* whose bodies contain object *e*.

A *UMG* $S = < a_0, R>$ will be called *variative* if its scheme *R* contains at least two unitary rules with the same header and different bodies:

$$a \to n_1 \cdot a_1, \ldots, n_m \cdot a_m,$$
$$a \to n'_1 \cdot a'_1, \ldots, n'_l \cdot a'_l, \tag{3.130}$$

or, equivalently, *R* contains at least one non-terminal object *a* such that

$$| R_a | > 1. \tag{3.131}$$

Such non-terminal objects are called *alternating*, while all *NTOs* with a one-element set R_a are called *non-alternating*.

As may be seen, the number of non-terminal objects occurring in *UR*s belonging to the scheme *R* of a variative *UMG* is less than the number of *UR*s in *R*:

$$| A_S^N | < | R | \, . \tag{3.132}$$

A *UMG* $S = \, < a_0, R>$ such that all its *NTOs* are non-alternating is called *non-variative*. In this case, evidently,

$$|A_S^N| = | R | \, . \tag{3.133}$$

From a practical point of view, any non-variative unitary multiset grammar usually describes some hierarchical entity or process (see Examples 3.11 and 3.13), while any variative *UMG* describes a set of entities/processes with alternative structures/interconnections. Non-variative *UMGs* are also called *1-multigrammars*, while variative *UMGs* are called *k-multigrammars* (Sheremet 2010, 2011a).

Example 3.28
Let *UMGS* $= \, < \{1 \cdot a\}, R>$, where the scheme *R* contains the following unitary rules:

$$r_1 : a \to 1 \cdot b, 2 \cdot d,$$

$$r_2 : b \to 3 \cdot c, 1 \cdot d,$$

$$r_3 : d \to 2 \cdot d.$$

As can be seen, this *UMG* is non-variative, and $|A_S^N| = |R| = 3$. If we join to *R* one more unitary rule

$r_4 : b \to 2 \cdot c, 3 \cdot d$,then the created *UMG* will be variative, and $|A_S^N| = 3 < |R| = 4$. Thus, the *NTO b* will become alternating.

Obviously, the notions of connectivity and variativity may be applied to filtering *UMGs* by a simple projection of these notions onto the core *UMGs* of *FUMGs*.

3.4.5 Properties of Metagrammatical Extensions

Let $S = \, < v_0, R, F>$ be a multiset metagrammar. We shall call *S cyclic* if there exists at least one cyclic *FMG* $S' \in S^*$, where S^* is defined by (3.81). Otherwise the *MMG* is *acyclic*.

Similarly, an *MMG* $S = \, < v_0, R, F>$ will be called *finitary* if all *FMGs* $S' = \, < v_0, R', F - F_\Gamma>$ such that $<v_0, R' > \, \in S^*$ are finitary. Otherwise *MMG* is *infinitary*.

The notion of usefulness in the case of *MMGs* is also defined in a natural way: if there exists at least one useful *FMG* $S' = \, < v_0, R', F - F_\Gamma>$ such that $<v_0, R' > \, \in S^*$, then the *MMGS* $= \, < v_0, R, F>$ is called *useful*; otherwise it is *useless*.

Consider now *unitary MMGs*. It is not difficult to extend to *UMMGs* the notion of variativity introduced for *UMGs*: we shall call a *UMMG* $S = < v_0, R, F>$ *non-variative* if for every *NTO* $a \in A_S^N$

$$|R_a| = 1. \tag{3.134}$$

As may be seen, in this case, every *MG*

$$S' = < v_0, R' > \in S^* \tag{3.135}$$

is also non-variative. Otherwise, i.e., if there exists at least one *NTO* $a \in A_S^N$ such that $|R_a| > 1$, the *UMMG* is called *variative*.

We shall apply a similar approach to define the integrity (connectivity) of a *UMMG*. We shall call a *UMMG* $S = < v_0, R, F>$ *integrated (connected)* if there exists at least one integrated *UMG* $S' = < v_0, R' > \in S^*$. If all $S' \in S^*$ are disintegrated, the *UMMG S* is also called *disintegrated (disconnected)*.

With this we finish our consideration of the syntax, semantics, and basic mathematical properties of basic classes of multiset grammars in the multigrammatical framework and move on to some useful descriptional extensions of the basic multigrammatical toolkit. Such extensions allow simplified representation of various sociotechnological systems and problems associated with these *STSs*.

3.5 Descriptional Extensions of the Basic Multigrammatical Toolkit

We shall consider four main descriptional extensions of the above-described basic multigrammatical toolkit which are very useful from the practical point of view:

1. Rational numbers as multiplicities
2. Negative numbers as multiplicities
3. Composite names of objects (composites)
4. Multifunctional manufacturing devices (*MFMDs*)

3.5.1 Rational Multiplicities

As it is known from (Petrovskiy 2002, 2003; Singh et al. 2007), there may be multisets with not only integer but also *rational multiplicities* of objects. This extension of the notion of a multiset does not change the syntax and semantics of any class of multiset grammars and multiset metagrammars (*MGs, FMGs, MMGs, UMGs, FUMGs, UMMGs, TMGs, FTMGs, TMMGs*) because all definitions, introduced and used above, are based on operations on numbers which may be rational as well as integer; the only required property of these multiplicities is their positivity. Therfore, any rule

$$\{n_1 \cdot a_1, \ldots, n_m \cdot a_m\} \rightarrow \{n'_1 \cdot a'_1, \ldots, n'_l \cdot a'_l\}, \tag{3.136}$$

any metarule

$$\{\mu_1 \cdot a_1, \ldots, \mu_m \cdot a_m\} \rightarrow \{\mu'_1 \cdot a'_1, \ldots, \mu'_l \cdot a'_l\}, \tag{3.137}$$

any unitary rule

$$a \rightarrow n_1 \cdot a_1, \ldots, n_m \cdot a_m, \tag{3.138}$$

any unitary metarule

$$a \rightarrow \mu_1 \cdot a_1, \ldots, \mu_m \cdot a_m, \tag{3.139}$$

any temporal rule

$$\{n_1 \cdot a_1, \ldots, n_m \cdot a_m\} \rightarrow \{n'_1 \cdot a_1, \ldots, n'_l \cdot a'_l, \Delta n \cdot \Delta t\}, \tag{3.140}$$

any temporal metarule

$$\{\mu_1 \cdot a_1, \ldots, \mu_m \cdot a_m\} \rightarrow \{\mu'_1 \cdot a'_1, \ldots, \mu'_l \cdot a'_l, \mu'_{l+1} \cdot \Delta t\}, \tag{3.141}$$

and any boundary condition

$$n \leq a \leq n', \tag{3.142}$$

or

$$n \leq \gamma \leq n', \tag{3.143}$$

may contain not only integers but also rational positive multiplicities and bounds. For practical applications, it is sufficient only to postulate the maximal lengths of the fractional parts of rational numbers used in schemes, filters, and generated multisets. Let us consider this case in more detail.

The only structural elements whose semantics critically depends on the number of digits in the fractional parts of the values used are variable declarations in *MMG's*, *UMMG's*, and *TMMG's* filters. This is due to the definition of the set S^* including all *MGs*, *UMGs*, and *TMGs* which may be created by substitution of variable values to *MRs*, *UMRs*, and *TMRs*, respectively. If the domains of these variables are sets of integers, then the number of created *MGs*, *UMGs*, and *TMGs*, i.e., an upper bound of $|S^*|$, can be estimated as the product of the cardinalities of the aforementioned domains. If we want to use sets of rational numbers as the domains of variables, then we must restrict a number of digits in the fractional parts of domains bounds; otherwise, every such domain would be infinite, and hence, the number of created *MGs* (*UMGs*, *TMGs*) would also be infinite. The only way to make variable domains finite is to fix the aforementioned number of digits by the use of values occurring in boundary conditions as patterns (Sheremet 2010, 2011a).

To implement this approach, it is sufficient to replace (3.81) by

$$S^* = \bigcup_{u_1 \in \varepsilon\left(n_1,\, n_1'\right)} \ldots \bigcup_{u_l \in \varepsilon\left(n_l,\, n_l'\right)} \{\langle v_0,\ R \circ \langle u_1,\ \ldots,\ u_l \rangle \rangle\}, \tag{3.144}$$

where for any $i = 1, \ldots, l$

$$\varepsilon\left(n_i, n_i'\right) = \bigcup_{j=0}^{k} \{n_i + j \cdot \Delta n_i\}, \tag{3.145}$$

and

$$k = \frac{n_i' - n_i}{\Delta n_i}, \tag{3.146}$$

where Δn_i is a discrete value describing the domain of the variable γ_i. This discrete value is determined by the numbers of digit after the decimal point in n_i and n_i'. If the aforementioned numbers in n_i and n_i' are different, then the maximal one is used. Formally,

$$\Delta n_i = 10^{-\,max\left\{\delta(n_i),\ \delta\left(n_i'\right)\right\}}, \tag{3.147}$$

where

$$\delta\left(d_1 \ldots d_p . d_1' \ldots d_q'\right) = q, \tag{3.148}$$

$$\delta\left(d_1 \ldots d_p\right) = 0, \tag{3.149}$$

and $d_i \in \{0, 1, \ldots, 9\}, d_i' \in \{0, 1, \ldots, 9\}$ are decimal digits; in the general case $p \geq 0$, $q \geq 0$.

Example 3.29
Let *FUMG* $S = \ < v_0, R, F>$, where $v_0 = \{1 \cdot a_0\}$, $R = \{r_1, r_2\}$,

$$r_1 : a_0 \rightarrow \gamma_1 \cdot a_1, 1 \cdot a_2,$$

$$r_2 : a_0 \rightarrow \gamma_2 \cdot a_2,$$

and

$$F = \{0.2 \le \gamma_1 \le 0.25,\ 1 \le \gamma_2 \le 1.5,\ a_2 \le 1.33\}.$$

According to (3.144)–(3.149),

$$S^* = \bigcup_{u_1 \in \varepsilon(0.2,0.25)} \bigcup_{u_2 \in \varepsilon(1,1.5)} \{\langle v_0,\ R \circ \langle u_1,\ u_2 \rangle \rangle\}$$

$$= \bigcup_{u_1 \in \{0.20, 0.21,\ \ldots,\ 0.25\}} \bigcup_{u_2 \in \{1.0,\ 1.1,\ \ldots,\ 1.5\}} \{\langle v_0,\ R \circ \langle u_1,\ u_2 \rangle \rangle\}$$

$$= \{ < v_0, \{ < a_0 \rightarrow 0.20 \cdot a_1,\ 1 \cdot a_2 >,\ < a_1 \rightarrow 1.0 \cdot a_2 > \} >,$$

$$< v_0, \{ < a_0 \rightarrow 0.20 \cdot a_1,\ 1 \cdot a_2 >,\ < a_1 \rightarrow 1.1 \cdot a_2 > \} >,$$

$$\ldots$$

$$< v_0, \{ < a_0 \rightarrow 0.25 \cdot a_1,\ 1 \cdot a_2 >,\ < a_1 \rightarrow 1.0 \cdot a_2 > \} >,$$

$$\ldots$$

$$< v_0, \{ < a_0 \rightarrow 0.25 \cdot a_1,\ 1 \cdot a_2 >,\ < a_1 \rightarrow 1.5 \cdot a_2 > \} > \}.$$

3.5.2 Negative Multiplicities

All previous definitions and considerations were based on the presumption that multiplicities are non-negative numbers. However, it is not too difficult to permit *negative multiplicities* in multiobjects occurring in rules, unitary rules, temporal rules, unitary metarules, temporal metarules, and boundary conditions as well as in generated multisets.

If we consider the basic definition (3.30)–(3.33) of the mathematical semantics of multigrammars, its kernel is the inclusion of the left part of the rule in the multiset to which this rule is applied. From the "physical" point of view, this means that the amounts of objects occurring in the left part are no greater than the amounts of the same objects in the multiset, so replacement may be done in accordance with (3.33). If we permit negative multiplicities in multisets, then, first of all, the relation of inclusion must be redefined on such *MSs*. However, an extension of the definition (3.30)–(3.33) on these *MSs* has no practical foundation and for this reason is meaningless. So we shall consider only multisets with positive integers and, if necessary, positive rational multiplicities. However, restricting our consideration to only multisets with positive multiplicities, we may permit negative multiplicities in the right parts of the rules, *URs*, *MRs*, *UMRs*, as well as in *TRs* and *TMRs* (in the last two cases except multiplicities of multiobject Δt), and also negative numbers in boundary conditions. Negative multiplicities in the aforementioned right parts allow representation of elimination (destruction) of the appropriate amounts of *ORs* from (in) their current collection (resource base). This technique will be applied below while considering resilience of *STSs* (Chap. 8) as well as resource-based games (Chap. 9). Let us redefine the mathematical semantics of rules regarding negative multiplicities.

Consider a *rule*

$$\begin{aligned}&\{n_1 \cdot a_1,\ \ldots,n_m \cdot a_m\}\\ &\quad\to\{n'_1 \cdot a'_1,\ \ldots,n'_l \cdot a'_l,\ -n'_{l+1} \cdot a'_{l+1},\ \ldots,\ -n'_{l+k} \cdot a'_{l+k}\},\end{aligned} \tag{3.150}$$

where $n_1, \ldots, n_m, n'_1, \ldots, n'_l, n'_{l+1}, \ldots, n'_{l+k}$ are positive numbers (so $-n'_{l+1}, \ldots, -n'_{l+k}$ are negative numbers).

We shall rewrite the rule in the form

$$v \to v'_+ - v'_-, \tag{3.151}$$

where

$$v = \{n_1 \cdot a_1,\ \ldots,n_m \cdot a_m\}, \tag{3.152}$$

$$v'_+ = \{n'_1 \cdot a'_1,\ \ldots,n'_l \cdot a'_l\}, \tag{3.153}$$

$$v'_- = \{n'_{l+1} \cdot a'_{l+1},\ \ldots,n'_{l+k} \cdot a'_{l+k}\}, \tag{3.154}$$

and redefine an application of this rule to a multiset $\overline{v}$ such that $v \subseteq \overline{v}$ as follows:

$$\overline{v}' = \overline{v} - v + v'_+ - v'_-. \tag{3.155}$$

As can be seen, this definition adequately represents the essence of negative multiplicities described verbally above: subtraction of *MS* v'_- provides the elimination of represented by it amounts of *ORs* from a result of an application of a rule $v \to v'_+$ to an *MS* $\overline{v}$.

Example 3.30
Let the *MS*

$$\overline{v} = \{1 \cdot a, 3 \cdot b, 4 \cdot c, 5 \cdot d\},$$

and the rule r be

$$\{2 \cdot b, 2 \cdot c\} \to \{1 \cdot a, \ -3 \cdot d\}.$$

Because

$$\{2 \cdot b, 2 \cdot c\} \subseteq \{1 \cdot a, 3 \cdot b, 4 \cdot c, 5 \cdot d\},$$

the rule r may be applied to the multiset $\overline{v}$, and according to (3.150),

$$v'_+ = \{1 \cdot a\},$$
$$v'_- = \{3 \cdot d\},$$

and the result of application of the rule r to the multiset $\overline{v}$ will be the *MS*

$$\overline{v}' = (\{1 \cdot a, \ 3 \cdot b, \ 4 \cdot c, \ 5 \cdot d\} - \{2 \cdot b, \ 2 \cdot c\} + \{1 \cdot a\}) - \{3 \cdot d\} =$$
$$(\{1 \cdot a, \ 1 \cdot b, \ 2 \cdot c, \ 5 \cdot d\} + \{1 \cdot a\}) - \{3 \cdot d\} =$$
$$\{2 \cdot a, 1 \cdot b, 2 \cdot c, 5 \cdot d\} - \{3 \cdot d\} =$$
$$\{2 \cdot a, 1 \cdot b, 2 \cdot c, 2 \cdot d\}.$$

Let us now consider negative multiplicities in *temporal rules*. In fact, having a basic technique regarding rules, it is not so difficult to apply it to *TRs*.

Namely, it is sufficient to represent a temporal rule whose right part contains negative multiplicities similarly to (3.151):

$$< r; v \to v'_+ - v'_- + \{\Delta n \cdot \Delta t\} >, \qquad (3.156)$$

which means that an application of this *TR* adds a collection of *ORs* v'_+ to the resource base and extracts a collection of *ORs* v'_- from this *RB*, both operations executed in Δn time units since the moment when the resource base contains the collection of *ORs* v . The order of the operations is essential: first an addition is performed and then a subtraction.

The appearance of negative multiplicities in the right parts of temporal rules implies a correction of the mathematical semantics of temporal multiset grammars defined by (3.53)–(3.61), but this correction is very local and boils down to the replacement of (3.57) by the following definition:

$$\boldsymbol{L}(v_R, n) =$$

$$\left(\sum_{\substack{r[n_1, l_1] \ldots [n_k, l_k] \in v_R \\ \langle r; v \rightarrow v'_+ - v'_- + \{\Delta n \cdot \Delta t\} \rangle \in R \\ \mu(v_R, n) \in \{l_1, \ldots, l_k\}}} v'_+ \right) - \left(\sum_{\substack{r[n_1, l_1] \ldots [n_k, l_k] \in v_R \\ \langle r; v \rightarrow v'_+ - v'_- + \{\Delta n \cdot \Delta t\} \rangle \in R \\ \mu(v_R, n) \in \{l_1, \ldots, l_k\}}} v'_- \right) \tag{3.157}$$

As can be seen, at the same moment n when some *MDs* finish their operation cycles and load to a resource base object-resources produced during these *OCs*, some *ORs*, predefined by multiobjects with negative multiplicities in the right parts of the *TRs*, are eliminated from the obtained *RB*.

The definition of a filtering multigrammar (as well as of a *FUMG* and of a *FTMG*) remains the same, because all corrections made above concern the core *MGs* of *FMGs* and do not require any transformations of the semantics of filters.

Similarly, the definition of a multiset metagrammar (as well as of an *UMMG* and of a *TMMG*) remains unchanged: core unitary multigrammars created by a substitution of values of multiplicity-variables may contain negative multiplicity-constants, which are processed according to (3.155).

There is no special effect of "negative bounds-containing" boundary conditions regarding multiplicities of terminal objects in generated *TMSs*—these conditions are tested as usual.

Concerning boundary conditions, which are variable declarations, it is obvious that all negative values from intervals defined by these declarations are substituted into the bodies of metarules (as well as into the bodies of *UMRs* and *TMRs*), making rules (as well as *URs* and *TRs*) with negative multiplicities. The only evident nuance in this context is that the resulting rules (*URs*, *TRs*) must not contain multiobjects with negative multiplicities in their left parts. If such rules (*URs*, *TRs*) occur, they are eliminated from the created schemes.

Example 3.31
Let $S = \, < \{1 \cdot a_0\}, R, F>$, where $R = \{r_1, r_2\}$, and

$$r_1 : a_0 \rightarrow \gamma_1 \cdot a_1, 2 \cdot a_2,$$

$$r_2 : a_1 \to \gamma_2 \cdot a_2,$$
$$F = \{-1 \le \gamma_1 \le 1,\ -2 \le \gamma_2 \le 0,\ a_2 < 1\}.$$

As can be seen, the set of unitary multigrammars created by the substitution of the variable values is nothing but

$$\{ < \{1 \cdot a_0\}, \{ < a_0 \to -1 \cdot a_1, 2 \cdot a_2 >,\ < a_1 \to -2 \cdot a_2 > \} >,$$
$$< \{1 \cdot a_0\}, \{ < a_0 \to 2 \cdot a_2 >,\ < a_1 \to -2 \cdot a_2 > \} >,$$
$$< \{1 \cdot a_0\}, \{ < a_0 \to 1 \cdot a_1, 2 \cdot a_2 >,\ < a_1 \to -2 \cdot a_2 > \} >,$$
$$< \{1 \cdot a_0\}, \{ < a_0 \to -1 \cdot a_1, 2 \cdot a_2 >,\ < a_1 \to -1 \cdot a_2 > \} >,$$
$$< \{1 \cdot a_0\}, \{ < a_0 \to 2 \cdot a_2 >,\ < a_1 \to -1 \cdot a_2 > \} >,$$
$$< \{1 \cdot a_0\}, \{ < a_0 \to 1 \cdot a_1, 2 \cdot a_2 >,\ < a_1 \to -1 \cdot a_2 > \} >,$$
$$< \{1 \cdot a_0\}, \{ < a_0 \to -1 \cdot a_1, 2 \cdot a_2 >,\ < a_1 \to \{\varnothing\} > \} >,$$
$$< \{1 \cdot a_0\}, \{ < a_0 \to 2 \cdot a_2 >,\ < a_1 \to \{\varnothing\} > \} >,$$
$$< \{1 \cdot a_0\}, \{ < a_0 \to 1 \cdot a_1, 2 \cdot a_2 >,\ < a_1 \to \{\varnothing\} > \} > \},$$

and, finally, after eliminating the *MGs* whose schemes contain unitary rules with negative multiplicities in their left parts,

$$S^* = \{ < \{1 \cdot a_0\}, \{ < a_0 \to 2 \cdot a_2 >,\ < a_1 \to -2 \cdot a_2 > \} >,$$
$$< \{1 \cdot a_0\}, \{ < a_0 \to 1 \cdot a_1, 2 \cdot a_2 >,\ < a_1 \to -2 \cdot a_2 > \},$$

$$<\{1 \cdot a_0\},\ \{<a_0 \to 2 \cdot a_2>, <a_1 \to\ -1 \cdot a_2>\}>,$$

$$< \{1 \cdot a_0\}, \{ < a_0 \to 1 \cdot a_1, 2 \cdot a_2 >,\ < a_1 \to -1 \cdot a_2 > \} >,$$
$$< \{1 \cdot a_0\}, \{ < a_0 \to 2 \cdot a_2 >,\ < a_1 \to \{\varnothing\} > \} >,$$
$$< \{1 \cdot a_0\}, \{ < a_0 \to 1 \cdot a_1, 2 \cdot a_2 >,\ < a_1 \to \{\varnothing\} > \} > \}.$$

Formally, it is sufficient to replace (3.87) by

$$\mu_i \circ \langle \overline{n_1}, \ \ldots, \overline{n_l} \rangle = \begin{cases} \mu_i, \text{ if } \mu_i \in [1,\ \infty], \\ \overline{n_\mathrm{i}}, \text{ if } \mu_i \in \Gamma \qquad \wedge \overline{n_\mathrm{i}} \ge 0, \end{cases} \tag{3.158}$$

leaving unchanged the rest of relations (3.80)–(3.86) defining the mathematical semantics of *MMGs*.

3.5.3 Composite Names of Objects

One more useful extension to the proposed multigrammatical toolkit will be so-called composite object names, or composites, which we have already introduced and applied in Sect. 3.2.2 when describing the mathematical semantics of temporal multiset grammars.

Until now, when considering any multigrammar S as well as any other class of multigrammars (*UMGs, FUMGs, UMMGs, TMGs, FTMGs, TMMGs*), we have used objects' names as atomic ("non-splittable") lexemes from a finite set A_S. However, when introducing self-generated *TMGs*, we have taken into consideration objects' names as lexemes with some predefined syntax. Such objects' names will be called below *composites*, and this technique will be very convenient not only in the particular case of *SG TMGs* but in a much broader sense.

We shall connect with composites not only syntax, but a new opportunity for constructing more refined and selective conditions in filters. For this purpose, we shall use subsets of the set A_S defined by so-called structures, each of which is a string containing symbols of a basic alphabet U, used in objects' names, and the divider "↔", not used in the aforementioned names.

Let a string a be an object name, and

$$\omega = a_1 \leftrightarrow a_2 \ldots \leftrightarrow a_i \leftrightarrow \ldots \leftrightarrow a_k \tag{3.159}$$

be a structure, where $a_1, \ldots, a_k$ are strings in an alphabet U. We shall say that an *object name a has a structure* ω, which is denoted

$$a \sim \omega, \tag{3.160}$$

if

$$a = a_1 w_1 a_2 w_2 \ldots w_{i-1} a_i w_{i+1} \ldots w_{k-1} a_k, \tag{3.161}$$

where $w_1, \ldots, w_{k-1}$ are strings in an alphabet U, i.e., in the usual notation

$$\{w_1, \ldots, w_{k-1}\} \subset U^*, \tag{3.162}$$

where U^* is the set of all strings in an alphabet U (including the empty string). In other words, ω is used as a string pattern.

Example 3.32
Let ω = *Head*↔, a_1 = *HeadDepartment*, a_2 = *HeadLaboratory*, a_3 = *Assistant*. Then $a_1 \sim \omega$ and $a_2 \sim \omega$, i.e., a_1 and a_2, have the structure ω, while a_3 does not have the structure ω.

Any structure may be used in conditions in a filter, as an object name, i.e., as a general feature of all objects' names having this structure. In other words, a structure

ω replaces all object names that have it, and a condition containing ω is tested upon the multiset created by this replacement. Let us define these techniques formally.

Let $v = \{n_1 \cdot a_1, \ldots, n_m \cdot a_m\}$ be a multiset and $\omega \theta n$ be a boundary condition. Let $v \dot{-} \omega$ be a multiset created by the aforementioned replacement:

$$v \dot{-} \omega = \{n \cdot \omega \mid n \cdot a \in v \& a \sim \omega\} \bigcup \{n \cdot a \mid n \cdot a \in v \& \neg(a \sim \omega)\}. \tag{3.163}$$

In (3.163) is covered the non-evident fact that a multiset cannot contain two or more multiobjects with the same object: if this situation occurs, the multiplicities of all such objects are summed. The introduced operation $\dot{-}$ is the basis of the definition of the semantics of a boundary condition $\omega \theta n$. Namely, a multiset v satisfies a boundary condition $\omega \theta \bar{n}$ if $n \cdot \omega \in v \dot{-} \omega$ and $n \theta \bar{n}$ is true. Similarly, v satisfies boundary condition $\bar{n} \theta \omega$ if $n \cdot \omega \in v \dot{-} \omega$ and $\bar{n} \theta n$ is true. Finally, v satisfies $\omega \theta \omega'$ if $n \cdot \omega \in v \dot{-} \omega$, $n' \cdot \omega' \in v \dot{-} \omega'$, and $n \theta n'$ is true.

Example 3.33
Let the multiset

$$v = \{1 \cdot HeadCompany, 3 \cdot HeadDepartment, 8 \cdot HeadLaboratory, 10 \cdot Analyst\},$$

and the boundary condition be $Head \leftrightarrow \; \geq 10$. If so, then $\omega = Head\leftrightarrow$, and $v \dot{-} \omega = \{12 \cdot Head \leftrightarrow\}$.

As can be seen, v satisfies this condition, because $12 \geq 10$. If the condition is $Head \leftrightarrow \; \leq 3$, then v does not satisfy it.

An optimizing condition $\omega = opt$, where ω is a structure and $opt \in \{min, max\}$, is applied to a multiset $v \in V$ in just the same manner. A multiset $v \in V$ satisfies a minimizing (maximizing) condition $\omega = min$ ($\omega = max$) if for every multiset $v' \in V$ such that $v \neq v'$, the multiplicity n in the multiobject $n \cdot \omega \in v \dot{-} \omega$, is not greater (not less) than the multiplicity n' in the multiobject $n' \cdot \omega \in v' \dot{-} \omega$.

Example 3.34
Let $V = \{v_1, v_2, v_3\}$, where

$$v_1 = \{3 \cdot CarBMW, 2 \cdot CarBentley, 4 \cdot TruckCaterpillar\},$$

$$v_2 = \{1 \cdot CarBMW, 3 \cdot CarBentley, 1 \cdot TruckBizon\},$$

$$v_3 = \{2 \cdot CarBMW517, 2 \cdot CarBMW707, 3 \cdot CarNissanMicra, 2 \cdot CarNissanJuke\}.$$

The result of the application of the optimizing condition

$$Car \leftrightarrow \; = min$$

to this set of multisets is $\{v_2\}$, while the result of the application of the OpC

$$\leftrightarrow BMW \leftrightarrow \; = max$$

is $\{v_3\}$. If we apply to this *SMS* the *OpC*

$$Truck \leftrightarrow \; = max\,,$$

the result will be $\{v_1\}$.

3.5.4 Multifunctional Manufacturing Devices

Multifunctional manufacturing devices are capable of processing various input collections of *ORs*, so the above-introduced technique of representation of monofunctional *MDs* by temporal rules is not directly applicable to *MFMDs*. However, it is not so difficult to make some very limited correction to this representation to cover the multifunctional case. We shall do it regarding impatient scheduling, but the proposed techniques may be easily carried over the general case covering also patient activation of *MFMDs*.

Let

$$V_r = \{ < v_1,\, v_1',\, \Delta n_1 >,\; \ldots,\; < v_m,\, v_m',\, \Delta n_m > \} \tag{3.164}$$

be a set representing capabilities of an *MFMD* r in such a way that v_j is an input collection which allows manufacture of an output collection v_j' during Δn_j time units, $j = 1, \ldots, m$. The only restriction here is that there does not exist more than one triple with the same input collection v_j. Otherwise it would be unclear what output collection would be manufactured by r given v_j.

Let us represent an *MFMD*, whose capabilities are defined by (3.164), by a set of temporal rules with one and the same name r:

$$< r; v_1 \rightarrow v_1' + \{\Delta n_1 \cdot \Delta t\} >,$$

$$\ldots$$

$$< r; v_m \rightarrow v_m' + \{\Delta n_m \cdot \Delta t\} >. \tag{3.165}$$

As may be seen, this representation allows correct application of an *MFMD* r in the sense that a new operation cycle of this device cannot begin until an existing cycle has finished. However, in this case, the multiobject

$$1 \cdot r[n_1, l_1] \ldots [n_k, l_k] \in v_R \tag{3.166}$$

created during *MST* generation may have an ambiguous interpretation: if there exist two different *TRs* in the set (3.165) with equal Δn_i and Δn_j, then it is unclear what kind of operation of the *MFMD* r has been implemented.

To avoid this ambiguity, it is sufficient to transform the representation of an *MFMD* r by inclusion in the right parts of temporal rules (3.165) of multiobjects $i \cdot \boldsymbol{a}$, where i is an ordinary number giving the number of the application rule of this device, while $\boldsymbol{a}$ is a special distinguished object used for assignment of the aforementioned number. So the set (3.165) will become

$$< r; v_1 \rightarrow v_1' + \{\Delta n_1 \cdot \Delta t,\ 1 \cdot \boldsymbol{a}\} >,$$

$$\dots$$

$$< r; v_m \rightarrow v_m' + \{\Delta n_m \cdot \Delta t,\ m \cdot \boldsymbol{a}\} >. \quad (3.167)$$

Another transformation will be applied to multiobjects (3.166), which will become

$$1 \cdot r[n_1, l_1, i_1] \dots [n_k, l_k, i_k], \quad (3.168)$$

where every triple $[n_j, l_j, i_j]$ replacing a couple $[n_j, l_j]$ includes a third component i_j representing the way of an *MFMD* r application (in fact, the number of the temporal rule from the set (3.167)).

Example 3.35

Consider a flexible assembly line capable of producing buses and pickups according to the order. This *MFMD* may be represented by the following two temporal rules:

$$< (asm - line); \{1 \cdot (order - bus),\ 1 \cdot (frame - bus),\ 1 \cdot (engine - bus)\} \rightarrow$$
$$\{1 \cdot (bus),\ 180 \cdot (mnt),\ 1 \cdot (\boldsymbol{way})\} >,$$

$$< (asm - line); \{1 \cdot (order - pickup),\ 1 \cdot (frame - pickup),\ 1 \cdot (engine - pickup)\}$$
$$\rightarrow$$
$$\{1 \cdot (pickup),\ 120 \cdot (mnt),\ 2 \cdot (\boldsymbol{way})\} >,$$

where the assignment of the choice of application of the assembly line is done by the multiplicity of the distinguished object (***way***) (1 corresponds to a bus and 2 to a pickup). As can be seen, if the resource base contains an order to produce a bus (i.e., the *MO* $1 \cdot (order - bus)$) as well as the necessary components (the frame and engine of a bus represented respectively by multiobjects $1 \cdot (frame - bus)$ and $1 \cdot (engine - bus)$), then a bus is produced by this assembly line in 180 min. Similarly, the second *TR* represents the alternative way of applying the considered assembly line: a pickup would be manufactured in 120 min in the case that the resource base contained a corresponding order as well as pickup components.

A possible schedule of this assembly line may be as follows:

$$1 \cdot < (asm - line) \\ \times [08.00, 11.00, 1][11.00, 13.00, 2][14.00, 16.00, 2][16.00, 19.00, 1] >,$$

which means that during the time interval from 08.00 to 19.00, two buses will be produced (the first during 08.00–11.00, the second during 16.00–19.00) as well as two pickups (the first during 11.00–13.00, the second during 14.00–16.00).

Relevant corrections of the expressions to extend the mathematical semantics of *TMGs* to multifunctional *MDs*, including the general case covering patient activation of *MFMDs*, may be done by the interested reader as an exercise.

To conclude this chapter, we may affirm that multiset generation by application of a specific multiset grammar of any class is, in fact, a *goal-driven and knowledge-driven computation*, providing as a side effect all necessary data. From another perspective, this computation is *ubiquitous* and naturally *parallel* in the sense that generation proceeds in all possible directions, and some of them may be terminated as unpromising, i.e., not containing multisets satisfying the filters which determine the purpose of this computation. *Early detection and cutoff of such deadlocks is the conceptual background of MGF algorithmics*, considered thoroughly in Chap. 4.

Chapter 4
Basic Algorithmics of Multiset Grammars and Metagrammars

The algorithmics of the multigrammatical framework is the *collection of algorithms providing effective implementation of all classes of multiset grammars and metagrammars* introduced in Chap. 3 and listed in Table 3.2.

In turn, an effective implementation of any class of *MGs/MMGs assumes a minimal number of generation steps*, which are executed inside a process of generation of sets of terminal multisets (or as in some classes of the *MG* family, also sets of non-terminal *MSs*).

Here we do not present algorithms for all 12 classes of multigrammars appearing in the aforementioned table and forming the *MGF* but describe some of the most representative of them in the sense that they illustrate the basic technique of reduction of redundant search during multiset generation.

We shall consider the aforementioned technique in a step-by-step manner beginning in Sect. 4.1 for the simplest class of the *MG* family—*unitary multiset grammars.* A basic algorithm of generation for *STMSs* defined by an *acyclic non-variative unitary multiset metagrammar* is considered in Sect. 4.2. This algorithm is a key construction of all currently developed *MGF* algorithmics allowing solution of nontrivial tasks in the *STS* planning area as well as in various associated areas. Both cases considered in Sects. 4.1 and 4.2 refer to the conventional sequential algorithmics. Section 4.3 is dedicated to the more complicated case of algorithmics of *filtering multiset grammars*, referred to as a *highly parallel computing environment*. Finally, key issues of algorithmics of *filtering temporal multiset grammars* are considered in Sect. 4.4.

I. A. Sheremet, *Multigrammatical Framework for Knowledge-Based Digital Economy*, https://doi.org/10.1007/978-3-031-13858-4_4

4.1 Unitary Multiset Grammars

4.1.1 *Acyclic Non-variative* UMGs

Let us consider an acyclic non-variative *UMG*, which since the first reference in this book (Sect. 3.4.4) we have called *1-multigrammars.* As a *1-MGS* = < v_0, *R*> defines a one-element set of terminal multisets $\bar{V}_s = \{v\}$, the mathematical semantics of this subclass of *UMGs* may be redefined by replacement of (3.31) by (4.1), providing a much more conveniently implemented logic of generation of *TMSs* (Sheremet 2010, 2011a):

$$V_{(i+1)} = \bigcup_{v \in V_{(i)}} \bigcup_{\substack{n \cdot a \exists \in v \\ a \rightarrow n_1 \cdot a_1, \ldots, n_m \cdot a_m \in R}} \{v - \{n \cdot a\} + n * \{n_1 \cdot a_1, \ldots, n_m \cdot a_m\}\}, \quad (4.1)$$

where the new symbol ∃∈ (a combination of ∃ and ∈) denotes a choice of *any one* multiobject $n \cdot a$ from a multiset *v*, and

$$\bar{V}_s = V_{(i)} \quad (4.2)$$

such that $V_{(i+1)} = \{\varnothing\}$, i.e., no *UR* may be applied to the *MS v*, which is the only element of the set $V_{(i)}$ (this means that all objects in *v* are terminal).

Example 4.1
Consider the 1-multigrammar

$$S = <\{1 \cdot a,\ 2 \cdot c\}, \{r_1,\ r_2\}>,$$

where

$$r_1 : 1 \cdot a \rightarrow 3 \cdot b, 1 \cdot c,$$

$$r_2 : 1 \cdot b \rightarrow 2 \cdot c.$$

According to (3.30)–(3.31), there is the following generation chain:

$$\{1 \cdot a,\ 2 \cdot c\} \overset{r_1}{\Rightarrow} \{3 \cdot b,\ 3 \cdot c\} \overset{r_2}{\Rightarrow} \{2 \cdot b,\ 5 \cdot c\} \overset{r_2}{\Rightarrow} \{1 \cdot b,\ 7 \cdot c\} \overset{r_2}{\Rightarrow} \{9 \cdot c\}.$$

At the same time, application of (4.1) results in the essentially shorter generation chain:

$$\{1 \cdot a,\ 2 \cdot c\} \overset{r_1}{\Rightarrow} \{3 \cdot b,\ 3 \cdot c\} \overset{r_2 * 3}{\Rightarrow} \{9 \cdot c\},$$

where $r_2 * 3$ means the application of the *UR* r_2 according to (4.1).

As can be seen, (4.1) allows *replacement of a sequence of multiset subtractions-additions by only one operation, namely, multiset subtraction-"multiplication-by-constant"-addition*. Moreover, an application of the introduced operation ∃∈,giving a choice of only one generation chain instead of a full search of the set of all possible generation chains, demonstrates the linear computational complexity of the generation of the one-element *STMS* defined by a *1-MG*. This radical reduction of the computational complexity is possible because, *due to the commutativity and associativity of multiplication and addition, any generation chain leads to one and the same resulting TMS* (Sheremet 2010, 2011a, 2018, 2019b).

However, extension of this simplifying and optimizing technique to variative *UMGs* is not correct. If there exists at least one non-terminal object a such that $m = |R_a| > 1$, and the current generated multiset contains a multiobject $n \cdot a$, then it is necessary to generate $N = m^n$ branches, as indicated by (3.31) (Sheremet 2010, 2011a). This extra-large computational complexity of a substitution with only one *NTO* makes practical application of variative *UMGs* really impossible for implementational reasons.

There may be two different practically useful exits from this situation.

The first one, called *imperative* and already used in all previous work on the development and implementation of the multigrammatical framework, is based on the definition of the mathematical semantics of variative *UMGs* by relations (3.30), (4.1), and (3.32) with left-to-right choice of the next *NTO* for generation. This approach presents to the knowledge engineer a *constraint programming language* with strictly defined logics different from the basic logics of *MGs*. As was argued in (Sheremet 2010, 2011a, 2018, 2019b), this decision is quite satisfactory from the practical point of view, and it provides an effective solution of some complicated practical problems in the systems analysis and operations research areas. But at the same time, to solve a major part of the classical problems in the operations research area and in *STS* planning, it is necessary to use a metagrammatical extension of such *UMGs*.

The second approach, introduced in this book in application to *acyclic variative UMGs*, is based on the construction of a *non-variative unitary multiset metagrammar*, equivalent to the initial variative *UMG* in the sense that both generate one and the same set of terminal multisets. This technique allows application of the unified effective algorithmics of non-variative *UMMGs* to variative *UMGs*, which, it follows from the content of Sect. 3.4.1, are a special case of variative *UMGs*.

4.1.2 Acyclic Variative UMGs

Consider an *acyclic variative UMG* $S = < v_0, R>$, called in this book (since the first reference in Sect. 3.4.4) a *k-multigrammar*. Let us create a *non-variative UMMG* equivalent to this *UMG* in the aforementioned sense. The created *UMMG* will be denoted $S_0 = < \{1 \cdot a_0\}, R_0, F_0>$, where a_0 is an auxiliary non-terminal object and the scheme R_0 along with the filter F_0 are constructed from the kernel v_0 and the scheme R of the *UMG S* in the following way.

First of all, R_0 will contain a *UR* $a_0 \to v_0$ (here and above for simplicity, we shall use multisets as bodies of unitary rules).

Second, R_0 will contain all *URs* with non-alternating *NTOs* (i.e., those $a \in A_S^N$ with $|R_a| = 1$).

After execution of the described steps, the most sophisticated transformations will be done regarding *alternating NTOs*.

Let us begin from the simplest case, where

$$R_a = \{a \to v_1, \; \ldots, a \to v_m\}, \tag{4.3}$$

and the scheme R_0 contains only one *UR*

$$a' \to v' + \{n \cdot a\}, \tag{4.4}$$

whose body contains a multiobject $n \cdot a$. In this case, we shall include in R_0 the unitary metarule

$$a \to \gamma_1 \cdot a'_1, \ldots, \gamma_m \cdot a'_m, \tag{4.5}$$

as well as a set of unitary rules

$$R'_a = \{a'_1 \to v_1 + \{1 \cdot \overline{a}\}, \; \ldots, a'_m \to v_m + \{1 \cdot \overline{a}\}\}, \tag{4.6}$$

where γ_i, $i = 1, \ldots, m$, are auxiliary variables, $a'_1, \ldots, a'_m$ are auxiliary non-terminal objects, and $\overline{a}$ is an auxiliary terminal object. Also, we shall include in the filter F_0 one boundary condition

$$\overline{a} = n, \tag{4.7}$$

and m boundary conditions

$$\begin{gathered} 0 \le \gamma_1 \le n, \\ \ldots \\ 0 \le \gamma_m \le n. \end{gathered} \tag{4.8}$$

As may be seen, by these techniques, we allow *generation of all combinations of integer numbers, being the values of the multiplicity-variables* $\gamma_1, \ldots, \gamma_m$, *whose sum is n*. In other words, this application of *UMMGs'* descriptional capabilities provides a *non-procedural representation of all possible ways of splitting integer number n into m non-negative summands*. The described operations are executed for all alternating *NTOs*.

The proposed logic is the basis of the following statement, which we introduce without proof because it is self-evident.

Statement 4.1 We have $\bar{V}_S = \bar{V}_{S_0}$.

Verbally, the created multigrammar defines the same set of terminal multisets as the initial one. Let us illustrate this with a simple example.

Example 4.2

Let us consider the acyclic variative unitary *MG* $S = \ < v_0, R>$ whose kernel is the *MS* $v_0 = \{8 \cdot a, 3 \cdot c\}$ and whose scheme is $R = \{r_1, r_2, r_3\}$, where

$$r_1 : a \to 2 \cdot b,$$

$$r_2 : a \to 1 \cdot c,$$

$$r_3 : b \to 2 \cdot c.$$

According to (4.4)–(4.8), the created unitary multiset metagrammar equivalent to this *UMG* is $S_0 = \ < \{1 \cdot a_0\}, R_0, F_0>$, whose scheme is $R_0 = \{r_0, r_a, r'_1, r'_2, r'_3\}$, the *URs* and the *UMRs* in this scheme are

$$r_0 : a_0 \to 8 \cdot a, 3 \cdot c,$$

$$r_a : a \to \gamma_1 \cdot a_1, \gamma_2 \cdot a_2,$$

$$r'_1 : a_1 \to 2 \cdot b, 1 \cdot \bar{a},$$

$$r'_2 : a_2 \to 1 \cdot c, 1 \cdot \bar{a},$$

$$br'_3 : b \to 2 \cdot c,$$

and the filter F_0 contains the following boundary conditions:

$$\bar{a} = 8,$$

$$0 \leq \gamma_1 \leq 8,$$

$$0 \leq \gamma_2 \leq 8.$$

As can be seen, the alternating non-terminal object a occurs in only one body of the unitary rule, namely, r_0. So the unitary metarule r_a is added to the scheme of the created unitary multiset metagrammar S_0,then the unitary rule r'_3 whose header is the non-alternating *NTO* b, and, finally, two unitary rules r'_1 and r'_2 are associated with the unitary metarule r_a.

However, *in the general case, an alternating NTO* $a \in A_S^N$ *may occur in the body of more than one UR*. Let R_a be the same as in (4.3), and suppose there exist l unitary rules, the body of each *UR* containing an object a:

$$a'_1 \to v'_1 + \{n_1 \cdot a\},$$

$$\dots$$

$$a'_l \to v'_l + \{n_l \cdot a\}. \tag{4.9}$$

Following the sense of (4.5) and (4.6), we shall include in the scheme R_0 l unitary metarules

$$\overline{a}'_1 \to \gamma^1_1 \cdot a^1_1, \dots, \gamma^1_m \cdot a^1_m,$$

$$\dots$$

$$\overline{a}'_l \to \gamma^l_1 \cdot a^l_1, \dots, \gamma^l_m \cdot a^l_m, \tag{4.10}$$

as well as l sets of unitary rules

$$R'_1 = \{a^1_1 \to v'_1 + \{1 \cdot \overline{a}_1\}, \ \dots, \ a^1_m \to v'_1 + \{1 \cdot \overline{a}_1\}\},$$

$$\dots$$

$$R'_l = \{a^l_1 \to v'_l + \{1 \cdot \overline{a}_l\}, \ \dots, \ a^l_m \to v'_l + \{1 \cdot \overline{a}_l\}\}, \tag{4.11}$$

where γ^i_j are auxiliary variables, similar to γ_i in (4.5); a^i_j are auxiliary objects, similar to a'_i in (4.5) and (4.6); and $\overline{a}_1, \dots, \overline{a}_l$ are auxiliary terminal objects, similar to $\overline{a}$ in (4.6). Following the sense of (4.7) and (4.8), we shall include in the filter F_0 l boundary conditions

$$\overline{a}_1 = n_1,$$

$$\dots$$

$$\overline{a}_l = n_l, \tag{4.12}$$

as well as boundary conditions

$$0 \le \gamma^1_1 \le n_1, \dots, 0 \le \gamma^1_{n_1} \le n_1,$$

$$\dots$$

$$0 \le \gamma^l_1 \le n_l, \dots, 0 \le \gamma^l_{n_l} \le n_l. \tag{4.13}$$

Evidently, introducing (4.9)–(4.13), we in a natural way expand the logic of (4.3)–(4.8) to the general case, applying this logic to all occurrences of a non-terminal object a to the bodies of l unitary rules. In fact, any such occurrence is interpreted as a distinguished variable.

As in the previous case, the described operations are executed for all alternating *NTOs* of the *UMG S* that occur in the body of *more than one UR*. After all such *NTOs* are processed, non-terminal objects that occur in the body of *one UR* may remain in the scheme of the created *UMMG*. So they are processed according to (4.3)–(4.8).

Let us illustrate the described logic by an example.

Example 4.3

Let us consider the acyclic variative unitary *MG* $S = \langle v_0, R \rangle$, whose kernel is the *MS* $v_0 = \{6 \cdot a, 5 \cdot c\}$, and whose scheme $R = \{r_1, r_2, r_3, r_4\}$, where

$$r_1 : a \to 7 \cdot b, 3 \cdot c,$$

$$r_2 : a \to 4 \cdot b,$$

$$r_3 : b \to 3 \cdot c,$$

$$r_4 : b \to 9 \cdot c.$$

First of all we shall include in the scheme of the created *UMMG* the unitary metarule

$$r_0 : a_0 \to 6 \cdot a, 5 \cdot c.$$

As can be seen, now there are two alternating *NTOs* in the *URs* $r_0 - r_4$, namely, a and b. As can be seen, the alternating *NTO* b occurs the bodies of two *URs* r_1 and r_2, while the alternating *NTO* a occurs in the body of one *UR* r_0.

We shall begin with the *NTO* b, and according to (4.9)–(4.13), we shall include in the scheme R_0 of the created *UMMG*, equivalent to the initial *UMG*, the following *URs*:

$$r'_1 : a \to 7 \cdot b_1, 3 \cdot c,$$

$$r'_2 : a \to 4 \cdot b_2,$$

$$r'_3 : b_1 \to 3 \cdot c,$$

$$r'_4 : b_1 \to 9 \cdot c,$$

$$r'_5 : b_2 \to 3 \cdot c,$$

$$r'_6 : b_2 \to 9 \cdot c.$$

Now each of the alternating *NTOs* b_1 and b_2 occur in the body of only one *UR* (r'_1 and r'_2), and thus (4.3)–(4.8) may be applied to the scheme containing r_0 and $r'_1 - r'_6$. The result of this application will be the following set of *URs* and *UMRs* joined to the already created part of the scheme of the *UMMG* R_0:

$$a \to \gamma_1 \cdot a_1, \gamma_2 \cdot a_2,$$

$$a_1 \rightarrow 7 \cdot b_1, 3 \cdot c, 1 \cdot \overline{a},$$

$$a_2 \rightarrow 4 \cdot b_2, 1 \cdot \overline{a}$$

$$b_1 \rightarrow \gamma_1^1 \cdot b_1^1, \gamma_2^1 \cdot b_2^1,$$

$$b_1^1 \rightarrow 3 \cdot c, 1 \cdot \overline{b_1},$$

$$b_2^1 \rightarrow 9 \cdot c, 1 \cdot \overline{b_1},$$

$$b_2 \rightarrow \gamma_1^2 \cdot b_1^2, \gamma_2^2 \cdot b_2^2,$$

$$b_1^2 \rightarrow 3 \cdot c, 1 \cdot \overline{b_2},$$

$$b_2^2 \rightarrow 9 \cdot c, 1 \cdot \overline{b_2}.$$

The filter F_0 associated with the created scheme R_0 will contain the following boundary conditions:

$$\overline{a} = 6,$$

$$0 \leq \gamma_1 \leq 6,$$

$$0 \leq \gamma_2 \leq 6,$$

$$\overline{b_1} = 7,$$

$$0 \leq \gamma_1^1 \leq 7,$$

$$0 \leq \gamma_2^1 \leq 7,$$

$$\overline{b_2} = 4,$$

$$0 \leq \gamma_1^2 \leq 4,$$

$$0 \leq \gamma_2^2 \leq 4.$$

The described and illustrated logic may be rather simply implemented by a recursive function creating an equivalent *UMMG* $S_0 = \ < \{1 \cdot a_0\}, R_0, F_0>$ for any acyclic variative unitary *MG* $S = \ < v_0, R>$ with any structure of mutual interconnections of *NTOs* in the bodies of its *URs*.

4.1.3 Acyclic Variative Filtering UMGs

Let $S = \ < v_0, R, F>$ be an acyclic variative *FUMG*. As may be seen, there is no difference in principle from the previous case except a non-empty filter F. In this case, the sequence of steps to create a non-variative *UMMG* $S_0 = \ < \{1 \cdot a_0\}, R_0, F_0>$ equivalent to the core variative *UMG* of S is repeated practically without any changes, except that the construction of the filter F_0 of the *UMMG* S_0 begins from the set $F \neq \{\emptyset\}$, and, therefore, F_0 includes all the conditions in F and also all conditions of the form (4.12) and (4.13) corresponding to occurrences of alternating *NTOs* in bodies of unitary rules in the scheme R.

Now we are ready to consider the algorithmics of unitary multiset metagrammars, which is, in fact, a *core element of the algorithmics of the multigrammatical framework as a whole.*

4.2 Unitary Multiset Metagrammars

We shall begin with the case of *non-variative UMMGs*, which will be considered in the following order:

1. The algorithm for a *direct implementation of the UMMGs' mathematical semantics*, defined by (3.79)–(3.86)
2. The *basic idea used as the foundation of the UMMGs' improved algorithmics*, giving a radical reduction of the computational complexity of generation of sets of terminal multisets
3. A *macroset representation of polynomials*, whose application is a key element of the aforementioned improved algorithmics, and basic operations on macroset-represented polynomials
4. An algorithm for *generation of the macroset-represented polynomial defined by the scheme of a UMMG and a polynomial-driven generation of an STMS reduced by a prefiltration*

After this, techniques for *generalization of the algorithmics of non-variative UMMGs to the case of variative UMMGs* will be considered.

4.2.1 *Algorithm for a Direct Implementation of* UMMGs' *Mathematical Semantics*

The simplest version of the *UMMG's* algorithmics is, in fact, a *direct procedural implementation of* (3.79)–(3.86). Its transformation to an algorithm is rather obvious:

A loop on all possible collections of values of *UMMG* variables

A substitution of every collection to unitary metarules

An application of the obtained *UMG* to the generation of *TMSs*

A filtration of the generated *TMSs* by boundary conditions accumulation of selected *TMSs* in a special storage

A filtration of the selected *TMSs*, accumulated in the aforementioned storage, by optimizing conditions

Let us represent the announced sequence in the form of the procedure-represented function *Ummg*.

We shall use the following variables in the body of *Ummg*:

A set *R* whose value is a *UMMG* scheme, i.e., a set of unitary rules and metarules

A set *F* whose value is a *UMMG* filter, including boundary and optimizing conditions as well as declarations of domains of *UMMG* variables in the form of chain boundary conditions $n \leq \gamma \leq n'$ (the set of these declarations is a subfilter $F_\Gamma \subseteq F$)

A set *V* whose value is the set of terminal multisets satisfying the filter *F* and accumulated during previous generation steps

A set *FT* whose value is a set of triples <*a*, *opt*, *l*>, each corresponding to an optimizing condition $a = opt \in F$, where *a* is an object, $opt \in \{max, min\}$, and *l* is the current value of the multiplicity of the object *a* obtained after previous generation steps

A set *Γ#* whose value is the set of current values of the *UMMG* variables in the form of couples $\langle \gamma, m \rangle$ (the initial value of this variable is the empty set)

A set *F#* whose value is the current set of declarations of the *UMMG* variables, which is used to create the set *Γ#* of values of these variables (the initial value of *F#* is the aforementioned subfilter $F_\Gamma \subseteq F$)

A set *R#* whose value is the current set of unitary rules created during the sequential processing of the *URs/UMRs* belonging to the scheme *R*

A couple $\langle a, w \rangle$ whose value is a representation of a unitary rule/metarule belonging to the scheme *R* of the *UMMG*, in such a way that *a* is an object (the header of the *UR/UMR*) and *w* is a multiset (the body of this *UR/UMR*)

A multiset *v* whose value is the multiset generated via the previous steps (the *Ummg* calls)

Inside the body of *Ummg*, the variables *R*, *F*, *V*, and *FT* are global variables, available from all calls during the execution of *Ummg*, while the variables Γ#, *R#*, and *F#* are local.

The body of *Ummg* is as follows:

```
Ummg: function (v, R, F) returns (V);
        variables R, F global;
        variables FT initial {∅}, V initial {∅} global;
        variables v, Γ# initial {∅}, F# initial (F_Γ) local;
        /* main section: call of Umg function */
        call Umg (R,F#,Γ#);
        /* Umg body */
    Umg: function (R#, F#, Γ# );
        variables R#, F#, Γ# local;
        /* UMMG declarations processing */
        do n ≤ γ ≤ n' ∃∈ F#;
            do m ∈ [n, n'];
                call Umg (R#, F# − {n ≤ γ ≤ n'}, Γ# ∪ {⟨γ, m⟩});
            end m;
        end γ;
        call SubstUmr (R#,Γ#);
        call G (v_0, R#, F#);
        end Umg;
        /* substitution of UMMG variables into unitary metarules */
    SubstUmr: function (R#, Γ#);
            variables R#, Γ# local;
            do ⟨a, w⟩ ∈ R#;
                do x · b ∈ w;
                    if x is variable
                        then do ⟨x, m⟩∃∈ Γ#;
                            R#: −{⟨a, w⟩}
                            ∪ {⟨a, w − {x · b} ∪ {m · b}⟩};
                            end Γ#;
                    else;
                end w;
            end R#;
    end SubstUmr;

end Ummg
```

Let us comment on this pseudocode.

Its main part is a call of the recursive function *Umg* (line 8) with three arguments—*R* (a scheme of an *UMMG*), *F#* (a set of declarations of *UMMG* variables whose initial value is, as mentioned above, F_Γ), and Γ# (a set of concrete values of these variables whose initial value is $\{\varnothing\}$).

Function *Umg* operates in the following way. The only declaration $n \leq \gamma \leq n'$ is randomly extracted from *F#* (line 11), and the loop on values from interval $[n, n']$ (lines 12–14) is executed in such a way that a recursive *Umg* call with the same first argument *R#* and the corrected second and third arguments is performed (line 13). The second argument is the set of declarations *F#* without $n \leq \gamma \leq n'$, while the third one is the set of *UMMG* variables joined to the couple $\langle \gamma, m \rangle$, where m is the current value of the variable γ. As may be seen, this chain of recursive calls will spread until all declarations in *F#* will be extracted, and exactly at this moment, the ∃∈ loop will fall out to line 16, which contains a call of the procedure *SubstUmr* (*Substitution to Unitary metarules*) with two arguments, the first being a set of *URs/UMRs* and the second a set of concrete values of *UMMG* variables, which is formed by the aforementioned chain of calls. After return from *SubstUmr*, the value of the variable *R#* will contain only unitary rules occurring in the original *UMMG* scheme as well as unitary rules created from unitary metarules by substitution of the aforementioned concrete values. This results in a correct call of function ***G*** (line 17) with the first argument v_0 (i.e., the *UMMG* kernel), the second the current set of unitary rules, which in the mathematical definition of *UMMGs* is referred to as $R \circ \langle \overline{n}_1, \ldots, \overline{n}_l \rangle$ (relation (3.80)), and the third a subfilter *F#*. As can be seen, the number of such *SubstUmr* and, respectively, function ***G*** calls will be exactly

$$\prod_{j=1}^{l} (n'_j - n_j + 1) \tag{4.14}$$

(the length of the chain of function *Umg* calls is l, i.e., the number of *UMMG* variables, while the number of concrete values of the j-th such variable is $n'_j - n_j + 1$). Every call of the function ***G*** leads to the generation of a set of *TMSs* and its filtration by a filter $F - F_\Gamma$ containing all boundary and optimizing conditions except variable declarations. So, as can be seen, after the final exit from *Ummg* (all l-ary tuples of *UMMG* variables have been processed), the values of variables *V#* and *FT#* will be respectively the set of *TMSs* defined by the *UMMG* and the set of multiplicities which satisfy the optimizing conditions (let us recall that both variables *V#* and *FT#* are global and by this feature their values are corrected inside every *Ummg* call).

To finish the consideration of the function *Ummg*, let us describe the function *SubstUmr*. Its main part is a loop on elements of the scheme *R#*, i.e., unitary rules and metarules (lines 22–30). (Let us recall that every such element is a couple whose first element a is the *UR/UMR's* header and the second w its body represented as a multiset.) An internal loop (lines 23–29) performs a search on elements of the current *UR/UMR* body. Every such element is a multiobject $x \cdot b$, and if x is a variable (i.e., $x \in \Gamma$), then the corresponding couple $\langle x, m \rangle$, where m is the current value of x, is selected from the set Γ#, and *R#* is corrected by replacement of $\langle a, w \rangle$ by $\langle a, w' \rangle$, such that w' will contain $m \cdot b$ instead of $x \cdot b$ (line 26). As can be seen, every *UMMG* variable is replaced by its current value, and that is done regarding every *UMR* and every variable occurring in its body, so really $R\circ\langle\overline{n}_1, \ldots, \overline{n}_l\rangle$ is created by a *SubstUmr* call.

We shall not explain function ***G***, because its logic is evident.

The correctness of function *Ummg* is affirmed by the following statement (Sheremet 2010, 2011a).

Statement 4.2 Let $S = \langle v_0, R, F\rangle$ be an acyclic non-variative *UMMG* and *V*= *Ummg* (v_0, *R*, *F*) the result of the function *Ummg* call.

Then $V = \overline{V_s}$.

Looking at the function *Ummg*, we may see that, although it is valid in the sense that it fully corresponds to the mathematical semantics of *UMMGs* defined by (3.78)–(3.85), it is *evidently inefficient*, because the only tools of elimination of redundant generation steps are hidden inside the function ***G***, while *the number of its calls may be extra large* (if we suppose there are 10 variables in a *UMMG*, and every such variable domain has 1000 values, then there would be $1000^{10} = 10^{30}$ function ***G*** calls during this *UMMG* application).

So a special *improved algorithmics of the unitary multiset metagrammars* has been developed to avoid the aforementioned shortcomings of the direct implementation of the *UMMGs'* mathematical semantics.

4.2.2 Basic Idea for the Improvement of UMMGs' *Algorithmics*

To explain the basic idea that is the foundation of the improved *UMMG* algorithmics, let us begin from the following example.

Example 4.4
Let *UMMG* *S* = < {1 · *a*}, *R*, *F*>, where the scheme $R = \{r_1, r_2, r_3\}$, the unitary metarules in this scheme are

$$r_1 : a \to \gamma_1 \cdot b, 3 \cdot c, 1 \cdot d,$$

$$r_2 : b \to \gamma_2 \cdot c, 2 \cdot d, 3 \cdot e,$$

$$r_3 : d \to 2 \cdot c, \gamma_3 \cdot e,$$

and the filter *F* includes the following conditions:

$$1 \le \gamma_1 \le 3,$$

$$0 \le \gamma_2 \le 2,$$

$$1 \leq \gamma_3 \leq 2,$$
$$c \leq 8,$$
$$e = max.$$

Following (3.30), (4.1), and (3.32), let us create the one-element set of terminal multisets $\bar{V}_S$, such that the multiplicities of the terminal objects will be polynomials of the variables occurring in the unitary metarules r_1, r_2, *and* r_3. In other words, applying (4.1), we shall substitute not *multiplicity-constants* but *multiplicity-variables*, and after a substitution, we shall open the brackets and simplify the obtained symbolic expressions according to conventional algebraic techniques:

$$\{1 \cdot a\} \overset{r_1}{\Rightarrow} \{\gamma_1 \cdot b, 3 \cdot c, 1 \cdot d\} \overset{r_2}{\Rightarrow} \{\gamma_1 * \{\gamma_2 \cdot c,\ 2 \cdot d,\ 3 \cdot e\}\} + \{3 \cdot c, 1 \cdot d\}$$
$$= \{(\gamma_1 \cdot \gamma_2 + 3) \cdot c, (2 \cdot \gamma_1 + 1) \cdot d, (3 \cdot \gamma_1) \cdot e\} \overset{r_3}{\Rightarrow}$$
$$\{(2 \cdot \gamma_1 + 1) * \{2 \cdot c,\ \gamma_3 \cdot e\}\} + \{(\gamma_1 \cdot \gamma_2 + 3) \cdot c, (3 \cdot \gamma_1) \cdot e\}$$
$$= \{(\gamma_1 \cdot \gamma_2 + 4 \cdot \gamma_1 + 5) \cdot c, (3 \cdot \gamma_1 + 2 \cdot \gamma_1 \cdot \gamma_3 + 1 \cdot \gamma_3) \cdot e\}.$$

Combining this *TMS* with the conditions in the filter *F*, we obtain the following problem: to search the set of collections of values of variables $\gamma_1, \gamma_2,$ *and* γ_3 such that

$$1 \leq \gamma_1 \leq 3,$$
$$0 \leq \gamma_2 \leq 2,$$
$$1 \leq \gamma_3 \leq 2,$$
$$\gamma_1 \cdot \gamma_2 + 4 \cdot \gamma_1 + 5 \leq 8,$$

and from all such collections to select those which maximize the value of

$$3 \cdot \gamma_1 + 2 \cdot \gamma_1 \cdot \gamma_3 + 1 \cdot \gamma_3.$$

As may be seen from this example, such natural techniques, based on the manipulation of multiplicities represented by polynomials from variable-multiplicities occurring in unitary metarules, reduce the problem of generating the set of terminal multisets defined by a non-variative *UMMG* to *a problem of integer polynomial programming*. The most general and effective method of solution of such problems was proposed by E. Hansen (Hansen 1979, 1992; Hansen and Walster 2004), who applied interval analysis to global optimization of functions called monotonic by inclusion. We shall combine Hansen's approach with the above-described techniques of symbolic computation of multiplicities during multiset generation, i.e., in fact, *interval analysis with mixed computation* (Ershov 1977; Bjørner et al. 1988; Itkin 1991; Lloyd and Shepherdson 1991; Shokin 1996).

Let us consider in detail the implementation of this briefly described basic idea, beginning with some necessary auxiliary tools.

4.2.3 Macroset Representation of Polynomials and Variable-Containing Multiplicities

As is known from (Sheremet 2010, 2011a) and as was illustrated in Sect. 4.2.2, any multiset generated by application of the unitary rules and metarules of a *UMMG* has the form

$$v = \{c_1 \cdot a_{i_1}, \ \ldots, c_m \cdot a_{i_m}\}, \tag{4.15}$$

where c_i, $i = 1, \ldots, m$, is a *variable-containing multiplicity* (*VCM*), which is a polynomial of variable-multiplicities occurring in the unitary metarules, applied during the generation of this *MS*. Multisets (4.15) will be called *V-multisets (VMS).* A *V*-multiset in which all objects are terminal is called an ***s***.

In the most general case

$$C = n + \sum_{i=1}^{N_C} n_i \cdot \gamma_{i,1}^{l_{i,1}} \cdot \ldots \cdot \gamma_{i,k_i}^{l_{i,k_i}}, \tag{4.16}$$

where N_C is the number of monomials, each being the product of all occurrences of multiplicity-variables and multiplicity-constants appearing in one generation branch leading to an object a. Such monomials in Example 4.4 are $\gamma_1 \cdot \gamma_2$, $4 \cdot \gamma_1$, $3 \cdot \gamma_1$, $2 \cdot \gamma_1 \cdot \gamma_3$, $1 \cdot \gamma_3$.

VCM and *VMS* manipulation in the *UMMG* algorithmics considered below is based on a unified representation of polynomials, written in the form (4.16), as macrosets. A *macroset* in its simplest form as applied here is a set of multisets, and *VCM C* from (4.16) is represented by the macroset

$$v_C = \left\{ \begin{array}{r} \{n \cdot \overline{\gamma}_0\}, \{n_1 \cdot \overline{\gamma}_0, \ l_{1,1} \cdot \overline{\gamma}_{1,1}, \ \ldots, \ l_{1,k_1} \cdot \overline{\gamma}_{1,k_1}\}, \ \ldots \\ \ldots, \{n_m \cdot \overline{\gamma}_0, \ l_{m,1} \cdot \overline{\gamma}_{m,1}, \ \ldots, \ l_{m,k_m} \cdot \overline{\gamma}_{m,k_m}\} \end{array} \right\}, \tag{4.17}$$

where $m = N_C$ and $\overline{\gamma}_{i,j}$ are objects corresponding to variables $\gamma_{i,\,j}$, while $\overline{\gamma}_0$ is an auxiliary object which is used for representation of *constants* occurring in the polynomial. The multiset $\{n_0 \cdot \overline{\gamma}_0\}$ will be called a *prefix of the macroset* (4.17), and a macroset (4.17) without a prefix will be called *prefix-free*.

Example 4.5
Let

$$C = 3 + 5 \cdot \gamma_1^2 \cdot \gamma_2^3 + \gamma_2^5 \cdot \gamma_3.$$

Then

$$v_C = \{\{3 \cdot \overline{\gamma}_0\}, \{5 \cdot \overline{\gamma}_0,\ 2 \cdot \overline{\gamma}_1,\ 3 \cdot \overline{\gamma}_2\}, \{1 \cdot \overline{\gamma}_0,\ 5 \cdot \overline{\gamma}_2,\ 1 \cdot \overline{\gamma}_3\}\}.$$

Here $\{3 \cdot \gamma_0\}$ is the prefix of v_C.

4.2.4 Operations on Macroset-Represented Polynomials and Variable-Containing Multiplicities

The proposed macroset representation is sufficiently flexible, and it is the *basis of an implementation of mixed computation in the multiset case*. The core of this implementation is three operations—addition of two *VCMs*, multiplication of a *VCM* and a multiset containing multiplicity-constants and multiplicity-variables (such multiplicities are called *primary*), and substitution of variable domains in a macroset-represented *VCM*. In order to define the last operation, a new notion—a *multiset with interval multiplicities* (*MSIM*)—will be introduced, and two operations, addition and intersection of such multisets, will be defined and applied below.

The operation of *addition of two variable-containing multiplicities* is, from one side, a generalization of the corresponding operation on multisets and, from the other, a macroset implementation of simplification of a sum of two polynomials.

Let v and v' be two *VCMs*. We shall introduce two operations on these *VCMs*: the operation of their *addition*, denoted by a special symbol $\hat{+}$, and the auxiliary operation of *addition of two prefix-free VCMs*, denoted by a special symbol $\breve{+}$. According to the sense of simplification of a sum of polynomials,

$$
v \hat{+} v' = \begin{cases}
\{\{(n_0 + n'_0) \cdot \overline{\gamma}_0\}\} \cup (\overline{v} \breve{+} \overline{v}'), \\
\qquad\qquad \text{if } v = \{n_0 \cdot \overline{\gamma}_0\} \cup \overline{v} \wedge v' = \{n'_0 \cdot \overline{\gamma}_0\} \cup \overline{v}', \\
\\
\{\{n_0 \cdot \overline{\gamma}_0\}\} \cup (\overline{v} \breve{+} v'), \\
\qquad\qquad \text{if } v = \{n_0 \cdot \overline{\gamma}_0\} \cup \overline{v} \ \wedge \overline{\gamma}_0 \notin v' \\
\\
\{\{n'_0 \cdot \overline{\gamma}_0\}\} \cup (v \breve{+} \overline{v}') \\
\qquad\qquad \text{if } v' = \{n'_0 \cdot \overline{\gamma}_0\} \cup \overline{v}' \ \wedge \overline{\gamma}_0 \notin v \\
\\
v \breve{+} v' \text{ otherwise,}
\end{cases}
\tag{4.18}
$$

$$
v \breve{+} v' = \begin{cases}
\{\{(n_0 + n'_0) \cdot \overline{\gamma}_0\} \cup \overline{v}\} \\
\cup [(v - (\{\{n_0 \cdot \overline{\gamma}_0\} \cup \overline{v}))\breve{+}(v' - (\{\{n'_0 \cdot \overline{\gamma}_0\} \cup \overline{v}))], \\
\text{if } \{n_0 \cdot \overline{\gamma}_0\} \cup \overline{v} \in v \ \ \{n'_0 \cdot \overline{\gamma}_0\} \cup \overline{v} \in v', \\
\\
v \cup v' \ \text{ otherwise.}
\end{cases}
\tag{4.19}
$$

Let us comment on the presented recursive definitions.

Expression (4.18) shows *four possible alternatives*.

The *first* alternative corresponds to the case when both added *VCMs* include sets $\{n_0 \cdot \overline{\gamma}_0\}$ and $\{n'_0 \cdot \overline{\gamma}_0\}$, i.e., constants in the usual representation of polynomials. In this case, the resulting macroset will be the join of the macroset $\{\{(n_0 + n'_0) \cdot \overline{\gamma}_0\}\}$, i.e., the constant $n_0 + n'_0$ in the usual representation of polynomials, and the result of $v \breve{+} v'$, where $\overline{v}$ and $\overline{v}'$ are macrosets containing all multisets occuring in v and v', except $\{n_0 \cdot \overline{\gamma}_0\}$ and $\{n'_0 \cdot \overline{\gamma}_0\}$, respectively.

The *second* alternative corresponds to the case when v includes a multiset $\{n_0 \cdot \overline{\gamma}_0\}$, while v' does not include $\{n'_0 \cdot \overline{\gamma}_0\}$ (i.e., a corresponding polynomial does not contain a constant as a summand). In this case, the result is the join of $\{n_0 \cdot \overline{\gamma}_0\}$ and $\overline{v} \breve{+} v'$ (i.e., v' as it is).

The *third* alternative is mirror to the second.

Finally, the *fourth* alternative corresponds to v and v' without $\{n_0 \cdot \overline{\gamma}_0\}$ and $\{n'_0 \cdot \overline{\gamma}_0\}$, i.e., *polynomials without constant summands*.

(4.19) defines the logic of a $v \breve{+} v'$ computation.

There are *two alternatives* in (4.19).

The *first* one corresponds to the case where v and v' include multisets $\{n_0 \cdot \overline{\gamma}_0\} \cup \overline{v}$ and $\{n'_0 \cdot \overline{\gamma}_0\} \cup \overline{v}$: both contain an identical submultiset $\overline{v}$ joined with constants represented by prefixes $\{n_0 \cdot \overline{\gamma}_0\}$ and $\{n'_0 \cdot \overline{\gamma}_0\}$, which means that the two monomials differ only by constant multipliers and all other multipliers of these monomials coincide. So, if the added monomials are $n_0 \cdot Z$ and $n'_0 \cdot Z$, then the result will be $(n_0 + n'_0) \cdot Z$, and exactly this is represented by the first alternative of (4.19).

The *second* alternative corresponds to the alternative case, when macrosets v and v' have no similar multisets and for this reason are simply joined.

Example 4.6
Let us consider two *VCMs*: $1 + 2\gamma_1\gamma_2^3 + 3\gamma_2^4\gamma_3$ and $3 + 4\gamma_1 + 5\gamma_2^4\gamma_3$. Their macroset representations are

$$v = \{\{1 \cdot \overline{\gamma}_0\}, \{2 \cdot \overline{\gamma}_0,\ 1 \cdot \overline{\gamma}_1,\ 3 \cdot \overline{\gamma}_2\}, \{3 \cdot \overline{\gamma}_0,\ 4 \cdot \overline{\gamma}_2,\ 1 \cdot \overline{\gamma}_3\}\},$$
$$v' = \{\{3 \cdot \overline{\gamma}_0\}, \{4 \cdot \overline{\gamma}_0,\ 1 \cdot \overline{\gamma}_1\}, \{5 \cdot \overline{\gamma}_0,\ 4 \cdot \overline{\gamma}_2,\ 1 \cdot \overline{\gamma}_3\}\}.$$

According to the first alternative of (4.18),

$$v \hat{+} v' = \{\{4 \cdot \overline{\gamma}_0\}\} \cup$$
$$\cup (\{\{2 \cdot \overline{\gamma}_0,\ 1 \cdot \overline{\gamma}_1,\ 3 \cdot \overline{\gamma}_2\},\ \{3 \cdot \overline{\gamma}_0,\ 4 \cdot \overline{\gamma}_2,\ 1 \cdot \overline{\gamma}_3\}\} \breve{+}$$
$$\{\{4 \cdot \overline{\gamma}_0,\ 1 \cdot \overline{\gamma}_1\}, \{5 \cdot \overline{\gamma}_0,\ 4 \cdot \overline{\gamma}_2,\ 1 \cdot \overline{\gamma}_3\}).$$

According to the first alternative of (4.19),

$$\{\{2 \cdot \overline{\gamma}_0,\ 1 \cdot \overline{\gamma}_1,\ 3 \cdot \overline{\gamma}_2\}, \{3 \cdot \overline{\gamma}_0,\ 4 \cdot \overline{\gamma}_2,\ 1 \cdot \overline{\gamma}_3\}\} \breve{+}$$
$$\{\{4 \cdot \overline{\gamma}_0,\ 1 \cdot \overline{\gamma}_1\}, \{5 \cdot \overline{\gamma}_0,\ 4 \cdot \overline{\gamma}_2,\ 1 \cdot \overline{\gamma}_3\}\}$$
$$= \{\{(5+3) \cdot \overline{\gamma}_0,\ 4 \cdot \overline{\gamma}_2,\ 1 \cdot \overline{\gamma}_3\}\}$$
$$\cup (\{\{2 \cdot \overline{\gamma}_0,\ 1 \cdot \overline{\gamma}_1,\ 3 \cdot \overline{\gamma}_2\}\} \cup \{\{4 \cdot \overline{\gamma}_0,\ 1 \cdot \overline{\gamma}_1\}\})$$
$$= \{\{8 \cdot \overline{\gamma}_0,\ 4 \cdot \overline{\gamma}_2,\ 1 \cdot \overline{\gamma}_3\}, \{2 \cdot \overline{\gamma}_0,\ 1 \cdot \overline{\gamma}_1,\ 3 \cdot \overline{\gamma}_2\}, \{4 \cdot \overline{\gamma}_0,\ 1 \cdot \overline{\gamma}_1\}\},$$

and, finally,

$$v \hat{+} v' = \{\{4 \cdot \overline{\gamma}_0\}, \{8 \cdot \overline{\gamma}_0,\ 4 \cdot \overline{\gamma}_2,\ 1 \cdot \overline{\gamma}_3\}, \{2 \cdot \overline{\gamma}_0,\ 1 \cdot \overline{\gamma}_1,\ 3 \cdot \overline{\gamma}_2\}, \{4 \cdot \overline{\gamma}_0,\ 1 \cdot \overline{\gamma}_1\}\},$$

which corresponds to the polynomial $4 + 8\gamma_2^4\gamma_3 + 2\gamma_1\gamma_2^3 + 4\gamma_1$, which is exactly what we would have obtained if we had tried to add the two considered *VCMs* directly.

The operation of *multiplication of a VCM and a multiset with primary multiplicities* is a generalization of multiplication of a constant and a multiset. Let us recall that

$$n * \{\mathrm{n}_1 \cdot \mathrm{a}_1,\ \ldots,\ \mathrm{n}_\mathrm{m} \cdot \mathrm{a}_\mathrm{m}\} = \{(\mathrm{n} \times \mathrm{n}_1) \cdot \mathrm{a}_1,\ \ldots,\ (\mathrm{n} \times \mathrm{n}_\mathrm{m}) \cdot \mathrm{a}_\mathrm{m}\}, \tag{4.20}$$

where $\times$ is the symbol of the ordinary operation of multiplication of integer numbers.

We shall denote the new operation by $\circledast$ and define it as follows:

$$v \circledast \{\boldsymbol{n}_1 \cdot a_1, \ \ldots, \boldsymbol{n}_m \cdot a_m\} = \{(v \otimes \boldsymbol{n}_1) \cdot a_1, \ \ldots, (v \otimes \boldsymbol{n}_{\mathrm{m}}) \cdot a_m\}, \tag{4.21}$$

where ⊗ is the symbol of multiplication of a *VCM* and a primary *VCM,* the latter being a constant or a variable $\gamma \in \Gamma(S)$, where $\Gamma(S)$ is the set of all variables occurring in unitary rules and metarules in the scheme *R* of the considered *UMMG* $S = < v_0$, *R*, *F*>, i.e.,

$$\boldsymbol{n}_i \in \Gamma(S) \cup \mathbf{N}, \tag{4.22}$$

where ***N*** is the set of positive integer numbers.

In turn, if $v = \{v_1, \ldots, v_m\}$,

$$v \otimes \boldsymbol{n} = \{\mathrm{v}_1 \otimes \boldsymbol{n}, \ \ldots, \mathrm{v}_{\mathrm{m}} \otimes \boldsymbol{n}\}, \tag{4.23}$$

where

$$\{n_0 \cdot \overline{\gamma}_0, n_1 \cdot \overline{\gamma}_1, \ \ldots, n_m \cdot \overline{\gamma}_m\} \otimes \boldsymbol{n} =$$

$$= \begin{cases} \{(n_0 \times \boldsymbol{n}) \cdot \overline{\gamma}_0, n_1 \cdot \overline{\gamma}_1, \ldots, n_m \cdot \overline{\gamma}_m\} \cup (\overline{v} + \overline{v}'), & \text{if } \boldsymbol{n} \in \boldsymbol{N}, \\ \{n_0 \cdot \overline{\gamma}_0, n_1 \cdot \overline{\gamma}_1, \ldots, (n_i + 1) \cdot \overline{\gamma}_i, \ldots, n_m \cdot \overline{\gamma}_m\}, & \text{if } \boldsymbol{n} \equiv \gamma_i, \\ \{n_0 \cdot \overline{\gamma}_0, n_1 \cdot \overline{\gamma}_1, \ldots, n_m \cdot \overline{\gamma}_m, 1 \cdot \boldsymbol{n}\}, & \text{if } \boldsymbol{n} \in \Gamma(S) - \{\gamma_1, \ldots, \gamma_m\}. \end{cases} \tag{4.24}$$

As can be seen, if $\boldsymbol{n}$ is a constant, multiplication of $\boldsymbol{n}$ and n_0 (a monomial constant multiplier) is performed; if $\boldsymbol{n}$ is a variable γ_i already occurring in the monomial, then the multiplicity of the object $\overline{\gamma}_i$ (i.e., the power of the variable γ_i) is incremented by 1; finally, if $\boldsymbol{n}$ is a new variable (not belonging to the set $\{\gamma_1, \ldots, \gamma_m\}$), then a multiobject $1 \cdot \boldsymbol{n}$, corresponding to a monomial of the new multiplier $\boldsymbol{n}$, is joined to the set $\{n_0 \cdot \overline{\gamma}_0, n_1 \cdot \overline{\gamma}_1, \ \ldots, n_m \cdot \overline{\gamma}_m\}$.

Example 4.7

Let us consider the polynomial $1 + 2\gamma_1^2\gamma_2^3$, whose macroset representation is

$$v = \{\{1 \cdot \overline{\gamma}_0\}, \{2 \cdot \overline{\gamma}_0, \ 2 \cdot \overline{\gamma}_1, \ 3 \cdot \overline{\gamma}_2\}\},$$

and the multiset with the primary *VCM* is

$$v' = \{3 \cdot a, \overline{\gamma}_1 \cdot b, \overline{\gamma}_2 \cdot c\},$$

corresponding to the body of the unitary metarule

$$d \rightarrow 3 \cdot a, \gamma_1 \cdot b, \gamma_2 \cdot c.$$

According to (4.20)–(4.24),

$$\begin{aligned}
v \circledast v' &= \{(v \otimes 3) \cdot a,\ (v \otimes \gamma_1) \cdot b,\ (v \otimes \gamma_2) \cdot c\}, \\
v \otimes 3 &= \{\{1 \cdot \overline{\gamma}_0\}, \{2 \cdot \overline{\gamma}_0,\ 2 \cdot \overline{\gamma}_1,\ 3 \cdot \overline{\gamma}_2\}\} \otimes 3 \\
&= \{\{1 \cdot \overline{\gamma}_0\} \otimes, \{2 \cdot \overline{\gamma}_0,\ 2 \cdot \overline{\gamma}_1,\ 3 \cdot \overline{\gamma}_2\} \otimes\} \\
&= \{\{3 \cdot \overline{\gamma}_0\}, \{6 \cdot \overline{\gamma}_0,\ 2 \cdot \overline{\gamma}_1,\ 3 \cdot \overline{\gamma}_2\}\}, \\
v \otimes \overline{\gamma}_1 &= \{\{1 \cdot \overline{\gamma}_0\}, \{2 \cdot \overline{\gamma}_0,\ 2 \cdot \overline{\gamma}_1,\ 3 \cdot \overline{\gamma}_2\}\} \otimes \overline{\gamma} \\
&= \{\{1 \cdot \overline{\gamma}_0,\ 1 \cdot \overline{\gamma}_1\}, \{2 \cdot \overline{\gamma}_0,\ 3 \cdot \overline{\gamma}_1,\ 3 \cdot \overline{\gamma}_2\}\}, \\
v \otimes \overline{\gamma}_2 &= \{\{1 \cdot \overline{\gamma}_0,\ 1 \cdot \overline{\gamma}_2\}, \{2 \cdot \overline{\gamma}_0,\ 2 \cdot \overline{\gamma}_1,\ 4 \cdot \overline{\gamma}_2\}\},
\end{aligned}$$

and, finally,

$$\begin{aligned}
v \circledast v' = \{&\{\{3 \cdot \overline{\gamma}_0\},\ \{6 \cdot \overline{\gamma}_0,\ 2 \cdot \overline{\gamma}_1,\ 3 \cdot \overline{\gamma}_2\}\} \cdot a, \\
&\{\{1 \cdot \overline{\gamma}_0,\ 1 \cdot \overline{\gamma}_1\}, \{2 \cdot \overline{\gamma}_0,\ 3 \cdot \overline{\gamma}_1,\ 3 \cdot \overline{\gamma}_2\}\} \cdot b, \\
&\{\{1 \cdot \overline{\gamma}_0,\ 1 \cdot \overline{\gamma}_2\}, \{2 \cdot \overline{\gamma}_0,\ 2 \cdot \overline{\gamma}_1,\ 4 \cdot \overline{\gamma}_2\}\} \cdot c\}.
\end{aligned}$$

After multiplying the polynomial and the multiplicities occurring in the body of the considered unitary metarule, we obtain the multiset

$$\begin{aligned}
&\{[(1 + 2\gamma_1^2\gamma_2^3) \cdot 3] \cdot a,\ [(1 + 2\gamma_1^2\gamma_2^3) \cdot \gamma_1] \cdot b,\ [(1 + 2\gamma_1^2\gamma_2^3) \cdot \gamma_2] \cdot c\} = \\
&\{(3 + 6\gamma_1^2\gamma_2^3) \cdot a,\ (\gamma_1 + 2\gamma_1^3\gamma_2^3) \cdot b,\ (\gamma_2 + 2\gamma_1^2\gamma_2^4) \cdot c\},
\end{aligned}$$

which is equivalent to the macroset representation obtained above.

As can be seen, a *macroset representation is a rather simple and flexible foundation for symbolic transformation of polynomials and, in this way, of variable-containing multiplicities*. This tool will be a basic one in our *UMMG* algorithmics.

4.2.5 *Operations on Multisets with Interval Multiplicities*

Before our consideration of the third of the above-announced operations on variable-containing multiplicities—substitution of variable domains in macroset-represented *VCMs*—let us introduce the notion of a *multiset with interval multiplicities.*

An *MSIM* is a set of multiobjects whose multiplicities are not integers or variables, but *integer intervals* represented in the usual form $[m, m']$, where $m' \geq m \geq 0$.

An *MSIM*

$$v = \{[m_1,\ m'_1] \cdot a_1,\ \ldots,\ [m_k,\ m'_k] \cdot a_k\} \tag{4.25}$$

defines a macroset

$$v^* = \{\{i_1 \cdot a_1,\ \ldots,\ i_k \cdot a_k\} | i_1 \in [m_1,\ m'_1] \& \ldots \& i_k \in [m_k,\ m'_k]\}. \tag{4.26}$$

Each multiset in v^* has the same basis as v, and every multiplicity i_j of a corresponding object a_j belongs to the domain $[m_j, m'_j]$. As can be seen,

$$|v^*| = \prod_{j=1}^{k} |[m_j,\ m'_j]| = \prod_{j=1}^{k} |m'_j - m_j + 1|. \tag{4.27}$$

Example 4.8
Let $v = \{[1,3] \cdot a_1, [0,2] \cdot a_2\}$. Then

$$\begin{aligned} v^* = \{&\{1 \cdot a_1\}, \{1 \cdot a_1,\ 1 \cdot a_2\}, \{1 \cdot a_1,\ 2 \cdot a_2\},\\ &\{2 \cdot a_1\}, \{2 \cdot a_1, 1 \cdot a_2\}, \{2 \cdot a_1, 2 \cdot a_2\},\\ &\{3 \cdot a_1\}, \{3 \cdot a_1,\ 1 \cdot a_2\}, \{3 \cdot a_1,\ 2 \cdot a_2\}\}. \end{aligned}$$

Before defining two basic operations on *MSIMs* which are an obvious generalization of the corresponding operations on multisets, we shall recall operations on integer intervals widely used in interval analysis (Hansen 1992; Shokin 1996). These operations—addition, multiplication, exponentiation, and intersection—are denoted the same as if their operands were integers and are defined as follows:

$$[m,\ m'] + [n,\ n'] = [m + n,\ m' + n'], \tag{4.28}$$

$$[m,\ m'] \cdot [n,\ n'] = [\,min\,\boldsymbol{M},\ max\,\boldsymbol{M}], \tag{4.29}$$

$$[m,\ m']^n = [m^n,\ (m')^n], \tag{4.30}$$

$$[m,\ m'] \cap [n,\ n'] = \begin{cases} \{\varnothing\}, & \text{if } m' < n \vee n' < m \\ [\,max\{m,\ \ n\},\ min\{m',\ \ n'\}] & \text{otherwise,} \end{cases} \tag{4.31}$$

where

$$\boldsymbol{M} = \{m \cdot n,\ m' \cdot n,\ m \cdot n',\ m' \cdot n'\}. \tag{4.32}$$

Everywhere below $m \equiv [m, m]$, i.e., an integer value («point») is equivalent to a singular interval in which the left and the right bounds coincide.

Example 4.9
Consider two intervals: $[0,3]$ and $[2,5]$. Then

$$[0,\ 3]+[2,\ 5]=[2,\ 8],$$
$$[0,\ 3]\cdot[2,\ 5]=[\mathit{min}\{0,\ 6,\ 15\},\ \mathit{max}\{0,\ 6,\ 15\}]=[0,\ 15],$$
$$[0,\ 3]^2=[0,\ 9],$$
$$[0,\ 3]\cap[2,\ 5]=[2,\ 3].$$

Now we may introduce two key operations on *MSIMs* which are used in *UMMG* algorithmics.

MSIM addition and *intersection*, denoted $[+]$ and $[\cap]$, respectively, are defined as follows (the intervals that are operands of these operations are denoted Δm and Δn):

$$\begin{aligned} v[+]v' = &\{\ (\Delta m + \Delta n)\cdot a \mid \Delta m\cdot a \in v \& \Delta n\cdot a \in v'\ \} \\ &\cup\{\ \Delta m\cdot a \mid \Delta m\cdot a\in v \& a \notin v'\ \} \\ &\cup\{\ \Delta n\cdot a \mid \Delta n\cdot a \in v'\ \& a\notin v\ \}, \end{aligned} \tag{4.33}$$

$$v[\cap]v' = \begin{cases} \{(\Delta m\cap\Delta m')\cdot a \mid \Delta m\cdot a\in v \& \Delta m'\cdot a\in v'\}, \\ \text{if } \beta(v)=\beta(v')\wedge(\forall a\in\beta(v))\ \Delta m\cap\Delta m' \neq [0,0] \\ \{\varnothing\} \text{ otherwise.} \end{cases} \tag{4.34}$$

According to (4.33), interval multiplicities (*IMs*) of objects occurring in both v and v' are summed, while an *IM* occurring in only one of the *MSIMs* remains unchanged (this is equivalent to the addition of interval $[0,0]$ to this *IM*).

Concerning (4.34), let us note that the result of an *MSIM* intersection is a non-empty *MSIM* if the basis $\beta(v)$ and the basis $\beta(v')$ of the multisets coincide, and the result of the intersection of the interval multiplicities Δm and $\Delta m'$ of every object a belonging to this basis is a non-empty interval. The resulting *MSIM* includes all objects belonging to $\beta(v)=\beta(v')$, with the interval multiplicities being the intersections of their *IMs* Δm and $\Delta m'$ in v and v'. In all other cases, the result of *MSIM* intersection is the empty multiset.

Example 4.10
Let

$$v=\{[0,\ 3]\cdot a,\ [1,\ 5]\cdot b,\ [3,\ 6]\cdot c\},$$
$$v'=\{[1,\ 2]\cdot a,\ [3,\ 8]\cdot b,\ [1,\ 5]\cdot c.\}$$

Then

$$v[+]v' = \{[1,\ 5] \cdot a,\ [4,\ 13] \cdot b,\ [4,\ 11] \cdot c\},$$
$$v[\cap]v' = \{[1,\ 2] \cdot a,\ [3,\ 5] \cdot b,\ [3,\ 5] \cdot c\}.$$

If

$$v' = \{[1,\ 2] \cdot a,\ [3,\ 8] \cdot b,\ [1,\ 2] \cdot c\},$$

then $v[\cap]v' = \{\varnothing\}$, because the interval multiplicities of object c in v and v' do not intersect:

$$[3, 6] \cap [1, 2] = \{\varnothing\}.$$

Now we may return to the third aforementioned operation on macroset-represented *VCMs*, namely, *substitution of variable domains in such VCMs*.

We shall represent a set of variable domains in the form of a multiset with interval multiplicities, so *MSIMw* will be

$$w = \{[m_1,\ m'_1] \cdot \gamma_1,\ \ldots,\ [m_k,\ m'_k] \cdot \gamma_k\}. \tag{4.35}$$

Let a *VCMC* in polynomial representation be

$$C = n + \sum_{i=1}^{N_c} \left(n_i \cdot \prod_{j=1}^{k_i} (\gamma_{i,j})^{l_{i,j}} \right), \tag{4.36}$$

so the result of substitution of the domains of variables given by the *MSIM w* would be the interval

$$C \circ w = n + \sum_{i=1}^{N_c} \left(n_i \cdot \prod_{j=1}^{k_i} [m_{i,j},\ m\prime_{i,j}]^{l_{i,j}} \right), \tag{4.37}$$

where $\circ$ is the symbol of the described operation of substitution.

Example 4.11
Let

$$C = 3 + 2\gamma_1\gamma_2^3 + \gamma_2^2,$$
$$w = \{[1,\ 2] \cdot \gamma_1,\ [0,\ 3] \cdot \gamma_2\}.$$

According to (4.37),

$$C \circ w = 3 + 2 \cdot [1, 2] \cdot [0, 3]^3 + [0, 3]^2$$
$$= 3 + 2 \cdot [1, 2] \cdot [0, 27] + [0, 9]$$

$$= [3, 12] + [0,108] = [3,120].$$

Let us describe the proposed algorithm of the considered operation in application to a macroset representation of variable-containing multiplicities. This algorithm is represented by the function *SubstIm* (*Substitution of Interval multiplicities*) on two input variables: v, a macroset-represented *VCM*, and w, a multiset-represented set of variable domains. Also in the *SubstIm* body, there are used the local variables Δ and Δ', whose current values are integer intervals, andΔ is the output value, i.e., the result of substitutions of the variable domains in the polynomial, while Δ' is the same result in application to one monomial.

The *SubstIm* body is as follows:

SubstIm: **function** (v, w) **returns** (Δ);

  variables v, w, Δ, Δ' **local;**

  if $\{n \cdot \gamma_0\} \in v$ /* Δ initial value setting */

   then $\{\Delta = [n, n]$;

     $v := v - \{n \cdot \gamma_0\}$;

     };

   else $\Delta := [0,0]$;

  do $\bar{v} \in v$; /*loop on monomials – elements of polynomial v */

   if $k \cdot \gamma_0 \in \bar{v}$ /* Δ' initial value setting for current monomial*/

    then $\Delta' := [k, k]$;

    else $\Delta' := [1,1]$;

   do $m \cdot \gamma \in \bar{v}$; /*loop on elements of monomial, represented by *MS* $\bar{v}$*/

    if $[l, l'] \cdot \gamma \in w$ /*extraction of a variable domain*/

     then $\Delta' := \Delta' \cdot [l, l']^m$; /*multiplication*/

   end monomial;

   $\Delta : +\Delta'$; /* summing values of the monomials */

  end polynomial;

  end ***SubstIm***

As can be seen, first of all, an initial value is assigned to the output variable Δ (lines 3–7). If there is a multiset $\{n \cdot \gamma_0\}$ in the macroset v, i.e., the constant n occurs in the polynomial, then the aforementioned initial value is the point $[n, n]$. Otherwise it is the point $[0, 0]$. For further processing of the macroset v, the multiset $\{n \cdot \gamma_0\}$ is eliminated from v (line 5). After that, the main section (lines 8–17) accumulates the output value Δ by a loop on multisets belonging to the macroset v, i.e., monomials belonging to the polynomial. The initial value of the variable Δ' is assigned similarly to the variable Δ case. If there exists an element $k \cdot \gamma_0$ in the multiset $\overline{v}$, i.e., a multiplier-constant k in the monomial (line 9), the initial value is $[k, k]$ (line 10); otherwise it is $[1, 1]$ (line 11). After setting this value, a loop on elements of the multiset $\overline{v}$ computes the value Δ' given that the variable domains and their powers are multiplicities of corresponding objects in the multiset $\overline{v}$ (lines 12–15). The monomial value Δ' is added to Δ (line 16), and after all monomials have been processed, i.e., the loop on $\overline{v} \in v$ is finished, an exit is performed with the current value of Δ returned as the result of the substitution.

With this, we finish the consideration of the auxiliary toolkit developed for implementation of the improved *Ummg#* generation algorithm and move to the main content of the present section of the book.

4.2.6 *Improved Algorithm for* STMS *Generation*

The function *Ummg#* considered below is a natural refinement and generalization of the above-described function *Ummg*. The function *Ummg#* is the result of unifying the generation algorithmics of both *UMGs* and *UMMGs*, the first being a subclass of the second (let us recall that *UMGs* are *UMMGs* without unitary metarules), but also the result of application of the tools developed and the described above for reduction of redundant search of *TMSs* by cutting off unpromising search branches. Let us confirm that practical application of the multigrammatical framework with very large knowledge bases becomes possible due to the development of *Ummg#*.

Summarizing all that was said in the previous sections about the *UMMG* algorithmics, let us present an oriented graph of calls of the described algorithmic complex. In this graph, the nodes are marked by function names, and an arc links a node which is marked by the name of the function calling another one, with the node which is marked by the name of the function being called (Fig. 4.1).

Key operations are performed by the function ***G#***, which generates a *V*-multiset by applying, as long as it is possible, *URs/UMRs* in the scheme of a *UMMG*. After a terminal *VMS* is created, further actions depend on whether the current domains of all variables occurring in multiplicity-variables are points or not. If they are all points, then the terminal *VMS* after substitution of values of variables becomes the terminal multiset, and the function *FilterTms#* (*Filtering Terminal multisets*) is applied, providing a check whether this *TMS* satisfies the boundary and optimizing conditions in the *UMMG* filter. If this *TMS* yields an improvement of at least one of the optimized multiplicities, then its new value (values) is (are) fixed, and the *TMS*

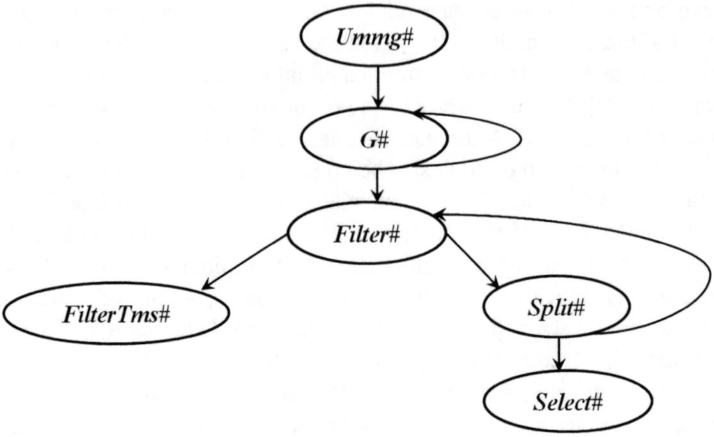

Fig. 4.1 Graph of calls of the algorithm for the reduced generation of a set of terminal multisets defined by an UMMG

becomes the first element of the accumulated set of terminal multisets, which may, finally, become the result of the application of the function *Ummg#*. If no improvement is possible, but a *TMS* satisfies the *UMMG* filter, this multiset is simply joined to the aforementioned accumulated current *STMS*. If at least one variable of a terminal *VMS* has a non-point current domain, then by application of the function *Split#*, one such domain is split into two subdomains, and the described actions, performed by the function *Filter#*, are implemented in both of them. Selection of the variable whose domain is to be split into subdomains is executed by the auxiliary function *Select#*.

Let us consider the variables used in the bodies of the aforementioned functions, in *Ummg#*. These variables are similar to those which were used in the function *Ummg*:

A set *v#*, whose current value is a generated *V*-multiset.

A set *R#*, whose static (unchangeable) value is the scheme *R*, containing unitary rules and metarules applied to *V*-multisets during generation.

A set *F#*, whose value, also static, is the filter *F*, containing boundary conditions of the form $k \leq a \leq k'$, optimizing conditions of the form $a = opt$ (*opt* may be *max* or *min*), and variable declarations of the form $n \leq \gamma \leq n'$, where the interval $[n, n']$ is the domain of the variable γ.

A set *V#*, whose current value is the set of couples $\langle v, w \rangle$ accumulated during the generation process, where v is a terminal multiset satisfying all boundary and optimizing conditions of the filter *F*, while w is the corresponding set of values of variables, represented in multiset form (i.e., $n \cdot \gamma \in w$ means n is the value of the variable γ), and, most importantly, *V#* is the final result of the ***Ummg#*** call.

A set *FT#*, whose current value is a set of triples $\langle a, opt, l \rangle$, each corresponding to one optimizing condition $a = opt \in F$, and l is the current (after the ***Ummg#*** stops—the final) value of the multiplicity of object a obtained as a result of the executed generation steps.

A couple $\langle a, w \rangle$, which is a representation of a unitary metarule $a \rightarrow \boldsymbol{n}_1 \cdot a_1, \ldots,$ $\boldsymbol{n}_m \cdot a_m$, where each $\boldsymbol{n}_i$ may be an integer number or a variable, so w is a *V*-multiset with a macroset-represented *VCM*.

A multiset *w#*, whose current value is a multiset with interval multiplicities, representing the set of current domains of variables, as defined by (4.35).

In the body of *Ummg#*, *R #*, *F #*, *FT#*, and *V#* are global variables, whose values are available to all other functions call while generation is executed. As in *Ummg*, where variables *R* and *F* are read-only, in *Ummg#*, variables *R#* and *F#* are treated similarly. At the same time, *FT#* and *V#*, like *FT* and *v* in *Ummg*, are updated (read-write) variables. All other variables used in *Ummg#* and the other functions forming together the improved *UMMG* algorithmics are local, and every new function call operates on its own values of these variables.

Ummg# and all other functions are described below similarly as in *Ummg*. The pseudocode contains primary comments which clarify the sense of its segments, and a verbal description containing references to the line numbers of the pseudocode follows.

The *Ummg#* body is as follows:

```
1        Ummg#: function (v#) returns (V#);

2               variables R#, F# external global;

3               variables FT# initial {∅}, V# initial {∅} global;

4               variables v#, w# initial {∅} local;

5               /* setting initial value of the variable FT# */

6               do a = opt ∈ F#;

7                  case opt:

8                     {"min": if k ≤ a ≤ k' ∈ F#

9                        then FT#:∪ {⟨a, "min", k'⟩};

10                       else  FT#:∪ {⟨a, "min",  MAX⟩};

11                    {"max": if k ≤ a ≤ k' ∈ F#

12                       then FT#:∪ {⟨a, "max", k⟩};

13                       else   FT#:∪ {⟨a, "max", 0⟩};

14                    };

15              end F#;

16              /* setting initial value of the variable w# */

17              do n ≤ γ ≤ n' ∈ F#;

18                 w#:∪ {[n, n'] · γ};

19              end γ;

20              /* generation function G# call */

21              call G# (v#, w#);

22              end Ummg#;
```

As can be seen, the initial part of *Ummg#* (lines 6–14) is similar to that in the *Ummg*. However, there are additional operations (lines 17–19) which allow an initial value of the variable *w#* to be set, which includes initial domains of all variables in the form of *MSIMs*.

Also similar to *Ummg* is the core part of *Ummg#*—a call of the *VMS* generating function ***G#*** (line 21), which, however, differs from ***G*** by the absence of input variables *R#* and *F#* (both are global, and there is no need to make them the arguments of the function ***G***), as well as by the presence of the additional input variable *w#*.

Let us consider the body of the function ***G#***.

```
1    G#: function (v#, w#);

2        variables R#, F#, FT#, V# global;
3        variables a, v, w, v#, w# local;

4        if v# is a terminal V-multiset
5           then {call Filter# (v#, w#);
6                 return;
7                };
8        do v · a ∃∈ v# where a is a non-terminal object;
9           do ⟨a, w⟩ ∈ R#; /* UR/UMR with header a */
10               call G# (v# − {v · a} +̂ v ⊛ w , w#);
11          end aw;
12       end va;

13       end G#;
```

Operation of ***G#*** begins with the test whether an input *V*-multiset *v#* is terminal (line 4), i.e., all objects occurring in *v#* are terminal. If so, then the function *Filter#*, applying the filter *F#*, is called (line 5), after which a return from ***G#*** is executed (line 6).

If *v#* is not a terminal *V*-multiset, i.e., there is at least one *NTO* in *v#*, then generation of new *V*-multisets may be continued. To continue generation, one multiobject *v* · *a*, where *a* is a non-terminal object, is selected from *v#* (line 8), and a *UR/UMR* with header a is applied to the *V*-multiset *v#* (lines 9–11), using a recursive call of the function ***G#*** with the first input value being the *V*-multiset created by a replacement of $v \cdot a$ by $v \circledast w$ (this operation is similar to the replacement of $n \cdot a$ by $n * a$ in the *UMG* case) and the second input value being the same *w#* (line 10).

Let us consider now the function *Filter#*, operating on terminal *V*-multisets generated by recursive ***G#*** call chains.

Filter# uses the same global variables *F#* and *FT#* as in the *Ummg#* and the ***G#*** bodies, as well as the local variables *v#* and *w#* input to this function.

The *Filter#* pseudocode is as follows.

```
Filter#: function (v#, w#);

        variables F#, FT# global;
        variables v#, w#, v local;
        v := SubstIm(v#, w#);
        if v is a terminal multiset /* all interval multiplicities in v are points */
           then {call FilterTms#(Point (v), w#); /*final filtering*/
                 return;
                 };

        /* at least one IM in v is not a point */
        /* prefiltration */

        /* boundary conditions */
           do k ≤ a ≤ k' ∈ F;
              if [n, n'] · a ∈ v
                 then {if n > k' ∨ n' < k
                          then return;    /* cut-off */
                      };
           end boundary;

         /* minimizing conditions */
           do < a, "min", l > ∈ FT#;
              if l = 0
                 then {if [n, n'] · a ∈ v
                          then {if n > 0
                                   then return;   /* cut-off */
                                   else;
                                };
                          else /* a ∉ v */
                      };
```

```
28                  else /* l > 0 */
29                      {if [n,n'] · a ∈ v
30                          then {if l < n
31                                  then return;  /* cut-off*/
32                               };
33                          else; /* a ∉ v */
34                      };
35          end minimizing;

36          /* maximizing conditions */
37           do < a,"max",l >∈ FT#;
38                  if [n,n'] · a ∈ v
39                  then { if n' < l
40                          then return;   /* cut-off */
41                          else;
42                       };
43                  else /* a ∉ v */
44                      { if l > 0
45                          then return;   /* cut-off */
46                          else;
47                      };
48           end maximizing ;
49           /* prefiltration successful */

50          /* splitting current domain of one selected variable into two */
51           call Split#(v#,w#);
52           end Filter#
```

Let us comment on the above pseudocode.

Filter# begins from a substitution of interval multiplicities from the *MSIM* *w#* (they are sets of possible values of the *UMMG* variables) into the *V*-multiset *v#* (line

4). If the result is a terminal multiset, i.e., all interval multiplicities in v are, in fact, points, then the function *FilterTms#*, which performs final filtering, is called (line 6). Here *Point* is the function transforming an *MSIM* with a point interval multiplicity into a one-element multiset:

$$\textbf{\textit{Point}}\ (v) = \{n \cdot a \mid [n,\ n] \cdot a \in v\}. \tag{4.38}$$

After a call of the function *FilterTms#*, an exit from *Filter#* is performed. The results of the *FilterTms#* operation are reflected in changes of the values of global variables *V#* and *FT#*.

Otherwise, i.e., if at least one multiplicity in the *MSIM* v is a non-point interval, so-called prefiltration is performed (lines 11–17).

Prefiltration includes three sequentially executed sections, performing tests on boundary, minimizing, and maximizing conditions, respectively.

Boundary conditions are checked inside the loop, appearing in lines 12–17. If the *MSIM* v includes a multiobject $[n, n'] \cdot a$ with the same object as a current chain boundary condition $k \le a \le k'$, then there may be three alternatives: $n > k'$ (the lowest value of the set of possible multiplicities of the object a is greater than the right bound of this condition), $n' < k$ (the largest value of the set of possible multiplicities of the object a is less than the left bound of this condition), or neither of these. In the first and the second cases, a cutoff is performed by a return (line 15); otherwise the loop is continued, because there exists a possibility that during further calls of the function ***G#*** (and thus the function *Filter#*) causing splitting of the interval $[n, n']$, some point values satisfying the boundary conditions may be obtained.

Minimizing conditions are processed in a similar way (lines 18–35). Every element $\langle a, "min", l\rangle \in FT\#$ is tested, first of all, to see whether $l = 0$ (line 20), i.e., the best possible case is already achieved or not. If so, and $[n, n'] \cdot a \in v$ where $n > 0$, there is no sense in further generation steps beginning from v; thus cutoff is performed by a return at line 23. Otherwise, i.e., if $l > 0$ (line 28), l is compared with the left bound n of the interval multiplicity $[n, n']$. If $n > l$, this means that further generation steps, which may result only in n increasing, are of no use, and a cutoff is performed (line 31). Otherwise the loop is continued.

Maximizing conditions are processed in practically the same manner (lines 36–48).

In the case all three prefiltration sections were passed successfully, the function ***Split#*** is called, implementing a splitting of the current domain of the selected variable occurring in the variable-containing multiplicities of terminal *V*-multiset *v#* (line 51). Note that the second input parameter in *FilterTms#* and in *Split#* is the same (*w#*), while the first input parameters are different.

The function *FilterTms#* implements a direct test of all boundary and optimizing conditions on multiplicities of terminal objects, as well as a correction of the set of *TMSs* accumulated at the previous steps. *FilterTms#* operates on the same global variables *FT #* , *F#*, and *V#* as all functions considered above in this section (i.e., *Ummg#*, *G#*, *Filter#*). Also *FilterTms#* operates on local variables v and *w#*, which are the same as in *Filter#*: v is a *TMS*, while $w\#$ is a *MSIM*, containing values of *UMMG* variables in *IM* form.

Also there is a Boolean variable x, which is a flag determining whether there is at least one replacement of optimized multiplicities. If there are no such replacements, $x = 0$, otherwise $x = 1$. The initial value of x is 0.

The *FilterTms#* pseudocode is as follows.

1   *FilterTms#*: **function** $(v, w\#)$;

2   **variables** $F\#, FT\#, V\#$ **global;**

3   **variables** $v, w\#, x$ initial (0) **local;**

4   /* boundary conditions test */

5   **do** $k \leq a \leq k' \in F$;

6   **if** $n \cdot a \in v$

7   **then** {**if** $(n > k') \vee (n < k)$ /* n is outside $[k, k']$ */

8   **then return**; /* cut-off */

9   **else**;

10   };

11   **else** /* zero multiplicity of a terminal object a, */

12   /* i.e., a does not occur in v */

13   {**if** $k \neq 0$

14   **then return**; /* cut-off */

15   **else**;

16   };

17   **end** *boundary#*;

18   /* minimizing conditions test */

19   **do** $\langle a, "min", l \rangle \in FT\#$;

20   **if** $n \cdot a \in v$

21   **then** {**if** $n > l$ /* already stored value l is less than n */

22   **then return**; /* cut-off */

```
                else;
            };
    end minimizing#;

  /* maximizing conditions test */
    do ⟨a, "max", l⟩ ∈ FT#;
      if n · a ∈ v
        then {if n < l  /* already stored value l is greater than n */
                then return;   /* cut-off */
                else;
             };

        else  /* zero multiplicity of a terminal object a, */
              /* i.e., a does not occur in v */
             {if l > 0
                then return;   /* cut-off */
                else;
             };
    end maximizing#;
  /* conditions test successful, so current min/max values of */
  /* multiplicities would be corrected by the TMS v */
```

```
42              do ⟨a, opt, l⟩ ∈ FT#;
43                  do n · a ∈ v;
44                      if l ≠ n   /* at least one multiplicity corrected */
45                          then {FT#:= FT# − {⟨a, opt, l⟩} ∪ {⟨a, opt, n⟩};
46                              x:= 1;   /* flag reset */
47                              };
48                  end na#;
49              end optimal#;

50          /* correction of the accumulated set V# */
51              if x = 0   /* no optimized multiplicities corrected /
52                  then V#:= V# ∪ {⟨v, w#⟩};   /* one more element joined */
53                  else V#:= {⟨v, w#⟩};   /* replacement by a one-element set */

54      end FilterTms#
```

Let us comment on the presented function, containing five functional sections.

The *first section* (lines 5–17) performs the boundary conditions check. Every such condition $k \le a \le k' \in F$ is analyzed in such a way that if there is a multiobject $n \cdot a$ in the multiset v and n is outside the interval $[k, k']$ (line 7), then the check fails, and an exit from the function is performed (line 8). If there is no object a in v, i.e., v contains $[0, 0] \cdot a$, the check $k \ne 0$ is executed (line 13). If $k \ne 0$, that means the condition $k \le 0 \le k'$ is not satisfied, and an exit is performed. In the case $k = 0$, the condition $0 \le 0 \le k'$ is true, and hence nothing is done (line 15).

If all boundary conditions are satisfied, i.e., no exits were done during the loop (lines 5–17) execution, the two following sections, performing checks of minimizing (lines 19–25) and maximizing (lines 27–39) conditions, are performed sequentially.

The loop on triples $\langle a, "min", l\rangle \in F\#$ performs, first of all, a check whether there is a multiobject $n \cdot a$ with non-zero multiplicity in the *MSIM* v (line 20). In the positive case, n and l are compared, and if $n > l$, an exit from this function is performed (n is already greater than an already achieved value l and may only increase). Otherwise ($n \le l$) nothing is done (line 31), because there exists a possibility that l may be replaced later by the lower value n (case $n < l$), and hence a generated *TMS* with the corresponding set of *UMMG* variable values may be joined to the already accumulated set *V#* (case $n = l$). The case in which there are no multiobjects with object a in the *MSIM* v, or, equivalently, $0 \cdot a \in v$, does not need particular consideration: if $l > 0$, the check fails (line 36), while if $l = n = 0$, nothing is done, as described above (line 37).

The loop on triples $\langle a, "max", l\rangle \in F\#$ is similar, except for the particular processing of the case in which the object a does not occur in *TMS* v, i.e., a multiplicity of a is zero (lines 35–38). If $l > 0$, then this means the already achieved value l is greater than the current multiplicity of object a, and thus exit is performed (line 36). Otherwise, i.e., if $l = n = 0$, nothing is done for the reasons mentioned above in the description of the "minimizing" section.

If all three considered sections are passed successfully, i.e., the *TMS v satisfies the filter F#*, then the *fourth section* (lines 42–49) is executed, performing a correction of the current min/max values by multiplicities occurring in multiobjects belonging to this *TMS* v. The external loop on all optimizing conditions $\langle a, opt, l\rangle \in FT\#$ (lines 42–49) contains an internal loop (lines 43–48) on the corresponding multiobjects $[n, n] \cdot a \in v$. If $l \neq n$, then this means $n > l$ (case "*min*"), and the current value l must be replaced by n. This replacement (line 45) is followed by resetting the flag x to the value 1, which means that the value of at least one multiplicity of the processed *TMS* is refined. If no such refinements are done, x continues to stay in the initial "zero" state.

Finally, *the fifth* section contains the only conditional operator (lines 51–53) and performs a correction of the current value of the accumulated set *V#*. If $x = 0$, i.e., no replacements of optimized multiplicities were done, the couple $\langle v, w\#\rangle$ is simply joined to *V#* (line 52). Otherwise, i.e., if at least one refinement was done, the current value of the variable *V#* is replaced by the one-element set $\{\langle v, w\#\rangle\}$, and the *FilterTms#* successfully terminates.

Let us consider now the function *Split#*, which is a key tool for *generation and selection of TMSs with the same basis as v but with interval multiplicities of objects, which become more and more "narrow" until they become points.*

The function *Split#* has a relatively short body, with the same input parameters v and *w#* as *Filter#* and *FilterTms#*:

```
1   Split#: function (v#, w#);
2       variables v#, w# local;
3       call Select#(v#, w#, n, n', γ);
4       m: = n + ent ((n' − n)/z);
5       call Filter# (v#, w# − {[n, n'] · γ} ∪ {[n, m] · γ});
6       call Filter# (v#, w# − {[n, n'] · γ} ∪ {[m + 1, n'] · γ});
7       end Split#
```

Here, first of all, a multiobject with an interval multiplicity to be split into two more "narrow" *IMs* is selected from all those belonging to the *MSIM* *w#*. This is done by a call of the function *Select#* (line 3) with the first two arguments *v#*and *w#*, which are known, and three more arguments n, n', and γ, which are unknown *before* a call and correspond to a multiobject $[n, n'] \cdot \gamma$ *after* a call. Some possible variants of logic which may be implemented by *Select#* will be considered below; until then, the only condition we may assume is $n' > n$, and the operation in line 4 computes the *middle point* of the interval $[n, n']$, i.e., the value m obtained by division of $n' - n$ by 2 and applying the standard function *ent* to the result (let us recall that *ent*(x), where x is a rational number, is the greatest integer number no greater than x, i.e., the "closest" to x from the left side).

After the value m is computed, two recursive calls of the function *Filter#* are executed, differing by the intervals of values of the selected variable γ (lines 5–6). The first one is $[n, m]$, i.e., the left subinterval of the interval $[n, n']$ (from the left bound n to the "middle point"), while the second one is $[m + 1, n']$, i.e., the right subinterval of $[n, n']$ (from the "middle point" to the right bound n'). Thus the interval $[n, n']$ is split, and both calls of *Filter#* will operate with the "narrowed" intervals of the current domain of the selected variable γ and unchanged intervals of all other variables.

Let us consider the function *Select#*, whose implementation in the general case has a *strong influence on the computational complexity* of *UMMG* algorithmics:

```
1    Select#: function (v#, w#, n, n', γ);
2        variables v#, w#, n, n', γ local;
3        do [m, m'] · γ' ∃∈ w#;
4          if γ' belongs to v#
5            then {  n: = m;
6                    n': = m';
7                    γ: = "γ'"
8                    return;
9                    };
10       end mmw;
11       end Select#
```

As can be seen, a key element of *Select#* is the arbitrary choice operator $\exists\in$ (line 3), which, obviously, is *decisive* for the aforementioned computational complexity. There may be at least three straightforward approaches to the implementation of this operator: random, “the narrowest,” and “the widest” interval $[n, m']$. Both non-random approaches are “adaptive” in the sense that variables are selected in a varying sequence, because the set of *UMMG* variables at any following step in the general case may be reordered by an increase of the lengths (cardinalities) of their domains.

There exists a more sophisticated but to an equal degree more effective method, based on the *improved algorithmics of solution of systems of polynomial equations on a set of integer values*. It will be considered in future publications.

Let us illustrate the described algorithmics by the following example, in which operation of the function *Ummg#* will be represented by an execution table with four columns:

The sequential number of the generation step associated with a function call

A function call, including its name and arguments

Operations performed during execution of the called function

The sequential number of the next generation step

Also inside column 1, there are labels of the form *i. j*, which are used when returning from functions call and for loop continuation (Table 4.1).

Table 4.1 Execution of the function *Ummg#*

Generation step number	Function call	Operations performed	Next generation step number
1 1.1	***Ummg#*** $(\{1 \cdot a\})$	$FT\# := \{<d, "min", 100>\};$ $w\# := \{[2,3] \cdot \gamma_1, [1,2] \cdot \gamma_2\};$ **call** $\boldsymbol{G}\#(\{1 \cdot a\}, \{[2,3] \cdot \gamma_1, [1,2] \cdot \gamma_2\});$ **exit** $V\# = \{<\{6 \cdot c, 2 \cdot d\},$ $\{[2,2] \cdot \gamma_1, [1,2] \cdot \gamma_2\} > \};$ $FT\# := \{<d, "min", 2>\};$	2
2	***G#***	$\{1 \cdot a\}$ is not a terminal *V*—multiset	
2.1	$(\{1 \cdot a\},$ $\{[2,3] \cdot \gamma_1, [1,2] \cdot \gamma_2\})$	**do** $1 \cdot a \ \exists \ \in \{1 \cdot a\};$ **do** $\langle a, \{2 \cdot b, \gamma_1 \cdot c, 1 \cdot e\}\rangle \in R\#;$ **call** $\boldsymbol{G}\#(\{2 \cdot b, \gamma_1 \cdot c, 1 \cdot e\}, \{[2,3] \cdot \gamma_1,$ $[1,2] \cdot \gamma_2\});$ **return** 1.1	3
3 3.1	***G#*** $(\{2 \cdot b, \gamma_1 \cdot c, 1 \cdot e\},$ $\{[2,3] \cdot \gamma_1, [1,2] \cdot \gamma_2\})$	$\{2 \cdot b, \gamma_1 \cdot c, 1 \cdot e\}$ is not a terminal *V*—multiset **do** $2 \cdot b \ \exists \ \in \{2 \cdot b, \gamma_1 \cdot c, 1 \cdot e\};$ **do** $\langle b, \{3 \cdot c, \gamma_2 \cdot d\}\rangle \in R\#;$ **call** $\boldsymbol{G\#}\left(\begin{array}{c} \{(6+\gamma_1) \cdot c, \ (2 \cdot \gamma_2) \cdot d, \ 1 \cdot e\}, \\ \{[2, \ 3] \cdot \gamma_1, \ [1, \ 2] \cdot \gamma_2\} \end{array} \right)$ **return** 2.1	4
4 4.1	$\boldsymbol{G\#}\left(\left\{ \begin{array}{c} (6+\gamma_1) \cdot c, \\ 2\gamma_2 \cdot d, \ 1 \cdot e \end{array} \right\},\right.$ $[2,3] \cdot \gamma_1, [1,2] \cdot \gamma_2)$	$\{(6+\gamma_1) \cdot c, (2 \cdot \gamma_2) \cdot d, 1 \cdot e\}$ is not a terminal *V*–multiset **do** $1 \cdot e \ \exists \ \in \{2 \cdot b, \gamma_1 \cdot c, 1 \cdot e\};$ **do** $<e, \{\gamma_1 \cdot c, 1 \cdot d\} > \ \in R\#;$ **call** $\boldsymbol{G}\#(\{(6+2 \cdot \gamma_1) \cdot c, (1+2 \cdot \gamma_2) \cdot d\},$ $\{[2,3] \cdot \gamma_1, [1,2] \cdot \gamma_2\})$ **return** 3.1	5
5 5.1		$\{(6+2 \cdot \gamma_1) \cdot c, (1+2 \cdot \gamma_2) \cdot d\}$ is a terminal *V*–multiset **call** ***Filter*** $\#(\{(6+2 \cdot \gamma_1) \cdot c, (1+2 \cdot \gamma_2) \cdot d\},$	6

(continued)

Table 4.1 (continued)

Generation step number	Function call	Operations performed	Next generation step number
	$\boldsymbol{G\#}\left(\left\{\begin{matrix}(6+2\cdot\gamma_1)\cdot c,\\(1+2\cdot\gamma_2)\cdot d\end{matrix}\right\}, \{[2,3]\cdot\gamma_1,[1,2]\cdot\gamma_2\}\right)$	$\{[2,3]\cdot\gamma_1,[1,2]\cdot\gamma_2\})$; **return** 4.1	
6 6.1	$\boldsymbol{Filter\#}\left(\left\{\begin{matrix}(6+2\cdot\gamma_1)\\ \cdot c,\\(1+2\cdot\gamma_2)\\ \cdot d\end{matrix}\right\}, \left\{\begin{matrix}[2,\ 3]\cdot\gamma_1,\\ [1,\ 2]\cdot\gamma_2\end{matrix}\right\}\right)$	$v := \{[10,12]\cdot c,[3,5]\cdot d\}$; v is not a terminal multiset: prefiltration; $\Delta := [10,12]\cap[10,17]\neq\{\emptyset\}$; $\Delta := [3,5]\cap[0,100]\neq\{\emptyset\}$; prefiltration successful; v is not a terminal multiset; **call** ***Split #*** $(\{(6+2\cdot\gamma_1)\cdot c,(1+2\cdot\gamma_2)\cdot d\}$, $\{[2,3]\cdot\gamma_1,[1,2]\cdot\gamma_2\})$; **return** 5.1	7
7 7.1 7.2	$\boldsymbol{Split\#}\left(\left\{\begin{matrix}(6+2\cdot\gamma_1)\\ \cdot\\ c,\\(1+2\cdot\gamma_2)\\ \cdot d\end{matrix}\right\}, \left\{\begin{matrix}[2,\ 3]\cdot\gamma_1,\\ [1,\ 2]\cdot\gamma_2\end{matrix}\right\}\right)$	**call** ***Select#*** $(\{(6+2\cdot\gamma_1)\cdot c,(1+2\cdot\gamma_2)\cdot d\}$, $\{[2,3]\cdot\gamma_1,[1,2]\cdot\gamma_2\},n,n',\gamma)$; /*result is $n=2$, $n'=3$, $\gamma=\gamma_1$*/ $m := 2 + \boldsymbol{ent}((3-2)/2) = 2$; **call** ***Filter#*** $(\{(6+2\cdot\gamma_1)\cdot c,(1+2\cdot\gamma_2)\cdot d\}$, $\{[2,2]\cdot\gamma_1,[1,2]\cdot\gamma_2\})$; **call** ***Filter#*** $(\{(6+2\cdot\gamma_1)\cdot c,(1+2\cdot\gamma_2)\cdot d\}$, $\{[3,3]\cdot\gamma_1,[1,2]\cdot\gamma_2\})$; **return** 6.1;	8 10
8 8.1	$\boldsymbol{Filter\#}\left(\left\{\begin{matrix}(6+2\cdot\gamma_1)\\ \cdot c,\\(1+2\cdot\gamma_2)\\ \cdot d\end{matrix}\right\}, \{[2,\ 2]\cdot\gamma_1,\ [1,\ 2]\cdot\gamma_2\}\right)$	$v := \{[10,10]\cdot c,[3,5]\cdot d\}$; v is not a terminal multiset: prefiltration; $\Delta := [10,10]\cap[10,17]\neq\{\emptyset\}$; $\Delta := [3,5]\cap[0,100]\neq\{\emptyset\}$; prefiltration successful;	9

		v is not a terminal multiset; **call** ***Split #*** $(\{(6+2\cdot\gamma_1)\cdot c,(1+2\cdot\gamma_2)\cdot d\},$ $\{[2,2]\cdot\gamma_1,[1,2]\cdot\gamma_2\})$; **return** 7.1	
9 9.1	***Split#*** $\left(\left\{\begin{matrix}(6+2\cdot\gamma_1)\cdot c,\\(1+2\cdot\gamma_2)\cdot d\end{matrix}\right\},\left\{\begin{matrix}[2,\ 2]\cdot\gamma_1,\\ [1,\ 2]\cdot\gamma_2\end{matrix}\right\}\right)$	**call** ***Select#*** $(\{(6+2\cdot\gamma_1)\cdot c,(1+2\cdot\gamma_2)\cdot d\},$ $\{[2,2]\cdot\gamma_1,[1,2]\cdot\gamma_2\},n,n',\gamma)$; /*result is $n=1, n'=2, \gamma=\gamma_2$*/ m := 1 + ***ent***((2 − 1)/2) = 1; **call** ***Filter#*** $(\{(6+2\cdot\gamma_1)\cdot c,(1+2\cdot\gamma_2)\cdot d\},$ $\{[2,2]\cdot\gamma_1,[1,1]\cdot\gamma_2\})$; **call** ***Filter#*** $(\{(6+2\cdot\gamma_1)\cdot c,(1+2\cdot\gamma_2)\cdot d\},$ $\{[2,2]\cdot\gamma_1,[2,2]\cdot\gamma_2\})$; **return** 8.1	10
10 10.1	***Filter#*** $\left(\left\{\begin{matrix}(6+2\cdot\gamma_1)\cdot c,\\(1+2\cdot\gamma_2)\cdot d\end{matrix}\right\},\left\{\begin{matrix}[2,\ 2]\cdot\gamma_1,\\ [1,\ 1]\cdot\gamma_2\end{matrix}\right\}\right)$;	v := $\{[10,10]\cdot c,[3,3]\cdot d\}$; v is a terminal multiset; **call** ***FilterTms#*** $(\{10\cdot c,3\cdot d\},$ $\{[2,2]\cdot\gamma_1,[1,1]\cdot\gamma_2\})$; **return** 9.1	11
11	***FilterTms#*** $(\{10\cdot c,3\cdot d\},\ \{[2,2]\cdot\gamma_1,[1,1]\cdot\gamma_2\})$;	multiplicity of object c=10: 10≤10 ≤ 17 filtration successful **return** 10.1	

Example 4.12
Let the unitary multiset metagrammar be $S = \ < \{1 \cdot a\}, R, F>$, where the scheme $R = \{r_1, r_2, r_3\}$,

$$r_1 : a \rightarrow 2 \cdot b, \gamma_1 \cdot c, 1 \cdot e,$$

$$r_2 : b \rightarrow 3 \cdot c, \gamma_2 \cdot d,$$

$$r_2 : e \rightarrow \gamma_1 \cdot c, 1 \cdot d,$$

and the filter

$$F = \{10 \leq c \leq 17,\, d = min, \ \ 2 \leq \gamma_1 \leq 3, \ \ 1 \leq \gamma_2 \leq 2\}.$$

In the *Ummg#* body, we shall use the maximal integer value $MAX = 100$. The execution table is as follows.

Execution starts from the *Ummg#* call with the argument $\{1 \cdot a\}$ (the sequential number of the generation step is 1). This sets the initial values of the variables *FT#* and *w#*, after which the function ***G#*** call with arguments $\{1 \cdot a\}$ and $\{[2,3] \cdot \gamma_1, [1,2] \cdot \gamma_2\}$ is performed (step 2). Because $\{1 \cdot a\}$ is not a terminal *V*-multiset, a loop on the elements of the multiset $\{1 \cdot a\}$ (i.e., the only element $1 \cdot a$) is performed, and inside it all, *URs/UMRs* with head *a* are searched (i.e., the only such unitary metarule $a \rightarrow 2 \cdot b, \gamma_1 \cdot c$), so the function ***G#*** call with the arguments $\{2 \cdot b, \gamma_1 \cdot c\}$ and $\{[2,3] \cdot \gamma_1, [1,2] \cdot \gamma_2\}$ is performed (step 3). (For simplification, we shall use everywhere below a natural *V*-multiset representation instead of a macroset one.) Here, because $\{2 \cdot b, \gamma_1 \cdot c\}$ is not a terminal *V*-multiset, prefiltration is executed. Firstly, the interval of the possible multiplicities of the object *c* is calculated by the function *SubstIm* call, and the result $[2, 3]$ is assigned to the variable Δ. Because the left bound of this interval, i.e., 2, is less than the right-bound value of the boundary condition $1 \leq c \leq 7$, i.e., 7, no cutoff is executed. For the same reason, the optimizing condition with the current value *<d, "min", 100>* does not produce a cutoff (for brevity, this test, implemented by ***G#*** lines 19–25, is not included in the execution table). So the generation branch is recognized as promising, and the loop on the non-terminal multiobjects belonging to the *V*-multiset $\{2 \cdot b, \gamma_1 \cdot c\}$, i.e., the only such multiobject $2 \cdot b$, is performed. That leads to a recursive ***G#*** call with the arguments $\{2 \cdot b, \gamma_1 \cdot c\} - \{2 \cdot b\}^{\smile} + 2 \circledast \{3 \cdot c, \gamma_2 \cdot d\} = \{(6 + \gamma_1) \cdot c, 2\gamma_2 \cdot d\}$ and, again, $\{[2,3] \cdot \gamma_1, [1,2] \cdot \gamma_2\}$. Execution of this new call (step 4) begins with the test, where the input *V*-multiset (the first argument) is terminal. Because both objects *c* and *d* occurring in this *V*-multiset are terminal, the test result is positive, and prefiltration is performed. The interval multiplicity calculated from the *VCM* $6 + \gamma_1$ and the current domain of the variable $\gamma_1 \in [2, 3]$ is $[8, 9]$, and because the left bound of this interval, i.e., 8, is greater than the right bound of the condition $1 \leq c \leq 7$, a cutoff is performed by a return to step 3.1, marking the continuation of the loop on *URs/UMRs* with the head *b*. This leads to another ***G#*** call, with the first argument $\{2 \cdot b, \gamma_1 \cdot c\} - \{2 \cdot 1\}^{\smile} + 2 \circledast \{\gamma_1 \cdot c, 1 \cdot d\} = \{3\gamma_1 \cdot c, 2 \cdot d\}$ and the same second argument (step 5). The new call execution follows in the same manner as described above. Because $\{3\gamma_1 \cdot c, 2 \cdot d\}$ is a terminal *V*-multiset, a *Filter#* call is performed with the same arguments as in the ***G#*** call in step 5. The *Filter#* (step 6) execution starts with creation of the *MSIM* $\{[6,9] \cdot c, [2,2] \cdot d\}$ and its assignment to the variable *v*. Because *v* now is not a terminal multiset (the interval [6, 9] is not a point), prefiltration is performed. Because both tests (on the boundary condition $1 \leq c \leq 7$ and the optimizing condition with the current state<*d*, "min", 100>) are successful, the obtained new value of the variable *v* is tested to see whether it is a terminal multiset. Because it is not a *TMS* (the interval [6, 7] is not a point), the function *Split#* is called, and the result of this call is the variable γ_1 with an interval of possible values [2, 3], which is split into two intervals: [2, 2] and [3, 3]. Afterward, another *Filter#* call is executed (step 8). Because the created *MSIM* $\{[6,6] \cdot c, [2,2] \cdot d\}$ is a terminal multiset, *FilterTms#* is called with the arguments $\{6 \cdot c, 2 \cdot d\}$ and $\{[2,2] \cdot \gamma_1, [1,2] \cdot \gamma_2\}$ (step 9). During its execution, both tests (on the boundary condition $1 \leq c \leq 7$ and the current state of the optimizing condition $d = min$, i.e., <*d*,"*min*", 100 > $\in$ *FT#*) are successful, so operations correcting the variables *V#* and *FT#* follow. Because the multiplicity 2 of the object *d* is less than the third component 100 in the current value of the variable *FT#*, this value is replaced by the new triple <*d*, "*min*", 2 > , and the value of the flag variable *x* becomes 1 ("a replacement was performed"). After the loop is finished, the flag *x* is tested, and because its value is 1, the value of the variable *V#* is replaced by the one-element set $\{<\{6 \cdot c, 2 \cdot d\}, \{[2,2] \cdot \gamma_1, [1,2] \cdot \gamma_2\}>\}$, and the return to step 8.1 (the *Filter#* continuation) is performed. This leads to the return to step 7.1, which is a continuation of the *Split#* by a new *Filter#* call (step 10) with the second argument corresponding to the right subinterval of the split interval [2, 3] of the domain of variable γ_1, i.e., [3, 3]. The first argument is the same: $\{3\gamma_1 \cdot c, 2 \cdot d\}$. Execution of this call creates the variable *v* value $\{[9,9] \cdot c, [2,2] \cdot d\}$, which is a terminal multiset, so the function *FilterTms#* is called with the first argument $\{9 \cdot c, 2 \cdot d\}$ and the second argument $\{[3,3] \cdot \gamma_1, [1,2] \cdot \gamma_2\}$ (step 11). Because the multiplicity 9 of the object *c* is greater than the upper bound of the condition $1 \leq c \leq 7$, a cutoff is executed by a return to step 10.1, and afterward a chain of returns—7.2, 6.1, 5.1, 3.2, 2.1, and, finally, 1.1—leads to an exit from *Ummg#* with the value of the variable *V#* being the one-element set of terminal multisets $\{<\{6 \cdot c, 2 \cdot d\}, \{[2,2] \cdot \gamma_1, [1,2] \cdot \gamma_2\}>\}$ and the value of the variable *FT#* being the one-element set {<*d*,"*min*",2>}. (Let us note that the values of some variables in the result of the *Ummg#* call may be non-point intervals, such as $[1,2] \cdot \gamma_2$, which means that any integer number from this interval may be used.)

The correctness of the described and illustrated algorithm is affirmed by the following statement, which is similar to Statement 4.2 regarding acyclic non-variative *UMGs*.

Statement 4.3 Let $S = \langle v_0, R, F \rangle$ be an acyclic non-variative UMMG, and

V= ***Ummg#*** (v_0, *R*, *F*) be the result of a call of the function ***Ummg#*** with the arguments v_0, *R*, *F*.

Then $V = \overline{V_s}$.

We do not provide a proof of this statement, because it would, in fact, be a retelling of the above description in other words.

Also we shall not estimate the computational complexity of the function Ummg# here, because it is quite evident: creation of a V-multiset by an application of a UMMG scheme has complexity linear in the number of URs/UMRs in the scheme, while generation of an STMS given this V-multiset has a logarithmically based complexity.

Let us pay some attention to a generalization of the proposed algorithmics in the case of acyclic and cyclic *variative* unitary multiset metagrammars.

4.2.7 Implementation of Acyclic and Cyclic Variative UMMGs

The technique used for reduction of acyclic *variative* unitary multiset *grammars* to acyclic *non-variative UMGs*, introduced in Sect. 4.1.2, may be extended without difficulty to variative unitary multiset *metagrammars*: however, only in cases where an object a which is the header of m *URs/UMRs* with bodies v_1 ,. . .,v_m occurs in l bodies of *URs/UMRs* such that multiobjects containing the object a in these bodies have multiplicity-constants. If there is at least one body which includes an object a with a multiplicity-variable γ, then the technique proposed above in Sect. 4.1.2 is *not applicable*, because its straightforward use would produce boundary conditions of the form $\bar{a} =\gamma$, which is outside the current syntax and semantics of *UMMGs*. However, such a solution may be introduced rather simply by a slight correction of the aforementioned syntax and semantics, as well as the *UMMGs* algorithmics.

Regarding *cyclic UMMGs* (both variative and non-variative), it is quite evident why the above-presented algorithmics, developed for acyclic *UMMGs*, would not be valid. It is not applicable to cyclic *UMMGs* because cyclicity produces an infinite chain of calls of the function ***G#*** generating the terminal *V*-multiset w, which is used by the functions *Filter#* and *FilterTms#* to search collections of values of *UMMG* variables. Thus there would be an infinite set of non-terminal *V*-multisets, excluding application of the function *FilterTms#* and, thus, search of the aforementioned collections satisfying the *UMMG* filter. However, it is possible to handle cyclic *UMMGs* by applying, as a foundation, the techniques already introduced in Sect. 3.4.2 for detecting infinite generation chains, as well as the algorithmics of acyclic *UMMGs*.

We shall study these interesting general cases, briefly described in this section, in future publications.

4.3 Filtering Multiset Grammars

In the previous sections, we have considered the developed *prefiltration-based early cutoff techniques for reduction of the computational complexity of generation of sets of terminal multisets defined by filtering unitary MGs/MMGs*. However, there is one more advantage of the multigrammatical framework: *the natural parallelism of generation of multisets due to the possibility of independent application of rules (URs, TRs) to currently available MSs.* This feature being supported by appropriate hardware is a very promising basis for an effective implementation of all classes of multiset grammars. However, a "brutal force" approach, demanding extensive parallelism, is not suitable here because of the evident cost restrictions. In this section, we shall consider an approach to an effective implementation of the filtering multiset grammars combining the above-considered cutoff techniques with maximal possible parallelization.

Developing the declared approach, we shall apply as a background multi-agent technology (*MAT*) (Mesbahi and Egerstedt 2010; Olfati-Saber et al. 2007; Shoham and Leyton-Brown 2009; Waldrop 2018; Wooldridge 2009). The main idea of the suggested technique (Sheremet 2020d) is the *representation of an FMG's application engine* (*AE*), *which generates multisets, as a multi-agent system* (*MAS*). This *MAS* operates in such a way that *every rule is applied* (i.e., the agent representing it becomes active) *as soon as there occurs a multiset matching this rule*. By applying this approach, a maximal possible degree of parallelism may be achieved.

4.3.1 Primary Multi-agent Implementation

We shall use the scheme depicted in Fig. 4.2, as the basis for a primary multi-agent implementation of an *FMG's AE*. The proposed multi-agent system implementing an *FMG* $S = \langle v_0, R, F\rangle$ contains $l = |R|$ agents $r_1, \ldots, r_l$, each corresponding to one rule from the set R (everywhere below, we shall denote rules and the agents implementing them by the same symbols); one agent F^*, implementing a filter F;

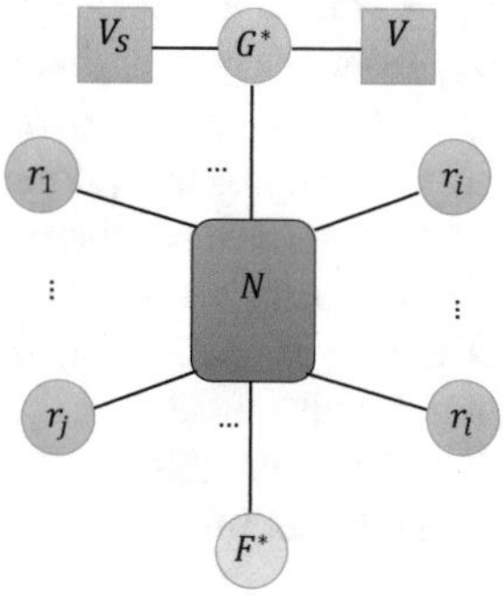

Fig. 4.2 A multi-agent system implementing an *FMG* application engine

Table 4.2 The set of messages circulating between the agents

N	Sender	Receiver	Message	Comment
1	G^*	r_i	$\langle j, v\rangle$	j—number of *MS*, $v \neq \{\emptyset\}$
2	r_i	G^*	$\langle j, v'\rangle$	v'—result of application of rule r_i to *MS* v
3	r_i	G^*	$\langle j, \{\emptyset\}\rangle$	Rule r_i is not applicable to v
4	G^*	F^*	$\langle j, v\rangle$	As 1
5	F^*	G^*	$\langle j, 1\rangle$	j-th *MS* satisfies filter F
6	F^*	G^*	$\langle j, 0\rangle$	j-th *MS* does not satisfy filter F

and one agent G^*, supervising interaction between agents (in fact, implementing ubiquitous generation by application of rules to multisets generated at previous steps). The agent G^* uses the storage V_S for an accumulation of all generated terminal multisets satisfying the filter F. Also G^* operates the storage V, containing the current set of generated multisets which have not yet been transferred to other agents. The initial state of V is $\{v_0\}$. Agents communicate via a network N.

The described multi-agent system operates according to the mathematical semantics of *FMGs* defined by (3.30)–(3.33) and (3.39). The set of messages which circulate between the agents is represented in Table 4.2.

All messages are couples, the first components of which are the numbers of multisets processed by the agents, and every *MS* has its unique number. Assigning numbers to multisets is performed by the agent G^*. The current maximal number is denoted J. Also the agent G^* uses the variable Z, whose value is a set of couples $\langle j, n\rangle$, where n is the number of agents r_i which until the current moment have not sent to G^* their message with the result of the application of the corresponding rule to the j-th *MS*.

The agent G^* sends the couple $\langle j, v\rangle$ to all agents r_i. After this, G^* joins to the current value of the variable Z a couple $\langle j, l\rangle$. Every agent r_i receiving a message $\langle j, v\rangle$ from the agent G^* tries to apply the rule r_i to the *MS* v. If this application is possible, then r_i sends to G^* the message $\langle j, v'\rangle$, where v' is the multiset which is the result of application of the rule r_i to the *MS* v. Otherwise r_i sends to G^* the message $\langle j, \{\emptyset\}\rangle$. The agent G^*, after receiving a message $\langle j, v\rangle$, where $v \neq \{\emptyset\}$, assigns J a new value $J + 1$ and sends a message $\langle J, v\rangle$ to all agents. Also a couple $\langle J, l\rangle$ is joined to Z, and a couple $\langle j, q\rangle \in Z$ is eliminated from Z, which means at least one rule was applied to the j-th *MS*, so this multiset is non-terminal. If the agent G^* receives a message $\langle j, \{\emptyset\}\rangle$ from an agent r_i, that means the rule r_i was not applied to the j-th multiset, and $\langle j, q\rangle \in Z$ is replaced by $\langle j, q - 1\rangle$. If it now occurs that $\langle j, 0\rangle \in Z$, then this means that no rule was applicable to the j-th *MS*, so it is terminal, and, according to the *FMG's* semantics, it must be filtered. So the agent G^* sends the couple $\langle j, v\rangle$ to the agent F^*, which performs testing, to test whether v satisfies the boundary subfilter $F_{\leq} \subseteq F$. If this testing is successful, the agent F^* sends to the agent G^* the message $\langle j, 1\rangle$, otherwise the message $\langle j, 0\rangle$. In the case $\langle j, 1\rangle$, the couple $\langle j, v\rangle$ is joined to the current value of the variable V_s. After no active agents r_i remain, the agent G^* applies an optimizing subfilter $F_{opt} \subseteq F$ to the aforementioned current value V_s, eliminating from it all multisets which do not satisfy F_{opt}. The final value of the variable V_s is exactly the set of terminal multisets defined by the *FMG* $S = \langle v_0, R, F \rangle$.

As may be seen, *the proposed multi-agent system performs generation and filtration of multisets by parallel operation of all agents belonging to this MAS.*

4.3.2 Reduction of Inter-agent Exchange

As may be seen, there are some evident bottlenecks limiting the speed of multiset generation by the described *FMG AE*. The most essential of them is the massive inter-agent exchange by multiple repeated sets of multiobjects via the *MAS* communication network *N*. To reduce this exchange, it is possible not to send *MOs* to agents r_i whose corresponding rules are not applicable to the multisets occurring in the storage *V*, and this non-applicability may be recognized directly by the agent G^*. If this idea were to be implemented, only those agents r_i which, possibly, might apply corresponding rules to a current *MS*, would receive it. By this cutoff technique, the traffic on the *MAS* communication network might be reduced sharply.

To implement the proposed logic, we shall introduce an auxiliary database *L*, whose elements would be couples $\langle a, \{\langle i_1, n_1 \rangle, \ldots, \langle i_k, n_k \rangle\}\rangle$, where a is the name of an object, and in the couples $\langle i, n \rangle$ forming the set which is the second component of a couple, the integer i is the number of a rule whose left part contains a multiobject $n \cdot a$.

Example 4.13
Let the scheme $R = \{r_1, r_2\}$, where r_1 is

$$\{9 \cdot a\} \to \{10 \cdot b\},$$

and r_2 is

$$\{5 \cdot a, 3 \cdot c\} \to \{7 \cdot b\}.$$

Then

$$L = \{<a, \{<1,\ 9>,\ <2,\ 5>\}>,\ <c, \{<2,\ 3>\}>\}.$$

The database *L* has an internal organization such that there exists an associative index allowing direct selection of any couple $\langle a, w \rangle$ by an object name a. To reduce search in the selected list of couples $\langle i, n \rangle$, it may be created and maintained as a list ordered by increase of multiplicities n. Let $v = \{n_1 \cdot a_1, \ldots, n_m \cdot a_m\}$ be a current multiset processed by an agent G^* and $Q = \{\langle a_1, \leq n_1 \rangle, \ldots, \langle a_m, \leq n_m \rangle\}$ a set of queries to a database *L*, each selecting couples $\left\langle a_i, \left\{ j_1^i, \ \ldots, \ j_{k_i}^i \right\} \right\rangle$ such that $j_p^i \leq n_i$, where $p = \{1, \ldots, k_i\}$. It is clear that in this only case, rules $r_{j_1^i}, \ldots, r_{j_{k_i}^i}$

may have an opportunity to be applied to the *MS* v, and, in total, only those rules may be applied all of whose multiobjects from their left parts have multiplicities not greater than those of the same objects in v. So there is an obvious criterion for the selection of rules which may be applicable to a current multiset v.

Statement 4.4
Let $\left\{\langle a_{i_1},\ N_1\rangle,\ \ldots,\ \left\langle a_{i_p},\ N_p\right\rangle\right\}$ be a set selected from a data base L by a query

$$Q=\{\langle a_1,\ \leq n_1\rangle,\ \ldots,\ \langle a_m,\ \leq n_m\rangle\}, \tag{4.39}$$

corresponding to a multiset $v=\{n_1\cdot a_1,\ldots,n_m\cdot a_m\}$, and $\left\{n_{j_1}\cdot a_{j_1},\ \ldots,\ n_{j_s}\cdot a_{j_s}\right\}$ be the left part of a rule r. Then r may be applicable to v if

$$\left\{a_{j_1},\ \ldots,\ a_{j_s}\right\}\subseteq\left\{a_{i_1},\ \ldots,\ a_{i_p}\right\}. \tag{4.40}$$

As can be seen, the proposed associative organization of the set of the left parts of rules belonging to scheme R permits fast selection of sets of rules which may be applied to a current multiset.

Let us consider *further enhancement* of an *FMG's* application engine, based on multi-agent technology.

First of all, it is evident that it is not necessary to send all multiobjects of a multiset v to which a rule r_i is applicable to the agent r_i, because replacement of the left part of this rule by its right part is a local operation, in the general case affecting a relatively small number of multiobjects in the processed multiset v, while all the other *MOs* remain unchanged. So it is sufficient to send to an agent r only tuples $\langle f_1 n_{i_1},\ \ldots, f_t n_{i_t}\rangle$ of multiplicities of objects $a_{i_1},\ldots,a_{i_t}$ in an *MS* v such that the tuple $A=\langle a_{i_1},\ \ldots,\ a_{i_t}\rangle$ is the lexicographically ordered set of the objects occurring in both the left and the right parts of the rule r, and the signs f_j before the multiplicities n_{i_j} of objects a_{i_j} occurring in the left part of the rule r are " – ", while all others are "+". On receiving such a tuple, the agent r computes a tuple $\langle \boldsymbol{n}_{i_1},\ \ldots,\ \boldsymbol{n}_{i_t}\rangle$, where $\boldsymbol{n}_j=n_j+\left(f_j n_{i_j}\right), j=1,\ldots,t$. (In the general case, objects may belong to both the left and the right parts of a rule r, but this case is simply handled by positioning in the j-th place of a tuple A the number $-n+n'$, where n is the multiplicity of the object a_{i_j} in the left part and n' is its multiplicity in the right part of this rule.)

Example 4.14
Let the multiset

$$v=\{5\cdot a,\ 7\cdot b,\ 10\cdot c,\ 18\cdot d\},$$

and the rule r be

$$\{3\cdot a,\ 2\cdot b\}\rightarrow\{5\cdot c\}.$$

Because the set of lexemes occurring in this rule is ordered lexicographically as $\langle a, b, c \rangle$, so the tuple sent by the agent G^* to agent r would be $\langle 5, 7, 10 \rangle$, and the agent r would send to the agent G^* the tuple $\langle 5-3, 7-2, 10+5 \rangle = \langle 2, 5, 15 \rangle$, and, thus, the result of $v \overset{r}{\Rightarrow} v'$ would be $v' = \{2 \cdot a, 5 \cdot b, 15 \cdot c, 18 \cdot d\}$. As can be seen, the multiplicity of the object d is not communicated to the agent r, because this object does not occur in the rule r.

The proposed technique *further reduces the traffic on the MAS communication network and, thus, the total time of generation of STMSs defined by an FMG.*

Application of the described *MAS*-based generation of *STMSs* is flexible in the following way: due to the granularity of the multigrammatical knowledge representation, local corrections of *FMG* schemes by replacement of one rule by another are easily reflected by a corresponding replacement of only the concerned agents without touching the others. The same may be done with *FMG* filters. *This flexibility allows the simplest implementation of the most practically useful "what-if" regimes of application of MG-centered STS schedulers.*

A further improvement of the above-considered techniques of parallelization of *STMS* generation is possible by regarding *FMGs* as even more "granulated" than described above. This is achieved by consideration of an *FMG* scheme as a *bipartite weighted oriented graph*, whose two non-intersecting sets of nodes are a set of objects A_S and a set A_R of names of rules of this *FMG*. The edge connecting a node-object and a node-name is marked by the integer value which is the multiplicity of this object in the left part of this rule. The edge connecting a node-name and a node-object is marked by the integer value which is the multiplicity of this object in the right part of this rule. *A MAS with the structure induced by this graph in such a way that every node of this graph is implemented by some specific agent allows an "objects-and-rules" parallelism which is more efficient in comparison with the described "rules" one.* A detailed consideration of this approach will be presented in the near future.

4.4 Key Issues of Algorithmics of Filtering Temporal *MGs*

To develop an algorithmics of filtering temporal multiset grammars, i.e., a set of algorithms generating *IS* schedules in a possibly minimal number of generation steps, we shall begin from the simplest case, which is described as follows:

An *IS* is in an initial state with a technological base R_0 and a resource base v_0.

A single order arrives at the *IS* at a time moment n_0.

An order defines a collection of resources q which must be manufactured by the *IS* no later than at a time moment $n > n_0$.

The problem is to generate a set of *IS* schedules satisfying these conditions.

Similar problems have been considered for a long time in scheduling theory under the name "resource-constrained project scheduling" (Klein 1999), and there are many algorithms providing effective solutions of such problems (Abdolshah

2014). However, *the MGF provides the most general formulation of the set of these problems which may be described by appropriate filtering self-generated temporal multigrammars and self-generated TMMGs*. Here we shall consider only a basic case and a core algorithmics whose application provides close to non-redundant search in the solution space. The general case and its comparison with known formulations of the *RCPSP* and relevant algorithmics will be the content of a future publication.

4.4.1 "Bottom-Up" Generation of Schedules

A direct solution of the formulated problem is to use as a foundation the definition of *FTMG's* mathematical semantics—namely, to construct an *FTMG* $S = < v_0, R, F>$, where $F = \{t \leq n\} \cup F_q$ and F_q is a set of boundary conditions defining the collection of resources whose manufacturing is the final objective of an incoming order. If $q = \{n_1 \cdot a_1, \ldots, n_m \cdot a_m\}$, then, evidently,

$$F_q = \{a_1 \geq n_1, \ \ldots, \ a_m \geq n_m\}, \tag{4.41}$$

i.e., at the end of an execution of any schedule, the *IS* resource base should contain no fewer than n_1 units of object-resource $a_1, \ldots, n_m$ units of object-resource a_m, so that all necessary amounts of the listed *ORs* may be delivered to the customer. As can be seen, V_S contains all *MSTs*

$$\{v_A + v_R + \{\, i \cdot t\}\} \tag{4.42}$$

that may be generated by S such that $n_0 \leq i \leq n$, and

$$v_R = \left\{r_1\tau_1, \ \ldots, \ r_p\tau_p\right\}. \tag{4.43}$$

From this, a generated set of schedules $\sum_S$ satisfying the aforementioned conditions 1–3, is

$$\sum_S = \bigcup_{v_A + v_R + \{\, n \cdot t\} \in V_s} \boldsymbol{Sch}(v_R), \tag{4.44}$$

where function *Sch* is defined by (3.48).

However, this direct approach is extremely inefficient and is not applicable in practice. The main reason is that the algorithm generates schedules in a "bottom-up" manner, beginning from the resource base. This creates a lot of redundant branches, which do not lead to necessary schedules satisfying the filter *F*.

4.4.2 "Top-Down" Generation of Schedules

Much more promising seems a "top-down" (or an objective-driven) generation from an order q to a resource base. As may be seen from Sects. 4.1 and 4.2, just such an approach has already been successfully applied to unitary *MGs* and *MMGs*. Let us consider the proposed technique of schedule generation, whose basic idea is to use the same logic of generation as in the definition of the mathematical semantics of *FTMGs*, but applied to the so-called *mirror (or dual) temporal multigrammars.*

The notion of a multiset grammar which is dual to another *MG* was defined in Sect. 3.1.3. Here we shall generalize the notion of duality first to *temporal MGs*.

Let us begin by considering the meaning of a transformation from a "bottom-up" to a "top-down" *MG*-represented *IS*.

Consider an *IS* with a manufacturing technological base, represented by a set of rules *R*, and a resource base v_0. As may be seen, an order $q = \{n_1 \cdot a_1, \ldots, n_m \cdot a_m\}$ may be completed by this *IS* if

$$(\exists v \in V_S)\; q \subseteq v, \tag{4.45}$$

where $S = \; < v_0, R>$. This criterion may be reformulated in the dual mode.

Given an *MG*$-S$ dual to an *MG S*, we shall introduce a criterion dual to (4.45) for the recognition of order feasibility: an order q may be completed by this *IS* if

$$(\exists v \in V_{-S})\; v \subseteq v_0. \tag{4.46}$$

As may be seen, the two criteria are equivalent from the substantial point of view. However, from a computational complexity point of view, the second criterion, allowing an order-driven generation of possible collections of resources which are sufficient for this order completion, is much preferable. This preference is especially evident if the manufacturing technological base of the *IS* consists only of one-output manufacturing devices, which are represented by "black boxes" and rules as in Fig. 4.3.

In this case, generation of multisets is executed in the aforementioned "top-down" manner (from an order to consumed resources) that fully corresponds to the efficient algorithmics of the filtering unitary multiset grammars and the unitary multiset metagrammars described above in Sects. 4.1 and 4.2. A radical reduction of the computational complexity of *TMS* generation in the *FUMGs/UMMGs* case is achieved by the possibility of cutting off unpromising branches at the very early generation steps by restricting the values of boundary conditions as well as the current values of minimized multiplicities.

Let us consider now the case of *filtering temporal multiset grammars*, having in mind the objective to apply to *FTMGs* the developed and described background of the improved algorithmics of *FUMGs/UMMGs*.

By analogy with the notion of duality in the multiset grammars, we shall call a filtering temporal multiset grammar $-S = \; < \boldsymbol{q}, - R, - F>$ *dual to an FTMG*

$$\{n_1 \cdot a_1, ..., n_l \cdot a_l\} \to \{1 \cdot a\}$$

a)   b)

$$\{1 \cdot a\} \to \{n_1 \cdot a_1, ..., n_l \cdot a_l\}$$

c)   d)

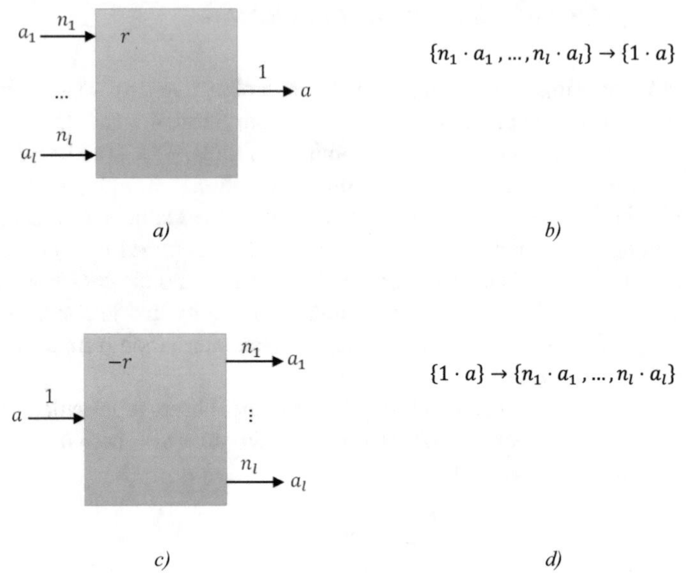

Fig. 4.3 Representation of one-output manufacturing devices: (**a**) "black box"; (**b**) rule; (**c**) mirror "black box"; (**d**) mirror rule

$S = <v_0, R, F>$, where $q = \{n_1 \cdot a_1, ..., n_m \cdot a_m\}$, $v_0 = \{n_1 \cdot a_1, ..., n_l \cdot a_l\}$, where q and v_0 have the same sense as above, $F = \{t \le n\}$, and

$$-R = \{\langle -r; v \to v' + \{-\Delta n \cdot \Delta t\}\rangle \mid$$

$$\langle r; v \to v' + \{\Delta n \cdot \Delta t\}\rangle \in R\}, \tag{4.47}$$

$$-F = \{t \ge -n\} \cup F_{v_0}, \tag{4.48}$$

$$F_{v_0} = \{a \le k \mid k \cdot a \in v_0\} \cup \{a = 0 \mid a \in \overline{A}_S \& a \notin v_0\}. \tag{4.49}$$

Elements of the set $-R$ will be called below *mirror temporal rules* (*MTRs*).

As can be seen, an application of a dual *FTMG* $-S$ generates temporal multisets $v_A + v_R + \{k \cdot t\}$ with negative multiplicities k. Due to (4.47) this generation is executed upon negative time scales, and application of any mirror temporal rule leads to the appearance of an interval $[n_i + \Delta n, n_i]$, where both n_i and Δn are negative numbers in the time scale corresponding to this *MTR* (Fig. 4.4). As can be seen, "individual pseudoschedules" (*IPSs*) are generated, containing "mirror" operation cycles of manufacturing devices.

A cutoff boundary condition $t \ge -n$, thus, selects those *MSTs* corresponding to schedules whose duration is no more than n. The only question now is to transform *IPSs* into schedules of *MDs*. As it is not difficult to see, any multiobject $1 \cdot r[n_1, l_1]...$ $[n_k, l_k]$ belonging to a multiset v_R would be replaced by $1 \cdot r[-n_1, -l_1]...[-n_k, -l_k]$, no matter that the left bounds of all these intervals are greater than the right ones.

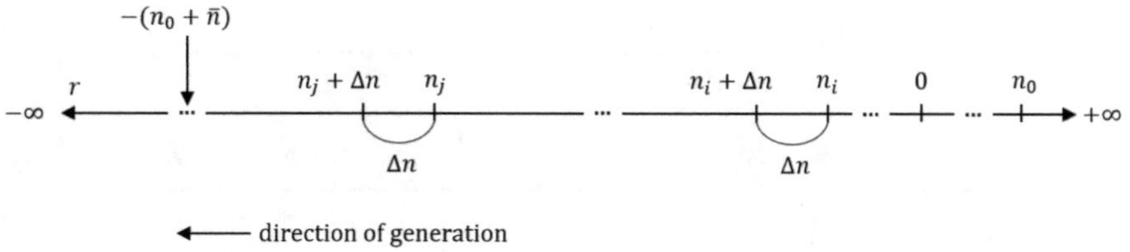

Fig. 4.4 Application of a mirror temporal rule with a multiplicity $\Delta n < 0$ of the object Δt

In fact, operation of an industrial system may begin at a moment n_0 by an application of a manufacturing technological base to an initial resource base v_0, so the kernel of the *FTMGS* would be $v_0 + \{n_0 \cdot t\}$, just as the kernel of the *FTMG* $-S$ would be $\boldsymbol{q} + \{n_0 \cdot t\}$. In this general case, evidently,

$$F_{v_0} = \{a \leq k \mid k \cdot a \in v_0\} \cup \{a = 0 \mid a \in \overline{A}_S \& a \notin v_0\}, \tag{4.49}$$

$$-F = \{t \geq -n + n_0\} \cup F_{v_0}. \tag{4.50}$$

The meaning of the described reverse transformation in the general case is illustrated by Fig. 4.5.

The correctness of the described logic of top-down generation of schedules is verified by Statement 4.5, in which the following relations (4.51)–(4.55) are used. Let

$$\bar{V}_S = \{ v_1, \ldots, v_l\}, \tag{4.51}$$

and, if so,

$$-\bar{V}_S = \{-v_1, \ldots, -v_l\}, \tag{4.52}$$

where

$$-\{v_A + v_R + \{i \cdot t\}\} = \{v_A + (-v_R) + \{i \cdot t\}\}, \tag{4.53}$$

$$-\{r_1\tau_1, \ldots, r_p\tau_p\} = \{r_1(-\tau_1), \ldots, r_p(-\tau_p)\}, \tag{4.54}$$

$$-[n_1, l_1]\ldots[n_k, l_k] = [-n_1, -l_1]\ldots[-n_k, -l_k]. \tag{4.55}$$

(Here in (4.54), brackets are used to denote the result of inverting individual schedules of *MDs*, not as dividers inside strings containing names of *MDs* and the aforementioned schedules.)

Statement 4.5 Let $S = \, < v_0, R, F>$ be a filtering temporal multiset grammar, and $-S = \, < \boldsymbol{q}, -R, -F>$ an *FTMG*, dual to *S*. Then

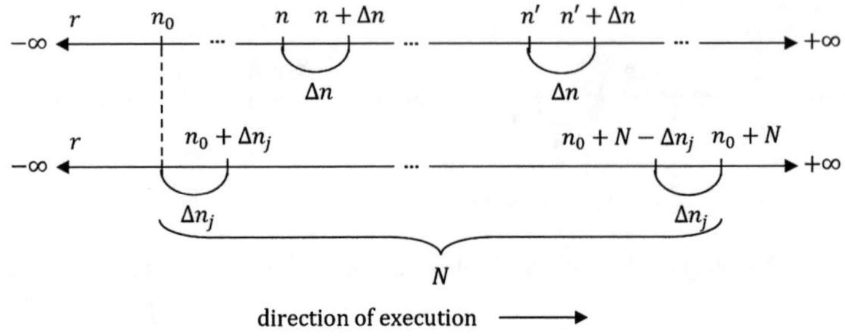

Fig. 4.5 Reverse transformation of individual pseudoschedules into schedules of manufacturing devices

$$\bar{V}_S = -\bar{V}_{-S} \tag{4.56}$$

The main result of the proposed technique for generation of a set of *ISs* schedules is the *possibility of applying the same algorithmics of MST generation as defined by the relations of the semantics of FTMGs*. In other words, there is *no necessity to develop special "top-down" algorithmics, when it is sufficient to use the "bottom-up" one with a minimal extension*. The substantial difference between them is that "bottom-up" generation operates on a resource base, while "top-down" operates on orders.

As is clear, the proposed approach is especially efficient for one-output manufacturing devices, by reason of the early cutoffs during generation allowed by the monotonic growth of values of multiplicities in generated *MSTs*. This case is very similar to the *FUMG/UMMG* one, where the aforementioned *monotonic growth of values of multiplicities in generated MSs along with information about upper bounds of these multiplicities, taken from filters, is the basis of a radical reduction of the number of redundant generation steps*.

However, there is one additional nuance concerning application of mirror temporal rules. A condition which must be satisfied for a direct application of a temporal rule is rather evident: the current resource base must be sufficient for activation of a manufacturing device, which may be explicated in multiset notation as $v \subseteq v_A$, where

$$v \to v' + \{\Delta n \cdot \Delta t\} \tag{4.57}$$

is a *TR* representing this *MD* and v_A is the aforementioned current *RB*. If we consider application of an *MTR*

$$v' \to v + \{-\Delta n \cdot \Delta t\}, \tag{4.58}$$

it is clear that the condition which must be satisfied for its application is more "soft." Namely, the *MTR* may be applied if in the currently created order (the analog of the current *RB* in "bottom-up" generation) there is at least one object that will be created by an application of this manufacturing device. Therefore, the activation condition for *MTRs* would not be $v \subseteq v_A$, but much "softer":

$$v \cap v_A \neq \{\varnothing\}. \tag{4.59}$$

From this, definition (3.53) of the function $\boldsymbol{\pi}$ for "top-down" generation becomes as follows:

$$\begin{aligned}
&\boldsymbol{\pi}(v_A + v_R + \{n \cdot t\}) = \\
&\{v_A \ - v + v_R \ - \{1 \cdot r\tau\} + \{1 \cdot r\tau[n, \ l]\} + \{n \cdot t\} \mid \\
&r\tau \in v_R \ \& \langle r; v \rightarrow v' + \{\Delta n \cdot \Delta t\}\rangle \in R \& v \cap v_A \neq \{\varnothing\} \& \\
&\boldsymbol{T}(\tau) \leq n \& l = n + \Delta n\}.
\end{aligned} \tag{4.60}$$

As is evident, in the case of one-output *MDs*, (3.53) and (4.60) are equivalent, but if $|v| > 1$, then there would be some "offset," including objects which were produced by the manufacturing device but not used at the current step of schedule creation. This "offset" is, obviously, $v - (v \cap v_A)$, and the question is what to do with it. Let us recall that during top-down generation, orders are created in a reverse mode, while the "offset" is a part of the resource base. So during real completion of the order, all these "offsets" would be considered to be included in the resource base.

In fact, an "offset" appears also in "bottom-up" generation: it includes multiobjects which appear in the generated multisets but do not appear in orders, which are submultisets of these *MSs*.

To finish this section, we may affirm that the algorithmics of *FUMGs/UMMGs* applied in the case of *ISs* whose *MTBs* include only one-output *MDs* may be additionally improved by *cutoffs implemented by the use of multiplicities of time markers of the generated MSTs*.

Now we have a recursive equational representation of an improved operational semantics of filtering temporal multiset grammars which may be used as a foundation on which to develop their procedural representations and is suitable for all possible special cases of *FTMGs* and all hardware environments used for *FTMG*-based *IS* schedulers.

4.4.3 Terminalization

The last enhancement of the algorithmics of *MGs* considered here concerns the opportunity to reduce the time to complete orders (and, in fact, the computational complexity of schedule creation) by use in rules (*URs*, *UMRs*, *TRs*) of *objects representing object-resources which might be already manufactured and stored in a resource base before an order arrives*. If any such an *OR* does exist, there is no need to construct and implement a manufacturing chain to produce it beginning from primary object-resources present in the *RB*; instead we can simply *include such ORs in the generated temporal multiset* as a result of any executed chain, without its creation and implementation.

Let us redefine a resource base $v_0 = \{n_1 \cdot a_1, \ldots, n_l \cdot a_l\}$ as the sum of two non-intersecting collections

$$v_0 = v_0^p + v_0^s, \tag{4.61}$$

where v_0^p is a collection of *primary ORs* used for manufacturing *secondary ORs*, which may be *produced from primary ORs and other secondary ORs*. The collection of secondary *ORs* in the *RB* v_0 is

$$v_0^s = \{n_1 \cdot (*a_1), \ \ldots, n_k \cdot (*a_k)\}, \tag{4.62}$$

where $*$ is an auxiliary symbol used as a selecting prefix in names of secondary object-resources occurring in an *RB*. The names of primary objects do not begin with the symbol $*$, so by renaming secondary *ORs* in this way, we distinguish them from primary *ORs*.

To provide an opportunity to use secondary *ORs* stored in a resource base in manufacturing chains, we shall join to the set of temporal rules *R* a set of so-called terminalizing TRs:

$$R_* = \cup_{(*a)\epsilon\beta\left(v_0^s\right)} \{\langle *a; \{1 \cdot (*a)\} \rightarrow \{1 \cdot a, \ \Delta n \cdot \Delta t\}\rangle \}, \tag{4.63}$$

where the name of a terminalizing temporal rule is similar to the name of the relevant secondary *OR* but differs from it by an absence of brackets (if they are equal, then the considered *FTMG* would be self-generating, which is incorrect), and Δn is the time interval necessary to take one *OR a*, by name ($*$ *a*), from the *RB* (in the simplest case, where $\Delta n = 0$, this operation is performed immediately). So any secondary *OR* *a* may be involved in a manufacturing process in two ways: by its production (this way is defined by *TRs* already present in the set *R*) or by its extraction from the resource base, if it appears in the *RB* (this way is defined by terminalizing *TRs*, joined to the set *R*).

We shall use terminalization not only as a simple and flexible extension of *FTMGs* but also as a similar extension of *UMGs* (see Sect. 8.1.1).

Example 4.15
Let the *IS MTB* contain three manufacturing devices represented by the following temporal rules:

$$< r_1; \{2 \cdot a, \ 1 \cdot b\} \rightarrow \{1 \cdot c, \ 1 \cdot mnt\} >,$$

$$< r_2; \{3 \cdot a, \ 1 \cdot c\} \rightarrow \{1 \cdot d, \ 2 \cdot mnt\} >,$$

$$< r_3; \{3 \cdot a, \ 2 \cdot c\} \rightarrow \{2 \cdot e, \ 5 \cdot mnt\} >,$$

and the resource base of this *IS* be $v_0 = \{6 \cdot a, 9 \cdot b\}$. Let the *RB* contain three secondary *ORs* c and ten secondary *ORs* d. So according to (4.61) and (4.62), this *RB* is represented by the multiset

$v_0 = \{6 \cdot a, 9 \cdot b, 3 \cdot \ * c, 10 \cdot \ * d\}$and additional temporal rules representing extraction of one secondary object c from the *RB* in 1 minute and one secondary object d from the *RB* in 2 minutes are

$$< *c; \{1 \cdot *c\} \to \{1 \cdot c,\ 1 \cdot mnt\} >,$$

$$< *d; \{1 \cdot *d\} \to \{1 \cdot d,\ 2 \cdot mnt\} >.$$

So the *TMG* representing the considered *IS* will permit creation of schedules including not only production of *ORs* c, d, and e but also extraction of already produced *ORs* c and d.

The more complicated case of the improved algorithmics of filtering *self-generating* multiset grammars will be considered in future publications.

Now we finish the consideration of two branches of the *MGF*—unitary *MGs* and temporal *MGs*—which are similar in some sense to the mathematical tools applied in two different segments of the classical mechanics, statics and dynamics, and move to the applications of the *MGF*.

Chapter 5
Resource-Consuming Sociotechnological Systems

Basic features of resource-consuming sociotechnological systems were described in Sect. 2.2. Here we shall consider a multigrammatical representation of *RCSs* and its application to *planning*, which is implemented via their analysis as well as via synthesis of needed systems. Unitary multiset grammars, applied for this purpose, provide an understandable and natural representation of *RCSs*' organizational structures, their equipment, and the resources necessary for their operation, as well as of interconnections between all these components of *RCSs*. The techniques of *UMG* application described below in Sect. 5.1 cover not only strictly hierarchical but also non-hierarchical *RCSs*.

The following tasks, usually solved by an *RCS*'*s* staff and which we wish to solve by various applications of different types of *UMGs/UMMGs*, are discussed in Sect. 5.2: assessment of sufficiency of amounts of resources available to an *RCS*, of the cost of an additional supply in the case of their insufficiency, and of the maximal period of time of *RCS* operation given available amounts of resources; evaluation of the increase and/or decrease of total amounts of various resources consumed by a system, in the case that its organizational structure and assigned equipment are transformed; and assessment of the feasibility of some decisions changing the system's structure and equipment. All the listed tasks are associated with the area of analysis (assessment) of sociotechnological systems. From the other side, the capabilities of filtering unitary multiset grammars and unitary multiset metagrammars provide an easy and natural representation of tasks commonly referred to as the area of *STS* synthesis; the relevant technique is considered in Sect. 5.3.

I. A. Sheremet, *Multigrammatical Framework for Knowledge-Based Digital Economy*, https://doi.org/10.1007/978-3-031-13858-4_5

5.1 Multigrammatical Representation of *RCSs*

Following the methodology of step-by-step increases in the sophistication of the studied *STSs*, we shall consider representations of *RCSs* in order of increasing complexity:

1. Organizational (*O*-)
2. Organizational-technological (*OT*-)
3. Resource-consuming organizational-technological (*ROT*-)

Below we shall refer to all listed representations of *RCSs* simply as *O*-, *OT*-, and *ROT*-systems (without repeating that they are resource-consuming).

5.1.1 Organizational Systems

Any *O-system* is identified by a hierarchical structure which reflects the subordination of its elements (organizational units, *OUs*) and their mutual nesting, if we speak about subsystems (down to elementary *OUs* consisting only of individual positions).

Example 5.1
Consider a company consisting of a director, three identical departments, and two laboratories. Each department, in turn, consists of its head and three laboratories, and each laboratory also consists of its head, three senior analysts, six analysts, and nine junior analysts. This organizational system may be represented by a tree, as in Fig. 5.1.

The corresponding *UMG S*= <{1· (*company*) }, *R*> is such that the scheme *R* includes the following unitary rules:

$$
\begin{aligned}
r_1 &: (\text{company}) \to 1\cdot(\text{director}), 3\cdot(\text{department}), 2\cdot(\text{laboratory}),\\
r_2 &: (\text{department}) \to 1\cdot(\text{head of department}), 3\cdot(\text{laboratory}),\\
r_3 &: (\text{laboratory}) \to 1\cdot(\text{head of laboratory}), 3\cdot(\text{senior analyst}),\\
&\qquad 6\cdot(\textit{analyst}), 9\cdot(\textit{junior analyst}).
\end{aligned}
$$

Evidently, the set of terminal multisets generated by this *UMG* is

$$
\begin{aligned}
\overline{V}_s = \{\{&1\cdot(\text{director}), 3\cdot(\text{head of department}), 11\cdot(\text{head of laboratory}),\\
&33\cdot(\text{senior analyst}), 66\cdot(\text{analyst}), 99\cdot(\text{junior analyst})\}\}.
\end{aligned}
$$

As can be seen, a multigrammatical representation differs from a tree one by its implicitness, which makes it much more compact (*there is no necessity for multiple repeats of the same subtrees if they re-occur*) and flexible (*it is as simple as possible*

r_1: (*company*)→ 1·(*director*), 3·(*department*), 2·(*laboratory*),

r_2: (*department*)→ 1·(*head of department*), 3·(*laboratory*),

r_3: (*laboratory*)→ 1·(*head of laboratory*), 3·(*senior analyst*),

6· (*analyst*), 9·(*junior analyst*).

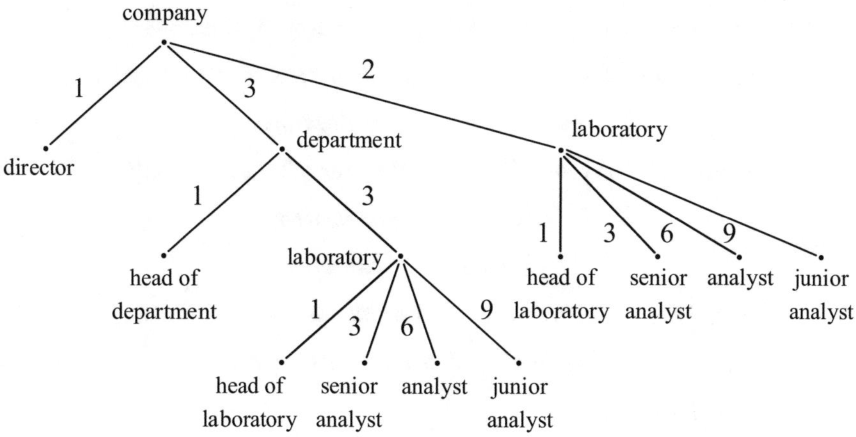

Fig. 5.1 A tree representation of an organizational system

to change the organizational structure of the company and any of its divisions by replacement of the appropriate unitary rules).

5.1.2 Organizational-Technological Systems

OT-systems are a further extension of *O*-systems by knowledge about the technological equipment (various devices and complexes) which is assigned to the organizational elements (down to the individual positions) and used (applied) by them for their operation. As was already said in Sect. 2.2, the number and types of such devices are defined by the proper supply standards. This is illustrated by the following example.

Example 5.2
Suppose the director of the company from the previous example is supplied with an auto, an i-phone, and a workstation, every head of department with an i-phone and a workstation, and the rest of the company personnel with workstations (one for every position). Also there is one server and one router enabling operation of the company network, as well as switches integrating the subnetworks of laboratories. All these devices are the property of the company, and knowledge about their assignment is

joined to the *UMG* representing this *O*-system by the obvious changes to the corresponding unitary rules and by joining to the scheme *R* some additional *URs*:

$$r_1 : (company) \to 1 \cdot (director), 3 \cdot (department), 2 \cdot (laboratory),$$
$$1 \cdot (company\ network\ equipment),$$
$$r_2 : (director) \to 1 \cdot (position\ director), 1 \cdot (equipment\ of\ director),$$
$$r_3 : (department) \to 1 \cdot (head\ of\ department), 3 \cdot (laboratory),$$
$$r_4 : (head\ of\ department) \to 1 \cdot (position\ head\ of\ department),$$
$$1 \cdot (equipment\ of\ head\ of\ department),$$
$$r_5 : (laboratory) \to 1 \cdot (head\ of\ laboratory), 3 \cdot (senior\ analyst),$$
$$6 \cdot (analyst), 9 \cdot (junior\ analyst),$$
$$1 \cdot (equipment\ of\ laboratory),$$
$$r_6 : (head\ of\ laboratory) \to 1 \cdot (position\ head\ of\ laboratory),$$
$$1 \cdot (equipment\ of\ head\ of\ laboratory),$$
$$r_7 : (senior\ analyst)$$
$$\to 1 \cdot (position\ senior\ analyst), 1 \cdot (equipment\ of\ senior\ analyst),$$
$$r_8 : (analyst) \to 1 \cdot (position\ analyst), 1 \cdot (equipment\ of\ analyst),$$
$$r_9 : (junior\ analyst)$$
$$\to 1 \cdot (position\ junior\ analyst), 1 \cdot (equipment\ of\ junior\ analyst),$$
$$r_{10} : (equipment\ of\ director) \to 1 \cdot (auto), 1 \cdot (i-phone), 1 \cdot (workstation),$$
$$r_{11} : (company\ network\ equipment) \to 1 \cdot (server), 1 \cdot (router),$$
$$r_{12} : (equipment\ of\ head\ of\ department) \to 1 \cdot (i-phone), 1 \cdot (workstation),$$
$$r_{13} : (equipment\ of\ laboratory) \to 1 \cdot (switch),$$
$$r_{14} : (equipment\ of\ head\ of\ laboratory) \to 1 \cdot (workstation),$$
$$r_{15} : (equipment\ of\ senior\ analyst) \to 1 \cdot (workstation),$$
$$r_{16} : (equipment\ of\ analyst) \to 1 \cdot (workstation),$$
$$r_{17} : (equipment\ of\ junior\ analyst) \to 1 \cdot (workstation).$$

As may be seen, this *UMG* defines the following one-element set of terminal multisets:

$$\overline{V}_s = \{\{1 \cdot (position\ director), 3 \cdot (position\ head\ of\ department),$$

$$11 \cdot (position\ head\ of\ laboratory), 33 \cdot (position\ senior\ analyst),$$

$$66 \cdot (position\ analyst), 99 \cdot (position\ junior\ analyst),$$

$$1 \cdot (auto),$$

$$4 \cdot (i - phone), 213 \cdot (workstation), 1 \cdot (server), 1 \cdot (router), 11 \cdot (switch)\}\}.$$

Thus, the proposed multigrammatical representation of an *OT*-system allows *natural recursive top-down computation of aggregated and at the same time detailed data* (in the form of a terminal multiset), including total numbers of all positions (as was already true in the case of an *O*-system) as well as total numbers of all types of technical equipment which is the property of the company (and this is a new feature compared to *O*-systems).

5.1.3 Resource-Consuming Organizational-Technological Systems

Let us note that every *OT*-system during its operation utilizes various resources necessary for both its human and technological components. An *ROT-system* also may easily be described by the relevant unitary multigrammar, to establish a time interval of resource consumption and to create a *UMG* whose unitary rules contain appropriate knowledge. In this way, it is quite simple to evaluate the resource amounts necessary for *ROT*-system operation during a time interval of any length. Let's illustrate this by the following example.

Example 5.3
Let us consider the *OT*-system from Example 5.2 and assume that the resources consumed by it are employees' monthly salaries, liters of petrol consumed by the director's auto, the electrical energy consumed by the computing equipment (the server, the router, switches, and workstations), as well as payments for the computers' and i-phones' traffic to the appropriate network providers. Electrical energy also is bought from the supplier. Sȯ we shall obtain the *ROT*-system whose consumption is related to a 1-month cycle of operation.

The scheme *R* of the relevant *UMG* $S = <\{1 \cdot (company)\}, R>$ contains, in addition to the *URs* in the previous example, the following unitary rules:

$$r_{18} : (position\ director) \rightarrow 10{,}000 \cdot (eur),$$

$$r_{19} : (position\ head\ of\ department) \rightarrow 5000 \cdot (eur),$$

$$r_{20} : (position\ head\ of\ laboratory) \rightarrow 3000 \cdot (eur),$$

$$r_{21} : (position\ senior\ analyst) \rightarrow 2000 \cdot (eur),$$

$$r_{22} : (position\ analyst) \rightarrow 1000 \cdot (eur),$$
$$r_{23} : (position\ junior\ analyst) \rightarrow 500 \cdot (eur),$$
$$r_{24} : (auto) \rightarrow 300 \cdot (l),$$
$$r_{25} : (server) \rightarrow 1000 \cdot (kwh),$$
$$r_{26} : (router) \rightarrow 100 \cdot (kwh),$$
$$r_{27} : (switch) \rightarrow 50 \cdot (kwh),$$
$$r_{28} : (workstation) \rightarrow 50 \cdot (kwh),$$
$$r_{29} : (i - phone) \rightarrow 200 \cdot (Mb).$$

So the total amount of resources consumed by this *ROT*-system may be evaluated by the one-element set of terminal multisets generated by this *UMG:*

$$\overline{V}_s = \{\{239{,}500 \cdot (eur),\ 300 \cdot (l),\ 12{,}300 \cdot (kwh),\ 800 \cdot (Mb)\}\}.$$

If the cost of 1 l of petrol is 1 euro, the cost of 1 kWh h of electrical energy is 1 euro, and the cost of one megabyte of network traffic is 1 eurocent, and it is necessary to evaluate the total company expenses for a month, it is sufficient to join to the scheme *R* of this *UMG* three additional unitary rules:

$$r_{30} : (l) \rightarrow 1 \cdot (eur),$$
$$r_{31} : (kwh) \rightarrow 1 \cdot (eur),$$
$$r_{32} : (Mb) \rightarrow 0.01 \cdot (eur).$$

From this

$$\overline{V}_s = \{\{252{,}108 \cdot (eur)\}\}.$$

This means the company's total expenses per month are 252,108 euros.

5.1.4 Alternative Ways to Apply UMGs *to the Representation of* RCSs

As may be seen from the above simple examples, *UMGs* are a perfect tool for representation of *RCS* organizational structures, their equipment, the resources necessary for their operation, and the nexus between any specific *RCS* components.

Of course, there may be different approaches to the application of *UMGs* to such *RCS* modeling. For example, technological equipment may be assigned to *RCS* positions by unitary rules like

$$(p) \rightarrow n_1 \cdot (d_1), \ldots, n_m \cdot (d_m), \tag{5.1}$$

where (p) is a position, while (d_1), . . ., (d_m) are devices assigned to it. In this case the *UMG*'s objects like (p) become non-terminals, and thus the terminal multiset generated by this *UMG* would not contain multiobjects like $n \cdot (p)$. This technique simplifies the transfer from the representation of *O*-systems to the representation of *OT*-systems: as is shown in (Sheremet 2010, 2011a), the application of *UMGs* in such a way is the basis of the following *equation, connecting the organizational and technological components of any OT-system*:

$$R_{OT} = R_O \cup R_T, \tag{5.2}$$

where R_O is a scheme representing the *O*-system, R_T is a set of *URs* representing an assignment of technical equipment to organizational positions, and R_{OT} is a scheme representing the *OT*-system. The same technique, in fact, was used in Example 5.3, where the set of *URs* denoted R_C (the lower index means "consumption") was joined to the scheme R_{OT}; thus in the general case (Sheremet 2010, 2011a),

$$R_{ROT} = R_{OT} \cup R_C = R_O \cup R_T \cup R_C. \tag{5.3}$$

Note that, due to the non-variativity of unitary multigrammars representing *O*-, *OT*-, and *ROT*-systems, these *UMGs* are 1-multigrammars: they generate only *one-element sets of terminal multisets*.

5.1.5 Application of UMGs *to Representation of Non-hierarchical* RCSs

It may seem that all that was said and illustrated above in this section concerns only corporate or governmental *RCSs* with strictly hierarchical structures. To demonstrate the applicability of a *UMG*-based representation of *RCSs* to all other resource-consuming systems, let us consider the case of some territory where a human socium lives and works, while at the same time persons belonging to this socium are not totally hierarchically ordered. In this case, its decomposition may be initiated from the unitary rule

$$(socium) \rightarrow 1\ (systems), 1 \cdot (persons), \tag{5.4}$$

where the non-terminal object (*systems*) is a denotation of the set of all business and state *RCSs*, while the non-terminal object (*persons*) is, similarly, a denotation of the set of all individuals not belonging to any of the aforementioned systems. The object (*systems*) is the header of a single unitary rule

$$(\text{systems}) \rightarrow m_1 \cdot (sys_1), \ldots, m_k \cdot (sys_k), \tag{5.5}$$

which means there are m_1 systems (of type) sys_1, …, m_k systems (of type) sys_k ; if any sys_i of sys_1, …, sys_k is unique, then $m_i = 1$.

Any system may be decomposed into its subsystems, down to the individual positions and *multiple access technological systems* (*MATS*) used by the personnel of this system and its subsystems, according to the techniques illustrated by Examples 5.1–5.3, where the *MATSs* are the server, the router, and the switches.

Concerning the multiobject (*persons*) from the body of the *UR* (5.4), it may be appropriate that the set of all individuals of the considered *RCS* (let it be for concreteness, some living house) may be divided into subsets (classes), each including individuals with similar sets of personal technical devices and consumed resources. This may be represented by a unitary rule

$$(person) \rightarrow n_1 \cdot (person_1), \ldots, n_l \cdot (person_l), \tag{5.6}$$

and unitary rules

$$(person_i) \rightarrow k_{i,1} \cdot (res_{i,1}), \ldots, k_{i,Ki} \cdot (res_{i,Ki}), m_{i,1} \cdot (dev_{i,1}), \ldots, m_{i,Li} \cdot (dev_{i,Li}), \tag{5.7}$$

which means each person belonging to the i-th class consumes $k_{i,\ 1}, \ldots, k_{i,\ Ki}$ units of resources $res_{i,\ 1}, \ldots, res_{i,\ Ki}$ and uses $m_{i,\ 1}, \ldots, m_{i,\ Li}$ devices $dev_{i,\ 1}, \ldots, dev_{i,\ Li}$, respectively, during the predefined period of time.

Henceforth, the resources $res_{i,\ j}$ consumed by $person_i$ may be detailed in the ordinary way described above by Example 5.3, i.e., by including in a *UMG* representing this *RCS* proper unitary rules defining amounts of food, water, electrical energy etc., required by the i-th *abstract person*.

It may be impossible in the general case to define a single collection of resources and devices used by any such "abstract individual," so in the general case, there may be alternative collections, and hence, the scheme of a *UMG* representing such an *RCS* may contain more than one *UR* with the same header ($person_i$) (Sheremet 2019c). Thus such a *UMG* may be variative and, hence, may be not a 1-*MG* but a k-multigrammar.

After introducing the basic techniques of multigrammatical representation of resource-consuming systems, we turn next to the consideration of various tasks of *RCS* analysis, listed in Sect. 2.2, which are relevant for such systems' managers and designers.

5.2 *RCS* Analysis

The everyday work of the management of any *RCS* includes online assessment and evaluation of the amounts of resources consumed by the system and its subsystems. *The larger the RCS is, the more difficult this work is for managers, and the more useful would be a knowledge-based software engine with multigrammatical knowledge representation, providing online analytics in a "what-if" mode.*

5.2.1 *Assessment of Sufficiency of Amounts of Resources Available to an* RCS

The *first* task usually considered by *RCS* managers is to assess whether the amounts of resources available to the system are sufficient for its normal (regular) operation. Let S=<{1·a_0}, R> be a 1–*MG* representing an *RCS* a_0, $\overline{V}_s = \{ v\}$, and

$$\overline{v} = \{n_1 \cdot a_1, \ \ldots, n_m \cdot a_m\}, \tag{5.8}$$

be a multiset representing amounts of the available resources (i.e., the *RCS* resource base). Evidently, such a system will operate normally if

$$v \subseteq \overline{v}, \tag{5.9}$$

i.e., the amounts of resources consumed by the *RCS* are not greater than the amounts of resources existing in its resource base.

Example 5.4
Let us consider the *ROT*-system (*company*) from Example 5.3, and let the resource base of this *RCS* be represented by the multiset $\overline{v} = \{300{,}000 \cdot (eur)\}$. Because -$v = \{252{,}108 \cdot (eur)\}$, and thus $v \subseteq \overline{v}$, this resource base is sufficient for the normal operation of this system during 1 month.

To assess the sufficiency of an *RCS* resource base during a time interval $n \cdot \Delta t$, where Δt is the basic interval of resource consumption chosen during the creation of the 1–*MG* S, it is reasonable to use a 1–*MG* S' =<{1·a'_0},R'>, where

$$R' = R \cdot \cup \{r\}, \tag{5.10}$$

and r is a unitary rule with a multiplicity-constant n

$$a'_0 \rightarrow n \cdot a_0. \tag{5.11}$$

As may be seen, in this case really

$$\overline{V}_s = \{\, n \times v\}. \tag{5.12}$$

The condition of *RB* sufficiency in this more general case is the same as (5.9); of course, this condition refers to the *RB* during a time period $n \cdot \Delta t$.

Example 5.5
Let us consider the *ROT*-system from Example 5.3; its interval of operation, as announced, is 1 month. Let there be available the sum of 3.6 million euros for the next year, the planning for which is a task to be solved by the system management. According to (5.10)–(5.12), to calculate similar expenses for the 1-year period, it is sufficient to apply the *UMG* $S' = < \{1 \cdot (company\text{-}a\text{-}year)\}, R' >$, where the scheme R' contains all unitary rules in the scheme *R* of the *UMG S*, and also the additional *UR*

$$r_0 : (company - a - year) \rightarrow 12 \cdot (company).$$

As may be seen,

$$\overline{V}_{s'} = \{\{3{,}025{,}296 \cdot (eur)\}\}.$$

So, because $\{\{3{,}025{,}296 \cdot (eur)\}\} \subseteq \{\{3{,}600{,}000 \cdot (eur)\}\}$, the resource base is sufficient for the normal operation of this system during 1 year.

5.2.2 Assessment of the Cost of an Additional Supply

In the case when an *RB* is not sufficient for normal operation of an *RCS*, it is evident that the missing resources which it is necessary to acquire are represented by the multiset

$$\Delta \upsilon = \upsilon - \overline{\upsilon}, \tag{5.13}$$

Let us consider the *task of acquiring these resources*.

We shall assume that a market allowing this acquisition is represented by a set of unitary rules *R*, each *UR* having the form

$$a \rightarrow 1 \cdot (a\text{—}splr), m \cdot (e\text{—}splr), m \cdot e, \tag{5.14}$$

where *a* is a resource name, *splr* the name of a supplier, and the cost of one item *a* is *m* units of currency *e*. Under antimonopoly regulations, the *i*-th supplier ($i=1,\ldots, K$) has permission to provide resources to a customer at a total cost no more than some maximal value M_i. Each supplier offers several types of resources in various amounts, and there is competition between the suppliers by reason of the intersection of the sets of types of items offered by them. The *RCS* resource base includes some money, represented by a multiobject $M \cdot e$ (M units of currency e). All this money may be spent for the acquisition of the missing resources, but the *RCS* management's objective is to minimize the acquisition cost.

This task may be solved by application of a *filtering unitary multiset grammar S* $= < \Delta v, R, F>$, where the filter F includes the following conditions:

$$e \leq M, \tag{5.15}$$

$$e = min, \tag{5.16}$$

$$\begin{gathered} (e\text{—}splr)_1 \leq M_1, \\ \ldots \\ (e\text{—}splr)_K \leq M_K. \end{gathered} \tag{5.17}$$

As can be seen, the result of application of the *UMG S* is a set of terminal multisets each having the form

$$\left\{m \cdot e, l_{i_1} \cdot (a\text{—}splr)_{i_1}, \ldots, l_{i_N} \cdot (a\text{—}splr)_{i_N}, m_1 \cdot (e\text{—}splr)_1, \ldots, m_K \cdot (e\text{—}splr)_K\right\}, \tag{5.18}$$

where m is the total cost of acquisition (minimal over all possible variants), l_{i_j} is the number of items a acquired from the i_j-th supplier (in the general case some i_j and i_k may be equal), and p_i is the money which the *RCS* must pay to the i-th supplier for all resources acquired from that supplier (i.e., p_i is in fact the income of the i-th supplier). Obviously, m is the sum of $m_1, \ldots, m_K$. In the general case, the set $\overline{V}_s$ may contain more than one multiset (5.18), but all these multisets will include a multiobject $m \cdot e$ with minimal value among all generated *TMSs*, as defined by (5.16).

5.2.3 *Assessment of the Maximal Period of Time of* RCS *Operation Given Available Amounts of Resources*

Let us consider now an inverse task: what would be the longest period during which the *RCS* would be able to operate with the resource base available to it. This task is simply represented by an application of *unitary multiset metagrammars*.

As in the previous section, let $S = <\{1 \cdot a_0\}, R>$ be a 1–*MG* representing an *RCS* a_0 and

$$\overline{v} = \{n_1 \cdot a_1, \ldots, n_m \cdot a_m\} \tag{5.19}$$

be a multiset representing the *RCS's* resource base. We shall construct a unitary multiset metagrammar $S' = <\{1 \cdot a'_0\}, R, F>$ such that

$$R' = R \cdot \cup \{r\}, \tag{5.20}$$

where r is the unitary metarule

$$a'_0 \rightarrow n \cdot a_0 \tag{5.21}$$

with the multiplicity variable n representing the required period as a number of basic intervals Δt. The filter F will contain two conditions concerning n:

$$n \leq \mathbf{N} \tag{5.22}$$

and

$$n = max, \tag{5.23}$$

where $\mathbf{N}$ is some fixed maximal possible value of n. Also F will contain m boundary conditions

$$a_i \leq n_i, \tag{5.24}$$

where $i=1,\ldots, m$, and each of these conditions provides a relevant restriction on the amount of a resource established by the available *RB*. As can be seen, the result of application of the constructed *UMMG* S' in a successful case will be a one-element set

$$\bar{V}_S = \{v \ + \{M \cdot n\}\}, \tag{5.25}$$

where M is the required value of the variable n, while $v \subseteq \overline{v}$ is the collection of resources necessary for *RCS* operation during n' basic intervals. Otherwise, i.e., when the available *RB* is not sufficient for *RCS* operation for even one such interval, the result will be the empty set.

Example 5.6

Consider the *ROT*-system from Example 5.3. Suppose there is available the sum of 7.5 million euros, and the system management requires an estimate of how many months this *RCS* will be able to operate given such a resource base. According to (5.20)–(5.25), to solve this task, it is sufficient to apply the *UMMG* $S' = <\{1 \cdot (company\text{-}time)\}, R', F>$, whose scheme R' contains all unitary rules belonging to the scheme R of the *UMG S* and also the additional unitary metarule

$$r_0 : (company - time) \rightarrow n \cdot (company),$$

while the filter *F* includes the following conditions: $n \leq 100$, $n = max$, and $(eur) \leq$ 7,500,000. (As can be seen, here N=100). The result is

$\overline{V}_{s'} = \{\{7{,}311{,}132 \cdot (eur), 29 \cdot n\}\}$.

So, the resource base is sufficient for the normal operation of this system during 29 months.

5.2.4 *Assessment of Variations of Amounts of Resources Consumed by an* RCS

The *third* of many everyday tasks to be considered by an *RCS's* management is to *evaluate the variation* (increase and/or decrease) *of the total amounts of resources consumed by the system, in the case of a change* (transformation) *to its organizational structure and assigned equipment.*

Let $S = < \{1 \cdot a_0\}, R>$ be a 1-multigrammar representing an *RCS* a_0, while the objective system which is the result of the aforementioned transformation is represented by a 1-multigrammar $S' = < \{1 \cdot a'_0\}, R'>$, where

$$R' = R - \Delta R^- \cup \Delta R^+, \tag{5.26}$$

and ΔR^- is the set of unitary rules deleted from scheme *R*, while ΔR^+ is the set of unitary rules joined to *R*. Also let $\overline{V}_s = \{v\}$, $\overline{V}_{s'} = \{v'\}$.

The resulting variation may be represented as a couple

$$\Delta v = \langle \Delta v^+, \Delta v^- \rangle, \tag{5.27}$$

where

$$\Delta v^+ = v' - v, \tag{5.28}$$

$$\Delta v^- = v - v', \tag{5.29}$$

so Δv^+ represents amounts of resources which would additionally be consumed after the changes to the *RCS*, while Δv^- represents amounts of resources which would be released as a result of this action.

Example 5.7

Consider the *RCS* from Example 5.3. Suppose the company director is assessing the consequences of a reduction of the number of junior analysts in every laboratory from nine to six and creating a new organizational unit—expert bureau—consisting of a head of the bureau and four analysts; also the director must assign to this new unit salaries and equipment equivalent to those existing in the laboratory. The expert bureau would be subordinated to the company director.

As may be seen,

$$\Delta R^- = \{r_1, r_5\},$$

$$\Delta R^+ = \{r'_1, r'_5, r_{33}, r_{34}, r_{35}\},$$

where

$$r'_1 : (company) \\ \rightarrow 1 \cdot (director), 3 \cdot (department), 2 \cdot (laboratory), 1 \cdot (expert\ bureau), \\ 1 \cdot (company\ network\ equipment),$$

$$r'_5 : (laboratory) \rightarrow 1 \cdot (head\ of\ laboratory), 3 \cdot (position\ senior\ analyst), \\ 6 \cdot (position\ analyst), 6 \cdot (position\ junior\ analyst), \\ 1 \cdot (equipment\ of\ laboratory),$$

$$r_{33} : (expert\ bureau) \rightarrow 1 \cdot (head\ of\ expert\ bureau), 4 \cdot (analyst), \\ 1 \cdot (equipment\ of\ expert\ bureau),$$

$$r_{34} : (head\ of\ expert\ bureau) \rightarrow 1 \cdot (position\ head\ of\ laboratory), \\ 1 \cdot (equipment\ of\ head\ of\ laboratory),$$

$$r_{35} : (equipment\ of\ expert\ bureau) \rightarrow 1 \cdot (switch).$$

According to (5.26)–(5.29),

$$\Delta \upsilon = < \{7300 \cdot (eur)\}, \{18,150 \cdot (eur)\} >,$$

i.e., the assessed change would reduce the company expenses by 10,850 euros per month.

Definitions (5.28) and (5.29) may be simplified in the case when $v' \subseteq v$ or $v \subseteq v'$:

$$\Delta \upsilon = < \{\phi\}, \upsilon - \upsilon' >, \tag{5.30}$$

$$\Delta \upsilon = < \upsilon' - \upsilon, \{\phi\} >. \tag{5.31}$$

respectively. Finally, the result of the assessment illustrated by Example 5.7 will be $<\{\varnothing\}, \{10{,}850 \cdot (eur)\}>$.

5.2.5 Assessment of Feasibility of Restructuring Decisions

The *fourth* task, which is a generalization of the previous one and is very often considered by an *RCS's* management, is an *assessment of the feasibility of some decisions changing the system's structure and its equipment*. This problem may be solved by more or less local changes to the scheme *R* representing the *RCS* and, if necessary, to its resource base.

The most typical case relates to a change to an *RCS's* structure, and this may be simply implemented by replacement of the scheme *R* by a new scheme R', as was proposed above by (5.26).

After this change, the *TMS* $v \in \overline{V}_{s'}$ would be compared with the resource base $\overline{v}$, and a conclusion about feasibility is reached in full accordance with (5.9): *if it is true, then a restructuring is feasible*.

Along with a change to an *RCS's* structure, there may be a change (usually a reduction) to the resource base caused by acquisition of some new equipment and related expenses for this action.

Example 5.8

Let us consider the *RCS* from Example 5.7. Its budget is 3.6 million euros per year, and the director of the company is studying the feasibility of elimination of two junior analysts and one analyst from every laboratory along with a 10% raise in the salary of the higher management (director, heads of departments, heads of laboratories). This may be done by

$$\Delta R^{-} = \{r_5, r_{18}, r_{19}, r_{20}\},$$

$$\Delta R^{+} = \{r'_5, r'_{18}, r'_{19}, r'_{20}\},$$

where

$$r'_5 : (laboratory) \rightarrow 1 \cdot (head\ of\ laboratory), 3 \cdot (senior\ analyst),$$

$$5 \cdot (analyst), 7 \cdot (junior\ analyst),$$

$$1 \cdot (equipment\ of\ laboratory),$$

$$r'_{18} : (position\ director) \rightarrow 11{,}000 \cdot (eur),$$

$$r'_{19} : (position\ head\ of\ department) \rightarrow 5500 \cdot (eur),$$

$$r'_{20} : (position\ head\ of\ laboratory) \rightarrow 3300 \cdot (eur).$$

After correction $R' = R - \Delta R^{-} \cup \Delta R^{+\cdot}$

$$\overline{V}_{s'} = \{\{3{,}007{,}446 \cdot (eur)\}\},$$

so

$$v = \{3{,}007{,}446 \cdot (eur)\} \subseteq \overline{v} = \{3{,}600{,}000 \cdot (eur)\},$$

and the change to the *RCS* structure is feasible.

Let us repeat that the four most often considered tasks, briefly described above, are solved by an *RCS's* management *in a "what-if" mode*. However, online analytics supporting decision-makers, to be practically useful, must produce not only *conclusions* about an *RCS's* current state but also *recommendations* of what to do *to achieve a new, more desirable state of an RCS* as an objective. Such tasks relate already to *RCS* synthesis, which is considered in the next section, and solution of these tasks is made possible by application of the full spectrum of the capabilities of the introduced multigrammatical framework, not only unitary multigrammars, as were demonstrated here.

5.3 *RCS* Synthesis

Unlike analysis, the *objective of RCS synthesis is the selection from a set of all possible variants of system structure and assigned equipment of those variants* (or most often one of them) *which satisfy some predefined conditions regarding integral and more or less detailed parameters of the created* (modified) *system*. Such conditions may define boundary values of the aforementioned parameters as well as their optimality in comparison with all other variants.

The plurality of *RCS* variants is a consequence of the *multialternativity* of their organizational structures and ways of equipment and resources supply.

5.3.1 Approaches to RCS *Synthesis*

In the general case, the problem of *RCS* synthesis may be solved in two different ways.

The *first* is *synthesis by analysis*: a decision-maker uses the capabilities of a software engine for an evaluation of possible variants as described in the previous section, while a sequential comparison of variants in order to select the best is done in a "handmade" manner.

The *second* way, fully automated and thus much more rational, is based on *creation of a filtering variative unitary multiset grammar or unitary multiset metagrammar whose scheme describes possible variants of RCS structure and supply, while the filter defines variants satisfying selecting conditions*. The filter may contain boundary as well as optimizing conditions. Generation and selection of possible variants are done by a software engine. Every selected variant of an *RCS* is represented by a 1-multigrammar generating a one-element set of terminal multisets, and this one *TMS* satisfies the filter. So synthesis is done, in fact, by creating one or

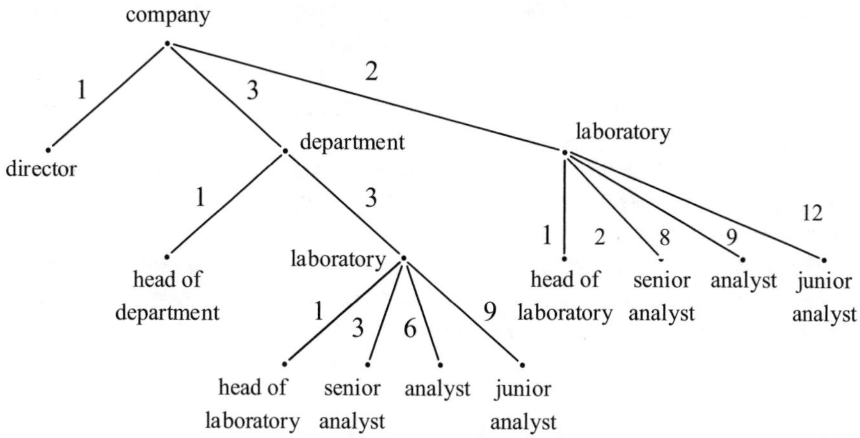

Fig. 5.2 A tree representation of an organizational system with variable structure of organizational units

more 1-*MGs* representing the system. Each such 1-*MG*, in turn, may be represented as a tree.

However, there is one hidden nuance concerning names and organizational structures in applications of *FUMGs*/*UMMGs*.

Consider the tree of the *O*-system *company* depicted in Fig. 5.1. As can be seen, there are two occurrences of the organizational unit *laboratory* in this tree, and in both cases, the organizational structure of this *OU* is the same. Now let the scheme of an *FUMG* representing the set of possible structures of this *O*-system contain two unitary rules representing different possible variants of structure of the *OU* *laboratory*:

$$(laboratory) \\ \to 1 \cdot (head\ of\ laboratory), 3 \cdot (senior\ analyst), 6 \cdot (analyst), 9 \cdot (junior\ analyst), \tag{5.32}$$

$$(laboratory) \\ \to 1 \cdot (head\ of\ laboratory), 2 \cdot (senior\ analyst), 8 \cdot (analyst), 12 \cdot (junior\ analyst), \tag{5.33}$$

the first one corresponding to the variant represented in Fig. 5.1. However, if we strictly follow the mathematical semantics of *FUMGs*, then there may be also variants like that depicted in Fig. 5.2,where laboratories belonging to departments have the organizational structure defined by the *UR* (5.32), while two laboratories directly belonging to the company have an organizational structure defined by the *UR* (5.33). And, of course, both of the latter two laboratories may have different structures, as well as laboratories belonging to departments (hence, departments also may have different structures), and so on. Thus, in fact, directly following the mathematical semantics of *FUMGs*, in the case of variative *FUMGs*, *we break a commonly used and substantially understood bijection between the name of an OU and its structure*. In other words, we make a structure of an organizational unit variable, thus creating some difficulties in understanding the results of synthesis.

There may be two approaches in this situation.

The first one, which will be illustrated by examples in Sect. 5.3.2, is based on a *direct application of FUMGs/UMMGs* but taking into account the described nuance concerning *OU* naming.

The second approach, presented in Sect. 5.3.3, is based on a *selective mathematical semantics of FUMGs developed specially in order to implement the aforementioned bijection between names and structures of OUs.*

5.3.2 *Direct* FUMG/UMMG *Application to* RCS *Synthesis*

Let us begin with an example that is similar to examples in Sect. 5.2.

Example 5.9

Suppose the company director is developing its structure, and the following are possible variants and other initial data:

1. Two four-laboratory departments and three stand-alone laboratories.
2. Three three-laboratory departments and two stand-alone laboratories.
3. A stand-alone laboratory may contain, except the head of the laboratory, four analysts and six assistants, while a laboratory inside a department has two analysts and five assistants.
4. The total monthly payroll of the company may be no more than 120,000 euros and should be minimal among all possible choices.
5. The salary of every person in a stand-alone lab is 10% greater than the salary of a person in the corresponding position in a laboratory inside a department.
6. The total number of persons working at the company may be no more than 110.

All the above may be represented by the variative filtering unitary multigrammar $S = <\{1 \cdot (company)\}, R, F>$, where the unitary rules in the scheme R are as follows:

$$r_1 : (company) \rightarrow 2 \cdot (dpt - 4 - labs), 3 \cdot (lab - alone), 1 \cdot (director);$$

$$r_2 : (company) \rightarrow 3 \cdot (dpt - 3 - labs), 2 \cdot (lab - alone), 1 \cdot (director);$$

$$r_3 : (dpt - 4 - labs) \rightarrow 4 \cdot (lab - dpt), 1 \cdot (head - dpt);$$

$$r_4 : (dpt - 3 - labs) \rightarrow 3 \cdot (lab - dpt), 1 \cdot (head - dpt);$$

$$r_5 : (lab - dpt) \rightarrow 2 \cdot (analyst - lab - dpt), 5 \cdot (assistant - lab - dpt),\ 1 \cdot (head - lab - dpt);$$

$$r_6 : (lab - alone) \rightarrow 4 \cdot (analyst - lab - alone),$$

$$6 \cdot (assistant - lab - alone),\ 1 \cdot (head - lab - alone);$$

$$r_7 : (director) \rightarrow 1 \cdot (person), 10{,}000 \cdot (eur);$$

$$r_8 : (head - dpt) \rightarrow 1 \cdot (person), 5000 \cdot (eur);$$

$$r_9 : (head - lab - dpt) \rightarrow 1 \cdot (person), 3000 \cdot (eur);$$

$$r_{10} : (head - lab - alone) \rightarrow 1 \cdot (person), 3300 \cdot (eur);$$

$$r_{11} : (analyst - lab - dpt) \rightarrow 1 \cdot (person), 1000 \cdot (eur);$$

$$r_{12} : (analyst - lab - alone) \rightarrow 1 \cdot (person), 1100 \cdot (eur);$$

$$r_{13} : (assistant - lab - dpt) \rightarrow 1 \cdot (person), 500 \cdot (eur);$$

$$r_{14} : (assistant - lab - alone) \rightarrow 1 \cdot (person), 550 \cdot (eur).$$

As may be seen, the following objects are used:

1. $(company)$, $(director)$, $(head - dpt)$, (eur), whose sense is the same as in Examples 5.1–5.8
2. $(dpt - 3 - labs)$ and $(dpt - 4 - labs)$, designating departments with three and four laboratories, respectively
3. $(lab - dpt)$ and $(lab - alone)$, designating a laboratory inside a department and a stand-alone laboratory, respectively
4. $(analyst - lab - dpt)$, $(analyst - lab - alone)$, $(assistant - lab - dpt)$, $(assistant - lab - alone)$, designating an analyst and an assistant at a laboratory inside a department and a stand-alone laboratory, respectively
5. $(person)$, representing any person of the company

The filter F representing the restrictions verbally described above will contain the following boundary and optimizing conditions:

$$c_1 : (person) \leq 110;$$

$$c_2 : (eur) \leq 120{,}000;$$

$$c_3 : (eur) = min\,.$$

As may be seen, the *FUMG S*, which is a k–multigrammar, generates two 1-multigrammars, whose schemes are

$$R_1 = \{r_1,\ r_3, r_5, r_6, r_7,\ r_8,\ r_9,\ r_{10},\ r_{11}\},$$

Table 5.1 Table representation of the tree corresponding to the scheme R_1

No.	Node	Edges and nodes–sons
1	*Company*	$2 \cdot (2), 3 \cdot (4), 1 \cdot (5)$
2	*Dpt–4–labs*	$4 \cdot (3), 1 \cdot (6)$
3	*Lab–dpt*	$1 \cdot (7), 2 \cdot (9), 5 \cdot (11)$
4	*Lab–alone*	$1 \cdot (8), 4 \cdot (10), 6 \cdot (12)$
5	*Director*	$1 \cdot (13), 10{,}000 \cdot (14)$
6	*Head–dpt*	$1 \cdot (15), 5000 \cdot (16)$
7	*Head–lab–dpt*	$1 \cdot (17), 3000 \cdot (18)$
8	*Head–lab–alone*	$1 \cdot (19), 3300 \cdot (20)$
9	*Analyst–lab–dpt*	$1 \cdot (21), 1000 \cdot (22)$
10	*Analyst–lab–alone*	$1 \cdot (23), 1100 \cdot (24)$
11	*Assistant–lab–dpt*	$1 \cdot (25), 500 \cdot (26)$
12	*Assistant–lab–alone*	$1 \cdot (27), 550 \cdot (28)$
13	*Person*	
14	*Eur*	
...	...	...
27	*Person*	
28	*Eur*	

Table 5.2 Table representation of the tree corresponding to the scheme R_2

No.	Node	Edges and nodes–sons
1	*Company*	$3 \cdot (2), 2 \cdot (4), 1 \cdot (5)$
2	*Dpt–3–labs*	$3 \cdot (3), 1 \cdot (6)$
3	*Lab–dpt*	$1 \cdot (7), 2 \cdot (9), 5 \cdot (11)$
4	*Lab–alone*	$1 \cdot (8), 4 \cdot (10), 6 \cdot (12)$
5	*Director*	$1 \cdot (13), 10{,}000 \cdot (14)$
6	*Head–dpt*	$1 \cdot (15), 5000 \cdot (16)$
7	*Head–lab–dpt*	$1 \cdot (17), 3000 \cdot (18)$
8	*Head–lab–alone*	$1 \cdot (19), 3300 \cdot (20)$
9	*Analyst–lab–dpt*	$1 \cdot (21), 1000 \cdot (22)$
10	*Analyst–lab–alone*	$1 \cdot (23), 1100 \cdot (24)$
11	*Assistant–lab–dpt*	$1 \cdot (25), 500 \cdot (26)$
12	*Assistant–lab–alone*	$1 \cdot (27), 550 \cdot (28)$
13	*Person*	
14	*Eur*	
...	...	...
27	*Person*	
28	*Eur*	

$$R_2 = \{r_2, r_4, r_5, r_6, r_7, r_8, r_9, r_{10}, r_{11}\},$$

each defining the corresponding variant of the synthesized *RCS*. The table-represented trees of these variants are shown as Tables 5.1 and 5.2.

As may be seen,

$$\bar{V}_{S_1} = \{\{100 \cdot (person),\ 113{,}000 \cdot (eur)\}\},$$

$$\bar{V}_{S_2} = \{\{98 \cdot (person),\ 114{,}500 \cdot (eur)\}\},$$

and after the filtration, the result is

$$\bar{V}_{S_1} = \{\{100 \cdot (person),\ 113{,}000 \cdot (eur)\}\},$$

which corresponds to the first variant, represented by scheme R_1 and Table 5.1.

More complicated cases may need unitary multiset metagrammars for description and selection of possible solutions. Let us illustrate this with the following example, concerning synthesis of an *O*-system.

Example 5.10

Suppose the company director is working on its structure, taking into account the following proposals:

1. The salary of the director must be twice the salary of a head of a department, which, in turn, must be 1.5 times the salary of a head of a laboratory, which, in turn, must be 1.5 times the salary of an analyst, which, in turn, must be twice the salary of an assistant, which must be between 300 and 500 euros.
2. A laboratory must consist of two to five groups, each including an analyst and two assistants, and all laboratories must be of the same size.
3. The company may contain "3–lab" as well as "4–lab" departments; the total number of departments must be not greater than 5.
4. The total number of persons in the company must be not greater than 100.
5. The monthly payroll must be not greater than 120,000 euros.
6. Every person should have the maximal possible salary.

All the above may be easily represented by the unitary multiset metagrammar S=<{1 · (*company*)}, *R*, *F*>, where the unitary rules and unitary metarules in the scheme *R* are as follows:

$$r_1 : (company) \to m \cdot (dpt - 3 - labs), n \cdot (dpt - 4 - labs), 1 \cdot (director);$$

$$r_2 : (dpt - 3 - labs) \to 3 \cdot (lab), 1 \cdot (dpt), 1 \cdot (head - dpt);$$

$$r_3 : (dpt - 4 - labs) \to 4 \cdot (lab), 1 \cdot (dpt), 1 \cdot (head - dpt);$$

$$r_4 : (lab) \to 1 \cdot (head - lab), l \cdot (group);$$

$$r_5 : (group) \to 1 \cdot (analyst), 2 \cdot (assistant);$$

$$r_6 : (director) \to 1 \cdot (person), 1 \cdot (salary - director);$$

$$r_7 : (head - dpt) \to 1 \cdot (person), 1 \cdot (salary - head - dpt);$$

$$r_8 : (head - lab) \to 1 \cdot (person), 1 \cdot (salary - head - lab);$$

$$r_9 : (analyst) \to 1 \cdot (person), 1 \cdot (salary - analyst);$$

$$r_{10} : (assistant) \to 1 \cdot (person), 1 \cdot (salary - assistant);$$

$$r_{11} : (salary - director) \to 2 \cdot (salary - head - dpt);$$

$$r_{12} : (salary - head - dpt) \to 1.5 \cdot (salary - head - lab);$$

$$r_{13} : (salary - head - lab) \to 1.5 \cdot (salary - analyst);$$

$$r_{14} : (salary - analyst) \to 2 \cdot (salary - assistant);$$

$$r_{15} : (salary - assistant) \to x \cdot (eur).$$

The filter *F* representing the conditions which implement the verbally described restrictions above contains the following boundary and optimizing conditions:

$$c_1 : \quad 300 \leq x \leq 500;$$

$$c_2 : \quad 2 \leq l \leq 5;$$

$$c_3 : \quad 0 \leq m \leq 5;$$

$$c_4 : \quad 0 \leq n \leq 5;$$

$$c_5 : \quad (person) \leq 100;$$

$$c_6 : \quad (euro) \leq 120{,}000;$$

$$c_7 : \quad (dpt) \leq 5;$$

$$c_8 : \quad x = max.$$

As can be seen, the unitary metarule r_1 defines the company, consisting of the director, *m* "3–lab" departments, and *n* "4–lab" departments. Because the total number of departments is not greater than five, there are three boundary conditions in the filter *F*: c_3 and c_4 define the domains of the variable-multiplicities *m* and *n*, while c_7 gives the aforementioned upper bound. The unitary rules r_2 and r_3 define the structure of the "3–lab" and the "4–lab" departments, respectively, each consisting of a head of department and an appropriate number of laboratories. Also there is a multiobject $1 \cdot (dpt)$ inside the bodies of both *URs*, thus providing a count of the total number of the departments in the company as a multiplicity of the terminal object (*dpt*), referred to by the aforementioned boundary condition c_7.

The unitary metarule r_4 defines the structure of a laboratory, which consists of a head and *l* groups, where *l* may be from 2 to 5 (the boundary condition c_2 defines the domain of the variable-multiplicity *l*). The *UR* r_5 defines a group as consisting of one analyst and two assistants.

The unitary rules $r_6 - r_{10}$ define all positions in the company, assigning them salaries and a unified property—the terminal object (*person*)—whose multiplicity in every generated *TMS* is, thus, the total number of persons in the company, which, in turn, is restricted to no more than 100 (the boundary condition c_5). The unitary rules

$r_{11} - r_{14}$ define the ratios between salaries of all positions, as was verbally described above. Finally, the unitary metarule r_{15} defines the basic salary of an assistant as an unknown number of euros (the variable-multiplicity x). The domain of the variable-multiplicity x is fixed by the boundary condition c_1.

The optimizing condition c_8 requires the maximization of all salaries, related as they are by the appropriate constant multipliers to a basic salary x euros.

As can be seen, the described unitary multiset metagrammar is a natural and compact description of this nontrivial combinatorial problem, whose formalization and solution by application of conventional tools and well-known techniques are non-obvious. So a direct *FUMG/UMMG* application in the area of *RCS* synthesis may be both simple and useful.

Let us move to an alternative approach to *RCS* synthesis, based on a selective mathematical semantics of variative *FUMGs* oriented specially to this problem.

5.3.3 *Variative* FUMGs *with a Problem-Oriented Mathematical Semantics and Their Application to* RCS *Synthesis*

An objective of correction of the mathematical semantics of *FUMGs*, introduced in (Sheremet and Zhukov 2016), is *to keep during MS generation the aforementioned bijection between names and structures of organizational units belonging to a constructed RCS*.

Let us begin from a variative *UMG* $S = \langle\{1 \cdot a_0\}, R\rangle$ and assume that this *UMG* generates a set X_S containing couples $\langle v, R^+\rangle$,where v is an *MS*, while R^+ is a set of unitary rules already applied during previous generation steps. So the result of a *UMG* application is, in fact, a *set of elements each corresponding to some variant of an organizational chart of a synthesized resource-consuming system*, which is represented by a set R^+, *along with* an *MS* v containing *the parameters of this variant*. Intermediate generated constructions are not multisets but triples $\langle v, R^+, \bar{R}\rangle$, where v and R^+ are the same as above, while $\bar{R}$ is a subset of the scheme R including all *URs* which may be applied to the *MS* v at the next generation step. An iterative representation of the proposed semantics of these *UMGs* which is a generalization of (3.30)–(3.33) is as follows:

$$X_{(0)} = \{\langle\{1 \cdot a_0\},\ R,\ \{\varnothing\}\rangle\}, \tag{5.34}$$

$$X_{(i+1)} = X_{(i)} \cup \left(\bigcup_{\langle v,\ \bar{R}\rangle \in X_{(i)}} \bigcup_{r \in \bar{R}} \bar{\pi}\,(v,\ R^+,\ \bar{R},\ r) \right), \tag{5.35}$$

$$X_S = \left\{\langle v,\ R^+\rangle \mid \langle v,\ R^+,\ \bar{R}\rangle \in X_{(\infty)}\right\}, \tag{5.36}$$

where

$$\bar{\pi}\,(v, R^{+}, \bar{R}, r) =$$

$$= \begin{cases} \langle v - \{n \cdot a\} + n{*}\mathbf{w},\ R^{+} \cup \{a \to w\},\ \bar{R} - alt(a \to w)\rangle, \text{ if } n \cdot a \in v, \\ \{\varnothing\} \text{ otherwise,} \end{cases} \tag{5.37}$$

$$alt(a \to w) = \{a \to w' \mid a \to w' \in \bar{R}\&\mathbf{w}' \neq \mathbf{w}\}, \tag{5.38}$$

r is the unitary rule

$$a \to n_1 \cdot a_1, \ldots, n_m \cdot a_m, \tag{5.39}$$

w and w' designate, respectively, $n_1 \cdot a_1, \ldots, n_m \cdot a_m$ and $n'_1 \cdot a'_1, \ldots, n'_k \cdot a'_k$, and also

$$\mathbf{w} = \{n_1 \cdot a_1,\ \ldots,\ n_m \cdot a_m\}, \tag{5.40}$$

$$\mathbf{w}' = \{n'_1 \cdot a'_1,\ \ldots,\ n'_k \cdot a'_k\}. \tag{5.41}$$

As may be seen from this definition, due to (5.37) and (5.38), an application of a *UR* r, $a \to n_1 \cdot a_1, \ldots, n_m \cdot a_m$, eliminates from the current set $\bar{R}$ all *URs* with the same header a and different bodies, so *after the first application of r no other UR with the same header will be applied during a generation branch beginning from the current triple* $\langle v, R^{+}, \bar{R}\rangle$. At the same time, the applied *UR* r is joined to the current set R^{+}. So, as can be seen, this generation permits the *accumulation of the organizational chart of the constructed RCS*, as well as the necessary bijection between names and structures of organizational units. Because, in essence, a *1-MG* with a scheme R^{+} is created by each generation branch, a replacement of a multiset $\{n \cdot a\}$ by a multiset $n * \mathbf{w}$ is performed similarly to (4.1), where this way of reducing the number of generation steps was introduced for 1-multigrammars. In order to exclude repeating generation steps as done by (4.1), we shall replace (5.35) by

$$X_{(i+1)} = \bigcup_{\langle v,\, \bar{R}\rangle \in X_{(i)}} \ \bigcup_{\substack{n \cdot a \exists \in v \\ \langle a \to w\rangle \in \bar{R}}} \{<v - \{n \cdot a\} + n{*}\mathbf{w},\ R^{+} \cup \{a \to w\},\ \bar{R} - alt(a \to w)>\}, \tag{5.42}$$

and define

$$\bar{X}_S = \{\langle v,\ R^{+}\rangle \mid \langle v,\ R^{+},\ \bar{R}\rangle \in X_{(i)}\&X_{(i+1)} = \{\varnothing\}\}. \tag{5.43}$$

To select only one variant of an *RCS* organizational chart, a filter may be applied, and the *problem-oriented mathematical semantics of FUMGs*, which are a generalization of (5.34)–(5.41), may be defined as follows.

Let $S = \langle\{1 \cdot a_0\}, R, F\rangle$ be a filtering variative *UMG* and $S' = \langle\{1 \cdot a_0\}, R\rangle$ its core *UMG*. Similarly to definition (3.39) of the semantics of filtering multiset grammars,

$$\bar{X}_S = \{\langle v,\ R^+\rangle \mid \langle v,\ R^+\rangle \in \bar{X}_{S'} \& \{v\} \downarrow F = \{v\}\}. \tag{5.44}$$

As can be seen, all multisets belonging to elements of the set $\bar{X}_{S'}$ generated by the core *UMG* S' of the variative *FUMG* S are filtered by the filter F and, if they satisfy it, elements containing them are included in the set $\bar{X}_S$.

Example 5.11
Let us consider the problem of synthesis of a resource-consuming system represented by the filtering variative unitary multiset grammar $S = \langle\{1 \cdot (company)\}, R, F\rangle$, whose scheme R contains the following unitary rules:

$$\begin{aligned}
r_1 &: (company) \to 1 \cdot (director), 3 \cdot (department), 2 \cdot (laboratory),\\
r_2 &: (department) \to 1 \cdot (head\ of\ department), 3 \cdot (laboratory),\\
r_3 &: (laboratory) \to 1 \cdot (head\ of\ laboratory), 3 \cdot (senior\ analyst),\\
&\qquad 6 \cdot (analyst), 9 \cdot (junior\ analyst),\\
r_4 &: (laboratory) \to 1 \cdot (head\ of\ laboratory), 2 \cdot (senior\ analyst),\\
&\qquad 8 \cdot (analyst), 12 \cdot (junior\ analyst),
\end{aligned}$$

and the filter $F = \{(analyst) = max\}$. As may be seen, there are two variants of the organizational chart of a laboratory defined by *URs* r_3 and r_4, corresponding to the trees depicted in Figs. 5.1 and 5.2. The objective of synthesis is to obtain an organizational chart of the company maximizing the number of analysts. The application of relations (5.34), (5.42), and (5.43) generates two couples:

$$x_1 = \left\langle \left\{ \begin{array}{l} 1 \cdot (director),\ 3 \cdot (head\ of\ department),\ 11 \cdot (head\ of\ laboratory), \\ 33 \cdot (senior\ analyst),\ 66 \cdot (analyst),\ 99 \cdot (junior\ analyst) \end{array} \right\}, \{r_1,\ r_2,\ r_3\} \right\rangle,$$

$$x_2 = \left\langle \left\{ \begin{array}{l} 1 \cdot (director),\ 3 \cdot (head\ of\ department),\ 11 \cdot (head\ of\ laboratory), \\ 22 \cdot (senior\ analyst),\ 88 \cdot (analyst),\ 132 \cdot (junior\ analyst) \end{array} \right\}, \{r_1,\ r_2,\ r_4\} \right\rangle,$$

and, after filtering, $\bar{X}_S = \{x_2\}$. Thus, the second variant defined by the set of *URs*, $R^+ = \{r_1, r_2, r_4\}$, gives the organization of the company with the maximal number of analysts.

There are already *a lot of methods applying FUMGs/UMMGs to various problems concerning analysis and synthesis of O-, OT-, and ROT- systems* (some of them concerning budget planning, and business planning are considered in sufficient detail in (Sheremet 2010, 2011a), while consideration of the others would require separate publications). In fact, this simple and flexible toolkit due to its relatively easy implementation and application may be disseminated at any level of state and corporate management.

As was mentioned in Sect. 2.2, one of the most difficult tasks associated with resource-consuming systems is *systems reverse engineering*, the essence of which is to determine a possible structure of an *RCS* given some primary knowledge about its possible structures, technological bases, and supply standards, as well as a fully or partly known collection of resources consumed by this *RCS* during some fixed period of time. Techniques of reverse engineering, in turn, provide effective solutions of various tasks from the *Business Intelligence* area (Munoz 2017; Rausch et al. 2013). The final purpose of any such task is, usually, to determine some unknown parameters of an *RCS* (elements of its structure, *TB*, and supply standards), which are monitored via observations by an external observer of integral amounts of some resources consumed by the system during some considered time period. Such data may be electrical power, water, money, food, fuels, etc. The technique of multigrammatical solution of such tasks is, in fact, the same as described above in this section and was described in detail in the aforementioned first books in which the *MGF* was introduced (Sheremet 2010, 2011a).

Now we may move to application of the multigrammatical framework to industrial systems which are associated with the most general *MGF* segment—temporal multiset grammars.

Chapter 6
Industrial Sociotechnological Systems

The main feature of ***industrial STSs*** is manufacturing new object-resources by means of their manufacturing technological bases (sets of manufacturing devices) and resource bases (collections of resources available to the aforementioned devices). The manufacturing process of any *IS* is driven by orders, whose sources are external subjects (systems), first of all, resource-consuming systems. As was mentioned above in Sect. 2.3, ***planning of IS operation*** covers, mainly, competitions implemented by *IS* staff and aimed to find the most rational collaborations able to complete incoming or expected orders. This activity includes, first of all, assessment of order feasibility; the goal of this assessment is to determine whether available resources are sufficient for order completion and, if not, then to assess what part of this order may be completed given the available *RB* (for ***closed*** *ISs*) or what additional resources must be acquired for full order completion (for ***open*** *ISs*).

The set of tasks associated with the aforementioned activities may be rather easily and fully adequately represented by means of ***filtering unitary multiset grammars and unitary metagrammars***, thus motivating further the development of *FUMG/UMMG* pragmatics, or application techniques. Just such a representation will be applied in Sect. 6.1. Also, taking into account the importance of ecological issues in connection with industrial activities, special attention is paid to *UMG*-based assessment of environmental damage, which may be caused by *ISs* during their operation. While considering the aforementioned proposed techniques of *UMG* application, *UMG* representations of such fundamentals of economical science as the "added value" of K. Marx, the Leontief model, and Pareto-optimal solutions will be briefly discussed.

In fact, all the key features concerning the algorithmics of creation of schedules as a subprocess of generation of temporal multisets were highlighted in Sects. 3.2 and 4.4 while considering ***temporal multiset grammars***. This is not by chance, because *TMGs* and all their extensions (*FTMGs*, *SG FTMGs*, *TMMGs*, *SG TMMGs*) were developed as a basic tool for the representation of manufacturing technological bases of industrial systems and, through the generation of sets of temporal multisets, for the creation of ***schedules*** associated with them to enable order completion. Here in

I. A. Sheremet, *Multigrammatical Framework for Knowledge-Based Digital Economy*, https://doi.org/10.1007/978-3-031-13858-4_6

this chapter, we shall consider *TMGs* from the point of view of their application to the implementation of such fundamental approaches to ***management of large-scale projects*** as "planning-programming-budgeting systems," "purposeful planning," and "output budgeting," all of which were referred to in the Introduction. This, the most valuable area of *TMG* application, is considered in detail in Sect. 5.2. Namely, for *ISs* manufacturing passive resources, different ways of assessment of order feasibility for *TMG*-based representation of *ISs MPR* are described. A method of assessment of additional amounts of resources to be acquired and the time period necessary for completion of initially non-feasible orders is described. The proposed method involves the creation of three interconnected schedules—input, supply, and output. Execution of these schedules leads to the coherent operation of supplying, producing, and consuming systems.

All the aforementioned tasks will be considered in the ***single-order*** case. The general case, concerning *IS* operation, is a ***flow of orders***, each new order arriving while previous orders are completed. So a problem of multi-order scheduling arises, which is considered thoroughly in Sect. 6.2.4.

As was mentioned in Sect. 2.2, all resources consumed by any *RCS* are produced by some industrial systems, and, thus, ***resource-consuming STSs are the main source of requests (orders) to industrial systems***. Various types of operational interconnections between resource-consuming systems and industrial systems will be considered in Sect. 6.3.

Let us consider the above-introduced tasks sequentially, beginning from the simplest form of multigrammatical representation of industrial systems—unitary multiset grammars. Before we move to the main content of this chapter, let us recall that, like any other *STSs*, industrial systems in the general case may be ***local*** or ***distributed***, ***open*** or ***closed***. Unless stated otherwise, we shall consider local *ISs*. Every specific case of interconnection of a particular *IS* with its environment (whether the system is open or closed) will be announced when we consider that system.

6.1 *UMG*-Based Representation of Industrial Systems and Associated Tasks

Let us repeat that the main objective of an ***assessment of order feasibility*** is to determine whether available resources are sufficient for order completion. If these resource amounts are not sufficient, there are two tasks to be considered:

1. What part of an order may be completed without additional resources?
2. What amounts of resources must be acquired additionally, if an order is not feasible?

The first task is relevant for ***closed ISs***, which do not have the opportunity to acquire missing resources from external systems; the second is for ***open*** industrial systems, which, on the contrary, have such opportunities.

6.1.1 UMG-*Based Representation of* ISs

Here, we shall consider only industrial systems manufacturing ***passive*** resources.

Following the intermediate "black box" representation introduced and applied in Sect. 2.3 (Fig. 2.5), we shall begin from an *IS* manufacturing technological base containing ***one-output manufacturing devices*** (Fig. 6.1). Any such "box" manufactures one object-resource a , consuming n_1 object-resources a_1, …, n_m object-resources a_m. The resource base of an *IS* contains object-resources, which are available to all manufacturing devices belonging to the *MTB* of this system. Every *MD* may extract all necessary *ORs* from the *RB*, manufacture the corresponding object-resource, and put (load) it into the *RB*. An *MD* operates only in the case when the *RB* contains all object-resources necessary for this *MD*, i.e., an operation cycle of this device starts only when the *RB* contains the aforementioned n_1 *ORs* a_1, …, n_m *ORs* a_m.

An ***order*** to be completed by the *IS* also is represented by a "black box" with $\boldsymbol{m}$ inputs, corresponding to $\boldsymbol{n_1}$ object-resources $\boldsymbol{a_1}$, …, $\boldsymbol{n_m}$ object-resources $\boldsymbol{a_m}$, which this system must produce (manufacture), and one output q, designating a request—an entity integrating the aforementioned $\boldsymbol{m}$ order elements.

Therefore, an *IS* may be represented as depicted in Fig. 6.1, where the resource base is the ellipse labeled "v_0" and the k manufacturing devices belonging to the *IS*

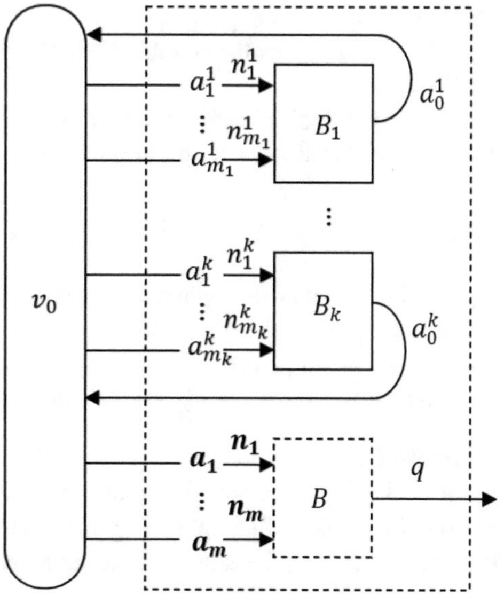

Fig. 6.1 A "black box" representation of an *IS* and an order

manufacturing technological base are boxes labeled "B_1," ..., "B_k," while the order is a box labeled "B." All inputs and outputs are marked in full accordance with the previous text. The resource base will be represented by a multiset $v_0 = \{n_1 \cdot a_1, \ldots, n_m \cdot a_m\}$, whose meaning is quite evident: the *RB* includes n_1 object-resources $a_1, \ldots, n_m$ object-resources a_m.

Similar to the representation of "black boxes" by rules, introduced above in Sect. 3.2.1, every manufacturing device B_i may be represented by a unitary rule r_i

$$a_0^i \rightarrow n_1^i \cdot a_1^i, \ldots, n_{m_i}^i \cdot a_{m_i}^i. \tag{6.1}$$

A set of rules (the scheme of a unitary multiset grammar) $R = \{r_1, \ldots, r_k\}$ is a multigrammatical representation of an *IS* manufacturing technological base and, unless stated otherwise, will be also called an *IS MTB*.

We shall use two substantially interconnected interpretations of unitary rules, applied to *IS* modeling and introduced in Sect. 3.1.3: the structural (*URs* like (3.37)) and the technological (*URs* like (3.38)).

Let us recall that according to the ***structural interpretation*** (see Example 3.10), the *UR* (6.1) means that a manufactured object-resource a_0^i consists of (is manufactured from) n_1^i object-resources $a_1^i, \ldots, n_{m_i}^i$ object-resources $a_{m_i}^i$.

Example 6.1
Let the scheme R contains two *URs* in the structural interpretation:

$$r_1 : (auto) \rightarrow 1 \cdot (frame), 1 \cdot (engine), 4 \cdot (door), 4 \cdot (wheel);$$

$$r_2 : (engine) \rightarrow 1 \cdot (motor), 1 \cdot (transmission), 1 \cdot (fuel\ tank).$$

So, according to these *URs*, an auto consists of one frame, one engine, four doors, and four wheels. An engine consists of one motor, one transmission, and one fuel tank. So r_1 and r_2 denote assembly lines B_1 and B_2, manufacturing autos and engines for them.

The ***technological interpretation*** of unitary rules (see Example 3.11) is an extension of the structural one in such a way that a *UR*

$$a_0^i \rightarrow n_1^i \cdot a_1^i, \ldots, n_{m_i}^i \cdot a_{m_i}^i, k_1^i \cdot b_1^i, \ldots, k_{l_i}^i \cdot b_{l_i}^i, \tag{6.2}$$

represents not only the structural components (spare parts) of an object-resource a_0^i, which are represented by multiobjects $n_1^i \cdot a_1^i, \ldots, n_{m_i}^i \cdot a_{m_i}^i$, but also resources necessary for assembling *OR* a_0^i from these components, which are represented by multiobjects $k_1^i \cdot b_1^i, \ldots, k_{l_i}^i \cdot b_{l_i}^i$ (k_1^i object-resources $b_1^i, \ldots, k_{l_i}^i$ object-resources $b_{l_i}^i$).

Example 6.2
Let the scheme R' contains the unitary rules r_1' and r_2' obtained from the *URs* r_1 and r_2 from the previous Example 6.1 by joining to their bodies multiobjects representing resources necessary for auto and engine assembling:

$$r'_1 : (auto) \rightarrow 1 \cdot (frame), 1 \cdot (engine), 4 \cdot (door), 4 \cdot (wheel), 400 \cdot (kwh),$$
$$50 \cdot (mnt : autos\ AL), 300 \cdot (eur);$$
$$r'_2 : (engine) \rightarrow 1 \cdot (motor), 1 \cdot (transmission), 1 \cdot (fuel\ tank), 100 \cdot (kwh),$$
$$80 \cdot (mnt : engines\ AL), 5 \cdot (mnt : computer\ test), 900 \cdot (eur).$$

As can be seen, the operation of assembling one auto requires, along with the structural components (a frame, an engine, doors, and wheels), also 400 kW h of electrical energy, money (300 euros) for payment of manufacturing personnel, as well as time (50 min)—the duration of operation of the "autos" assembly line. Similarly, the operation of assembling one engine for this auto, along with its structural components (a motor, a transmission, a fuel tank), requires also 100 kWh h of electrical energy, 900 euros, as well as time resources—80 min of operation of the "engines" assembly line and 5 min of computer testing of the manufactured engine.

As may be seen, unitary multigrammars provide a description of *IS* manufacturing technological bases in an easily understood ***top-down*** manner, reflecting the structures of manufactured objects. Let us note that in the general case, there may be more than one way of manufacturing any object, so a scheme *R* representing an *IS MTB* may contain unitary rules with the same header and different bodies; thus, a *UMG S*, whose scheme is *R*, may be ***variative***. One more nuance is the representation of identical "black boxes"—manufacturing devices which are identical material objects thus represented by identical unitary rules. But because the scheme *R* is a ***set*** of *URs*, all such devices would "stick together." To exclude such incorrectness, in the body of every *UR* r_i representing a "black box" B_i, we shall place a multiobject representing the duration of an operation cycle of this *MD* in the form $n_i \cdot (\Delta t : B\ i)$, where Δt is a time measurement unit (in the above Example 6.2, as in Example 3.11, it is a minute, represented by the lexeme *mnt*). So the string *B i* is a unique identifier of an *MD*, and instead of a single time axis, we shall use individual time axes each corresponding its own *MD*. Any *MD* B_i operates in its individual ("*own*") time, so to produce *n* objects a_0^i, this device would execute *n* such cycles, and for this reason, unlike *IS* time, these ***individual times are additive values***. *IS* time, due to the possibility of parallel operation of *MDs*, is not an additive value, as was illustrated in detail in Sect. 3.2 when we introduced the temporal multiset grammars.

Example 6.3
Let the scheme R'' contain the following unitary rules r''_1, r''_2, and r''_3:

$$r''_1 : (auto) \rightarrow 1 \cdot (frame), 1 \cdot (engine), 4 \cdot (door), 4 \cdot (wheel),$$
$$400 \cdot (kwh), 50 \cdot (mnt : autos\ AL), 800 \cdot (eur);$$
$$r''_2 : (engine) \rightarrow 1 \cdot (motor), 1 \cdot (transmission), 1 \cdot (fuel\ tank),$$

$$100 \cdot (kwh), 80 \cdot (mnt : engines\ AL\ 1), 900 \cdot (eur);$$

$$r_3'' : (engine) \rightarrow 1 \cdot (motor), 1 \cdot (transmission), 1 \cdot (fuel\ tank),$$

$$100 \cdot (kwh), 80 \cdot (mnt : engines\ AL\ 2\ ), 900 \cdot (eur).$$

As can be seen, the *IS* manufacturing technological base represented by this scheme contains engine assembly lines, represented by the *URs* r_2'' and r_3'', whose bodies include different multiobjects $(mnt : engines\ AL\ 1)$ and $(mnt : engines\ AL\ 2\ )$, representing the durations of operation cycles of two identical but independently operating assembly lines producing engines.

The introduction of regular identifiers of manufacturing devices and multiobjects representing durations of *MD* operation cycles naturally leads us to a change to the multiset representation of a resource base, which henceforth will contain, along with other object-resources, also *ORs* like $n \cdot (\Delta t : B\ i)$, which represent the ***operational resource of manufacturing devices*** available during operation of an *IS*.

An order to be completed by an industrial system with an *MTB* represented by a scheme *R*, and an *RB* represented by a multiset v_0, will be represented by a multiset

$$q = \{\boldsymbol{n}_1 \cdot \boldsymbol{a}_1, \ \ldots, \boldsymbol{n}_m \cdot \boldsymbol{a}_m\}, \tag{6.3}$$

which means the objective of the order q is to get from the *IS* $\boldsymbol{n}_1$ object-resources $\boldsymbol{a}_1, \ldots, \boldsymbol{n}_m$ object-resources $\boldsymbol{a}_m$.

From now, we shall call a triple $I = \ < q, R, v_0 >$ an "***assigned to order q industrial system with manufacturing technological base R and resource base*** v_0" (for short *AIS*) and a triple $I = \ < \{\emptyset\}, R, v_0 >$ a "***free industrial system***" (for short *FIS*). Let us consider the techniques of *IS* analysis, based on the introduced notions and constructions.

6.1.2 Assessment of Order Feasibility

Let us consider a *UMG* $S = \ < q, R >$ corresponding to an *AIS* $I = \ < q, R, v_0 >$. This *UMG* in the most general case is ***variative*** (there may be many ways of manufacturing some object-resources) and ***disintegrated*** (disconnected) (there may be objects that are headers of unitary rules and do not occur in the body of any another *UR*; such objects represent "*final object-resources*," which are not components of any other *ORs* and, most often, belong to orders, which define the final *ORs*). (Note also that this *UMG* may also be ***cyclic***, which reflects circulation of some resources in a manufacturing cycle—e.g., electrical energy may be used as a resource enabling operation of devices which, in turn, generate electrical energy. This feature will be considered in Chap. 8, which is dedicated, among other things, to the resilience of critical infrastructures including multiple such cycles.)

As may be seen, a set of terminal multisets $\overline{V}_s$ is a set of collections of resources, each necessary for the completion of an order q in one of many possible ways. The following obvious statement defines the ***condition of order feasibility by AISs***.

Statement 6.1 Let $I = \ < q, R, v_0>$ be an industrial system assigned to an order q, with an *MTB R* and an *RB* v_0,and $S = \ < q, R >$. Then the order q may be completed by this *AIS* if

$$(\exists v \in \bar{V}_S) \cdot v \subseteq v_0. \tag{6.4}$$

Otherwise, the order q cannot be completed by the *AIS I*.

Speaking informally, an order may be completed if there exists at least one way of completing it that requires a resource collection which is part of the *IS* resource base.

Example 6.4
Let us consider an *FIS* $I = \ < \{\emptyset\}, R, v_0>$, whose manufacturing technological base R is the same as the *MTB* R'' from the previous Example 6.3 and whose resource base is

$$v_0 = \{3 \cdot (\mathit{frame}), 10 \cdot (\mathit{door}), 16 \cdot (\mathit{wheel}), 4 \cdot (\mathit{motor}), 3 \cdot (\mathit{transmission}),$$
$$4 \cdot (\mathit{fuel\ tank}), 150 \cdot (\mathit{mnt} : \mathit{autos\ AL}), 300 \cdot (\mathit{mnt} : \mathit{engines\ AL\ 1}),$$
$$200 \cdot (\mathit{mnt} : \mathit{engines\ AL\ 2}), 1800 \cdot (\mathit{kwh}), 10{,}000 \cdot (\mathit{eur})\}.$$

As can be seen, the *IS* resource base contains 3 frames, 10 doors, 16 wheels, 4 motors, 3 transmissions, and 4 fuel tanks. Its available operation resource base is 150 min of operation of the autos assembly line, 300 min of operation of the first engines assembly line, and 200 min of operation of the second engines assembly line. The electrical power resource is 1800 kWh h, and, finally, the financial resource is 10,000 euros.

Let the order $q = \{2 \cdot \mathit{auto}\}$, and, hence,

$$\bar{V}_q = \{\bar{v}_1, \bar{v}_2, \bar{v}_3\},$$

where

$$\bar{v}_1 = \bar{v} + \{160 \cdot (\mathit{mnt} : \mathit{engines\ AL\ 1})\},$$
$$\bar{v}_2 = \bar{v} + \{160 \cdot (\mathit{mnt} : \mathit{engines\ AL\ 2})\},$$
$$\bar{v}_3 = \bar{v} + \{80 \cdot (\mathit{mnt} : \mathit{engines\ AL\ 1})\} + \{80 \cdot (\mathit{mnt} : \mathit{engines\ AL\ 2})\},$$
$$\bar{v} = \{2 \cdot (\mathit{frame}), 2 \cdot (\mathit{engine}), 8 \cdot (\mathit{door}), 8 \cdot (\mathit{wheel}), 2 \cdot (\mathit{motor}), 2 \cdot (\mathit{transmission}),$$
$$2 \cdot (\mathit{fuel\ tank}), 100 \cdot (\mathit{mnt} : \mathit{autos\ AL}), 1000 \cdot (\mathit{kwh}), 3400 \cdot (\mathit{eur})\}.$$

As can be seen, $\bar{v}_1, \bar{v}_2$, and $\bar{v}_3$ are submultisets of multiset v_0, so the order q may be completed by the *AIS* $I = \ < q, R, v_0>$.

At the end of this section, let us make a short remark regarding ***UMG's capabilities*** to represent some basic notions and interconnections from the area of ***classical economics***. As may be seen, the technological interpretation of *URs* provides a natural representation of the basic economical category "***added value***" (Marx 2018), which is expressed by multiplicities of objects occurring in *UR* bodies and representing monetary units, used for the calculation of payments for operations executed by *MDs* (in the *URs* from Examples 6.2–6.4, this object is (*eur*), representing euros). On the other hand, *UMGs'* capabilities provide a no less natural representation of the ***Leontief model*** of the producing economy and its associated problems (Bjerkholt and Kuzz 2006; Haimes and Jiang 2001), including computation of amounts of input resources necessary for the production of needed collections of output resources. Until now, such interconnections were described by the application of operations and relations on matrices, and the aforementioned computation was done based on the algorithmics of matrix equations (Horn and Johnson 1991).

6.1.3 Assessment of Amounts of Resources to Be Acquired to Allow Completion of Initially Non-feasible Orders

There is a trivial solution to the question of what collection of additional resources must be acquired to complete an order which cannot be completed by reason of insufficiency of one or another resource in the current *RB*. The set of such collections is, evidently,

$$\Delta V = \{v - v_0 \mid v \in \bar{V}_S\}. \tag{6.5}$$

Example 6.5

Consider an *AIS* $= \ < q, R, v_0>$, where q and R are the same as in the previous Example 6.4, while v_0 differs from the *RB* in the aforementioned example and is as follows:

$$\begin{aligned} v_0 = \{ & 1 \cdot (\textit{frame}), 9 \cdot (\textit{door}), 20 \cdot (\textit{wheel}), 5 \cdot (\textit{motor}), 5 \cdot (\textit{transmission}), \\ & 2 \cdot (\textit{fuel tank}), 120 \cdot (\textit{mnt : autos AL}), 250 \cdot (\textit{mnt : engines AL } 1), \\ & 300 \cdot (\textit{mnt : engines AL } 2), 1200 \cdot (\textit{kwh}), 2000 \cdot (\textit{eur})\}. \end{aligned}$$

As can be seen, this resource base is insufficient for the completion of order q, and, according to (6.5),

$$\Delta V = \{\{1 \cdot (frame),\ 1400 \cdot (eur)\}\},$$

i.e., to complete order q, it is necessary to add to the initial resource base one frame and 1400 euros. (Note that ΔV is a one-element set despite $\bar{V}_S$ being two-element; this means that the resource base would be extended by one and the same collection of resources independent of how the order is completed.)

However, in the general case, $|\Delta V| > 1$, so there must be some technique to select only one multiset from this multi-element set of multisets, so that this *MS* would be the objective of the acquisition activities. If such a multiset is known, then the techniques to solve this task proposed in Sect. 5.2.2 may be directly applied to this *MS*.

But in the most general case, the acquisition of the missing collections of *ORs* is an exchange of one set of resources, belonging to one subject, for another such set, belonging to another subject. Speaking more generally, ***acquisition concerns the notion of property***, which is not an industrial but an economical category. Such problems inter alia form the area of ***economical combinatorics*** and, for this reason, will be rather deeply considered in the next Chap. 7.

Here, we shall only present techniques to make the set ΔV ***more compact***. Namely, if there are two multisets $v \in \Delta V$ and $v' \in \Delta V$ such that $v \subset v'$, it is reasonable to exclude v' from ΔV, because the ***"smaller" additional MS*** v ***is already sufficient for order completion***. This idea is generalized by the following obvious statement.

Statement 6.2 Let $I = \, < q, R, v_0 >$ be assigned to order q industrial system with *MTB* R and *RB* v_0, *UMG* $S = \, < q, R >$,and

$$\left(\exists v \in \bar{V}_q\right) \cdot v_0 \subseteq v \,. \tag{6.6}$$

Then $\Delta V = \{v_0 - v \mid v \in \bar{V}_q\}$ may be replaced by (reduced to) $\Delta V' = \mathbf{min}\ \Delta V$.

We shall call min ΔV the compressed representation of ΔV.

A more complicated "reverse" task associated with *AIS* resource base insufficiency is to determine *what part of the order may be completed given an insufficient RB* (below, this case is called ***partial order completion***). From the practical point of view, this case is relevant when there is no opportunity to acquire the resources necessary for full completion of an order (e.g., one frame and 1400 euros in the previous Example 6.5). This situation is typical for closed *ISs*.

6.1.4 Assessment of Partial Order Completion in the Case of an Insufficient Resource Base

Consider an *AIS* $I = \, < \boldsymbol{q}, R, v_0 >$,where $\boldsymbol{q} = \{\boldsymbol{n_1} \cdot \boldsymbol{a_1}, \ldots, \boldsymbol{n_m} \cdot \boldsymbol{a_m}\}$.

Let us introduce an auxiliary object q and create a *UMMG* $S = \ < \{1 \cdot q\}, R', F >$, where $R' = R \cup \{r_q\}$, r_q is a unitary metarule

$$q \rightarrow \gamma_1 \cdot \boldsymbol{a_1}, \ldots, \gamma_m \cdot \boldsymbol{a_m}, \tag{6.7}$$

and the filter

$$F = \bigcup_{j=1}^{m} \{0 \leq \gamma_j \leq \boldsymbol{n}_j\} \cup \{a \leq n \mid n \cdot a \in v_0\} \cup \{a = 0 \mid a \in \overline{A}_S \& a \notin v_0\}. \tag{6.8}$$

Let us comment on (6.7)–(6.8).

The *UMR* (6.7) and the first component of the definition of the filter F determine that partial completion of the order q is represented by use of variable-multiplicities $\gamma_1, \ldots, \gamma_m$, whose values in the generated *TMSs* in set $\bar{V}_S$ are amounts of manufactured objects $\boldsymbol{a_1}, \ldots, \boldsymbol{a_m}$ (of course, $\boldsymbol{a_1}, \ldots, \boldsymbol{a_m}$ in the general case can be non-terminal objects, representing items assembled by the *AIS* from primary resources). Every boundary condition $0 \leq \gamma_j \leq \boldsymbol{n}_j$ in the aforementioned component defines that the value of the variable γ_j may be any integer in the interval $[0, \boldsymbol{n}_j]$.

The second component of the definition of the filter F is induced by the resource base of the *AIS* v_0 in such a way that in every *TMS* in $\bar{V}_S$, the amount of resource a is not greater than its amount n in the *RB*. Only variants of a partial completion of the order are selected that satisfy all boundary conditions $a \leq n$ such that $n \cdot a \in v_0$.

However, if we define the filter F by only the two considered components, then every *TMS* satisfying these components and containing at least one multiobject $n \cdot a$ such that $a \notin v_0$ (i.e., the *RB* does not contain a at all) will appear in $\bar{V}_S$. This is unacceptable. To exclude such cases, the third component is joined to the filter F; it contains boundary conditions $a = 0$ for all objects a which do not occur in the resource base of the considered *AIS*. Such conditions eliminate *TMSs* with "out-of-*RB*" multiobjects.

The constructed unitary multiset metagrammar S (especially the filter F) perhaps looks too bulky, but it demonstrates the essence of application of *UMMGs* to the creation of a set of possible variants of partial order completion given an insufficient resource base.

According to the mathematical semantics of *UMMGs*, the set of terminal multisets $\bar{V}_S$ includes all *TMSs* which contain multiobjects of two types:

$m_i \cdot \gamma_i$, where the multiplicity $m_i \in [0, \boldsymbol{n}_i]$ is the number of objects $\boldsymbol{a_i} \in q$ which may be manufactured by the *AIS* $I = \ < \boldsymbol{q}, R, v_0 >$ when the resource base v_0 is insufficient for full order completion

$k \cdot a$, where the multiplicity $k \in [0,n]$ is the number of objects a, where $n \cdot a \in v_0$, which would be spent for this purpose

So each $v \in \bar{V}_S$ defines

$$v_{\boldsymbol{q}} = \{m_i \cdot a_i \mid m_i \cdot \gamma_i \in v\} \subseteq \boldsymbol{q} \tag{6.9}$$

as a part of the order $\boldsymbol{q}$ which may be manufactured by the *AIS* $I = \ <\boldsymbol{q}, R, v_0>$ with resource base v_0, which is insufficient for full order completion. At the same time, the multiset $v - v_{\boldsymbol{q}}$ is the part of the resource base necessary for the completion of this part of the order $\boldsymbol{q}$.

Example 6.6
Let the assigned industrial system $I = \ <\boldsymbol{q}, R, v_0>$, where R and $\boldsymbol{q}$ are the same as in Example 6.4 and the resource base v_0 is the same as in Example 6.5. According to (6.7)–(6.8), r_q is the unitary metarule $q \rightarrow \gamma \cdot (auto)$, and the filter

$$F_q = \{0 \le \gamma \le 2,\ (frame) \le 1, (door) \le 9, (wheel) \le 20, (motor) \le 5,$$
$$(transmission) \le 5, (fuel\ tank) \le 2,$$
$$(mnt : autos\ AL) \le 120, (mnt : engines\ AL\ 1) \le 250,$$
$$(mnt : engines\ AL\ 2) \le 30,\ (kwh) \le 1200,\ (eur) \le 2000\}.$$

So, according to the mathematical semantics of *UMMGs*, the set $\bar{V}_S$ contains two terminal multisets, corresponding to two values of the variable γ—namely, 0 and 1:

$$\bar{V}_q = \{\{\varnothing\},\quad \{1 \cdot \bar{\gamma}, 1 \cdot (frame), 4 \cdot (door), 4 \cdot (wheel), 1 \cdot (motor), 1(transmission),$$
$$1 \cdot (fuel\ tank), 500 \cdot (kwh), 1700 \cdot (eur), 50 \cdot (mnt : autos\ AL),$$
$$80 \cdot (mnt : engines\ AL\ 1)\ \}\ \}.$$

As can be seen, the first element of $\bar{V}_S$, which is the empty set, corresponds to $\gamma = 0$, while the second element corresponds to $\gamma = 1$, and there are two different ways of partially completing order $\boldsymbol{q}$. The first one is to produce nothing, while the second is to produce one auto.

Let us note, however, that the proposed solution of the task of assessment of the possibility of partial completion of an order by an *AIS* with an insufficient resource base is not fully suitable from the practical point of view. As seen from the presented Example 6.6, the set $\bar{V}_S$ along with practically sensible *TMSs* may include also multisets with small (including zero) object multiplicities, which are recognized intuitively as practically useless. So it is reasonable to make a solution set ***free of the aforementioned useless variants***.

This objective may be simply achieved by the application of the function **max** introduced in Sect. 3.1.2. The set of practically reasonable ("***best***") solutions is defined by the following obvious statement.

Statement 6.3 The set of "the best" parts of an order $\boldsymbol{q}$ which may be completed by an *AIS* $I = \ <\boldsymbol{q}, R, v_0>$ with an insufficient resource base v_0 is **max** $\bar{V}_S$.

As may be seen, the set **max** $\bar{V}_S$ contains only non-dominated terminal multisets; this set is similar to ***Pareto-optimal sets*** in classical optimization (Harrington 2014; Hillier and Lieberman 2014; Lasdon 2013).

Example 6.7
Let us consider the set $\bar{V}_S = \{\{\emptyset\}, v_1\}$ from the previous Example 6.6. As can be seen, **max** $\bar{V}_S = \bar{V}_S - \{\{\emptyset\}\} = \{v_1\}$, because $\{\emptyset\} \subseteq v_1$.

6.1.5 Assessment of Ecological Damage Which May Be Caused by ISs *During Their Operation and Order Feasibility Under Ecological Restrictions*

One of the most demanded and important features of modern and future industrial systems is their ***ecological cleanness***. So it would be very reasonable to develop techniques for the multigrammatical representation of *ISs'* impact on the environment, i.e., finally, the scale of environmental pollution, which is directly proportional to the amounts of various harmful substances emitted by industry. Let us consider a rather simple *UMG*-based technique of assessment of the aforementioned amounts.

Let R be a set of unitary rules in a technological interpretation, representing the manufacturing technological base of some *IS*. Let us also recall that according to (3.38), the body of any such *UR*

$$a \to n_1 \cdot a_1, \ldots, n_m \cdot a_m, k_1 \cdot b_1, \ldots, k_l \cdot b_l \tag{6.10}$$

contains two sets of *ORs*: $\{a_1, \ldots, a_m\}$, representing structural components of a produced *OR* a, and $\{b_1, \ldots, b_l\}$, representing *ORs* applied for manufacturing (assembling) an *OR* a from n_1 *ORs* a_1, …, n_m *ORs* a_m. The unitary rule as a whole, in fact, represents some manufacturing device or an entire industrial facility producing object-resources (of a type) a.

From the ecological point of view, it is a common fact that any operation cycle of an *MD* (facility) may cause some environmental pollution as a side result of physical, chemical, biological, and other effects occurring during *OCs*. To take such results into account, it is sufficient to join to the *UR* body a set of multiobjects representing the amounts of the aforementioned substances affecting the environment (*SAEs*) created during an operation cycle which produces the *OR* a. We shall denote the set of *ORs* representing substances created during an *OC* by V_C, so the total set of names of *ORs* present in *URs* belonging to the set R is

$$V = V_{A,B} \cup V_C, \tag{6.11}$$

where $V_{A,\ B}$ is the set of names of *ORs* occurring in bodies of *URs* (6.10) and denoting *ORs* used for the production of other *ORs* as well as durations of operation cycles of *MDs*, while V_C is the set of names of *ORs* representing measurement units of *SAEs* during the aforementioned operation cycles. Naturally, sets $V_{A,\ B}$ and V_C do not intersect. In this case, any *UR* representing an *MD* would be as follows:

$$a \rightarrow n_1 \cdot a_1, \ldots, n_m \cdot a_m, k_1 \cdot b_1, \ldots, k_l \cdot b_l, i_1 \cdot c_1, \ldots, i_p \cdot c_p, \tag{6.12}$$

where multiobjects $i_1 \cdot c_1, \ldots, i_p \cdot c_p$ such that $\{c_1, \ldots, c_p\} \subseteq V_C$ represent amounts of the substances $c_1, \ldots, c_p$ affecting the environment during an operation cycle which are created by the production of an *OR* a.

Now, it is quite simple to assess the total amounts of *SAEs* created as a result of completion of an order $q = \{\boldsymbol{n_1} \cdot \boldsymbol{a_1}, \ldots, \boldsymbol{n_m} \cdot \boldsymbol{a_m}\}$.

Let *AIS* $I = \ < q, R, v_0 >$ be such that all *URs* belonging to R have the form (6.11) (in the general case, the list of *MOs* $i_1 \cdot c_1, \ldots, i_p \cdot c_p$ may be empty; this corresponds to clean *MDs*). Also, we shall extend the resource base v_0 of this *AIS* by joining to it a multiset

$$v_C = \{\boldsymbol{i_1} \cdot \boldsymbol{c_1},\ \ldots,\ \boldsymbol{i_p} \cdot \boldsymbol{c_p}\} \subseteq v_0\ , \tag{6.13}$$

representing ***ecological restrictions***—the maximum amounts of *SAEs* allowed for this order. After this inclusion, it is possible to apply criterion (6.4) of an order's feasibility to this *AIS* and, thus, to determine whether order q is feasible by an *AIS* with manufacturing technological base R and resource base v_0 under ecological restrictions v_C.

Statement 6.4 Let $I = \ < q, R, v_0 >$ be assigned to order q industrial system with *MTB* R and *RB* v_0, $\boldsymbol{v_C}$ be a representation of ecological restrictions established for this order, and $= < q, R >$. Then order q may be completed by this *AIS* respecting these ecological restrictions if

$$(\exists v \in \bar{V}_S) \cdot v \subseteq v_0 \cup v_C. \tag{6.14}$$

Otherwise, order q cannot be completed by the *AIS I*.

In practice, ecological restrictions are usually established not for every incoming order, but to apply to some time interval of operation of the considered *IS* during which an unpredictable number of orders may arrive. In this case, the multiset v_C is corrected after every period of order completion by the subtraction of a multiset representing the real amounts of *SAEs* created by the *IS* during this period, and the remaining multiset is applied as the ecological restrictions for the following incoming order. If this approach is implemented, then after some order, all those following it may be declared unfeasible because all predefined limits would be already exhausted. This way seems not rational, and more refined logics may be applied—for example, by some kind of scaling of total *IS* ecological restrictions to any order.

To implement ecological cleanness of a future industry, it is necessary to take ecological restrictions into account when conducting competitions between industrial actors who participate in it. Let us move now to a consideration of the extremely important area of *UMG*-based modeling of competitions, to which the technique introduced above in this section for treating ecological restrictions may be applied without any difficulties.

6.1.6 Technique of Competition Modeling

A competition (tender) is a standard tool for rational selection of the best of multiple possible contractors or to distribute a total amount of work among several contractors.

The main tools of multiset modeling of competitions are variative filtering unitary multigrammars and unitary multiset metagrammars. *FUMG/UMMG* variativity provides a representation of ***ISs containing different subjects able to complete one and the same order by manufacturing identical objects each in its own way, resulting in cooperation.***

There are two different approaches to an implementation of competitions. We shall name them cooperative and selective.

The ***cooperative approach to competitions*** ensures the most rational distribution of total work among all possible contractors, especially when the capabilities of none of the competitors alone are sufficient for total order completion. In this case, the most rational way may be to split the order between different subjects, thus integrating their capabilities.

The selective approach usually called "***winner takes all***" is applied, perhaps, even more often than the cooperative one, due to the practical simplicity of its implementation. The one winner is selected, and he gets the whole work, though he may create a collaboration by his own decisions.

Let us begin with the ***cooperative approach***.

Consider an *AIS* $I = \ < \boldsymbol{q}, R, v_0 >$, where the scheme R contains for every type of produced object-resource a $k \geq 1$ unitary rules with the same header a and different bodies:

$$a \to 1 \cdot r_1^a, n_1^1 \cdot a_1^1, \ldots, n_{m_1}^1 \cdot a_{m_1}^i,$$

$$\ldots$$

$$a \to 1 \cdot r_k^a, n_1^k \cdot a_1^k, \ldots, n_{m_k}^k \cdot a_{m_k}^k, \tag{6.15}$$

where the i-th alternative corresponds to the subject r_i^a of the manufacturing technological base of the *AIS I*. This subject allows manufacturing of one *OR* a by utilizing for this purpose n_1^i object-resources a_1^i, …, $n_{m_i}^i$ object-resources $a_{m_i}^i$ and doing this work during one operation cycle. If $k = 1$, then the subject r_1^a has a monopoly on manufacturing *ORs* of type a.

Let us consider now the *UMG* $S = \ < \boldsymbol{q}, R >$ and assume that the resource base v_0 is sufficient for order $\boldsymbol{q}$ completion, which means there exist $M \geq 1$ ways of doing this work. Each such way is associated with a specific resource collection v_j, $j = 1,\ldots, M$, which is necessary and sufficient for this order completion, so every $v_j \in \bar{V}_S$ satisfies a condition

$$v_j \subseteq v_0. \tag{6.16}$$

The set of all such collections is

$$\bar{V} = \{v_1, \ldots, v_M\} \subseteq \bar{V}_S, \tag{6.17}$$

where for every $j = 1,\ldots, M$

$$v_j = v'_j + v''_j, \tag{6.18}$$

where, in turn, v'_j is a *TMS* containing all multiobjects of the form $m \cdot r_i^a$ (m is the number of operation cycles of the *MTB* subject r_i^a involved in the j-th way of order completion) and v''_j is a *TMS* containing multiobjects of the form $l \cdot a$ (l is the amount of the object-resource a spent during the implementation of the considered j-th way). In fact, v'_j defines the cooperation and degree of participation of all its members in completion of the order $\boldsymbol{q}$, while v''_j defines the total of passive resources spent by this cooperation during this process.

The ***objective of a competition is to select the best way of order completion***, i.e., **the one TMS** $v_j \in \bar{V}$. This task may be done by constructing a filtering unitary multiset grammar $S_{\boldsymbol{q}} = \ <\boldsymbol{q}, R, F>$, where

$$F = F_{v_0} \cup F_c, \tag{6.19}$$

and

$$F_{v_0} = \{a \leq n \mid n \cdot a \in v_0\} \cup \{a = 0 \mid a \in \bar{A}_S \& a \notin v_0\} \tag{6.20}$$

is a subfilter performing filtration of the set $\bar{V}$, while F_c is a subfilter ***selecting a unique TMS v*** $\in \bar{V}$ ***and thus determining a winner*** (in fact, a winning collaboration). F_c **is a formal representation of a set of substantial rules verbally defining the notion "the best way of order completion."** It may contain conditions reflecting natural restrictions on the available operating resources of manufacturing devices (e.g., the number of their operation cycles), as well as limitations on some material resources consumed by such *MDs*. Of course, the best way of order completion may be defined by inclusion in a subfilter F_c of some optimizing conditions relevant to minimal amounts of consumed resources and/or maximal values of some parameters describing the quality of the manufactured items.

Finally, the result of a competition is represented by a multiset $\boldsymbol{v} \in \bar{V}_{S_q}$.

Example 6.8
Let the scheme R contain the following unitary rules r_1, r_2, r_3,and r_4:

$$r_1: \quad (auto) \to 1 \cdot (auto\ east), 1 \cdot (frame), 1 \cdot (engine), 4 \cdot (door), 4 \cdot (wheel),$$
$$80 \cdot (mnt : autos\ AL\ east), 300 \cdot (kwh), 600 \cdot (eur);$$

$$r_2: (auto) \to 1 \cdot (auto\ west), 1 \cdot (frame), 1 \cdot (engine), 4 \cdot (door), 4 \cdot (wheel),$$
$$60 \cdot (mnt : autos\ AL\ west), 500 \cdot (kwh), 800 \cdot (eur);$$

$$r_3: (engine) \to 1 \cdot (engine\ south), 1 \cdot (motor), 1 \cdot (transmission), 1 \cdot (fuel\ tank),$$
$$70 \cdot (mnt : engines\ AL\ south), 120 \cdot (kwh), 900 \cdot (eur);$$

$$r_4: (engine) \to 1 \cdot (engine\ north), 1 \cdot (motor), 1 \cdot (transmission), 1 \cdot (fuel\ tank),$$
$$90 \cdot (mnt : engines\ AL\ north\ ), 200 \cdot (kwh), 500 \cdot (eur).$$

These *URs* represent a set of industrial facilities manufacturing autos and engines for these autos. There are two facilities with assembly lines capable of manufacturing autos from frames, engines, doors, and wheels and two facilities with assembly lines capable of manufacturing engines from motors, transmissions, and fuel tanks. The assembly line at the facility located in the east region and represented by the *UR* r_1 manufactures one auto during 80 min (the *MO* 80 · (*mnt* : *autos AL east*)), consuming 300 kWh h of electrical energy (the multiobject 300 · (*kwh*)), and the cost of this operation cycle is 600 euros (the *MO* 600 · (*eur*)). The assembly line at the facility located in the west region and represented by the *UR* r_2 manufactures one auto during 60 min (the multiobject 60 · (*mnt* : *autos AL west*)), consuming 500 kWh h of electrical energy (the multiobject 500 · (*kwh*)), and the cost of this operation cycle is 800 euros (the *MO* 800 · (*eur*)). The assembly line at the facility located in the south region and represented by the *UR* r_3 manufactures one engine during 70 min (the multiobject 70 · (*mnt* : *engines AL south*)), consuming 120 kWh h of electrical energy (the multiobject 120 · (*kwh*)), and the cost of this operation cycle is 900 euros (the *MO* 900 · (*eur*)). Finally, the assembly line at the facility located in the north region and represented by the *UR* r_4 manufactures one engine during 90 min (the multiobject 90 · (*mnt* : *engines AL north*)), consuming 200 kWh h of electrical energy (the multiobject 200 · (*kwh*)), and the cost of this operation cycle is 500 euros (the *MO* 500 · (*eur*)).

Comparing these *URs* with (6.10), we see that the multiobjects $1 \cdot r_i^a$ identifying ways of manufacturing *ORs* *a* are 1 · (*auto east*), 1 · (*auto west*), 1 · (*engine south*), and 1 · (*engine north*).

As can be seen, there is a competitive production environment at the level of final manufactured objects (autos), as well as at the level of their components (engines).

Let the resource base of the considered *IS* be

$$v_0 = \{6 \cdot (frame), 13 \cdot (door), 20 \cdot (wheel), 5 \cdot (motor), 7 \cdot (transmission),$$
$$8 \cdot (fuel\ tank), 160 \cdot (mnt : autos\ AL\ east), 120 \cdot (mnt : autos\ AL\ west),$$
$$250 \cdot (mnt : engines\ AL\ south), 300 \cdot (mnt : engines\ AL\ north),$$
$$1200 \cdot (kwh), 2500 \cdot (eur)\}.$$

It is assumed that all the spare parts necessary for auto manufacturing are available to all four producing facilities from some common storage, as well as electrical energy produced by some power plant; mutual payments, if necessary, are implemented by some common banking office.

Let the order be two autos, so

$$\boldsymbol{q} = \{2 \cdot (auto)\},$$

and an optimal completion of this order would consume no more than 1000 kWh h of electrical energy and minimize the cost of manufacturing, so the selecting subfilter is

$$F_c = \{(kwh) \le 1000,\ (eur) = min\ \}.$$

The subfilter representing the limits relevant to the current state of the industrial system is

$$F_{v_0} = \{(frame) \le 6, (door) \le 13, (wheel) \le 20, (motor) \le 5,$$
$$(transmission) \le 7, (fuel\ tank) \le 8,$$
$$(mnt : autos\ AL\ east) \le 160, (mnt : autos\ AL\ west) \le 120,$$
$$(mnt : engines\ AL\ south) \le 250, (mnt : engines\ AL\ north) \le 300,$$
$$(kwh) \le 1200, (eur) \le 2500\}.$$

As can be seen,

$$\bar{V}_S = \{v_{2,0,2,0}, v_{2,0,1,1}, v_{2,0,0,2}, v_{1,1,2,0}, v_{1,1,1,1}, v_{1,1,0,2}, v_{0,2,2,0}, v_{0,2,1,1}, v_{0,2,0,2}\},$$

where $v_{i,j,k,l}$ is a *TMS* corresponding to the way of order completion which produces i autos at the east facility, j autos at the west facility, k engines at the south facility, and l engines at the north facility (in any case $i + j = k + l = 2$), so

$$v_{2,0,2,0} = v + \{2 \cdot (auto\ east), 160 \cdot (mnt : autos\ AL\ east),$$
$$2 \cdot (engine\ south), 140 \cdot (mnt : engines\ AL\ south),$$
$$840 \cdot (kwh), 3000 \cdot (eur)\},$$

$$v_{2,0,1,1} = v + \{2 \cdot (auto\ east), 160 \cdot (mnt : autos\ AL\ east),$$
$$1 \cdot (engine\ south), 1 \cdot (engine\ north),$$
$$70 \cdot (mnt : engines\ AL\ south), 90 \cdot (mnt : engines\ AL\ north\ ),$$
$$920 \cdot (kwh), 2600 \cdot (eur)\},$$

$$v_{2,0,0,2} = v + \{2 \cdot (auto\ east), 160 \cdot (mnt : autos\ AL\ east),$$
$$2 \cdot (engine\ north), 180 \cdot (mnt : engines\ AL\ north\ ),$$
$$1000 \cdot (kwh), 2200 \cdot (eur)\},$$

$$v_{1,1,2,0} = v + \{1 \cdot (auto\ east), 1 \cdot (auto\ west),$$
$$80 \cdot (mnt : autos\ AL\ east), 60 \cdot (mnt : autos\ AL\ west),$$
$$2 \cdot (engine\ south), 140 \cdot (mnt : engines\ AL\ south),$$
$$1040 \cdot (kwh), 3200 \cdot (eur)\},$$

$$v_{1,1,1,1} = v + \{1 \cdot (auto\ east), 1 \cdot (auto\ west),$$
$$80 \cdot (mnt : autos\ AL\ east), 60 \cdot (mnt : autos\ AL\ west),$$
$$1 \cdot (engine\ south), 1 \cdot (engine\ north),$$
$$70 \cdot (mnt : engines\ AL\ south), 90 \cdot (mnt : engines\ AL\ north\ ),$$
$$1120 \cdot (kwh), 2800 \cdot (eur)\},$$

$$v_{1,1,0,2} = v + \{1 \cdot (auto\ east), 1 \cdot (auto\ west),$$
$$80 \cdot (mnt : autos\ AL\ east), 60 \cdot (mnt : autos\ AL\ west),$$
$$2 \cdot (engine\ north), 180 \cdot (mnt : engines\ AL\ north\ ),$$
$$1200 \cdot (kwh), 2400 \cdot (eur)\},$$

$$v_{0,2,2,0} = v + \{2 \cdot (auto\ west), 120 \cdot (mnt : autos\ AL\ west),$$
$$2 \cdot (engine\ south), 140 \cdot (mnt : engines\ AL\ south),$$
$$1240 \cdot (kwh), 3400 \cdot (eur)\},$$

$$v_{0,2,1,1} = v + \{2 \cdot (auto\ west), 120 \cdot (mnt : autos\ AL\ west),$$
$$1 \cdot (engine\ south), 1 \cdot (engine\ north),$$
$$70 \cdot (mnt : engines\ AL\ south), 90 \cdot (mnt : engines\ AL\ north\ ),$$
$$1320 \cdot (kwh), 3000 \cdot (eur)\},$$

$$v_{0,2,0,2} = v + \{2 \cdot (auto\ west), 120 \cdot (mnt : autos\ AL\ west),$$
$$2 \cdot (engine\ north), 180 \cdot (mnt : engines\ AL\ north\ ),$$
$$1400 \cdot (kwh), 2600 \cdot (eur)\},$$

and the constant summand representing amounts of spare parts consumed for order completion

$$v = \{4 \cdot (frame), 8 \cdot (door), 8 \cdot (wheel), 4 \cdot (motor), 4 \cdot (transmission),$$
$$4 \cdot (fuel\ tank)\}.$$

As can be seen, the terminal multiset that is the only element of the *STMS* $\bar{V}_{S_q}$ is $\boldsymbol{v} = v_{2,\ 0,\ 0,\ 2}$, which means that the order will be completed by cooperation between the east auto manufacturing facility, which will produce both autos, and the north engine manufacturing facility, which will produce both engines for these autos. This way of order completion satisfies the restriction on the consumed amount of electrical energy (1000 kWh h is not greater than the established upper bound 1000 kWh h) and has the minimal cost of manufacturing (2200 euros) compared to all other variants also satisfying the aforementioned restriction (namely, $v_{2,\ 0,\ 2,\ 0}$ and $v_{2,\ 0,\ 1,\ 1}$).

Concerning (6.18), it is evident that

$$\boldsymbol{v} = \boldsymbol{v}' + \boldsymbol{v}'',$$

where the multiset

$$\boldsymbol{v}' = \{2 \cdot (auto\ east),\ 2 \cdot (engine\ north)\}$$

represents the created cooperation, while the multiset

$$v'' = \{160 \cdot (mnt : autos\ AL\ east), 180 \cdot (mnt : engines\ AL\ north\ ),$$
$$1000 \cdot (kwh), 2200 \cdot (eur)\}$$

represents the collection of resources necessary to this cooperation for the completion of the order.

Let us consider now the ***selective approach*** to competitions. To model this approach, we shall use a technique similar to that applied above in Sect. 6.1.4 for the ***partial completion of orders by an IS with an insufficient resource base***.

Let *AIS* $I = < \boldsymbol{q}, R, v_0>$ and order $\boldsymbol{q} = \{n \cdot \boldsymbol{a}\}$. Consider a filtering unitary multiset grammar $S_q = < q, R_q, F>$, where $q = \{n \cdot a\}$, where a is an introduced auxiliary object, the filter F is the same as (6.19), and

$$R_q = R - R_{\boldsymbol{a}} \cup R_a, \tag{6.21}$$

where the set $R_{\boldsymbol{a}} \subseteq R$ contains $k>1$ unitary rules with the same header $\boldsymbol{a}$ and different bodies $w_1, \ldots, w_k$, as defined by (6.15), while the set R_a contains $2k$ unitary rules

$$a \rightarrow 1 \cdot r_1^a, n \cdot a_1,$$
$$\ldots$$
$$a \rightarrow 1 \cdot r_k^a, n \cdot a_k, \tag{6.22}$$

$$a_1 \rightarrow w_1,$$
$$\ldots$$
$$a_k \rightarrow w_k, \tag{6.23}$$

where $a_1, \ldots, a_k$ are specially introduced auxiliary objects. As can be seen from (6.21) to (6.23), the terminal multiset generated by the constructed *FUMG* S_q and corresponding to the competition winner is $\{1 \cdot r_i^a\} + v$, where i is its number and v is a multiset representing all the rest of the cooperation participating in order completion, as well as the amounts of resources consumed by this cooperation. So due to (6.21)–(6.23), the winner would get the right to produce all n *ORs* $\boldsymbol{a}$. Note that the ***principle "winner takes all" is implemented only at the top level of a competition***, i.e., among competitors able to manufacture the final *ORs* $\boldsymbol{a}$, n of which are defined as the objective of the tender. The set of other participants in order completion is created by the cooperative approach defined above by (6.15)–(6.20). However, the winner, by including in the filter F special substantial conditions, may implement any desired logic of selection of the necessary participants and, hence, form his preferred cooperation.

Example 6.9
Let us consider the same *IS* as in the previous example, and let us construct the filtering unitary multiset grammar implementing the selective approach regarding the same order. According to (6.21)–(6.23), the scheme of the constructed *FUMG* will be as follows:

$$r'_1: \ (order) \to 1 \cdot (\ east), 2 \cdot (auto\ east);$$

$$r'_2: \ (order) \to 1 \cdot (\ west), 2 \cdot (auto\ \ west);$$

$$r_1: \ (auto\ east) \to 1 \cdot (frame), 1 \cdot (engine), 4 \cdot (door), 4 \cdot (wheel),$$
$$80 \cdot (mnt : autos\ AL\ east), 300 \cdot (kwh), 600 \cdot (eur : east);$$

$$r_2: (auto\ west) \to 1 \cdot (frame), 1 \cdot (engine), 4 \cdot (door), 4 \cdot (wheel),$$
$$60 \cdot (mnt : autos\ AL\ west), 500 \cdot (kwh), 800 \cdot (eur : west);$$

$$r_3: (engine) \to 1 \cdot (engine\ south), 1 \cdot (motor), 1 \cdot (transmission),$$
$$1 \cdot (fuel\ tank), 70 \cdot (mnt : engines\ AL\ south), 120 \cdot (kwh),$$
$$900 \cdot (eur : south);$$

$$r_4: (engine) \to 1 \cdot (engine\ north), 1 \cdot (motor), 1 \cdot (transmission), 1 \cdot (fuel\ tank),$$
$$90 \cdot (mnt : engines\ AL\ north\ ), 200 \cdot (kwh), 500 \cdot (eur : north).$$

Here, (*order*) is the auxiliary object *a* from (6.22), and the *URs* r'_1 and r'_2 are a concretization of (6.22), while r_1 and r_2 are a concretization of (6.23). All the rest is the same as in the previous example. In the generated terminal multisets, the introduced multiobjects 1 · (*east*) and 1 · (*west*) serve as markers of the competition winners, while the introduced objects (*auto east*) and (*auto west*) are, in fact, the names of the final objects to be manufactured by them.

As may be seen, the proposed technique allows not only the implementation of competitions but also the ***distribution of work and resources inside collaborations spearheaded by winners of a competition***. However, due to natural limitations of the applied *UMG/UMMG*-based toolkit, which may operate on only ***additive*** resources, there is no opportunity to assess and to use the ***total time of order completion*** in the process to select a winner. Time is not a fully additive resource; as was already mentioned above, it is additive only regarding a separate manufacturing device: to produce *n* items, each produced by this device during an interval Δt, it is necessary to spend an interval $n \cdot \Delta t$. So this feature may generally be used for the elimination of variants of order completion which imply a workload of manufacturing devices exceeding their real capabilities. But concerning the entire manufacturing technological base of an industrial system, time is not an additive resource due to the possibility of parallel operation of all *MDs* which were supplied by the *RB* at the initial moment of their current operation cycles: it is clear that n identical devices, supplied with the necessary resources and working simultaneously, may produce n items during the same interval Δt. So ***to solve all tasks considered previously in this section regarding the real durations of order completion, it is necessary to use schedules, which, in turn, may be created by use of temporal multiset grammars and their various extensions***. Let us consider this direction of *MGF* application.

6.2 *TMG*-Based Representation of Industrial Systems and *IS* Scheduling

Let us begin with the one-order case, where an ***industrial system manufacturing passive resources permits the completion of some separate order***. Various issues concerning this case will be considered in Sects. 6.2.1–6.2.3. The general case concerning scheduling of multiple orders incoming to an *IS MPR* at unpredictable time moments will be considered in Sect. 6.2.4. Consideration of scheduling issues concerning *ISs* manufacturing not only passive but also *active* resources is beyond the scope of this book.

6.2.1 TMG*-Based Representation of* ISs MPR*, Bottom-Up Assessment of Order Feasibility, and Ecological Issues Modeling*

It is not necessary to repeat the representation of *ISs MPR* by temporal multiset grammars introduced in Sect. 3.2. Let us only recall that the state of an *IS* at any moment n of its operation is

$$v = v_A + v_R + \{n \cdot t\}, \tag{6.24}$$

where v_A is a multiset representing amounts of passive resources available to manufacturing devices (active resources) at this moment, and the current state of the *IS's* manufacturing technological base is

$$v_R = \{1 \cdot r_1\tau_1, \ \ldots, \ \ 1 \cdot r_N\tau_N\}, \tag{6.25}$$

where, in turn, τ_j is the schedule of the *MD* r_j, i.e., a string $[n^j_1, l^j_1] \ldots \left[n^j_{k_j}, l^j_{k_j}\right]$, including moments n^j_i of activation of *MD* r_j and moments l^j_i at which it finishes an operation cycle ($n^i_j \le l^j_i$). All $r_i\tau_j$ are composite names.

A *TMG* $S = < v_0, R>$ represents an *IS MPR*, so the set of all possible states of this *IS* is nothing but V_S. But until we have defined a representation of orders, there is no substantial and formal interconnection between the basic knowledge representation (namely, *TMGs*) and the considered objects (*ISs MPR*).

In the previous Sect. 6.1, where we have used unitary multiset grammars as a tool for a representation of industrial systems, a triple $I = < \boldsymbol{q}, R, v_0>$was called an "*IS* with *MTB R* and *RB* v_0, assigned to order $\boldsymbol{q}$," while a triple $I = < \{\emptyset\}, R, v_0>$ was called a free *IS*; an order $\boldsymbol{q}$ was introduced as a collection of object-resources $\{\boldsymbol{n_1} \cdot \boldsymbol{a_1}, \ldots, \boldsymbol{n_m} \cdot \boldsymbol{a_m}\}$ to be manufactured by this *IS*.

Because *TMGs* generate temporal multisets, there occurs an opportunity to include in an order desired restrictions on the duration of order completion. Following the terminology introduced in Sect. 6.1, we shall call a triple $I = < \{\emptyset\}, R, v_0>$ a ***free industrial system manufacturing passive resources*** (*FIS MPR*), where

$$v_0 = v^0_A + v^0_R + \{n_0 \cdot t\}. \tag{6.26}$$

So v_0 is the initial state of this *IS MPR* at a moment n_0 when an order arrives at this system.

Similarly, a triple $I = < \boldsymbol{Q}, R, v_0>$ will be called an ***IS MPR with MTB R and RB** v_0, **assigned to order** $\boldsymbol{Q}$=<q,t>* (*AIS MPR*), where, as above, $\boldsymbol{q} = \{\boldsymbol{n_1} \cdot \boldsymbol{a_1}, \ldots, \boldsymbol{n_m} \cdot \boldsymbol{a_m}\}$, which is the ***substantial part of the order***, defines a collection of object-resources to be manufactured by this *IS* by a time moment defined by the ***temporal part of an order t,*** which may be any set of conditions introduced in Sect. 3.2.1 ($t \le n, t = n, n \le t \le n', t = min$,etc.). However, we shall use the most practical form of the temporal part of an order

$$\boldsymbol{t} = \{n \le t \le n', t = min\ \}, \tag{6.27}$$

defining the earliest of all possible time moments in an interval $[n, n']$.

As may be seen now, the set of all possible schedules of an *AIS MPR I* allowing the completion of an order $\boldsymbol{Q}$ is nothing but V_s, where $S = < v_0, R, F>$ is the filtering temporal multiset grammar whose filter, containing the time subfilter $\boldsymbol{t}$, is

$$F = F_q \cup \boldsymbol{t}, \tag{6.28}$$

and

$$F_q = \{\boldsymbol{a_1} \ge \boldsymbol{n_1},\ \ldots, \boldsymbol{a_m} \ge \boldsymbol{n_m}\}. \tag{6.29}$$

The *FTMG S* is said to be ***associated with the AIS** I.* Verbally, *S*, beginning from kernel v_0 (representing the initial state of the resource base and manufacturing technological base of this *AIS MPR* at time moment n_0), generates temporal multisets (representing all achievable states of the resource base and manufacturing technological base of the *AIS MPR*), each *MST* satisfying both subfilters F_q and $\boldsymbol{t}$ of filter F:

The amounts of object-resources $\boldsymbol{a_1}, \ldots, \boldsymbol{a_m}$ in the resource base of the *IS* are not less than $\boldsymbol{n_1}, \ldots, \boldsymbol{n_m}$, respectively, thus including the submultiset $\boldsymbol{q} = \{\boldsymbol{n_1} \cdot \boldsymbol{a_1}, \ldots, \boldsymbol{n_m} \cdot \boldsymbol{a_m}\}$ in this *MST* (i.e., in the *IS's RB*) (1).

The time moment when this *MST* occurs inside the *RB* is the minimal of all such moments belonging to interval $[n, n']$ (2).

Condition (1), represented by (6.29), ensures the completion of the ***substantial part of the order*** (all needed amounts of object-resources are manufactured), while condition (2) ensures the timely manufacturing of the aforementioned amounts of *ORs*, i.e., the ***temporal part of the order***.

Example 6.10

Let the manufacturing technological base of the considered *IS* be represented by the set *R*, including the following temporal rules r_1 and r_2:

$$< r_1; \{1 \cdot (frame),\ 1 \cdot (engine),\ 4 \cdot (door),\ 4 \cdot (wheel),\ 400 \cdot (kwh),\ 300 \cdot (eur)\}. \\ \rightarrow \{1 \cdot (auto),\ 50 \cdot (mnt)\} >,$$

$$< r_2; \{1 \cdot (motor),\ 1 \cdot (transmission),\ 1 \cdot (fuel\ tank),\ 100 \cdot (kwh),\ 900 \cdot (eur)\} \\ \rightarrow \{1 \cdot (engine),\ 80 \cdot (mnt)\} >,$$

and the resource base of this *IS* be

$$v_0 = \{12.00 \cdot t,\ 3 \cdot (frame), 10 \cdot (door), 16 \cdot (wheel), 4 \cdot (motor), 3 \cdot (transmission), \\ 4 \cdot (fuel\ tank), 1800 \cdot (kwh), 10{,}000 \cdot (eur)\},$$

that means that at 12.00, the *RB* contains object-resources represented by the multiobjects in the temporal multiset v_0 (12.00 $\cdot$ t is its time marker).

Let the order incoming to the considered *IS* at 12.00 be

$$\boldsymbol{Q} = < \{2 \cdot (auto)\}, \{14.00 \leq t \leq 16.00,\ t = min\ \} >,$$

which means the objective of this order is to produce two autos at the earliest possible time between 14.00 and 16.00.

According to (6.28) and (6.29), the filter F of the filtering temporal multiset grammar $S = < v_0, R, F>$ associated with this *IS* is

$$F = \{(auto) \geq 2,\ 14.00 \leq t \leq 16.00,\ t = min\}.$$

Given this *TMG*-based representation of an *AIS MPR*, we may consider the same three tasks as in the previous Sect. 6.1, concerning the assessment of:

Order feasibility

Additional amounts of resources necessary for the completion of an initially non-feasible order

Partial completion of a non-feasible order

The simplest is ***the first task***, which may be easily solved on the basis of bottom-up generation of *MSTs*.

Statement 6.5 Let an *AIS MPR* $I = \langle \boldsymbol{Q}, R, v_0 \rangle$ and the *FTMG* associated with this *IS* be $S = \langle v_0, R, F \rangle$. Then the order $\boldsymbol{Q}$ may be completed by this *IS* if

$$V_s \neq \{\varnothing\}. \tag{6.30}$$

Otherwise, the order $\boldsymbol{Q}$ is not feasible**.**

This statement is in fact a formal definition of order feasibility regarding a *TMG*-based representation of *AISs MPR*. However, it is not applicable to the other two tasks, which were considered above in Sects. 6.1.3 and 6.1.4 in the context of a *UMG*-based representation of *ISs*. This consideration was possible due to top-down (beginning from an order) generation of multisets representing collections of resources needed for order completion. Let us apply the same techniques to a *TMG*-based representation.

To finish this section, let us note that the proposed TMG representation of ISs MPR permits the most natural and the easiest modeling of ***ecological issues*** *by inclusion in the right parts of temporal rules of multiobjects representing amounts of substances appearing as a side result of one operating cycle—for example,* 1 · *m3CO2 (one cubic meter of carbon dioxide). As a result of operation of an entire TB of an IS, some integral amounts of such substances, represented by the respective multiobjects, would be present in any generated temporal multiset; also some temporal rules may be included in the KB to describe interactions of the aforementioned substances with other substances and material objects.* By *including in a filter of an FTMG some boundary and/or optimizing (apparently, minimizing) conditions, it is quite easy to reject variants of IS operation which cause ecological damage exceeding thresholds established by current regulations. So order feasibility in the nature-friendly case would be easily assessed by the application of the described simple technique.*

6.2.2 *Top-Down Assessment of Order Feasibility for a* TMG-*Based Representation of* ISs MPR

In fact, top-down generation was introduced regarding temporal multiset grammars and filtering *TMGs* in Sect. 4.4 in order to reduce the computational complexity of

MST generation. Here, we shall try to formulate a criterion of order feasibility regarding a *TMG*-based representation of *AISs MPR*, by adapting the criterion introduced by Statement 6.1. For this purpose, we shall use the toolkit of dual *FTMGs*.

Statement 6.6 Let an *AIS MPRI* = $<\boldsymbol{Q}, R, v_0>$, where

$$\boldsymbol{Q} = <\boldsymbol{q}, \boldsymbol{t}>, \tag{6.31}$$

$$\boldsymbol{q} = \{\boldsymbol{n_1} \cdot \boldsymbol{a_1}, \; \ldots, \boldsymbol{n_m} \cdot \boldsymbol{a_m}\}, \tag{6.32}$$

$$\boldsymbol{t} = \{n \leq t \leq n', t = min\}, \tag{6.33}$$

and the *FTMG* dual to the *FTMG S* = $< v_0, R, F>$associated with this *AIS* is $-S = <\boldsymbol{q}, -R, -F>$, where

$$-R = \{< -r; v' \rightarrow v + \{-\Delta n \cdot \Delta t\} > | < r; v \rightarrow v' + \{\Delta n \cdot \Delta t\} > \in R\}, \tag{6.34}$$

$$-F = F_{v_0} \cup -\boldsymbol{t}, \tag{6.35}$$

$$F_{v_0} = \{a \leq n \mid n \cdot a \in v_0\} \cup \{a = 0 \mid a \in \overline{A}_S \& a \notin v_0\}, \tag{6.36}$$

$$\begin{aligned} -\boldsymbol{t} = \{ -n' + n_0 \leq t \leq -n + n_0 \mid n \leq t \leq n' \in \boldsymbol{t}\} \\ \cup \{ t = max \mid t = min \in \boldsymbol{t}\}. \end{aligned} \tag{6.37}$$

Then the order $\boldsymbol{Q}$ may be completed by this *IS* if

$$V_{-S} \neq \{\emptyset\}. \tag{6.38}$$

Let us comment on the formulated statement, paying attention to the points that are of principal importance for its validation.

Equation (6.34) defines a set of mirror temporal rules generating *MSTs* representing collections of object-resources necessary for order completion; it is clear that all terminal *MSTs* generated by the dual *FTMG* $-S$ represent collections of *ORs* which would be taken from the *IS* resource base. The subfilter F_{v_0} defined by (6.36) is just the same as in (6.20); it includes boundary conditions restricting the amounts of *ORs* available for order completion to those that exist in the resource base, and no other *ORs* may be used. Finally, the subfilter $-\boldsymbol{t}$ defined by (6.37) includes the chain boundary condition $-n' + n_0 \leq t \leq -n + n_0$ obtained from the *CBC* $n \leq t \leq n'$ in the subfilter $\boldsymbol{t}$ by subtraction of the value n_0 (the moment the order arrives at the *IS*); this operation allows the correct reverse computation of time markers of *MSTs* during the application of mirror temporal rules during *MST* generation. In fact, this transformation leads to the replacement of an absolute time by a relative one, counted from the end of order completion. The optimizing condition $t = min$, if it appears in the subfilter $\boldsymbol{t}$, selects *MSTs* with minimal multiplicities in the time marker. So, taking into account the reverse direction of computation of such multiplicities, this condition is replaced by $t = max$, which is included in the subfilter $-\boldsymbol{t}$.

Now, it is clear that the order $\boldsymbol{Q}$ is feasible if the set of *MSTs* generated by the dual *FTMG* $-S$ is not empty, i.e., if there exists at least one temporal multiset satisfying the filter $-F$ and generated by the application of the mirror temporal rules in the scheme $-R$.

Example 6.11
Let us consider the *TMG* $-S = \, <\boldsymbol{q}, -R, -F>$, mirror to the *TMGS* $= \, <\boldsymbol{q}, R, F>$ associated with the *AIS I* from the previous Example 6.10. According to (6.34)–(6.37),

$$\begin{aligned}-R = \{ &< -r_1; \{1\cdot(auto)\} \rightarrow \{1\cdot(frame), 1\cdot(engine), 4\cdot(door), 4\cdot(wheel),\\ &\quad 400\cdot(kwh), 300\cdot(eur), -50\cdot(mnt)\} >,\\ &< -r_2; \{1\cdot(engine)\} \rightarrow \{1\cdot(motor), 1\cdot(transmission), 1\cdot(fuel\ tank),\\ &\quad 100\cdot(kwh), 900\cdot(eur), -80\cdot(mnt)\} > \},\end{aligned}$$

$$\begin{aligned}-F = \{ &(frame) \le 3, (door) \le 10, (wheel) \le 16, (motor) \le 4,\\ &(transmission) \le 3, (fuel\ tank) \le 4,\\ &(kwh) \le 1800,\ (eur) \le 10{,}000,\ -4.00 \le t \le -2.00,\ t = max\ \},\end{aligned}$$

while = $\{2 \cdot (auto)\}$, as in the previous example. As may be seen, the created mirror *FTMG* is context-free, because the left parts of all temporal rules are one-element multisets.

Statement 6.6 may be used as a basis for the assessment of amounts of resources to be acquired for the completion of initially non-feasible orders.

6.2.3 *Assessment of Additional Amounts of Resources to Be Acquired and the Time Period Necessary for Completion of Initially Non-feasible Orders*

The simplest way to assess the additional amounts of resources to be acquired for the completion of initially non-feasible orders may be to apply relation (6.5) and

Statement 6.2, adapted to the general case of a *TMG*-based representation of *ISs MPR*. The resulting relation is, evidently,

$$\Delta V = \{v_A - v_0 \mid v_A + v_R + \{n \cdot t\} \in V_{-s}\}, \tag{6.39}$$

and a compressed representation of ΔV is **min** ΔV.

However, this solution, which is rational for a *UMG*-based representation of *ISs*, in the more general case of a *TMG*-based representation of *ISs MPR* may be far from rationality. The main reason for this is that there is no necessity to accumulate and to store all the resources which are necessary only at the moments of activation of manufacturing devices using these resources, from the starting moment of order completion (or, just the same, of *IS* schedule execution). Following this approach, a lot of resources would be out of use for part of the time of *AIS MPR* operation—they would simply occupy places in storages without any profit. It is clear that it would be much more reasonable to supply an *AIS MPR* with resources in such a way that the *ORs* necessary for some operation cycle of any manufacturing device would be transferred to its inputs directly at the moment when they are really necessary, i.e., upon its activation. By this alternative approach, we minimize redundant stockpiling inside an *AIS MPR*, but its implementation requires generation not only of an ***IS schedule*** but also of an ***IS supply schedule***, i.e., a sequence of time-fixed operations supplementing the *IS* resource base by amounts of object-resources necessary for *MDs* whose operation cycles according to the *IS* schedule could begin if the aforementioned amounts were available to these *MDs*.

So the task is ***to create an IS supply schedule***. Let us consider in detail a solution of the considered task.

Recall that an *IS* schedule, as it was introduced in Sect. 3.2.2, is nothing but a set of couples $\langle i, R(i)\rangle$, where i is a time moment and $R(i) \subseteq R$ is the subset of the manufacturing technological base, including manufacturing devices, which are activated at this moment, i.e., receive a start message from the *IS* controller.

Let $I = \ < \boldsymbol{Q}, R, v_0>$ be, as above, an *IS MPR* with resource base v_0 and manufacturing technological base *R*, assigned to an order $\boldsymbol{Q}$=<$\boldsymbol{q,t}$>, where, as above, $\boldsymbol{q} = \{\boldsymbol{n_1} \cdot \boldsymbol{a_1}, \ldots, \boldsymbol{n_m} \cdot \boldsymbol{a_m}\}$, $\boldsymbol{t}$= $\{n \leq t \leq n', t = min\}$, and

$$\Sigma_S = \{\boldsymbol{Sch}(v_R) \mid v_A + v_R + \{n \cdot t\} \in V_S\} \tag{6.40}$$

is a set of *IS* schedules permitting the completion of this order. Consider a separate *IS* schedule

$$\boldsymbol{\sigma} = \{<n_0, R(n_0)>, \ \ldots, \ <n_l, R(n_l)>\} \in \Sigma_S, \tag{6.41}$$

where $R(n_0) \subseteq R, \ \ldots, \ R(n_l) \subseteq R$are sets of manufacturing devices activated, respectively, at moments $n_0, \ldots, n_l$. As may be seen, the multiset

$$\Delta v_0 = \bar{v_0} - v_0, \tag{6.42}$$

where

$$\bar{v_0} = \sum_{<r; v \to v'> \in R(n_0)} v, \tag{6.43}$$

is nothing but the collection of resources which must be added to the initial resource base to activate all *MDs* $r \in R(n_0)$.

Similarly, for a time moment n_1

$$\Delta v_1 = \bar{v_1} - v_1, \tag{6.44}$$

where

$$\bar{v_1} = (v_0 - \bar{v_0}) + \sum_{<r; v \to v'> \in R(n_1)} v \tag{6.45}$$

and v_1 is the resource base after finishing all operation cycles which end at the moment n_1.

By induction,

$$\Delta v_i = \bar{v_i} - v_i, \tag{6.46}$$

where

$$\bar{v_i} = (v_{i-1} - \bar{v}_{i-1}) + \sum_{<r; v \to v'> \in R(n_i)} v \ . \tag{6.47}$$

Finally, the set,

$$\sigma_{in} = \bigcup_{i=1}^{I} \{ <n_i, \Delta v_i> \mid \Delta v_i \neq \{\emptyset\} \}, \tag{6.48}$$

is nothing but the required ***IS MPR supply schedule*** which permits the completion of the order ***Q***.

If all Δv_i are empty multisets, then, evidently, the initial *RB* v_0 is sufficient for the completion of order $\boldsymbol{Q}$, and the order is feasible.

Example 6.12
Let the manufacturing technological base of the considered *IS* be represented by the set *R*, including the following temporal rules r_1 and r_2:

$$< r_1; \{1 \cdot (frame), 1 \cdot (engine), 4 \cdot (door), 4 \cdot (wheel), 300 \cdot (eur)\} \\ \rightarrow \{1 \cdot (auto), 50 \cdot (mnt)\} >,$$

$$< r_2; \{1 \cdot (motor), 1 \cdot (transmission), 1 \cdot (fuel\ tank), 900 \cdot (eur)\} \\ \rightarrow \{1 \cdot (engine), 80 \cdot (mnt)\} >,$$

the resource base of this *IS* be

$$v_0 = \{12.00 \cdot t, \quad 1 \cdot (door), 16 \cdot (wheel), 1 \cdot (transmission), 1 \cdot (fuel\ tank)\ \},$$

and the *IS* schedule be

$$\boldsymbol{\sigma} = \{ < 12.00, \{r_2\} >, < 13.20, \{r_1, \ r_2\} >, < 14.40, \{r_1\} > \}.$$

According to (6.42)–(6.48), the supply schedule of this *IS* will be created as follows.

Firstly, the collection of *ORs* which must be added to the initial *RB* at the moment 12.00 is determined:

$$\Delta v_0 = \bar{v_0} - v_0 = \\ \{1 \cdot (motor), 1 \cdot (transmission), 1 \cdot (fuel\ tank), 900 \cdot (eur)\} - \\ \{12.00 \cdot t, \quad 1 \cdot (door), 16 \cdot (wheel), 1 \cdot (transmission), 1 \cdot (fuel\ tank)\ \} = \\ = \{1 \cdot (motor), 900 \cdot (eur)\},$$

after which the remaining *RB* will be

$$v_1 = v_0 - \bar{v_0} = \\ \{12.00 \cdot t, \quad 1 \cdot (door), 16 \cdot (wheel), 1 \cdot (transmission), 1 \cdot (fuel\ tank)\ \} \\ - \{1 \cdot (motor), 1 \cdot (transmission), 1 \cdot (fuel\ tank), 900 \cdot (eur)\} \\ = \{12.00 \cdot t, \quad 1 \cdot (door), 16 \cdot (wheel)\ \}.$$

At the next step, corresponding to the moment 13.20 when the first operation cycle of the *MD* r_2 will finish and the second *OC* of this *MD* will begin, as well as the first *OC* of the *MD* r_1, the collection of *ORs* which must be added to the current *RB* (apart from the result of the first operation cycle of the *MD* r_2, $1 \cdot (engine)$, which is added at this moment) is determined:

$$\Delta v_1 = \bar{v_1} - v_1 = \{1 \cdot (frame), 1 \cdot (engine), 4 \cdot (door), 4 \cdot (wheel),$$
$$1 \cdot (motor), 1 \cdot (transmission), 1 \cdot (fuel\ tank), 1200 \cdot (eur)\} -$$
$$\{12.00 \cdot t, \quad 1 \cdot (door), 16 \cdot (wheel), 1 \cdot (engine)\} =$$
$$\{1 \cdot (frame), 3 \cdot (door),$$
$$1 \cdot (motor), 1 \cdot (transmission), 1 \cdot (fuel\ tank), 1200 \cdot (eur)\},$$

and the remaining *RB* will be

$$v_2 = v_1 - \bar{v_1} = \{12.00 \cdot t,\ 12 \cdot (wheel)\}.$$

Finally, at the third step, corresponding to the moment 14.40 when the second operation cycle of the *MD* r_2 will finish and the second *OC* of *MD* r_1 will begin, the collection of *ORs*, which must be added to the current *RB* (apart from the result of the first operation cycle of the *MD* r_1, $1 \cdot (auto)$, which is added at the moment 14.10, and the result of the second operation cycle of the *MD* r_2, $1 \cdot (engine)$, which is added at the moment 14.40), is determined:

$$\Delta v_2 = \bar{v_2} - v_2 = \{1 \cdot (frame),\ 1 \cdot (engine),\ 4 \cdot (door),\ 4 \cdot (wheel),\ 300 \cdot (eur)\} -$$
$$\{12.00 \cdot t,\ 12 \cdot (wheel),\ 1 \cdot (auto),\ 1 \quad \cdot (engine)\ \}$$
$$= \{1 \cdot (frame),\ 4 \cdot (door),\ 300 \cdot (eur)\}.$$

So the supply schedule of the considered *IS* is as follows:

$$\sigma_{in} = \{ < 12.00, \{1 \cdot (motor),\ 900 \cdot (eur)\} >,$$
$$< 13.20, \left\{ \begin{array}{l} 1 \cdot (frame),\ 3 \cdot (door),\ 1 \cdot (motor), \\ 1 \cdot (transmission),\ 1 \cdot (fuel\ tank),\ 1200 \cdot (eur) \end{array} \right\} >,$$
$$< 14.40, \{1 \cdot (frame),\ 4 \cdot (door),\ 300 \cdot (eur)\} > \}.$$

The remaining *RB* at the moment 14.40 will be

$$v_3 = v_3 - \bar{v_3} = \{12.00 \cdot t,\ 8 \cdot (wheel),\ 1 \cdot (auto)\}.$$

Now it is possible to define one more schedule, called ***output***, which permits sequential transfer of object-resources manufactured by an *AIS MPR*, and defined by a completed order, to the external system which is the source of this order, i.e., the consumer of these *ORs*.

An output schedule may be created in the following simple way:

$$\boldsymbol{\sigma}_{out} = \bigcup_{i=1}^{l} \{ < n_i, \Delta v_i > \mid \Delta v_i \neq \{\emptyset\}\}, \tag{6.49}$$

where

$$\Delta v_i = \boldsymbol{q} \cap \sum_{< r; v \to v' + \{\Delta n \cdot \Delta t\} > \in R(n_i - \Delta n)} v'. \tag{6.50}$$

As can be seen, an output schedule is created by the intersection of an order and the sum of the collections of all new *ORs* which will appear at a moment n_i as a result of operation cycles which finish at this moment. This definition is valid if no manufactured *OR* is used by some manufacturing device as input at some moment $n_j > n_i$. Otherwise, such an *OR* should not be included in the resulting collection. However, this condition is practically always satisfied: orders define only final *ORs*.

Example 6.13
Let the manufacturing technological base and the resource base of the considered *IS* as well as the order to be completed be the same as in the previous example.

Then according to (6.49) and (6.50), the output schedule of this *IS* will be as follows:

$$\boldsymbol{\sigma}_{out} = \{ < 14.10, \{1 \cdot (auto)\} >, < 15.30, \{1 \cdot (auto)\} > \}.$$

As may be seen, both *IS* supply and output schedules are created simply by a direct execution of an *IS* schedule.

Let us note that there is one more task concerning a *TMG*-based *IS's* representation that was not considered in the case of a *UMG*-based one due to the lack of time in the unitary rules. Namely, as was announced at the beginning of Sect. 6.2.3, the set of variants of the *RB* supplementation ΔV defined by (6.39) may be the one-element set $\Delta V = \{\{\emptyset\}\}$. This means that the amounts of resources in the *IS RB* are sufficient. However, despite this sufficiency, the order is not feasible, because the durations of all possible ways of order completion do not fit the boundary condition $n \leq t \leq n'$. So the task is ***to assess the minimal possible time for order completion***.

This task may be solved in an obvious way. In the initial *AIS* $I = < \boldsymbol{Q}, R, v_0 >$ with the order $\boldsymbol{Q} = <\boldsymbol{q},\boldsymbol{t}>$, where $= \{n \leq t \leq n', t = min\}$, we shall replace $\boldsymbol{t}$ by

$$\boldsymbol{t}' = \{t = min\} \tag{6.51}$$

and apply the *FTMG* $-S' = < \boldsymbol{q}, -R, -F' >$, where in the filter $-F'$ obtained from the filter $-F$ defined by (6.35), the subfilter $-\boldsymbol{t}$ instead of (6.37) is as follows:

$$-\mathbf{t} = \{\, t = max \mid t = min \in \mathbf{t}\}. \tag{6.52}$$

So the result of a solution of such a task will be a set $V_{-s'} \neq \{\emptyset\}$ including all variants of order completion with the minimal duration of *IS* schedules even if all of them do not satisfy the boundary condition $n \leq t \leq n'$.

We shall not consider an assessment of partial order completion in the case of an insufficient resource base, because, in fact, there are no principal distinctions in a solution of this task in the case of a *TMG*-based representation of industrial systems if we use as a basis the techniques described in Sect. 6.1.4 for solving a similar task in a *UMG*-based representation. The same may be said about *TMG*-based modeling of competitions.

6.2.4 Multi-order Scheduling

Until now, we have considered tasks with the common feature that only one order came in to the *IS MPR*. The general case is a ***flow of orders***, any of which may arrive at the *IS MPR* at an unpredictable time moment, i.e., in such a way that successive orders arrive while at least one of the previous orders is being completed according to an already created and executed *IS* schedule. To achieve the maximal possible efficiency of the *IS*, it would be reasonable to implement multi-order operation, excluding the necessity for any of new incoming orders to await the completion of all previous orders and permitting their completion in parallel to the extent that it may be possible. So at a moment of arrival of a new order, the current *IS* schedule would be corrected in such a way that all previous orders as well as the new one would be completed simultaneously. There may be two different approaches to a solution of this problem. ***The first*** one simply presumes a supplement of the current schedule by operation cycles allowing the completion of the new order without any changes to the preplanned *OCs* allowing the completion of the previous orders. ***The second*** approach is more general and flexible due to the possibility of correction of the current schedule (of course without making already processed orders uncompleted), thus making available some additional resources which would not be accessible in the first case. Here, we shall consider only the first approach.

We shall begin from a ***generalization of the representation of the current state of an IS MPR***, which, let us recall once more, according to (3.44) is

$$v = v_A + v_R + \{n \cdot t\}, \tag{6.53}$$

where v_A is the current resource base, and

$$v_R = \{1 \cdot r_1\tau_1, \ \ldots, \ \ 1 \cdot r_N\tau_N\} \tag{6.54}$$

is the current state of the *MTB* at moment n. A string τ_j, $j = 1, \ldots, N$, is a representation of the schedule of the manufacturing device r_j in such a way that $\tau_j = \left[n_1^j, l_1^j\right] \ldots \left[n_{k_j}^j, l_{k_j}^j\right]$, where n_i^j and l_i^j ($n_j^i \leq l_i^j$) are, respectively, moments of activation of the *MD* r_j and moments at which it finishes an operation cycle. All $r_j\tau_j$ are composite names.

To cover the general case of multi-order scheduling, we shall introduce a representation of an *MD* schedule as a string

$$\tau_j = \left[n_1^j, l_1^j : \boldsymbol{i_1}\right] \ldots \left[n_{k_j}^j, l_{k_j}^j : \boldsymbol{i_{k_j}}\right], \tag{6.55}$$

where ":" is a divider, while $\boldsymbol{i_l}$, $l = 1, \ldots, k_j$, is a unique identifier (number) of an order whose completion is contributed to by the operation cycle of the *MD* r_j beginning at the moment n_l^j. Thus, *any* manufacturing device may take part in the completion of *any* order, and this possibility allows maximal concurrency of the process of order completion. However, the schedule of any *MD* r_j satisfies restriction (3.46), fixing the sequential execution of the *MD* operation cycles (or the inadmissibility of their intersection).

Let $v = v_A + v_R + \{\boldsymbol{n} \cdot t\}$ be the current state of the *IS MPR* at the moment $\boldsymbol{n}$, when order $\boldsymbol{Q} = \ <\boldsymbol{q,t}>$ has arrived at the *IS* and has obtained the unique identifier $\boldsymbol{i}$distinguishing it from all other orders in the *IS's* current schedule $\boldsymbol{\sigma}$. To generalize from the one-order case to allow construction of an *IS* schedule in the multi-order case, we shall make one very local correction of definition (3.53) of the function $\boldsymbol{\pi}$, which henceforth will be defined as

$$\begin{gathered}\boldsymbol{\pi}(v_A + v_R + \{n \cdot t\}) = \\ \{v_A - v + v_R - \{1 \cdot r\tau\} + \{1 \cdot r\tau[n,\ l : \boldsymbol{i}]\} + \{n \cdot t\} \mid \\ r\tau \in v_R \ \&\langle r; v \rightarrow v' + \{\Delta n \cdot \Delta t\}\rangle \in R\&v \subseteq v_A \ \&\boldsymbol{T}(\tau) \leq n\&l = n + \Delta n\}.\end{gathered} \tag{6.56}$$

As may be seen, this correction is the replacement of $[n, l]$ by $[n, l : \boldsymbol{i}]$, which supplements the current *IS* schedule $\boldsymbol{\sigma}$ with the identifier $\boldsymbol{i}$ of the order $\boldsymbol{Q}$ to which each operation cycle of an *MD* r contributes. So application of the function $\boldsymbol{\pi}$ during the generation of a new *MST* serves as a basis for the creation of a schedule $\boldsymbol{\sigma}'$ allowing, if it is possible, the completion of order $\boldsymbol{Q}$ as well as all previous orders. Also we shall redefine the function $\boldsymbol{T}$, which, according to (3.54), selects the last element of the list $[m_1, n_1]\ldots[m_l, n_l]$, i.e., n_l. Henceforth, to fit (6.55), this function will be defined as follows:

$$\boldsymbol{T}([m_1,\ n_1 : \boldsymbol{i_1}] \ldots [m_l,\ n_l : \boldsymbol{i_l}]) = n_l. \tag{6.57}$$

After the introduced corrections, we may define the logic of a supplement of an *IS MPR* schedule, allowing the completion of an order $\boldsymbol{Q}$=<$\boldsymbol{q,t}$> whose identifier is $\boldsymbol{i}$, as follows. The set of possible states of this *IS* is nothing but the set of temporal multisets defined by the filtering temporal multiset grammar

$$S = < v_A + v_R + \{\boldsymbol{n} \cdot t\}, R, F >, \tag{6.58}$$

where

$$F = F_{\boldsymbol{q}} \cup \boldsymbol{t}, \tag{6.59}$$

and $F_{\boldsymbol{q}} = \{\boldsymbol{a_1} \geq \boldsymbol{n_1}, \ldots, \boldsymbol{a_m} \geq \boldsymbol{n_m}\}$, as defined by (6.29); $\boldsymbol{q} = \{n \leq t \leq n', t = min\}$, as defined by (6.33); v_A is the resource base of the *IS* at the moment $\boldsymbol{n}$ at which order $\boldsymbol{Q}$ arrives at the *IS*; and v_R is the state of the *IS* manufacturing technological base at this moment. According to (3.66),

$$V_S = \boldsymbol{\pi}^* \left(v_0 + v_R^0 + \{n_0 \cdot t\}, F_{\leq}\right) \downarrow F_{opt}, \tag{6.60}$$

where

$$F_{\leq} = F_{\boldsymbol{q}} \cup \{n \leq t \leq n'\}, \tag{6.61}$$

$$F_{opt} = \{t = min\}. \tag{6.62}$$

So, selecting one multiset $v \in V_S$, if $| V_S | \geq 1$, we obtain a ***new IS schedule which supplements the current schedule and allows the completion of order Q*** as well as all previous orders which were already being completed at the moment $\boldsymbol{n}$ at which $\boldsymbol{Q}$ arrived. In the case $|V_S| = 0$, i.e., $V_S = \{\emptyset\}$, an order $\boldsymbol{Q}$ is ***non-feasible***.

As can be seen, supplementing of an already executing *IS MPR* schedule by a schedule additionally allowing the completion of order $\boldsymbol{Q}$ is implemented by the application of an *FTMG S* whose kernel $v_A + v_R + \{\boldsymbol{n} \cdot t\}$ matches the current state of the resource base and the manufacturing technological base of the *IS* at moment $\boldsymbol{n}$. In this way, generation of *MSTs* corresponding to possible states of the *IS* during the completion of order $\boldsymbol{Q}$ allows the sequential creation of its possible schedules within the restrictions on available active and passive resources imposed by the previous orders and reflected by the aforementioned kernel of the *FTMG S*.

Let us consider now the ***top-down logic*** of schedule creation, which is a core element of the one-order case and provides valuable reduction of redundant search during the application of temporal rules; this follows from the well-known results concerning the minimization of computational complexity of logical inference in various versions of Prolog and augmented Post systems (Sheremet 1994, 2013; Bratko 2012), where, namely, top-down inference (from goal to subgoals until ground facts) enables practically useful implementation of any knowledge representation model. Below, we shall construct the necessary relations for applying the top-down logic to the multi-order case, which is not difficult if we use the techniques introduced in Sect. 4.4.2.

We shall begin with the construction of an *FTMG* $-S$ dual to the *FTMG S* defined by (6.58):

$$-S = < \boldsymbol{q} + (-v_R) + \{\boldsymbol{n} \cdot t\}, -R, -F >, \tag{6.63}$$

where

$$\boldsymbol{q} = \{\boldsymbol{n_1} \cdot \boldsymbol{a_1}, \ldots, \boldsymbol{n_m} \cdot \boldsymbol{a_m}\}, \tag{6.64}$$

$$v_R = \{1 \cdot r_1\tau_1, \ldots, 1 \cdot r_N\tau_N\}, \tag{6.65}$$

$$-v_R = \{1 \cdot r_1(-\tau_1), \ldots, 1 \cdot r_N(-\tau_N)\}, \tag{6.66}$$

$$-([n_1, l_1 : \boldsymbol{i_1}] \ldots [n_k, l_k : \boldsymbol{i_k}]) = [-n_1, -l_1 : \boldsymbol{i_1}] \ldots [-n_k, -l_k : \boldsymbol{i_k}], \tag{6.67}$$

$$-R = \{\langle -r; v' \to v + \{-\Delta n \cdot \Delta t\}\rangle \mid \langle r; v \to v' + \{\Delta n \cdot \Delta t\}\rangle \in R\}, \tag{6.68}$$

$$F = F_{\boldsymbol{q}} \cup \boldsymbol{t}, \tag{6.69}$$

$$F_{\boldsymbol{q}} = \{\boldsymbol{a_1} \geq \boldsymbol{n_1}, \ldots, \boldsymbol{a_m} \geq \boldsymbol{n_m}\}, \tag{6.70}$$

$$\boldsymbol{t} = \{n \leq t \leq n', t = min\}, \tag{6.71}$$

$$-F = F_A \cup -\boldsymbol{t}, \tag{6.72}$$

$$F_A = \{a \leq k \mid k \cdot a \in v_A\} \cup \{a = 0 \mid a \in \overline{A}_S \& a \notin v_A\}, \tag{6.73}$$

$$-\boldsymbol{t} = \{-n' + \boldsymbol{n} \leq t \leq -n + \boldsymbol{n}, t = max\}. \tag{6.74}$$

After these changes, the following statement verifying the top-down logics in the multi-order case and close in meaning to Statement 4.5 is obvious.

Statement 6.7 The following holds:

$$V_S = -V_{-S} \tag{6.75}$$

Let us now consider the relations defining ***IS schedules for the multi-order case***. Let

$$v = v_A + v_R + \{n \cdot t\} \in V_S \tag{6.76}$$

be an *MST* generated by an *FTMG S* and, following (6.49)–(6.52),

$$v_R = \bigcup_{r \in R} \{r\tau\}, \tag{6.77}$$

where

$$\tau = [n_1, l_1 : \boldsymbol{i}_1] \ldots [n_k, l_k : \boldsymbol{i}_k], \tag{6.78}$$

and, as above, $\boldsymbol{i}_1, \ldots, \boldsymbol{i}_k$ are identifiers of orders to be completed by the considered *IS*.

Now, we may define an ***IS schedule for the multi-order case***. Evidently, it is sufficient to generalize (3.48) in the following way:

$$\boldsymbol{Sch}(v_R) = \bigcup_{n \in \boldsymbol{N}(v_R)} \left\{ < n, \bigcup_{r[n_1,\ l_1:\boldsymbol{i}_1\ ]\ldots[n_k,\ l_k:\boldsymbol{i}_k\ ] \in v_R n \in \{n_1,\ \ldots,\ n_k\}} \{r\} > \right\}, \tag{6.79}$$

where, as in the one-order case,

$$\boldsymbol{N}(v_R) = \bigcup_{r[n_1,\ l_1:\boldsymbol{i}_1\ ]\ldots[n_k,\ l_k:\boldsymbol{i}_k\ ] \in v_R} \{n_1, \ldots, n_k\} \tag{6.80}$$

is the set of time moments when at least one *MD* r is activated.

So an *IS* schedule in the multi-order case is as follows:

$$\boldsymbol{\sigma} = \{ < n_0, \boldsymbol{R}(n_0) >, \ \ldots, \ < n_p, \boldsymbol{R}(n_p) > \}, \tag{6.81}$$

where for any $n \in \{ n_0, \ldots, n_p\}$,

$$\boldsymbol{R}(n) = \left\{ r_1^n / \boldsymbol{i}_1^n, \ \ldots, \ r_j^n / \boldsymbol{i}_j^n \right\}, \tag{6.82}$$

is the set of *MDs* r_j^n activated at time moment n with the objective to execute its operation cycle for the assigned order $\boldsymbol{i}_j^n$; in other words,

$$r_j^n/\boldsymbol{i}_j^n \in \boldsymbol{R}(n) \tag{6.83}$$

means *MD* r_j^n will be activated at moment n to execute its *OC* to contribute to the completion of the order with identifier $\boldsymbol{i}_j^n$. Here, in (6.82) and (6.83), "/" is a divider between an *MD* name and an order identifier.

Definition (6.81) also provides an opportunity to redefine *IS* supply and output schedules for the multi-order case.

We shall begin from the redefinition of Δv_i, which in (6.42)–(6.47) is nothing but a multiset representation of the collection of *ORs* which would be added to the *IS* resource base. Representation (6.83) entails the following definition of $\bar{v}_i$:

$$\bar{v}_i = (v_{i-1} - v_{i-1}^-) + \sum_{<r/i;\, v \to v' + \{\Delta n \cdot \Delta t\}> \in R(n_i)} v, \tag{6.84}$$

i.e., the *RB* must be replenished with the *ORs* necessary for the completion of all *OCs* which would be activated at time moment n_i. By application of (6.85), the definition (6.48) of an ***IS supply schedule*** remains unchanged for the multi-order case.

Regarding an ***IS output schedule***, some additional operations are necessary. First of all, we shall generalize (6.49) in such a way that for any moment n_i, there are defined collections of *ORs* which are delivered at this moment to the consumers which are the sources of orders. So we shall define

$$\boldsymbol{\sigma}_{out} = \bigcup_{j=1}^{l} \{ < n_j, \boldsymbol{\Delta v}_j > \mid \boldsymbol{\Delta v}_j \neq \{\emptyset\} \}, \tag{6.85}$$

where

$$\boldsymbol{\Delta v}_j = \bigcup_{i \in \boldsymbol{I}_j} \left(\left\{ < \boldsymbol{q}_i \cap \left(\sum_{< r/\boldsymbol{i}; v \to v' + \{\Delta n \cdot \Delta t\} > \in \boldsymbol{R}(n_j - \Delta n)} v' \right), \boldsymbol{i} > \right\} - \{ < \{\emptyset\}, \boldsymbol{i} > \} \right). \tag{6.86}$$

As can be seen, (6.85) is just the same as (6.49), but Δv_j is replaced by $\boldsymbol{\Delta v}_j$, which, according to (6.86), is a join of couples $<\Delta v, \boldsymbol{i}>$, such that the non-empty multiset Δv is a collection of *ORs* whose *OC* is finished at the moment n_j, and which, at the same time, complete a substantial part $\boldsymbol{q}_i$ of an order with an identifier $\boldsymbol{i}$. The join is done over all orders being completed at the moment n_j (the set of their identifiers is $\boldsymbol{I}_j$). Henceforth, as can be seen, an *IS* output schedule allows the sequential delivery of parts of *OR* collections defined by simultaneously completed substantial parts of orders, to their sources (*OR* consumers).

Thus, we have generalized all three types of *IS* schedules (***input***, ***output***, and ***supply***) to the multi-order case.

Now, we may move to a consideration of operational interconnections between resource-consuming and industrial systems.

6.3 Operational Interconnections Between Resource-Consuming and Industrial Systems

In practice, industrial systems in most cases manufacture equipment and consumables necessary for the establishment, improvement, and operation of *OT* and *ROT* resource-consuming systems. Henceforth, we shall call any such industrial system an "***RCS supplier***" if it completes orders allowing the aforementioned activities. In turn, any such *RCS* will be called a "***source of orders to the IS***." In this environment, we may consider multiple variants of operational interconnections between an *RCS* and its supplier or between an *IS* and a source of orders to it:

1. A ***created OT-system*** with a known structure and assigned technological equipment, which is manufactured by an *IS*; so an order, which must be completed by an *IS*, defines amounts of various types of devices necessary for *OT*-system creation.

2. An ***existing (operating) OT-system which is modified (adapted) by a change of its structure***, as well as of its technological equipment, in which some items are replaced to create a new structure and assignments; an order which must be completed by the *IS* defines amounts of various types of new equipment which must be added to already operating devices and systems or replace some of them.
3. ***An existing (operating) ROT-system with exactly known amounts of consumable resources necessary for the operation of its human and technological elements during a predefined time interval***; an order which must be completed by the *IS* defines amounts of various types of consumables necessary for this *ROT*-system's full supply for the aforementioned time interval.

A practical task in any of the three listed variants of an *RCS* is to create from specific collections of resources, needed for separate elements of a considered system, an integrated order to be completed by an *IS*, and this task is usually rather difficult to carry out by reason of its high dimensionality for more or less valuable *RCSs*. However, this task becomes simple if we apply the multigrammatical tools of *RCS* analysis already developed and described in the previous chapter. Let us consider basic techniques of such application.

Let us begin with the case of ***creation of an OT-system***, which is represented by a non-variative unitary multiset grammar (i.e., 1-*MG*) $S=<\{1\cdot a_0\},R_{OT}>$, constructed in such a way that any position is connected with assigned technological equipment by unitary rules like (6.1), and for this reason, the scheme R_{OT} is a join of sets of unitary rules R_O and R_T, representing, respectively, the organizational structure and the aforementioned *TE* of this system. Then the set of technological devices to be manufactured by the *IS* for the creation of the *OT*-system *S* is nothing but the multiset $\boldsymbol{v} \in V_S$ such that $V_S=\{\boldsymbol{v}\}$. Hence, the order to be completed is $\boldsymbol{Q} = <\boldsymbol{v},\boldsymbol{t}>$, where $\boldsymbol{v}$ is the substantial part of this order (above it was denoted $\boldsymbol{q}$) and $\boldsymbol{t}$ is its temporal part, defining the time interval when the industrial system $\boldsymbol{S} =< \boldsymbol{Q}, \boldsymbol{R}, \boldsymbol{v}_0>$ assigned to this order completes it. These are the following evident subtasks in this case:

An assessment of the minimal time necessary for the creation of the *OT*-system *S* (i.e., necessary for the completion of the order $\boldsymbol{Q}$ by the *IS* $\boldsymbol{S}$)

An assessment of amounts of resources to be acquired for the completion of the order $\boldsymbol{Q}$ by the *IS* $\boldsymbol{S}$

A computation of the input and output schedules of the *IS* $\boldsymbol{S}$ allowing the creation of the *OT*-system *S* and meeting the time restrictions defined by the temporal part $\boldsymbol{t}$ of the order $\boldsymbol{Q}$

The techniques for the solution of these tasks were considered in detail in the previous Sect. 6.2.

Let us note that in the general case, there may be some collection $\boldsymbol{v}'$ of technological equipment already available to the created *OT*-system *S*, and thus it would not be necessary to manufacture it by means of the *IS* $\boldsymbol{S}$. If so, then evidently $\boldsymbol{Q} = <\boldsymbol{v} - \boldsymbol{v}',\boldsymbol{t}>$.

The case of a ***modified*** (adapted) ***OT-system*** is similar to the case already considered, and the only difference is the application of a 1-*MG* $S' =<\{1\cdot a_0\}, R'>$ representing an objective system obtained from the initial one represented by a 1-*MG*

$S = <\{1 \cdot a_0\}, R>$ by elimination from R of its subset $\Delta R^- \subseteq R$ and supplementing of the rest of R the set ΔR^+ according to (5.26). So it is clear that the substantial part of an order $\boldsymbol{Q}$ to be completed for the modification of a considered *OT*-system would be $\boldsymbol{\Delta v}^+ = v^+ - v$, as defined by (5.28), or if some part $\boldsymbol{v}'$ of the *TE* is already available, then, as above, $\boldsymbol{Q} = \ <\boldsymbol{\Delta v}^+ - \boldsymbol{v}',\boldsymbol{t}>$.

The most complicated case is an ***ROT-system, which consumes resources necessary for the operation of its human and technological elements during some predefined time interval of*** $\boldsymbol{\Delta n}$ ***time units.*** In this case, Δn is the duration of an *IS* operation cycle, i.e., time period which is necessary for manufacturing all resources which will be consumed by the *ROT*-system during the next Δn time units. That is, while a *ROT*-system is consuming an available collection of *ORs*, the *IS* must produce the collection of resources which will be consumed by the *ROT*-system in the next cycle of its operation. So, the aforementioned collection must be ready for consumption not later than the end of the current *ROT*-system cycle.

Let a 1-*MG* $S = <\{1 \cdot a_0\}, R_{ROT}>$ represent a *ROT*-system and $\boldsymbol{v} \in V_S$, such that $V_S = \{\boldsymbol{v}\}$ is a multiset representing the aforementioned collection of *ORs* which, from one side, is consumed by S and, from the other side, is produced by an *AIS* $\boldsymbol{S} = <\boldsymbol{Q}, \boldsymbol{R}, \boldsymbol{v_0}>$, where $\boldsymbol{Q} = \ <\boldsymbol{v},\boldsymbol{t}>$, and

$$\boldsymbol{t} = \{t < n + \Delta n\}, \tag{6.87}$$

where n is the time moment of the beginning of the current operation cycle of the *IS* and the *RCS* supplied by it, so $n + \Delta n$ is the time moment of the beginning of the next *OC*. The boundary condition (6.87) ensures readiness of all *ORs* necessary to the *RCS* for the next *OC* not later than this cycle begins.

Of course, this synchronized operation of the *RCS* and the *IS* supplying it is possible if $|V_S| \geq 1$, i.e., if there exists at least one way of completing the order $\boldsymbol{Q}$. Otherwise, it is necessary to engage some external system (or several systems) for supplying the *IS* within its supply schedule, which may be easily created according to (6.41)–(6.48).

A practically useful generalization of the described *RCS-IS* interaction presumes an opportunity for a consumer to represent an order as a multiset whose objects are non-terminal objects of a 1-*MG* $S = <\{1 \cdot a_0\}, R_{ROT}>$ or $S = <\{1 \cdot a_0\}, R_{OT}>$. Namely, an initial order $\boldsymbol{Q} = \ <\boldsymbol{q},\boldsymbol{t}>$ is, in fact, replaced by an order $\boldsymbol{Q} = \ <\boldsymbol{v},\boldsymbol{t}>$, where $\boldsymbol{v} \in V_{S'} = \{\boldsymbol{v}\}$, and $S' = \ <\boldsymbol{q}, R_{ROT}>$ or $S' = \ <\boldsymbol{q}, R_{OT}>$, respectively. This replacement means that ***an order is formulated by the RCS management in some organizational units***, while the ***IS obtains an order formulated in the manufactured technological devices and consumables***, whose amounts strictly correspond to supplies of the aforementioned *OUs*.

The only task to be solved in this context is to redefine the *IS* output schedule in such a way that it contains ***not multiobjects representing technological devices and consumables but multiobjects representing OUs***.

Concerning ***OT-systems***, this task may be solved by the application of a specially constructed filtering temporal multiset grammar $\boldsymbol{S}_{OT} = \ <\boldsymbol{q}, -\boldsymbol{R} \cup \boldsymbol{R}_{OT}, \boldsymbol{F}_{OT}>$, whose scheme is the join of the set of mirror temporal rules $-\boldsymbol{R}$, representing the manufacturing technological base of the *IS*, and the scheme $\boldsymbol{R}_{OT}$, containing temporal rules constructed from the unitary rules in the scheme R_{OT} of the 1-*MG S* in the following obvious way:

$$\boldsymbol{R}_{OT} = \{<r;\{1\cdot a\}\rightarrow\{n_1\cdot a_1,\ \ldots,\ n_m\cdot a_m\}>|<r;a\rightarrow n_1\cdot a_1,\ \ldots,\ n_m\cdot a_m> \in R_{OT}\}. \tag{6.88}$$

As can be seen, the "duration of operation cycle" of "device" r is zero, so one unit a "is manufactured" from n_1 units $a_1, \ldots, n_m$ units a_m immediately. The filter $\boldsymbol{F}_{OT}$ is composed of the given *IS* resource base $\boldsymbol{v_0}$ and the set of boundary conditions $\boldsymbol{t}$ in full accordance with (6.35)–(6.37). Thus, the *IS* output schedule $\boldsymbol{\sigma}_{out}$ obtained from any *MST* $\boldsymbol{v} \in V_{S_{OT}}$ would contain only multiobjects containing a substantial part $\boldsymbol{q}$ of the order $\boldsymbol{Q}$, i.e., names of organizational units, and $<n_i, \Delta v_i> \in \boldsymbol{\sigma}_{out}$, where $a \in \Delta v_i$ means that one *OU* a would be ready (being supplied with all necessary manufactured devices) beginning at time moment n_i.

In the case of ***ROT-systems***, the approach is practically the same; the only difference is that a scheme R_{ROT} is applied instead of R_{OT}, and also a relevant *FTMG* $\boldsymbol{S}_{ROT} = <\boldsymbol{q}, -\boldsymbol{R} \cup \boldsymbol{R}_{ROT}, \boldsymbol{F}_{ROT}>$. Naturally, an output schedule of the *IS* ensures the readiness of all *OUs* belonging to the *ROT*-system in the broader sense: not only the necessary manufactured devices but also all necessary produced consumables are available to any *OU* beginning at the time moment of its appearance in the schedule, so this *OU* is ready for one operation cycle of the *ROT*-system (amounts of the aforementioned consumables are computed regarding this cycle).

Until now, we have limited our scope to local industrial systems. To consider ***distributed*** *ISs*, it is sufficient to apply the technique of representation of object-resources located at some places (points), which was announced in Sect. 2.5 in the context of any sociotechnological systems. Namely, if it is necessary to consider distributed *ISs*, then only the syntactic extension by the ordinary representation of *OR* locations is applied. We shall introduce a set of locations (positions) *Z*, where object-resources of both kinds may be present and each such location will be denoted by $z \in Z$. (In the general case, z may be defined as a point in some coordinates, as well as an area of any form—circle, quadrilateral, ellipse, or any shape defined by boundary points.) This extension provides the simplest representation of located *ORs*: instead of primary names of *ORs*, constructions (a,z) are used, where z is the name of a point (place, area) where *OR* a is located and "(", ",", and ")" are dividers.

Let us note that, as in the case of resource-consuming systems, an important class of tasks related to ***reverse engineering of industrial systems*** is highly relevant. As in the case of *RCSs*, techniques for the solution of these tasks would, in turn, make available solutions of a lot of tasks related to ***business intelligence***. Concerning *UMG*-based representation of *ISs*, these techniques are similar to those already developed for *RCSs* and allow the solution by application of *UMMGs*. If *TMG*-based representations are considered, similarly, *TMMGs* will be applied. This task will be considered in future publications.

Now, after our consideration of *MGF*-represented resource-consuming and resource-producing (industrial) sociotechnological systems, we may move on to resource-distributing, or economical, *STSs*.

Chapter 7
Economical Sociotechnological Systems and Economical Combinatorics

The most essential thing in the above-considered *MGF* modeling of ***resource-producing (industrial) systems*** is that all resources—belonging to an *IS* resource base, produced by the application of an *IS* manufacturing technological base, as well as transferred to customers—***do not belong to any holder***. In other words, no unit of any resource is the property of any person or group of persons: all resources are for common use.

As was already said in Sect. 2.4, an ***economy begins when we assign resources to some holders***—so ***operation of a sociotechnological system as a "side effect" results in the permanent redistribution (motion) of property between holders***. This chapter is dedicated to *MGF* modeling of various kinds of such motion, as well as classes of economical *STSs* associated with some common features originating from operations on property which are performed by these *STSs*.

As introduced in Sect. 2.4, we shall consider ***three basic types of operations on property, differing by the way of acquisition of resources by holders***:

1. Exchange
2. Lending
3. Producing (manufacturing)

These operations are the foundation of relevant and similarly named classes of economical systems.

Let us recall that the main feature of ***exchange economical systems*** is that ***no new resources are produced by them during their life cycle*** and all that is done by *EES* subjects is the ***mutual exchange of resources which are their property***. (Such an economy since the beginning of time has been implemented by ***merchants***.)

In ***lending economical systems***, an active subject (***lender***) "produces" (or "acquires") some amount of his property by the ***removal of this property*** from a passive subject (***borrower***), using the latter's lack of needed resources at a specific time moment as a tool for the implementation of this removal at some later moment, when the borrower is expected to possess amounts of resources sufficient to return all the borrowed as well as some additional amounts which are the benefit of the lender

I. A. Sheremet, *Multigrammatical Framework for Knowledge-Based Digital Economy*, https://doi.org/10.1007/978-3-031-13858-4_7

for his favor. (Such an economy has for centuries been implemented by ***bankers***.) As in *EESs*, ***no new resources are produced inside any LES***.

On the contrary, ***producing economical systems*** are able to ***produce (manufacture) new resources*** by means of their manufacturing technological bases, thus launching produced assets into local, regional, and/or global processes of circulation of property. (Such an economy from the beginning of time was implemented by agrarians and ***workers***, who in Marx's era of early capitalism became ***proletarians*** and now to an increasing extent are transforming into the ***cybertarians*** and ***cognitarians*** mentioned in the Introduction.)

We shall consider all three basic classes sequentially in Sects. 7.1–7.3, respectively, while Sect. 7.4 will be dedicated to the general case of economical systems integrating elements of some two or all of these classes.

To consider economical systems, we shall make a small change to the representations of *RBs* and *MTBs* used in Chaps. 5 and 6: namely, instead of multiobjects $n \cdot a$ appearing in multisets and rules (including temporal rules), we shall use multiobjects $n \cdot (a{:}h)$, where a is the name of an object-resource, h is the name of a holder, and ":" is a divider, necessary for unambiguous interpretation of the string $(a{:}h)$. Thus, if the current resource base of an *ES* is v, then $n \cdot (a{:}h) \in v$ means that n object-resources a are the property of a holder h. (Such a structure of objects was introduced in Sect. 2.5. Let us recall also that a similar structure of composite object names $(\Delta t : x)$ was introduced in Sect. 3.1.3 for the representation of durations of operation cycles of manufacturing devices; this fully matches the introduced representation of property, because time intervals of operation of any *MD* x measured in units Δt may be interpreted, in fact, as a specific kind of property of x, i.e., a time interval belonging to an *MD* x.) This local transformation does not exclude common resources, which are represented, as above, by multiobjects like $n \cdot a$, i.e., without any holder (not less than two such common resources would appear in any *MST* or *TR*—namely, t and Δt, respectively).

Example 7.1
Let the resource base of some economical system at the initial moment of its operation be

$$v_0 = \left\{ \begin{array}{l} 3 \cdot (eur : Alex), 2 \cdot (usd : Alex), 4 \cdot (rur : Alex), \\ 7 \cdot (eur : Bob), 5 \cdot (rur : Bob), \\ 3 \cdot (usd : Charlie) \end{array} \right\},$$

which means that Alex holds 3 euros, 2 dollars, and 4 rubles and Bob holds 7 euros and 5 rubles, while Charlie holds 3 euros.

Like industrial systems, economical systems may be ***local*** or ***distributed***, ***closed*** or ***open***. Unless it is said otherwise, we shall consider local *ESs*, announcing their nexus with the environment in any specific case.

Having developed a representation of ownership, we take the opportunity to describe logics of the aforementioned redistribution of property inside *ESs*. Before we consider ***exchange economical systems***, let us recall that they do not produce any resources, and thus they ***do not possess*** *MTBs* (***active resources***) but only resource bases (***passive resources***), ***whose elements are assigned to holders*** and, if necessary, ***redistributed among them***.

7.1 Exchange Economical Systems

7.1.1 Basic Scheme of an Exchange and Its Representation by Filtering Multiset Grammars

An exchange is the simplest way of acquisition of necessary resources. Every holder *A* may declare his offer, including couples$<w, w'>$, where *w* is a collection of resources which he is willing transfer to another holder *B*, who, in turn, would transfer to the initiator of this exchange a collection w', which before the exchange belonged to *B*. A ***set of such offers*** is accumulated by an *EES* scheduler, as well as a ***set of criteria*** established by holders and defining ***objectives of their participation in this exchange***. It is assumed that ***any chain of exchanges*** between holders ***is implemented promptly and simultaneously***, i.e., no time is spent for exchanges. So an *EES* scheduler's task is ***to construct one or more chains*** leading from the initial resource base of the *EES* to a final one satisfying the objectives of all holders who have announced their participation in an exchange or ***to prove*** that ***no such chain exists***.

Let us repeat once more that the most essential feature of the described basic operation on the resource base of an *EES* is that ***no new resources are produced (manufactured), so the total amount of any resource in the RB before and after any exchange chain are the same***. Nevertheless, the amounts of some resources which are the property of some owners may change. This feature will be called ***resource-safekeeping*** (*RSK*). It is formally defined by the following obvious equality:

$$(\forall a)(\forall i > 0) \sum_{n \cdot a : h \in v_0} n = \sum_{n \cdot a : h \in v_i} n, \tag{7.1}$$

where v_i is an *EES* resource base at any *i*-th moment of its operation.

As may be seen from this verbal description, the set of all possible exchange chains and their results may be precisely represented by a ***filtering multiset grammar*** $S = \ < v_0, R, F>$, where

$$v_0 = \{n_1 \cdot (a_{i_1} : h_{i_1}), \ \ldots, n_m \cdot (a_{i_m} : h_{i_m})\} \tag{7.2}$$

is ***the resource base of an EES at an initial moment*** of its operation. This *RB* contains n_1 object-resources a_{i_1} belonging to a holder $h_{i_1}, \ldots, n_m$ object-resources a_{i_m} belonging to a holder h_{i_m}; *R* is a set of rules of the form

$$v \to v', \tag{7.3}$$

each defining an ***offer of some holder or group of holders*** to exchange in such a way that a collection v with its distribution of property would be replaced by a collection v' with another distribution of property (of course, only in the case when v is a submultiset of the *RB*); *F* is a set of criteria established by all holders participating in an exchange, each criterion defining ***conditions which the RB must satisfy after an exchange***, so a filter

$$F = \bigcup_{h \in H} F_h, \tag{7.4}$$

where *H* is a set of identifiers of holders belonging to the *EES* and participating in exchanges, while subfilter F_h is a set of conditions established by a holder *h* and representing his objective in the exchanges.

It is clear that the set of multisets

$$V_s = V_{s'} \downarrow F, \tag{7.5}$$

where $S' = \langle v_0, R \rangle$ is the core multigrammar of the *FMG S* is nothing but the ***set of all possible RBs that may be created by all possible exchange chains***, each represented by some sequence of applied rules generating a resulting *RB*. If $V_s = \{\varnothing\}$, no such chain exists. Let us take into account that definition (7.5) differs from the definition of a set of terminal multisets generated by an *FMG S* (3.35) by use of $V_{s'}$ instead of $\overline{V}_{s'}$. This feature reflects a feature of the modeled activity, namely, that not only exchange chains ending in a terminal multiset (i.e., a resource base to which no offer may be applied) may be acceptable, but ***any such chain generating a non-terminal, as well as terminal, multiset satisfying filter*** *F*.

Following (7.1), ***no offer***, represented by a rule $v \to v'$, ***may be included*** in a set of rules *R* unless it satisfies the ***RSK feature*** of the considered *ES*, thus making it an *EES*, i.e., $v \to v' \in R$ only if

$$(\forall a) \sum_{n \cdot a : h \in v} n = \sum_{n \cdot a : h \in v'} n. \tag{7.6}$$

It is evident that if all rules belonging to the scheme *R* of an *FMG S* satisfy the *RSK* property of the represented *ES*, then it is, in fact, an *EES*. The inverse property needs additional study.

Example 7.2
Let $R = \{r_1, r_2\}$, where r_1 is

$$\{2\cdot(usd : Alex),\, 1\cdot(eur : Bob)\} \rightarrow \{1\cdot(eur : Alex),\, 2\cdot(usd : Bob)\},$$

and r_2 is

$$\{1\cdot(usd : Alex),\, 2\cdot(eur : Alex),\, 2\cdot(usd : Bob),\, 2\cdot(rur : Bob)\}$$
$$\{1\cdot(usd : Bob),\, 2\cdot(eur : Bob),\, 2\cdot(usd : Alex),\, 2\cdot(rur : Alex)\}.$$

Both r_1 and r_2 satisfy (7.6) and thus the *RSK*. At the same time, the rule r_3

$$\{2\cdot(usd : Alex),\, 3\cdot(eur : Bob)\} \rightarrow \{1\cdot(eur : Alex),\, 2\cdot(usd : Bob)\}$$

and rule r_4

$$\{1\cdot(usd : Alex),\, 4\cdot(eur : Bob)\} \rightarrow \{4\cdot(eur : Alex),\, 2\cdot(usd : Bob)\}$$

do not satisfy (7.6).

Now, we may investigate more deeply the introduced basic scheme of an exchange. We shall consider the following basic variants of exchange and *EESs* of the same name: ***direct***, ***cooperative***, and **broker.**

7.1.2 Direct Exchange ESs

The simplest case of an exchange is a ***direct exchange***, which is performed directly between holders belonging to an *EES*. Offers of holders willing to participate in an exchange may be represented by rules like

$$\left\{n_1\cdot\left(a_{i_1} : h\right),\, \ldots,\, n_m\cdot\left(a_{i_m} : h\right),\, n'_1\cdot\left(b_{j_1} : h'\right),\, \ldots,\, n'_l\cdot\left(b_{j_l} : h'\right)\right\}$$
$$\left\{n_1\cdot\left(a_{i_1} : h'\right),\, \ldots,\, n_m\cdot\left(a_{i_m} : h'\right),\, n'_1\cdot\left(b_{j_1} : h\right),\, \ldots,\, n'_l\cdot\left(b_{j_l} : h\right)\right\} \quad (7.7)$$

i.e., holder h would transfer n_1 object-resources a_{i_1}, …, n_m object-resources a_{i_m} to holder h', while holder h', in turn, would transfer to holder h n'_1 object-resources b_{j_1}, …, n'_l object-resources b_{j_l}.

Every holder h may propose several exchanges to another one, and there may be several holders h'_1, …, h'_k to which h may make his proposals. Every holder may (or may not) declare the objective of his participation in an exchange; this objective is represented by a set of boundary and optimization conditions $(a{:}h)\ \theta\ n$ and $a{:}h = opt$, respectively, where $\theta \in \{<, >, \leq, \geq, =\}$, $opt \in \{min, max\}$. If all conditions declared as objectives of holder h have the form $a{:}h\ \theta\ n$ and/or $a{:}h = opt$

(i.e., concern amounts of his own property), such a direct exchange *ES* is called ***self-objective for holder*** h; if a direct exchange *ES* is self-objective for all participating holders, it is called ***self-objective***; otherwise (i.e., if at least one holder h by his objective declares some conditions on amounts of property of at least one other holder), such a direct exchange *ES* is called ***cross-objective***.

Example 7.3
Let $R = \{r_1, r_2\}$, where r_1 is

$$\{3 \cdot (usd : Alex), 2 \cdot (eur : Bob)\} \rightarrow \{2 \cdot (eur : Alex), 3 \cdot (usd : Bob)\},$$

and r_2 is

$$\{1 \cdot (usd : Alex), 3 \cdot (eur : Alex), 2 \cdot (usd : Bob), 5 \cdot (rur : Bob)\}$$
$$\{1 \cdot (usd : Bob), 3 \cdot (eur : Bob), 2 \cdot (usd : Alex), 5 \cdot (rur : Alex)\}.$$

As can be seen, the direct exchange *ES* represented by these rules is resource-safekeeping.

Let the objective of holder *Alex* be
$F_1 = \{\ (usd : Alex) \geq 10,\ (rur : Alex) < 7\}$,while the objective of holder *Bob* is
$F_2 = \{\ (usd\text{: } Bob) \geq 12,\ (rur : Bob) = min\}$.

In this case, the direct exchange *ES* is self-objective, because it is self-objective for both holders (F_1 includes conditions regarding only *Alex,* and F_2 includes conditions regarding only *Bob*). If

$$F_1 = \{\ (usd : Bob) < 5,\ (usd : Alex) = max\ \},$$

then the considered direct exchange *ES* is cross-objective, because *Alex* is setting a condition (namely, a restriction) on *Bob's* property: F_1 includes the condition regarding *Bob*: $(usd : Bob) < 5$.

7.1.3 Cooperative Exchange ESs

A more general case is a ***cooperative exchange***, which is implemented by groups of holders belonging to an *EES*. Offers of such groups of holders willing to participate in an exchange may be represented by rules like

$$\{n_1 \cdot (a_{i_1} : h_{i_1}),\ \ldots,\ n_m \cdot (a_{i_m} : h_{i_m})\}$$
$$\{n'_1 \cdot (b_{j_1} : h_{j_1}),\ \ldots,\ n'_l \cdot (b_{j_l} : h_{j_l})\} \tag{7.8}$$

i.e., some amounts of property of holders $h_{i_1}, \ldots, h_{i_m}$ would be extracted from the resource base and, being redistributed as defined by (7.8), would become the property of holders $h_{j_1}, \ldots, h_{j_l}$. As above, all rules of the form (7.8) must satisfy the *RSK* property (7.6).

If all rules in the scheme R of a filtering multiset grammar $= < v_0, R, F>$ representing a cooperative *EES* satisfy the condition

$$\{h_{i_1}, \ldots, h_{i_m}\} = \{h_{j_1}, \ldots, h_{j_l}\}, \tag{7.9}$$

then such a cooperative exchange is called ***closed***; i.e., the holders belonging to the sets in the left and right parts of equality (7.9) simply redistribute their property among themselves. Otherwise, the exchange is called ***open***: as a result of application of rules like (7.8), some persons may lose their property without any compensation, while other persons may acquire new property without any expense.

Similar to a direct exchange, a cooperative exchange may be ***self-objective*** or ***cross-objective*** according to the form of the conditions declared by the exchange participants.

So there may be four types of cooperative exchange *ESs* (***self-objective closed***, ***self-objective open***, ***cross-objective closed***, ***cross-objective open***).

Example 7.4
Let $R = \{r_1, r_2\}$, where r_1 is

$$\{3 \cdot (usd : Alex), 4 \cdot (eur : Bob), 7 \cdot (eur : Bob), 5 \cdot (usd : Charlie)\}$$
$$\{10 \cdot (eur : Alex), 7 \cdot (usd : Bob), 1 \cdot (usd : Charlie), 1 \cdot (eur : Charlie)\},$$

and r_2 is

$$\{5 \cdot (usd : Alex), 4 \cdot (eur : Alex), 3 \cdot (usd : Bob), 5 \cdot (rur : Charlie)\}$$
$$\{2 \cdot (usd : Alex), 3 \cdot (rur : Alex), 6 \cdot (usd : Bob), 4 \cdot (eur : Bob), 2 \cdot (rur : Charlie)\}.$$

As can be seen, both rules satisfy *RSK* and form a closed cooperative exchange. If the rule r_3

$$\{2 \cdot (usd : Alex), 7 \cdot (eur : Alex), 3 \cdot (usd : Bob), 5 \cdot (eur : Charlie)\}$$
$$\{1 \cdot (usd : Alex), 3 \cdot (eur : Alex), 4 \cdot (usd : Bob), 9 \cdot (eur : Bob)\}$$

were to be added to the set *R*, this cooperative *EES* would become open, because as a result of application of this rule, *Charlie* would lose 5 euros, which would become the property of *Bob,* as well as 4 euros belonging to *Alex*.

It is clear that both considered types of exchange are based on ***complete awareness of all holders*** willing to participate in an exchange ***about one another*** and a *FMG* representation of an *EES* permits various ways of exchange, composing all possible exchange chains and computing their results. Implementation of both types is rather simple based on modern Internet technologies, allowing online trade and merchandizing. However, to follow this approach, it is necessary for any holder to replicate his offers to all potential participants in the possible exchange, which is not possible in cases where the exchange initiator does not have full information about all other holders of various properties. To avoid this inconvenience, the third approach, called a ***broker exchange***, may be applied.

7.1.4 Broker Exchange ESs

This approach is based on some intermediate entity (person or *STS*), usually called a ***broker***, who serves as an assigned exchange center of an *EES*. So all exchange activities of holders belonging to the *EES* are carried out through interaction with a broker, which is why such systems are called below ***broker exchange ESs***. In this case, it is not necessary for holders to be aware about one another's proposals and possibilities, because all such work is done by a broker.

Following this verbal description, we may assume that all rules representing exchange proposals are of the form

$$\left\{n_1 \cdot (a_{i_1} : h),\ \ldots,\ n_m \cdot (a_{i_m} : h),\ n'_1 \cdot \left(b_{j_1} : B\right),\ \ldots,\ n'_l \cdot \left(b_{j_l} : B\right)\right\}$$
$$\left\{n_1 \cdot (a_{i_1} : B),\ \ldots,\ n_m \cdot (a_{i_m} : B),\ n'_1 \cdot \left(b_{j_1} : h\right),\ \ldots,\ n'_l \cdot \left(b_{j_l} : h\right)\right\}, \tag{7.10}$$

where B denotes a broker. As can be seen, (7.10) is practically the same as (7.7); the difference is in the dedication of one and the same intermediate person (or *STS*) for all property transfers.

Filters defining holders' objectives in an exchange are accepted by a broker, as well as rules defining holders' proposals. So a broker creates a filtering multiset grammar necessary for the assessment of a possible exchange and for the selection of a chain ensuring, if possible, the achievement of the objectives established by holders willing to participate in the exchange. If no chains are possible due to lack of some resources in the ownership of participants, a broker may add some resources which he holds himself.

Of course, a broker's involvement requires some payment, which is not necessary in the previous cases, where holders connect with one another directly. This payment is expressed in proper amounts of some resources, which are transferred from their holders to the broker and become his property.

Example 7.5
Let $R = \{r_1, r_2\}$, where r_1 is

$$\{3 \cdot (usd : Alex), 4 \cdot (eur : Alex), 10 \cdot (pound : Broker)\} \to$$
$$\{10 \cdot (pound : Alex), 3 \cdot (usd : Broker), 4 \cdot (eur : Broker)\},$$

and r_2 is

$$\{5 \cdot (usd : Bob),\ 3 \cdot (pound : Bob),\ 10 \cdot (rur : Broker)\}$$

$$\{10 \cdot (rur : Bob),\ 5 \cdot (usd : Broker),\ 3 \cdot (pound : Broker)\}.$$

As can be seen, both rules satisfy *RSK* and form a closed cooperative exchange. The rule r_1 enables an exchange between *Alex*, who is offering 3 dollars and 4 euros for 10 pounds, and *Broker*. Similarly, the rule r_2 enables an exchange between *Bob*, who is offering 5 dollars and 3 pounds for 10 rubles, and *Broker*. The objectives of all participants may be as follows:

$F_1 = \{(pound : Alex) \geq 10\}$,

$F_2 = \{ (rur : Bob) = max\}$,

$F_3 = \{ (usd : Broker) \geq 12,\ (eur : Broker) = max\}$,which means that after an exchange, *Alex* wants to hold no less than 10 pounds and *Bob* wants to hold the maximal sum of rubles, while *Broker* wants to hold no less than 12 dollars as well as the maximal sum of euros.

In the general case, a broker may join to the scheme of an *FMG*, applied for exchange planning, some additional rules involving new participants and also join to its filter some additional conditions expressing their objectives if they were to be involved.

Of course, in practical cases, ***mixed approaches may be used, including all the described basic ways of exchange***.

The described techniques of multigrammatical representation of an exchange may be used to represent games in which participants are trying to eliminate resources of the opposing side by expending for this action their own resources. We shall call such games "***resource-based games***." By application of the *RBG* toolkit, real potential conflicts, including military and economic ones, may be rather simply and adequately modeled. We shall consider *RBGs* thoroughly in Chap. 9.

Now, we shall consider the more complicated class of economical *STSs* whose basic operation on property is ***lending***.

7.2 Lending Economical Systems

Lending is a way of property redistribution, which, unlike an exchange, which is performed ***promptly***, is implemented ***during a period of time agreed by its participants***. Such an agreement (***loan***) between persons (or *STSs*) borrowing some amounts of resources and persons (or *STSs*) lending these amounts requires return at a predefined moment of time of these amounts from the borrower (debtor) to the lender as well as some additional amounts, which, in essence, are nothing but a ***payment for the favor done by the lender***. In the general case, a loan may predefine not only the moment of repayment of a debt but a sequence of such moments, each assigned to some part of this debt. Multigrammatical modeling of this kind of relation between property holders is based on use of ***temporal multisets*** for representation of an *LES* resource base and application of ***filtering multiset grammars*** for representation of loans.

7.2.1 Representation of Loans

We shall assume that an *LES* resource base at a moment k includes an *MST*

$$v = \left\{ \begin{array}{l} n_1 \cdot (a_{i_1} : D), \ \ldots, n_m \cdot (a_{i_m} : D), \\ n'_1 \cdot (b_{j_1} : L), \ \ldots, n'_l \cdot (b_{j_l} : L), k \cdot t \end{array} \right\}, \tag{7.11}$$

which means that at this moment, a debtor D holds n_1 object-resources $a_{i_1}, \ldots, n_m$ object-resources a_{i_m}, while a lender L holds n'_1 object-resources $b_{j_1}, \ldots, n'_l$ object-resources b_{j_l}.

The simplest kind of loan between a debtor D and a lender L may be represented by two rules, the first r_1 representing the ***borrow step*** of the loan and involving a transfer of needed by D amounts of *ORs* from L to D and the second r_2 representing a ***debt recovery step***, i.e., return of mutually agreed amounts of resources from D to L at some predefined time moment.

A rule r_1 has the form

$$\begin{gathered} \{\boldsymbol{n_1} \cdot (\boldsymbol{a_1} : L), \ \ldots, \boldsymbol{n_l} \cdot (\boldsymbol{a_l} : L), k \cdot t\} \rightarrow \\ \{\boldsymbol{n_1} \cdot (\boldsymbol{a_1} : D), \ \ldots, \boldsymbol{n_l} \cdot (\boldsymbol{a_l} : D), 1 \cdot (A : L), k \cdot t\}, \end{gathered} \tag{7.12}$$

which means that at the moment k of the beginning of the loan implementation, $\boldsymbol{n_1}$ object-resources $\boldsymbol{a_1}$, ..., $\boldsymbol{n_l}$ object-resources $\boldsymbol{a_l}$, which until this moment were the property of the lender L, after this moment become the property of the debtor D. At the same moment, it is recorded that at a predefined moment, some mutually agreed amounts of object-resources will be returned to the lender, and this is represented by the multiobject $1 \cdot (A : L)$ and the rule r_2. As can be seen, the first possible application of the rule r_1 would be at the moment k.

The rule r_2, whose left and right parts contain a multiobject $k' \cdot t$, where k' is the moment at which the loan period is completed, is as follows:

$$\begin{gathered} \{\boldsymbol{m_1} \cdot (\boldsymbol{b_1} : D), \ \ldots, \boldsymbol{m_g} \cdot (\boldsymbol{b_g} : D), 1 \cdot (A : L), k' \cdot t\} \\ \{\boldsymbol{m_1} \cdot (\boldsymbol{b_1} : L), \ \ldots, \boldsymbol{m_g} \cdot (\boldsymbol{b_g} : L), k' \cdot t\}. \end{gathered} \tag{7.13}$$

As can be seen, the first application of this rule, which occurs at a mutually agreed moment k', transfers $\boldsymbol{m_1}$ object-resources $\boldsymbol{b_1}, \ldots, \boldsymbol{m_g}$ object-resources $\boldsymbol{b_g}$ from the debtor D to the lender L. However, this operation is only possible if the *LES RB* contains a multiobject $1 \cdot (A : L)$, confirming that there is a loan between L and D.

In the case when a loan presumes return of agreed amounts of object-resources in several parts, it is sufficient to define each part by a rule like (7.13), and every such rule includes its own time marker $k'_i \cdot t$, where i is the ordinary number of a partial transfer. Moreover, the same sequential partial transfer may be from the lender to the borrower, so there may be several rules like (7.12), and, in general, a loan may be represented by any necessary set of both types of rules, each including its specific time marker. Of course, such a chain must be integrated by multiobjects like $1 \cdot (A : L)$constructing sequences of mutual transfers of property between the lender and the debtor and in the inverse direction.

A further generalization of this approach may be the participation of several lenders and debtors in such activity, as well as inclusion of direct exchange operations.

7.2.2 Formal Definition of Loan Implementation

Let us formally define loan implementation, i.e., the logic of generation of temporal multisets by application rules like (7.12) and (7.13), as well as rules (3.23) applied to multisets.

In fact, the only nuance which appears when we try to apply rules like (7.12) and (7.13) to some *MST* is connected with time markers. According to (3.33), a rule $v \to v'$ may be applied to a multiset $\overline{v}$ if $\overline{v} \subseteq v$. We shall define that the relation $\subseteq$ is applicable only to ***temporal multisets with identical time markers***, so *MST* $\boldsymbol{v} = v + \{k \cdot t\}$ is a submultiset of (is included in) *MST* $\boldsymbol{v}' = v' + \{k \cdot t\}$, which is denoted as in the case of multisets,

$$\boldsymbol{v} \subseteq \boldsymbol{v}', \tag{7.14}$$

if and only if $v \subseteq v'$. Similarly, $\boldsymbol{v}$ is a strict submultiset of (is strictly included in) *MST* $\boldsymbol{v}'$, i.e.,

$$\boldsymbol{v} \subset \boldsymbol{v}', \tag{7.15}$$

if and only if $v \subset v'$. Both definitions do not contradict the similar definitions of inclusion and strict inclusion of *MSs* since addition of one and the same multiset (in this case $\{k \cdot t\}$) to two multisets already subordinated by either of these relations does not change it.

Now, let $\boldsymbol{v_0} = v_0 + \{\boldsymbol{n_0} \cdot t\}$ be an *LES* resource base at an initial moment $\boldsymbol{n_0}$, *R* be a set of rules representing all the above-described lending and exchange operations, *F* be a set of conditions representing objectives of holders participating in the considered activity, $\boldsymbol{S_0} = \langle \boldsymbol{v_0}, R, F \rangle$ be a filtering multiset grammar operating on temporal multisets, and $\boldsymbol{V}(\boldsymbol{v_0}, R, F)$ be the set of *RBs* which may be achieved by the *LES* represented by $\boldsymbol{v_0}$, *R*, and *F*. This set may be defined similar to (3.30)–(3.33):

$$V_{(0)} = V_{S_0}, \tag{7.16}$$

$$V_{(i+1)} = V_{(i)} \cup \left(\bigcup_{v \in V_{(i)}} V_{S_{i+1}} \right), \tag{7.17}$$

$$\boldsymbol{V}(\boldsymbol{v_0}, R, F) = V_{(\infty)}, \tag{7.18}$$

where

$$\boldsymbol{S_{i+1}} = < v - \{i \cdot t\} + \{(i+1) \cdot t\}, R, F > . \tag{7.19}$$

As can be seen, (7.16)–(7.19) generate the set of all *MSTs*, which represent *LES* resource bases, that may be obtained by chains of lending and exchange operations between holders. All such operations are represented by appropriate rules in the set R. Generation is implemented sequentially for all time moments $i=\boldsymbol{n_0}$, $\boldsymbol{n_0} + 1, \ldots$, in such a way that the set of *MSTs* with time marker $i \cdot t$ is created by joining the sets of *MSTs* generated by the *FMG* with kernel $v + \{i \cdot t\}$, scheme R and filter F. In the general case, the set V_S is infinite due to the infiniteness of the set of values of the aforementioned variable i. However, there is a simple criterion for the recognition of whether $\boldsymbol{V}(\boldsymbol{v_0}, R, F)$ is finite in the sense that no new multiobjects different from time markers appear in generated *MSTs* after some moment. This criterion is similar to the one introduced in Sect. 3.1.3 regarding multiset grammars.

Statement 7.1 Let

$$\boldsymbol{V}_{(i)} = \{ v \mid v + \{n \cdot t\} \in V_{(i)} \}, \tag{7.20}$$

$$\boldsymbol{V}_{(i+1)} = \{ v \mid v + \{l \cdot t\} \in V_{(i+1)} \}, \tag{7.21}$$

$$\boldsymbol{k} = max \ \{ k \mid v + \{k \cdot t\} \rightarrow v' + \{k \cdot t\} \in R \}, \tag{7.22}$$

and

$$\boldsymbol{V}_{(i)} = \boldsymbol{V}_{(i+1)} \, . \tag{7.23}$$

Then

$$(\forall j > i \geq \boldsymbol{k})\ V_{(j)} = V_{(i)}. \tag{7.24}$$

As can be seen, $V_{(j)}$and $V_{(i)}$contain all *MSs* obtained from the *MSTs* belonging, respectively, to $\boldsymbol{V}_{(j)}$ and $\boldsymbol{V}_{(i)}$ by elimination of time markers. So if no additional multiobject other than time markers appears in the set of such *MSs* corresponding to the time moment $i + 1$ compared to the similar set corresponding to the time moment i, then no such additional multiobject will appear at any moment $j > i$. Due to (7.22), after the moment k, when the last rule finalizing some loan is applied, due to (7.23), all following moments produce no additional generated multiobject.

Naturally, all generated *MSTs* must satisfy a ***resource-safekeeping*** feature, which in this case is as follows:

$$(\forall a)(\forall i)\left(\forall v \in \boldsymbol{V}_{(i)}\right) \quad \sum_{n \cdot a : h \in v_o} n = \sum_{n \cdot a : h \in v} n. \tag{7.25}$$

All introduced notions and definitions concerning lending economical systems are illustrated by the following example.

Example 7.6
Consider a loan defined by the following rules r_1 and r_2:

$$r_1 : \{3 \cdot (usd : Alex)\} \rightarrow \{3 \cdot (usd : Bob),\ 1 \cdot (AlexBobAgreement)\},$$
$$r_2 : \{5 \cdot (usd : Bob),\ 1 \cdot (AlexBobAgreement),\ 202010311200 \cdot t\}$$
$$\{5 \cdot (usd : Alex),\ 202010311200 \cdot t\}.$$

As may be seen, the application of the rule r_1 performs the borrow step of the loan implementation, namely, the transfer of 3 dollars from Alex, who is the lender, to Bob, who is the debtor. At the same moment, a reference to the agreement between Alex and Bob is included in the resource base. The rule r_2 may be applied only at the moment defined by the multiplicity of the object t, i.e., October 31, 2020, at 12.00. As soon as this moment occurs, 5 dollars are transferred from Bob to Alex, the reference to the agreement is extracted from the resource base, and the time marker, allowing implementation of the debt recovery step and also being extracted, is returned to the *RB*. Of course, the described operations may be implemented only if Bob has not less than 5 dollars in his property at the aforementioned moment.

Let us note that from the practical point of view, there are two main objectives of loan modeling:

Combinatorial, i.e., selection of a chain of exchanges between lenders and debtors whose implementation at a certain time creates an *LES* resource base satisfying the conditions comprising the filter *F*

Supervisory, i.e., monitoring of a property exchange between holders by an application of some compact representation of a selected chain of exchanges, aggregated into an entity similar to an *IS* schedule

The aforementioned entity, called an ***LES schedule***, will be formally defined in the next section.

7.2.3 *Formal Definition of an* LES *Schedule*

From a substantial point of view, the closest thing we have seen to an *LES* schedule is the notion of an ***IS supply schedule***, defined by (6.48), which operates on ***collections of object-resources to be transferred to an IS at predefined time moments***. On the other hand, ***an exchange of property presumes not only transfer of some ORs to some holders but also elimination of similar ORs from other holders***. In this way, the background of *LES* schedules is similar to the assessment of the consequences of changes of resource-consuming systems defined by (5.26). Let $V(v_0, R, F) = \{v_0, \ldots, v_i, \ldots, v_l\}$ be a set of *MSTs* representing a sequence of *LES RBs* occurring at time moments $\boldsymbol{n}_0, \ldots, \boldsymbol{n}_i, \ldots, \boldsymbol{n}_l = \boldsymbol{k}$, respectively, where $\boldsymbol{k}$ is defined by (7.22).

Following these premises, we shall define a schedule of a lending economical system, represented by v_0, R, and F, as a ***sequence of triples*** $<\boldsymbol{n}_i, v_i - v_{i-1}, v_{i-1} - v_i>$. As can be seen, the first component of this triple is a ***time moment***, and the second is the collection of object-resources which at this moment ***obtain new holders***, while the third component is the collection of resources which, on the contrary, ***lose their previous holders***. Evidently, $v_i \cap v_{i-1}$ is the collection of object-resources which at the moment $\boldsymbol{n}_i$ do not change their holders. Denoting an *LES* schedule by $\boldsymbol{\rho}$ and eliminating from the considered sequence elements which contain empty second and third components (such elements correspond to time moments when no exchange is performed, so there is no necessity to include them in a schedule), we may define it as follows:

$$\rho = \bigcup_{i=0}^{l} \{ <\boldsymbol{n}_i, v_i - v_{i-1}, v_{i-1} - v_i > | \; \{v_i, \; v_{i-1}\}$$

$$\subseteq V(v_0, \; R, \; F) \& (v_i - v_{i-1} \neq \{\emptyset\} \vee v_{i-1} - v_i \neq \{\emptyset\})\}, \tag{7.26}$$

and henceforth it is possible to implement any monitoring activity regarding the considered time interval $[\boldsymbol{n}_0, \boldsymbol{k}]$.

However, an *LES* schedule represents ***dynamics of change of ownership regarding the entire set of holders***. But very often it is necessary to monitor such dynamics regarding only the holder h. It is not difficult to define a ***private schedule of a holder*** h. Such a schedule will be denoted ρ_h and be defined as follows:

$$\boldsymbol{\rho}_h = \bigcup_{i=0}^{l} \left\{ \begin{array}{c} <\boldsymbol{n}_i, \boldsymbol{v}_i^h - \boldsymbol{v}_{i-1}^h, \boldsymbol{v}_{i-1}^h - \boldsymbol{v}_i^h > | \\ \boldsymbol{v}_i^h - \boldsymbol{v}_{i-1}^h \neq \{\emptyset\} \vee \boldsymbol{v}_{i-1}^h - \boldsymbol{v}_i^h \neq \{\emptyset\} \end{array} \right\}, \tag{7.27}$$

where for every $j \in [\boldsymbol{n}_0, \boldsymbol{k}]$, we have

$$\boldsymbol{v}_j^h = \left\{ n \cdot (a : h) \mid n \cdot (a : h) \in \boldsymbol{v}_j \right\}. \tag{7.28}$$

As can be seen, a ***private schedule of a particular holder** h **defines increments and decrements of his property over time***. Such a schedule may be effectively used for both forecasting and planning of the holder's activity, as well as for property monitoring. Of course, such a "slice" from an *LES* schedule may be constructed for any group of holders, thus providing an opportunity for monitoring changes to their property.

Example 7.7
Let $\boldsymbol{V}(\boldsymbol{v}_0, R, F) = \{\boldsymbol{v}_0, \boldsymbol{v}_1\}$, where

$$\boldsymbol{v}_0 = \{3000 \cdot (eur : Alex), 20000 \cdot (usd : Alex), 1 \cdot (ToyotaPrado : Alex),$$
$$1 \cdot (TownHouse : Alex), 7000 \cdot (eur : Bob), 5000 \cdot (usd : Bob), 1 \cdot (NissanJuke : Bob),$$
$$1 \cdot (Flat : Bob), 3000 \cdot (usd : Charlie), 1 \cdot (VWPassat : Charlie),$$
$$1 \cdot (VillageHouse : Charlie), 202010310000 \cdot t\}$$
$$\boldsymbol{v}_1 = \{3000 \cdot (eur : Alex), 15000 \cdot (usd : Alex), 1 \cdot (ToyotaPrado : Alex),$$
$$1 \cdot (NissanJuke : Alex), 1 \cdot (TownHouse : Alex), 7000 \cdot (eur : Bob), 10000 \cdot (usd : Bob),$$
$$1 \cdot (Flat : Bob), 3000 \cdot (usd : Charlie), 1 \cdot (VWPassat : Charlie),$$
$$1 \cdot (VillageHouse : Charlie), 202012310000 \cdot t\}.$$

According to (7.26),

$$\boldsymbol{\rho} = \{ < 202012310000, \{1 \cdot (NissanJuke : Alex), 5000 \cdot (usd : Bob)\},$$
$$\{1 \cdot (NissanJuke : Bob), 5000 \cdot (usd : Alex)\} > \},$$

which represents the property schedule that is the result of an exchange between two property holders—Alex and Bob—in fact, Alex buys a Nissan Juke car which was the property of Bob for 5000 dollars.

The private schedules of the holders belonging to the considered *STS* and changing their property are as follows:

$$\boldsymbol{\rho}_{Alex} = \{ < 202012310000, \{1 \cdot (NissanJuke : Alex)\}, \{5000 \cdot (usd : Alex)\} > \},$$
$$\boldsymbol{\rho}_{Bob} = \{ < 202012310000, \{5000 \cdot (usd : Bob)\}, \{1 \cdot (NissanJuke : Bob)\} > \}.$$

Let us note at the end of this section that common to both exchange and lending economical systems is ***the resource-safekeeping feature***, which is implied by their inability to produce new resources due to a lack of manufacturing devices in their technological bases. Now, we shall move to the most advanced class of *ESs*, which is able to produce new resources and is a ***property-based generalization*** of the ***industrial systems*** we already considered in Chap. 6.

7.3 Producing Economical Systems

Every producing economical system utilizes resources that are the property of multiple holders by application of its manufacturing technological base and distributes produced resources among the holders. Manufacturing devices belonging to a *PES MTB* are also somebody's property.

As is known from Sect. 2.4, producing economical systems are divided into *ESs* operating on passive resources and *ESs* operating on passive and active resources. In this book, we shall study only ***producing ESs OPR***. Let us begin their consideration from the techniques of representation of their *RBs* and *MTBs*, which are the same for all *PESs*.

7.3.1 *Representation of* PES *Resource Bases and Manufacturing Technological Bases*

To provide a unified representation of *PESs RBs*, we shall use as a basis the representation which was introduced for lending *ESs* for the following obvious reason: in contrast with *EESs*, which implement exchange chains promptly, both *LESs* and *PESs* operate continuously over a time scale, and thus it is reasonable to apply in both cases similarly represented *RBs*. Following this basic assumption, we shall represent a *PES RB* as a temporal multiset

$$v = \{n_1 \cdot (a_{i_1} : h_{i_1}),\ \ldots,\ n_m \cdot (a_{i_m} : h_{i_m}),\ k \cdot t\}, \tag{7.29}$$

where multiplicities and composite names of objects have the same sense as in (7.2), so (7.29) means that the resource base of a producing *ES* at a moment k includes n_1 object-resources a_{i_1} belonging to a holder h_{i_1}, …,n_m object-resources a_{i_m} belonging to a holder h_{i_m}.

Now, it is quite reasonable to represent the manufacturing technological base of an *ES OPR* as a set of temporal rules

$$\langle r : h; v \to v' \rangle, \tag{7.30}$$

where r is the name of a manufacturing device belonging to a holder h, and the multiset

$$v = \{\boldsymbol{n_1} \cdot (\boldsymbol{a_{i_1}} : \boldsymbol{h_{i_1}}),\ \ldots,\ \boldsymbol{n_m} \cdot (\boldsymbol{a_{i_m}} : \boldsymbol{h_{i_m}})\} \tag{7.31}$$

represents an input collection of passive resources necessary for one operation cycle of this *MD* and including n_1 object-resources a_{i_1}, which are supplied by holder h_{i_1}, …, n_m object-resources a_{i_m}, which are supplied by a holder h_{i_m}, and, finally, the multiset

$$v' = \left\{\boldsymbol{n'_1} \cdot \left(\boldsymbol{a'_{i_1}} : \boldsymbol{h'_{i_1}}\right),\ \ldots,\ \boldsymbol{n'_l} \cdot \left(\boldsymbol{a'_{i_l}} : \boldsymbol{h'_{i_l}}\right), \boldsymbol{\Delta n} \cdot \boldsymbol{\Delta t}\right\} \tag{7.32}$$

represents the output collection of passive resources produced during one operation cycle of this *MD*, including $\boldsymbol{n'_1}$ object-resources $\boldsymbol{a'_{i_1}}$ which will be supplied to a holder $\boldsymbol{h'_{i_1}}$, …, $\boldsymbol{n'_l}$ object-resources $\boldsymbol{a'_{i_l}}$ which will be supplied to a holder $\boldsymbol{h'_{i_l}}$; all recipients of the assigned *ORs* will get them at the end of the *MD's OC*, whose duration is $\boldsymbol{\Delta n}$ time units.

The manufacturing technological base of an *ES OPAR* differs from the *MTB* of an *ES OPR* by ***one additional feature***: a multiset $\boldsymbol{v'}$ which is the right part of a *TR* may include multiobjects like $\boldsymbol{n'} \cdot (\boldsymbol{r'} : \boldsymbol{h'})$, which means that one operation cycle of an *MD* r which is the property of a holder h produces $\boldsymbol{n'}$ ***manufacturing devices*** $\boldsymbol{r'}$ and supplies them to a holder $\boldsymbol{h'}$.

Example 7.8

Let the manufacturing technological base of some *ES OPR* be represented by the set R, including the following temporal rules r_1 and r_2:

$$< r_1 : Alex; \{1 \cdot (frame : Bob), 1 \cdot (engine : Charlie), 4 \cdot (door : David),$$
$$4 \cdot (wheel : Elvis), 300 \cdot (eur : Alex)\} \rightarrow \{1 \cdot (auto : Alex), 50 \cdot (mnt)\} >,$$

$$< r_2 : Charlie; \{1 \cdot (motor : Charlie), 1 \cdot (transmission : Charlie), 1 \cdot (fuel\ tank : Fred),$$
$$900 \cdot (eur : Charlie)\} \rightarrow \{1 \cdot (engine : Charlie), 80 \cdot (mnt)\} >,$$

and let the resource base of this *ES OPR* be

$$v_0 =$$
$$\{12.00 \cdot t, 5 \cdot (frame : Bob), \quad 12 \cdot (door : David), 16 \cdot (wheel : Elvis), 8 \cdot (wheel : David),$$
$$2 \cdot (motor : Charlie), 3 \cdot (transmission : Charlie), 3 \cdot (fuel\ tank : Fred),$$
$$9000 \cdot (eur : Alex), 5000 \cdot (eur : Charlie), 3000 \cdot (eur : David)\ \}.$$

As can be seen, the *TR* r_1 : *Alex* represents an assembly line belonging to Alex and manufacturing autos, which become his property, from frames supplied by Bob, engines supplied by Charlie, doors supplied by David, and wheels supplied by Elvis. Also, Alex spends 300 euros to perform one operation cycle of his *AL*. The *TR* r_2 : *Charlie* represents an assembly line belonging to Charlie and manufacturing engines, which become his property, from his own motors and transmissions, as well as from fuel tanks supplied by Fred. Charlie spends 900 euros to perform one operation cycle of his *AL*. The resource base of this *ES OPR* at 12.00 contains 9000 euros belonging to Alex; 5 frames belonging to Bob; 2 motors, 3 transmissions, and 5000 euros belonging to Charlie; 12 doors, 8 wheels, and 3000 euros belonging to David; 16 wheels belonging to Elvis; and 3 fuel tanks belonging to Fred.

Now, having introduced a representation of *PESs*, *RBs*, and *MTBs*, we may consider the key issues of modeling *ESs OPR*. *ESs OPAR* as well as *ISs MPAR* will be considered in future publications.

7.3.2 Schedules of Economical Systems Operating on Passive Resources

Let us take into account that the above-proposed change of *RB* representation from the general to a special case allows rather easy application to *PESs* of ***the same algorithmics of schedule creation which was developed for industrial systems***. This is because the aforementioned algorithmics operates on objects' names without splitting them into structural components, so, although these names are composites, there is ***no difference between processing privately owned ORs and ORs that are common property***. Thus, there is ***no need to develop a special PES algorithmics***.

The only issue that is decisively necessary for any producing *ES* is a *PES* schedule, which from the property motion point of view is very similar to the one we have introduced regarding ***lending ESs***. The principal difference between *LES* and *PES* schedules is that in the first case, no new *ORs* appear and no existing *ORs* are lost during system operation (recall that ***LESs satisfy the resource-safekeeping feature***), while in the second case, *ORs* are permanently produced (appear) and utilized (disappear), so the ***total collection of object-resources belonging to any PES RB***, i.e., its resource base, ***changes at moments of beginning and ending of operation cycles***. As a result, the ***property of holders also changes***, but the foundation of this dynamics is not exchange or lending, but ***production and utilization of resources and distribution of produced resources among holders***. In fact, property motion in this case is implemented in the form of mutual supply of manufacturing devices belonging to various holders, according to a *PES* schedule created for completing an order initiated by some holder.

So the only new task associated with producing *ESs* is to define the notion of a ***PES schedule***, by integrating basic features of *LES* and *IS* schedules: namely, because *PESs* are in fact the result of an "intersection" of *LESs* and *ISs*, inheriting from *LESs* assignment of resources to holders and from *ISs* production of new resources by application of the manufacturing technological base to available resources already in the *PES* resource base.

We shall apply denotations which we have already introduced in Sect. 6.2 regarding industrial systems to producing economical systems. So we shall introduce the notion of an ***ES OPR assigned to an order*** $\boldsymbol{Q} = \langle \boldsymbol{q,t} \rangle$ (for short *AES OPR*) and represent it as a triple $E = \langle \boldsymbol{Q}, R, v_0 \rangle$, where $\boldsymbol{q,t}, R$, v_0 have the same sense as in the *AIS MPR* definition, except for the structure of names of object-resources, which in this case are composites $(a : h)$,as well as names of manufacturing devices, which are also composites $(r : h)$. Similar to *ISs*, we shall call a triple $E = \langle \{\emptyset\}, R, v_0 \rangle$ a ***free ES OPR*** (for short *FES OPR*).

We shall call a set

$$v\sigma = \{ \langle n_0, R(n_0) \rangle, \ \ldots, \ \langle n_l, R(n_l) \rangle \}, \tag{7.33}$$

created as described in Sect. 3.2.2 regarding *ISs* and represented according to (6.41), a ***manufacturing schedule of an AES OPR*** *E*, while the set of resource bases of this *ES*

$$\boldsymbol{\rho} = \{ \langle n_0, \boldsymbol{v}_0 \rangle, \ \ldots, \ \langle n_l, \boldsymbol{v}_l \rangle \} \tag{7.34}$$

assigned to time moments $n_0, \ldots, n_i, \ldots, n_l$ will be called *E*. Let us define a resource schedule as a sequence of the following relations, corresponding to a direct execution of a manufacturing schedule:

$$\boldsymbol{v}_0 = v_0 - \left(\sum_{\langle r; v \rightarrow v' \rangle \, \in R(n_0)} v \right) - \{n_0 \cdot t\}, \tag{7.35}$$

$$\ldots$$

$$\boldsymbol{v}_i = \boldsymbol{v}_{i-1} + \left(\sum_{\langle r; v \rightarrow v'' + \{\Delta n \cdot \Delta t\} \rangle \, \in R(n_i - \Delta n)} v'' \right) - \left(\sum_{\langle r; v \rightarrow v' \rangle \, \in R(n_i)} v \right), \tag{7.36}$$

$$\ldots$$

$$\boldsymbol{v}_l = \boldsymbol{v}_{l-1} + \left(\sum_{< r; v \rightarrow v'' + \{\Delta n \cdot \Delta t\} > \in R(n_l - \Delta n)} v'' \right). \tag{7.37}$$

As can be seen from (7.36), every $\boldsymbol{v}_i$ is the *RB* at the moment n_i before the *ES* would move to the moment $n_i + 1$. It is formed from the *RB* which existed at the moment n_{i-1}, by adding to it all collections of *ORs* which were manufactured by *MDs* whose operation cycles ***finish*** at this moment and subtracting from the obtained result all collections of *ORs* which are necessary to *MDs* whose operation cycles ***begin*** at this moment. There are two exceptions to this regular logic. According to (7.35), $\boldsymbol{v}_0$ is formed by only the aforementioned subtraction, because at the initial moment, there are no *MDs* whose *OCs* have finished. Similarly, $\boldsymbol{v}_l$ is formed by only addition, because there are no *MDs* whose *OCs* begin at the final moment of a whole schedule, when order completion ends.

A resource schedule is the basis for a so-called property schedule of an AES OPR E, representing property motion at any given time and substantially similar to an *LES* schedule. For this reason, its formal definition is very similar to (7.26):

$$\boldsymbol{\rho} = \bigcup_{i=0}^{l} \left\{ \begin{array}{l} < \boldsymbol{n}_i, \boldsymbol{v}_i - \boldsymbol{v}_{i-1}, \boldsymbol{v}_{i-1} - \boldsymbol{v}_i > | \\ \qquad \boldsymbol{v}_i - \boldsymbol{v}_{i-1} \neq \{\emptyset\} \vee \boldsymbol{v}_{i-1} - \boldsymbol{v}_i \neq \{\emptyset\} \end{array} \right\}. \tag{7.38}$$

Similar to (7.27), we may define the ***private property schedule of a holder*** p:

$$\boldsymbol{\rho}_h = \bigcup_{i=0}^{l} \left\{ \begin{array}{c} < \boldsymbol{n}_i, \boldsymbol{v}_i^h - \boldsymbol{v}_{i-1}^h, \boldsymbol{v}_{i-1}^h - \boldsymbol{v}_i^h > | \\ \boldsymbol{v}_i^h - \boldsymbol{v}_{i-1}^h \neq \{\emptyset\} \vee \boldsymbol{v}_{i-1}^h - \boldsymbol{v}_i^h \neq \{\emptyset\} \end{array} \right\}, \tag{7.39}$$

where $\boldsymbol{v}_j^h$ for every $j \in [0, l]$ is defined by (7.28).

In practice, ***the behavior of any economical system involves the execution of all three considered basic types of operations on property: exchange, lending, and producing***. So it would be practically valuable to develop techniques of modeling such "mixed" *ESs*, integrating features of *EESs*, *LESs*, and *PESs*. Such techniques will be considered in the next section.

7.4 Mixed Economical Systems

We shall consider mixed *ESs* in order from simple to complex:

1. *EES* + *LES*
2. *EES* + *PES*
3. *LES* + *PES*
4. *EES* + *LES* + *PES*

As above, when speaking about producing economical systems, we shall limit our scope only to *ESs* operating passive resources.

7.4.1 Exchange and Lending Economical Systems

This case is in fact the simplest one, and it was already introduced in Sect. 7.2.2. Now, let us recall only that above we defined $V(\boldsymbol{v_0}, R, F)$ as the set of *RBs* that are created by an *LES* $E = <\boldsymbol{v_0}, R, F>$, where $\boldsymbol{v_0}$ is its initial resource base, *F* is a set of conditions representing objectives of holders participating in the lending process, and *R* is a set of rules. Thus, in fact, we have described a technique of an application of such rules to a unified representation of both lending and exchange operations. As a result, (7.16)–(7.19) generate the set of all *MSTs* representing *EES* + *LES* resource bases that may be obtained by ***chains of lending and exchange operations***. All such operations are represented by the relevant rules belonging to the set *R* and differing by the presence of time markers in rules representing loans and their absence in rules representing exchanges. So the last may be applied to any *MST* while the first only to *MSTs* with the same time markers, as in the rule.

7.4.2 Exchange and Producing Economical Systems

This case is not more complicated than the previous one. As it is easy to see, every exchange operation represented by a proper rule may be considered as a virtual manufacturing device with ***an operation cycle of zero duration***, i.e., the zero multiplicity Δn of the object Δt in a temporal rule representing such a virtual *MD*. So, because rules representing exchange operations are singular *TRs* with $\Delta n=0$, they may be included in the set of *TRs*, representing a *PES* manufacturing technological base, and, thus, be applied on the regular algorithmic basis developed for *PESs*.

7.4.3 Lending and Producing Economical Systems

This case is the most complicated, and there may be two ways of ***integrating LESs and PESs to give a unified model***:

1. Redefining the mathematical semantics of temporal multiset grammars and filtering *TMGs* introduced in Sects. 3.2.2 and 3.2.3, by including in it rules representing loans, i.e., containing *MSTs* in both the left and the right parts of rules
2. Developing a technique to represent loans by temporal rules, semantically equivalent to rules allowing replacement of *MSTs*

We shall follow the second way, trying to solve ***the task of transition from time markers*** $n \cdot t$, which are elements of *MSTs*, to ***time intervals*** Δn, which are multiplicities of the object Δt in temporal rules.

Let $R = R_M \cup R_L$ be a set representing the capabilities of an *ES* integrating manufacturing and lending in such a way that R_M, including temporal rules, represents the manufacturing technological base of this *ES*, while R_L, including rules on *MSTs*, represents loans established by holders belonging to this *ES*:

$$R_L = \bigcup_{i=1}^{l} \{< r_i; v_i + \{n_i \cdot t\} \to v'_i + \{n_i \cdot t\} > \}, \tag{7.40}$$

Let $\boldsymbol{v_0} = v_0 + \{n_0 \cdot t\}$ be the initial resource base of the considered *ES*. We shall construct an auxiliary initial resource base $\boldsymbol{v'_0}$ and a set $\boldsymbol{R_L}$ of temporal rules which, being joined with R_M, correctly represent this *ES* in the sense that the set of *MSTs* generated by the constructed *TMG* $\boldsymbol{S}$=<$\boldsymbol{v'_0}$,$R_M \cup \boldsymbol{R_L}$> is the same as the set of resource bases which are obtained by an application of the manufacturing and lending capabilities of this *ES* to the aforementioned initial *RB* $\boldsymbol{v_0}$. Namely, we shall define

$$\boldsymbol{v'_0} = \boldsymbol{v_0} + \bigcup_{i=1}^{l} \{1 \cdot \boldsymbol{b_i}\}, \tag{7.41}$$

where each auxiliary object $\boldsymbol{b_i}$ corresponds to rule r_i from (7.40). Also we shall define

$$\boldsymbol{R_L} = R_L^{start} \cup R_L^{loan}, \tag{7.42}$$

where

$$R_L^{start} = \bigcup_{i=1}^{l} \{< \boldsymbol{r_i}; \{1 \cdot \boldsymbol{b_i}\} \to \{1 \cdot b_i, (n_i - n_0) \cdot \Delta t\} > |$$

$$< r_i; v_i + \{n_i \cdot t\} \to v'_i + \{n_i \cdot t\} > \in R_L\}, \tag{7.43}$$

$$R_L^{loan} = \bigcup_{i=1}^{l} \{ <\boldsymbol{r}_i^+; v_i + \{1 \cdot b_i\} \rightarrow v'_i>, \quad <\boldsymbol{r}_i^-; \{1 \cdot b_i\} \rightarrow \{\emptyset\}> \}. \qquad (7.44)$$

As may be seen, at the initial moment n_0, due to $1 \cdot \boldsymbol{b}_i \in \boldsymbol{v}'_0$, an application of temporal rule $\boldsymbol{r}_i \in R_L^{start}$, whose left part is $\{1 \cdot \boldsymbol{b}_i\}$, to $\boldsymbol{v}'_0$ will add to the generated temporal multiset at a moment $n_0 + (n_i - n_0) = n_i$ the multiobject $1 \cdot b_i$, "manufactured" at this moment by $\boldsymbol{r}_i$.

At the moment n_i, the multiobject $1 \cdot b_i$, belonging to the generated *MST* and the left part of a rule $\boldsymbol{r}_i^+$, in the case that the multiset v_i is a submultiset of the generated *MST*, enables the replacement of v_i by v'_i as a result of application of the rule $\boldsymbol{r}_i^+$. So each of the two loan steps (borrowing and debt recovery) associated with the predefined moment n_i is correctly implemented. In this case, the multiobject $1 \cdot b_i$ is eliminated from the generated *MST*, and, hence, the rule $\boldsymbol{r}_i^-$ is not applied at the next generation step.

A more complicated case arises if the multiset v_i is not a submultiset of the generated *MST*. Then the rule $\boldsymbol{r}_i^+$ cannot be applied to the generated multiset, but there is a possibility that at some moment $n_j > n_i$, the generated multiset would include the *MS* v_i, and then this rule would be applied, and thus a step of the loan would be implemented at the wrong time. To avoid this possibility, the rule $\boldsymbol{r}_i^-$ is included in the set R_L^{loan}, and its application at the next generation step causes elimination of the multiobject $1 \cdot b_i$ from the generated *MST*, if this *MO* remains in this *MST* because of non-applicability of the rule $\boldsymbol{r}_i^+$.

Thus, as can be seen, lending and producing economical systems may be integrated without any major difficulties.

7.4.4 Exchange, Lending, and Producing Economical Systems

Having the technique of integrating all possible couples of basic types of *ESs* as a background, it is not difficult to integrate all three basic types.

Let:

R_E be a set of rules representing possible exchanges between subjects of an *ES*

R_L be a set of rules representing possible loans inside this *ES*

R_M be a set of temporal rules representing its manufacturing capabilities

Constructing a temporal multiset grammar

$$\boldsymbol{S} = <\boldsymbol{v}'_0, R_E \cup \boldsymbol{R}_L \cup R_M>, \qquad (7.45)$$

where $\boldsymbol{v}'_0$ and $\boldsymbol{R}_L$ are defined, respectively, by (7.41) and (7.42)–(7.44), we achieve the needed integration of the capabilities of all three basic types of *ESs* into one economical system representing the most general case of this class of *STSs*.

To end this chapter, let us note that all considered cases were related to interactions between *ES* components regardless of their locations, so, in fact, limiting the scope to local *ESs*. To handle ***distributed*** *ESs*, it would be sufficient, as was already done in the case of industrial systems, to apply the representation of object-resources located at some places (points) that was announced in Sect. 2.5 regarding any sociotechnological systems. Namely, if it is necessary to consider distributed *ESs*, then the minimal syntactic extension to the ordinary representation of *OR* locations already used in Chap. 6 will be applied. As above, we use a set of locations (positions) Z, where object-resources may be present and every such location will be denoted by $z \in Z$. Then instead of primary names of *ORs* $(a : h)$, we use constructions $(a : h, z)$, where z is the name of a point (place, area) where an *OR* a belonging to a holder h is located and "(", ",", and ")" are delimiters. (As was mentioned in Chap. 6, in the general case, z may be defined as a point in some coordinates, as well as an area of any form—circle, quadrilateral, ellipse, or any shape defined by boundary points.) So all results obtained and described above regarding local *ESs* may be easily generalized to the case of distributed *ESs*.

With this, we finish our consideration of applications of the multigrammatical framework to economical systems and, as was announced above, move to consideration of the basic issues of *MGF* application to various tasks from the area of the ***resilience and recovery of sociotechnological systems***.

Chapter 8
Resilience and Recovery of Sociotechnological Systems

As was declared in Sect. 2.5, the following kinds of destructive impacts on sociotechnological systems are possible:

1. On an *STS* resource base
2. On an *STS RB* and its technological base
3. Impacts on an *STS RB* and *TB*, multiplied by cascade (chain) effects

Any such impact eliminating some parts of the *RB* and/or the *TB* of an *STS* processing an order (or multiple orders at the same time) may (or may not) transpose the system to a state in which it is not able to complete this order (or some of these orders) (Sharkey and Pinkley 2019). In this case, there are two groups of tasks to be solved by the *STS* scheduler, depending on whether the system is ***closed*** (i.e., not connected by resources with any external *STS* that is able to aid it to complete currently processed orders) or ***open*** (i.e., connected with one or more external *STSs* that are able to recover the *RB* and *TB* of the affected system or to support the latter by completing some more or less valuable parts of the orders being processed at the moment of the impact).

Concerning both types of sociotechnological systems, there is a common task to be solved by the *STS* scheduler, namely, ***assessment of the STS's ability to complete the current order (orders) after an impact***. If the affected system is able to complete the current order (orders), it is considered ***resilient*** to this impact, and no additional actions from the *STSS* side, except establishing a new schedule enabling current order (orders) completion with a reduced resource base and technological base, are required in this case (Sheremet 2021a).

If a ***closed STS*** is ***vulnerable*** to an impact, i.e., it is not able to complete the current order applying the resource base and technological base remaining after this impact, then the *STSS* must assess ***what part of the order may be completed*** by the application of this *RB* and *TB* and ***create a schedule maximizing the aforementioned part***.

Concerning a ***vulnerable open STS***, the following tasks must be solved by the *STS* scheduler for recovery and support of this *STS*:

I. A. Sheremet, *Multigrammatical Framework for Knowledge-Based Digital Economy*, https://doi.org/10.1007/978-3-031-13858-4_8

1. ***Assessment of additional amounts of resources*** which it would be necessary to deliver to the affected *STS* for order completion
2. ***Creation of interconnected (correlated) schedules of affected and external systems*** allowing order completion

The last task may, in turn, have two versions:

1. An external system ***recovers*** all or some of the resources eliminated by an impact from the *RB* and *TB* of the affected system, which, applying these resources during implementation of a created new schedule, manages to complete the current order.
2. An external system ***supports*** the affected system by completing some part of the current order in such a way that it is completed as a whole as a result of the collaboration of the two systems implemented by the correlated orders created by their interconnected and interacting schedulers.

The whole set of listed tasks to be solved by an *STS* scheduler to ensure the resilience of its system will be considered with respect to the following set of initial conditions:

1. The affected *STS* is industrial or economical.
2. One order is considered or multiple orders are considered following one another at unpredictable time moments.
3. One impact is considered or multiple impacts are considered following one another at unpredictable time moments.
4. The affected *STS* is local or distributed.
5. Impacts are or are not multiplied by cascade effects.

There exists a ***large-scale research program*** targeting the creation of *MGF*-based schedulers enabling smart and resilient operation of a wide range of sociotechnological systems. In this book, from all $2^5=32$ variants of initial conditions, we shall study only local *ISs* and *ESs*, one order being interrupted by one impact, the last being in both possible implementations (without and with cascade effects). All the remaining variants will be considered briefly by describing the main techniques for their solution.

We shall consider issues concerning ***closed industrial systems*** in Sect. 8.1, ***open ISs*** in Sect. 8.2, and ***modeling cascade effects*** in Sect. 8.3.

The most important and valuable area of application of the proposed *MGF*-based tools for the assessment of *STS* resilience is critical infrastructures, discussed as a whole in Sect. 2.5. Below in Sect. 8.4, we shall consider the most important of all *CIs*—***energy infrastructures*** (*EIs*)—which enable the operation of all other *CIs* by electrical power (*EP*) and fuel supply. Let us note that *EIs* are essentially ***distributed*** *STSs*.

We shall not consider resilience and recovery of ***economical systems*** here, because all techniques concerning this topic will, in fact, be represented in Chap. 9, dedicated to ***resource-based games***.

8.1 Closed Industrial Systems

As is known from Chap. 6, there are two ways of representing industrial systems—*UMG*-based and *TMG*-based. We shall apply the same two approaches below when considering tasks to be solved by *IS* schedulers in the area of analysis of resilience and vulnerability of closed *ISs*.

8.1.1 UMG-*Based Analysis of* IS *Vulnerability*

Let us consider an industrial system $I = < q, R, v_0>$ assigned to an order q, with a manufacturing technological base R and a resource base v_0 , which contains only primary (not manufactured by this industrial system) object-resources, i.e., *ORs* represented by terminal objects of a unitary multiset grammar $S= < q, R >$. The simplest way of modeling a destructive impact applied to this *AIS* ***before the beginning of processing of an order*** is to represent it by a multiset Δv, which is nothing but the collection of primary object-resources eliminated from the *RB*. So the affected *AIS* is

$$I' = < q,R,v_0 - \Delta v > , \tag{8.1}$$

and the criterion of vulnerability of this *AIS* I to an impact Δv is a simple application of the criterion of feasibility of orders established by Statement 6.1.

Statement 8.1 An *AIS* I is vulnerable to an impact Δv if

$$(\forall v \in \bar{V}_S)\; v - (v_0 - \Delta v\,) \neq \{\emptyset\}. \tag{8.2}$$

Verbally, if an *AIS* resource base, reduced by an impact, is not sufficient for any of the possible ways of order completion, then this system is vulnerable to this impact. An alternative formulation of this statement, which may be applied directly, is as follows:

$$(\forall v \in \bar{V}_S)\;\; (v_0 - \Delta v\,) \cap v \neq v. \tag{8.3}$$

Example 8.1 Let the manufacturing technological base of an *IS* be represented by the scheme R containing the following two unitary rules in the technological interpretation:

$$r_1 : (auto) \to 1 \cdot (frame), 1 \cdot (engine), 4 \cdot (door), 4 \cdot (wheel),$$

$$400 \cdot (kwh), 50 \cdot (mnt : autos\ AL), 300 \cdot (eur);$$

$$r_2 : (engine) \rightarrow 1 \cdot (motor), 1 \cdot (transmission), 1 \cdot (fuel\ tank),$$
$$100 \cdot (kwh), 80 \cdot (mnt : engines\ AL), 900 \cdot (eur).$$

Let the resource base of the considered *IS* be as follows:

$$v_0 = \{3 \cdot (frame), 10 \cdot (door), 16 \cdot (wheel), 4 \cdot (motor), 3 \cdot (transmission),$$
$$4 \cdot (fuel\ tank), 150 \cdot (mnt : autos\ AL), 300 \cdot (mnt : engines\ AL),$$
$$1800 \cdot (kwh), 10{,}000 \cdot (eur)\}.$$

and, finally, let the order q be the multiset $\{2 \cdot (auto)\}$. As we have seen, this order is feasible by the considered *AIS* because $\bar{V}_S = \{v\}$, where

$$v = \{2 \cdot (frame), 8 \cdot (door), 8 \cdot (wheel), 2 \cdot (motor), 2 \cdot (transmission),$$
$$2 \cdot (fuel\ tank), 100 \cdot (mnt : autos\ AL), 1000 \cdot (kwh), 2400 \cdot (eur),$$
$$160 \cdot (mnt : engines\ AL)\} \subset v_0.$$

Let the impact be

$$\Delta v = \{1 \cdot (frame), 1 \cdot (door), 2 \cdot (wheel), 300 \cdot (kwh), 1000 \cdot (eur)\},$$

so

$$v_0 - \Delta v = \{2 \cdot (frame), 9 \cdot (door), 14 \cdot (wheel), 4 \cdot (motor), 3 \cdot (transmission),$$
$$4 \cdot (fuel\ tank), 150 \cdot (mnt : autos\ AL), 300 \cdot (mnt : engines\ AL),$$
$$1500 \cdot (kwh), 9000 \cdot (eur)\},$$

and, thus, $v \subset v_0 - \Delta v$, so the considered *AIS* does not satisfy criterion (8.2) and, for this reason, *may be* resilient to the impact Δv.

On the other hand, in the case

$$\Delta v = \{1000 \cdot (kwh),\ 2000 \cdot (eur)\}$$

the result of the impact will be

$$v_0 - \Delta v = \{3 \cdot (frame), 10 \cdot (door), 16 \cdot (wheel), 4 \cdot (motor), 3 \cdot (transmission),$$

$$4 \cdot (\textit{fuel tank}), 150 \cdot (\textit{mnt} : \textit{autos AL}), 300 \cdot (\textit{mnt} : \textit{engines AL}),$$
$$800 \cdot (\textit{kwh}), 8000 \cdot (\textit{eur})\},$$

and because v is not a submultiset of $v_0 - \Delta v$,the considered *AIS* is vulnerable to the impact Δv.

Let us note that we do not affirm that an AIS which does not satisfy the formulated criterion (8.2) *is* ***resilient****. From a substantial point of view, it is obvious that without incorporating the* ***duration of order completion*** *into the criterion of AIS resilience, it is not possible to assess whether an order would be completed or not. However, if the AIS resource base after impact is not sufficient for any way of application of the AIS MTB, it is evident that, regardless of the aforementioned duration, the order is* ***non-feasible*** *and, thus, the AIS is vulnerable to the impact.*

Let us consider a more complicated case, where an impact is applied not strictly before the beginning of processing of an order by the *AIS*, but ***within the time interval of processing of this order***. The main difference of this case compared with the previous one is that during a subinterval preceding the moment of an impact, some secondary object-resources might be manufactured by the application of the *MTB* and joined to the *RB*.

There may be two approaches to modeling such a possibility.

The ***first*** one, introduced in Sheremet (2019c), is based on an application of the so-called unitary multiset grammars with reduced generation.

The ***second***, described in Sect. 4.4.3 in the context of filtering temporal multiset grammars, is much simpler and is based on ***terminalization***, defined in that section, which in the case of *UMGs* is no more than the inclusion in a scheme *R*, representing the manufacturing technological base of an *IS*, of $\mid A_S - \bar{A}_S \mid$ additional unitary rules

$$a \rightarrow 1 \cdot \boldsymbol{a} \tag{8.4}$$

each corresponding to one non-terminal object (i.e., a ***secondary OR***) $a \in A_S - \bar{A}_S$. Here, $\boldsymbol{a}$ is an additional terminal object, representing a produced *OR* a. (In Sect. 4.4.3, an alternative way of representing manufactured objects stored in a resource base was used, i.e., concatenating a prefix "*" to their names in temporal rules, to represent manufacturing devices.) This makes it possible to include in an *IS* resource base a multiobject $n \cdot \boldsymbol{a}$ in the case that there are n already manufactured *ORs* a, and, as we have seen, this inclusion collects together all known alternative ways of manufacturing one more *OR* a, which is simply an ***extraction of this OR from the resource base*** (of course, if it occurs there). Such *URs* will be called ***terminalizing***, and a *UMG* whose scheme includes terminalizing *URs* for all non-terminal objects is called ***terminalized***. A unitary multiset metagrammar may be treated the same way, because there is no difference between unitary rules and unitary metarules regarding additional *URs* (8.2).

Let us note that since the definition terminalization, we may extend criterion (8.2) to the general case of an *IS RB*, which before the beginning of order processing may contain not only primary but also secondary *ORs*. Moreover, this technique may be applied to the ***criterial base of order feasibility***. Hence, there is no necessity for an extension of the mathematical and algorithmical background of assessment of *IS* vulnerability in the considered general case of impact application. The criterion of *AIS* vulnerability to an impact, applied within a time interval of order processing, may be obtained by the following very local correction of criterion (8.2):

$$(\forall v \in \bar{V}_S) v' - \Delta v \subset v, \tag{8.5}$$

where v' is the resource base at the moment of impact application. At any moment before the beginning of order processing, $v' = v_0$.

Example 8.2 Let the manufacturing technological base of some *IS* be represented by the scheme *R* containing the same two unitary rules r_1 and r_2 in the technological interpretation as in the previous Example 8.1 and also two terminalizing *URs* r_3 and r_4:

$$r_1 : (auto) \rightarrow 1 \cdot (frame), 1 \cdot (engine), 4 \cdot (door), 4 \cdot (wheel),$$

$$400 \cdot (kwh), 50 \cdot (mnt : autos\ AL), 300 \cdot (eur);$$

$$r_2 : (engine) \rightarrow 1 \cdot (motor), 1 \cdot (transmission), 1 \cdot (fuel\ tank),$$

$$100 \cdot (kwh), 80 \cdot (mnt : engines\ AL), 900 \cdot (eur);$$

$$r_3 : (auto) \rightarrow 1 \cdot (\boldsymbol{auto});$$

$$r_4 : (engine) \rightarrow 1 \cdot (\boldsymbol{engine}),$$

where (***auto***) and (***engine***) are terminal objects representing, respectively, an already manufactured auto and engine.

Let the resource base of the considered *IS* be as follows:

$$v_0 = \{1 \cdot (\boldsymbol{auto}), 2 \cdot (\boldsymbol{engine}), 1 \cdot (frame), 5 \cdot (door), 7 \cdot (wheel), 2 \cdot (motor),$$

$$3 \cdot (transmission), 2 \cdot (fuel\ tank),$$

$$120 \cdot (mnt : autos\ AL), 200 \cdot (mnt : engines\ AL),$$

$$1000 \cdot (kwh), 3000 \cdot (eur)\},$$

which means that the resource base contains one already produced auto and two engines.

Let the order q be the multiset $\{2 \cdot (auto)\}$. So the set of terminal multisets will be

$$\bar{V}_S = \{v_1, v_2, v_3, v_4\},$$

where

$$v_1 = \{2 \cdot (\boldsymbol{auto})\},$$

$$v_2 = \{1 \cdot (\boldsymbol{auto}), 1 \cdot (\mathit{frame}), 1 \cdot (\boldsymbol{engine}), 4 \cdot (\mathit{door}), 4 \cdot (\mathit{wheel}),$$
$$50 \cdot (\mathit{mnt} : \mathit{autos\ AL}), 400 \cdot (\mathit{kwh}), 300 \cdot (\mathit{eur})\},$$

$$v_3 = \{1 \cdot (\boldsymbol{auto}), 1 \cdot (\mathit{frame}), 4 \cdot (\mathit{door}), 4 \cdot (\mathit{wheel}),$$
$$1 \cdot (\mathit{motor}), 1 \cdot (\mathit{transmission}), 1 \cdot (\mathit{fuel\ tank}),$$
$$50 \cdot (\mathit{mnt} : \mathit{autos\ AL}), 80 \cdot (\mathit{mnt} : \mathit{engines\ AL}),$$
$$500 \cdot (\mathit{kwh}), 1200 \cdot (\mathit{eur})\},$$

$$v_4 = \{2 \cdot (\mathit{frame}), 8 \cdot (\mathit{door}), 8 \cdot (\mathit{wheel}),$$
$$2 \cdot (\mathit{motor}), 2 \cdot (\mathit{transmission}), 2 \cdot (\mathit{fuel\ tank}),$$
$$100 \cdot (\mathit{mnt} : \mathit{autos\ AL}), 160 \cdot (\mathit{mnt} : \mathit{engines\ AL}),$$
$$1000 \cdot (\mathit{kwh}), 2400 \cdot (\mathit{eur})\}.$$

As may be seen, this order is feasible regarding the considered *AIS* because there exists $v_2 \in \bar{V}_S$ such that $v_2 \subseteq v_0$.

Let the impact

$$\Delta v = \{1 \cdot (\boldsymbol{engine}), 1 \cdot (\mathit{door}), 2 \cdot (\mathit{wheel}), 300 \cdot (\mathit{kwh}), 1000 \cdot (\mathit{eur})\},$$

so

$$v_0 - \Delta v = \{1 \cdot (\boldsymbol{auto}),\ 1 \cdot (\boldsymbol{engine}),\ 1 \cdot (\mathit{frame}), 4 \cdot (\mathit{door}), 5 \cdot (\mathit{wheel}), 2 \cdot (\mathit{motor}),$$
$$3 \cdot (\mathit{transmission}), 2 \cdot (\mathit{fuel\ tank}),$$
$$120 \cdot (\mathit{mnt} : \mathit{autos\ AL}), 200 \cdot (\mathit{mnt} : \mathit{engines\ AL}),$$
$$700 \cdot (\mathit{kwh}), 2000 \cdot (\mathit{eur})\},$$

and, because there exists $v_3 \in \bar{V}_S$ such that $v_3 \subset v_0 - \Delta v$, the considered *AIS*, according to criterion (8.2), may be resilient to the impact Δv.

On the other hand, in the case

$$\Delta v = \{1 \cdot (\boldsymbol{auto}),\ 100 \cdot (kwh)\}$$

the result of the impact will be

$$\begin{aligned} v_0 - \Delta v = \{\ & 2 \cdot (\boldsymbol{engine}),\ 1 \cdot (frame), 5 \cdot (door), 7 \cdot (wheel), 2 \cdot (motor), \\ & 3 \cdot (transmission), 2 \cdot (fuel\ tank), \\ & 120 \cdot (mnt : autos\ AL), 200 \cdot (mnt : engines\ AL), \\ & 900 \cdot (kwh), 3000 \cdot (eur)\}, \end{aligned}$$

and because no $v \in \bar{V}_S$ is a submultiset of $v_0 - \Delta v$,the considered *AIS* is vulnerable to the impact Δv.

All that has been said above in this Sect. 8.1.1 concerned impacts on *IS* ***resource bases***. However, impacts may be applied also to manufacturing technological bases of industrial systems. This case may be similarly modeled by a set ΔR of unitary rules, representing manufacturing devices eliminated from an *IS MTB*. Thus, the result of an impact would be an *AIS*

$$I' = <q, R - \Delta R, v_0>, \tag{8.6}$$

if the impact was applied before the beginning of processing of an order, and

$$I' = <q, R - \Delta R, v'>, \tag{8.7}$$

if the impact was applied within a time interval of processing of an order.

Taking into account all that has been said in this section, we may represent the most general case, when an impact is applied at some moment within a time interval of order processing to an *RB* and an *MTB* ***simultaneously,*** by an *AIS*

$$I' = <q, R - \Delta R, v' - \Delta v>; \tag{8.8}$$

thus, the most general criterion of *AIS* vulnerability is as follows.

Statement 8.2 An *AIS* $I = <q, R, v_0>$ is vulnerable to an impact eliminating a set of manufacturing devices ΔR and a collection of object-resources Δv if

$$(\forall v \in \bar{V}_{S'})\ v' - \Delta v \subset v, \tag{8.9}$$

where $S' = <q, R - \Delta R>$ is a terminalized *UMG* and v' is the resource base of this *AIS* at the moment of impact.

It is supposed, as above, that before the beginning of order processing, $v' = v_0$.

Let us note that, as was proposed in Sheremet (2018, 2019c, 2021a) and illustrated in Sect. 3.1.3 (see Example 3.11), an impact on an *MTB* may be modeled by an inclusion in *MSs* v and Δv, as well as in bodies of *URs* belonging to R, of multiobjects like $n \cdot (\Delta t : x)$, where Δt is a time unit, by which the duration of operation of a manufacturing device x is measured. Any such multiobject represents the so-called operation resource of an *MD*: number of time units spent by it for manufacturing one object-resource from consumables (if $n \cdot (\Delta t : x)$ enters the body of the *UR* representing this *MD*) or number of time units of operation of this *MD* available during the considered operation cycle of the *IS* which technological base includes this *MD* (if $n \cdot (\Delta t : x)$ enters the resource base of this *IS* and thus value n implicitly determines the number of output *ORs* which the *MD* may produce being supplied by necessary collection of input *ORs*). If an *MS* Δv includes an *MO* $\boldsymbol{N} \cdot (\Delta t : x)$, where $\boldsymbol{N}$ is a maximal integer number in the current implementation of the *IS* scheduler, then the presence of an *MO* $\boldsymbol{N} \cdot (\Delta t : x)$ in an *MS* Δv means that the impact destroys the manufacturing device x completely, because for any $n \neq \boldsymbol{N}$

$$\{n \cdot (\Delta t : x)\} - \{\boldsymbol{N} \cdot (\Delta t : x)\} = \{0 \cdot (\Delta t : x)\} = \{\varnothing\}. \tag{8.10}$$

The consequence are the same if Δv includes $n' \cdot (\Delta t : x)$, where $n' \geq n$. If $n' < n$, then the impact produces some damage to the affected manufacturing device, and thus its operation resource is reduced from $n \cdot (\Delta t : x)$ to $(n - n') \cdot (\Delta t : x)$. So henceforth, ***it is sufficient to represent two different impacts—one on the resource base and the other on the manufacturing technological base of an IS—by only one impact concerning the resource base***.

As may be seen, the described technique simplifies and unifies the criterial base of a *UMG*-based assessment of the vulnerability of industrial systems, reducing it to only one criterion (8.2).

In introducing objects like $(\Delta t : x)$, we assume the additivity of time regarding any separate manufacturing device. However, let us recall that ***regarding any IS as a whole, time is not an additive value, because manufacturing devices may operate in parallel***. For this reason, an *IS* which, according to (8.2), is not vulnerable may be vulnerable from a substantial point of view, because the duration of completion of an order may exceed an upper bound declared by the time subfilter of the order. A fully adequate assessment of *IS* sustainability requires the application of a ***TMG-based*** representation of industrial systems and will be done in Sect. 8.1.3.

8.1.2 UMG-Based Assessment of Partial Order Completion by a Vulnerable IS

This task is similar to the task of a *UMG*-based assessment of partial order completion in the case of an insufficient resource base, considered in Sect. 6.1.4. So we shall apply the technique proposed there to the new task concerning vulnerable industrial systems.

Consider an *AIS* $I = \langle \boldsymbol{q}, R, v' \rangle$, where, as in the aforementioned Sect. 6.1.4,

$$\boldsymbol{q}=\{\boldsymbol{n_1}\cdot\boldsymbol{a_1},\ \ldots,\ \boldsymbol{n_m}\cdot\boldsymbol{a_m}\},\tag{8.11}$$

while $v'$is the resource base of the considered affected *IS* after impact.

We shall create a ***terminalized UMMG*** $= <\{1\cdot q\}, R', F>$, where $R' = R\cup\{r_q\}$, r_q is a unitary metarule

$$q\rightarrow\gamma_1\cdot\boldsymbol{a_1},\ldots,\gamma_m\cdot\boldsymbol{a_m},\tag{8.12}$$

and the filter

$$F=\bigcup_{j=1}^{m}\{0\leq\gamma_j\leq\boldsymbol{n_j}\}\cup\{a\leq n\mid n\cdot a\in v'\}\cup\{a=0\mid a\in\overline{A}_S\&a\notin v'\}.\tag{8.13}$$

According to the semantics of *UMMGs* and (8.11)–(8.13), each multiset $v\in\bar{V}_S$ determines

$$v_{\boldsymbol{q}}=\{m_i\cdot\boldsymbol{a_i}\mid m_i\cdot\bar{\gamma}_i\in v\}\subseteq\ \boldsymbol{q}\tag{8.14}$$

as the part of an order $\boldsymbol{q}$ which may be manufactured by an *AIS* $I = <\boldsymbol{q}, R, v'>$ with resource base v' remaining after an impact and being insufficient for the completion of this order. At the same time, the multiset $v - v_{\boldsymbol{q}}$ is the part of the resource base sufficient for the completion of this part of order $\boldsymbol{q}$.

Applying Statement 6.3 to the considered case, we may affirm that the set of "***the best***" parts of order $\boldsymbol{q}$ which may be completed by an affected vulnerable *AIS* $I = <\boldsymbol{q}, R, v'>$ with a resource base v' reduced by an impact is **max** $\bar{V}_S$.

Let us repeat that by applying *UMG*-based representations of *ISs*, we may assess their vulnerability in such a way that the modeled industrial system is considered vulnerable if it satisfies the declared criterion. In the case an *AIS* does not satisfy this criterion, it is not correct to consider it resilient, so it would be necessary to continue the assessment by an application of a *TMG*-based representation of this *AIS*. Let us move to this part of analysis of closed industrial systems affected by destructive impacts.

8.1.3 TMG-*Based Analysis of* IS *Resilience*

We shall use here the techniques developed in Sect. 6.2.3 for the assessment of additional amounts of resources to be acquired and time necessary for the completion of initially non-feasible orders. Let $I = <\boldsymbol{Q}, R, v_0>$ be an *IS MPR* with resource base v_0 and a manufacturing technological base R, assigned to an order $\boldsymbol{Q}$=<$\boldsymbol{q}$,$\boldsymbol{t}$>, where

$$\boldsymbol{q} = \{\boldsymbol{n}_1 \cdot \boldsymbol{a}_1, \ \ldots, \boldsymbol{n}_m \cdot \boldsymbol{a}_m\}, \tag{8.15}$$

$$\boldsymbol{t} = \{n \leq t \leq n', t = min\}. \tag{8.16}$$

We shall also use a value $\bar{n}$, which is the time moment of completion of an order $\boldsymbol{q}$, i.e., the moment when manufacturing of the last object-resource $\boldsymbol{a} \in \boldsymbol{q}$ is finished.

We shall represent an impact on this *AIS* as a triple $\boldsymbol{\delta} = \ <\boldsymbol{n}, \Delta R, \Delta v>$, where $\boldsymbol{n} \in [n_0, \bar{n}]$ is the moment of application of the impact, eliminating a set of *MDs* ΔR from the *MTB* of this *AIS* and a collection of resources Δv from its *RB*.

Consider an *FTMG* $S(\boldsymbol{\delta}) = \ < v(\boldsymbol{n}) - \Delta v, R - \Delta R, F(\boldsymbol{n})>$, associated with the affected *AIS*

$$I = \ <\boldsymbol{Q}(\boldsymbol{n}), R - \Delta R, v(\boldsymbol{n}) - \Delta v>, \tag{8.17}$$

where the multiset $v(\boldsymbol{n})$ is the *RB* at moment $\boldsymbol{n}$,

$$\boldsymbol{Q}(\boldsymbol{n}) = \ <\boldsymbol{q} - \Delta \boldsymbol{q}(\boldsymbol{n}), \boldsymbol{t}>, \tag{8.18}$$

where the multiset $\Delta \boldsymbol{q}(\boldsymbol{n})$ is the part of order $\boldsymbol{q}$ completed up to the moment $\boldsymbol{n}$, so

$$\boldsymbol{q} - \Delta \boldsymbol{q}(\boldsymbol{n}) = \{\boldsymbol{n}'_1 \cdot \boldsymbol{a}'_1, \ \ldots, \boldsymbol{n}'_{m'} \cdot \boldsymbol{a}'_{m'}\} \tag{8.19}$$

is nothing but the remaining part of the order to be completed, and, finally, $F(\boldsymbol{n})$ is a filter corresponding to $v(\boldsymbol{n})$ and $\boldsymbol{q} - \Delta \boldsymbol{q}(\boldsymbol{n})$ in the sense of (6.22)–(6.23), i.e.,

$$F(\boldsymbol{n}) = F_{\boldsymbol{q} - \Delta \boldsymbol{q}(\boldsymbol{n})} \cup \boldsymbol{t}, \tag{8.20}$$

$$F_{\boldsymbol{q} - \Delta \boldsymbol{q}(\boldsymbol{n})} = \{\boldsymbol{a}'_1 \geq \boldsymbol{n}'_1, \ \ldots, \boldsymbol{a}'_{m'} \geq \boldsymbol{n}'_{m'}\}. \tag{8.21}$$

In this way, we may introduce a criterion of *IS* resilience which is, in fact, an implication of Statement 6.5.

Statement 8.3 An AIS $I = \ <\boldsymbol{Q}, R, v_0>$ is resilient to an impact $\boldsymbol{\delta} = \ <\boldsymbol{n}, \Delta R, \Delta v>$ if

$$V_{S(\boldsymbol{\delta})} \neq \{\emptyset\}. \tag{8.22}$$

Otherwise, this *AIS* is vulnerable to this impact.

This criterion may be transformed into a ***top-down form*** similar to the top-down criterion of order feasibility established by Statement 6.5.

Let us consider now the case of a vulnerable closed *IS* and the task of partial order completion associated with such an *IS*.

8.1.4 TMG-*Based Assessment of Partial Order Completion by a Vulnerable* IS

Let an *AIS* $I = \; <\boldsymbol{Q}, R, v_0>$ be vulnerable to an impact $\boldsymbol{\delta} = \; <\boldsymbol{n}, \Delta R, \Delta v>$, i.e., $V_{S(\boldsymbol{\delta})} = \{\emptyset\}$. To assess what part of an order may be completed by this *AIS*, we shall apply a simple technique. Consider a filtering temporal multiset grammar

$$S'(\boldsymbol{\delta}) = \; < v(\boldsymbol{n}) - \Delta v, R - \Delta R, \{t \leq n'\} >, \tag{8.23}$$

whose filter does not contain lower restrictions on the amounts of manufactured object-resources declared by an order and does not demand that the time for their production should be minimized. Then the set of temporal multisets $V_{S'(\boldsymbol{\delta})}$ represents nothing but the set of collections of *ORs* which may be produced by the affected vulnerable *AIS* during the time interval $[\boldsymbol{n}, n']$, and finally, the subset

$$V(I, \boldsymbol{\delta}) = \{ v \mid v \in V_{S'(\boldsymbol{\delta})} \& \cdot \boldsymbol{q} - \Delta \boldsymbol{q}(\boldsymbol{n}) \cap v \neq \{\emptyset\}\} \tag{8.24}$$

of this set consists of temporal multisets representing collections of object-resources, each of which contains at least one *OR*, belonging to the remaining part of the order to be completed. This subset may be compressed by the application of the function *max* similar to the way it was applied in Statement 6.3.

Let us introduce a set

$$V_{max} = \mathbf{max}\{ (\boldsymbol{q} - \Delta \boldsymbol{q}(\boldsymbol{n})) \cap v \mid v + \{m \cdot t\} \in V_{S'(\boldsymbol{\delta})}\}, \tag{8.25}$$

which includes all "maximal" parts of collections of *ORs* which may be produced by the considered *AIS* after the moment of impact. However, in the case when a multiset v belongs to the set V_{max}, there may be several temporal multisets $v + \{m_1 \cdot t\}$, …, $v + \{m_l \cdot t\}$, belonging to the set $V_{S'(\boldsymbol{\delta})}$. Evidently, "***the best***" of them would be the *MST* with a ***minimal*** multiplicity of the object t, which corresponds to an optimizing condition $t = min$, which was excluded from the filter when we constructed the *FTMG* $S'(\boldsymbol{\delta})$. Such collections would be produced by an *AIS* in the minimal time. Let us define

$$\boldsymbol{V}_{max} = \{v + \{m \cdot t\} \mid v + \{m \cdot t\} \in V_{max} \& (\forall (v + \{m' \cdot t\}) \in V_{max})\, m' \geq m\}, \tag{8.26}$$

selecting in the set $\boldsymbol{V}_{max}$ "***maximal***" parts of collections of *ORs* which may be produced by the considered *AIS* in ***minimal*** time since the moment of the impact. Finally, the set of "the best" parts of the order $\boldsymbol{q}$ which may be completed by an *AIS* $I = \; <\boldsymbol{q}, R, v_0>$ which is vulnerable to an impact $\boldsymbol{\delta} = \; <\boldsymbol{n}, \Delta R, \Delta v>$is, evidently, nothing but

$$V(I,\boldsymbol{\delta}) = \{\Delta\boldsymbol{q}(\boldsymbol{n}) + v + \{m\cdot t\} \mid v + \{m\cdot t\} \in \boldsymbol{V}_{max}\}. \tag{8.27}$$

So we have formulated logics of ***partial order completion by a vulnerable AIS in the case of a TMG representation***.

Having constructed a set $\boldsymbol{V}(I,\boldsymbol{\delta})$, we may obtain the set of schedules of a vulnerable *AIS* by applying the relations introduced in Sect. 6.2.3.

With this, we finish the consideration of the key issues of resilience and vulnerability of ***closed*** *ISs* affected by destructive impacts and move to ***open*** industrial systems.

8.2 Open Industrial Systems

The criterial base for an assessment of resilience and vulnerability of both closed and open industrial systems is, in fact, the same. However, existence of a ***helpful external industrial system*** which may recover the reduced resource base of an affected vulnerable *IS* and/or complete some part of an order whose completion by this *IS* was interrupted by an applied destructive impact makes relevant questions whose solution would allow ***rational participation of an EIS in the mitigation of the negative consequences of this impact***.

We shall consider these tasks in the following way. A *UMG*-based analysis of recoverability and supportability of a vulnerable *IS* will be described in Sects. 8.2.1 and 8.2.2. A *TMG*-based analysis of the same issues will be considered in Sects. 8.2.3 and 8.2.4, where we shall assume that ***all manufacturing devices which were not affected by an impact begin their new operation cycles after the impact only on being freshly supplied with all necessary ORs***, i.e., ***all ORs which were supplied to MDs before the moment of an impact are considered lost***. The more general case of so-called non-interrupted MDs, which, not being affected by an impact, continue their current *OCs* after the moment of the impact and do not require refreshment of their input *ORs*, will be considered in Sect. 8.2.5.

8.2.1 UMG-*Based Analysis of Recoverability of Vulnerable* ISs

Consider a case where a destructive impact Δv is applied to an *AIS* $I = \langle q, R, v_0\rangle$ before the beginning of processing of an order, and the result of this impact is that this *AIS* becomes incapable of completing an order q with a reduced *RB* $v_0 - \Delta v$. An external industrial system, in turn, may or may not be capable of recoving the reduced *RB*, and there is a set

$$Q = \{\overline{v} - (v_0 - \Delta v) | \, \overline{v} \in \overline{V}_S\} \tag{8.28}$$

of requirements of the affected *IS* which is the background for a set of orders to the *EIS*, each, if completed by this system, recovering some part of the *RB* of the affected *AIS*, so that the latter becomes capable of completing the assigned order q. In (8.28), the multiset $\overline{v} - (v_0 - \Delta v)$ is nothing but the collection of object-resources to be added to the resource base of the affected *AIS I* reduced by the impact Δv, to make it capable of completing the order q. We assume here, as above in Sects. 8.1.1 and 8.1.2, that the *UMG* $S = \langle q, R \rangle$ is terminalized. The set of orders to the *EIS*, $\bar{Q}$, is created by replacing additional terminal objects $\boldsymbol{a} \in q \in Q$ by initial non-terminal objects a, which are the headers of terminalizing unitary rules $a \to 1 \cdot \boldsymbol{a}$: it is evident that orders to the *EIS* would contain *NTOs* representing *ORs* to be manufactured by the *EIS*. If any such *OR* occurs in the *RB* of the *EIS*, it is represented as in the *RB* of the affected *AIS*, i.e., as a terminal object $\boldsymbol{a}$.

Let $\boldsymbol{I} = \langle \{\emptyset\}, \boldsymbol{R}, \boldsymbol{v_0} \rangle$ be a ***free external industrial system***. Obviously, the resource base of an *AIS I* may be recovered if at least one order $\bar{q} \in \bar{Q}$ may be completed by the assigned *EIS*

$$\boldsymbol{I}_{\bar{q}} = \langle \bar{q}, \boldsymbol{R}, \boldsymbol{v_0} \rangle, \tag{8.29}$$

i.e.,

$$(\exists \bar{q} \in \bar{Q})(\exists \overline{v} \in \overline{V}_{S_{\bar{q}}})\overline{v} \subseteq \boldsymbol{v_0}\, , \tag{8.30}$$

where

$$\boldsymbol{S}_{\bar{q}} = \langle \bar{q}, \boldsymbol{R} \rangle \tag{8.31}$$

is a terminalized *UMG*. The same condition in a set-theoretic representation is as follows:

$$\bar{Q} = \bigcup_{\bar{q} \in Q} \{\overline{v} \mid \overline{v} \in \overline{V}_{S_{\bar{q}}} \& \overline{v} \subseteq \boldsymbol{v_0}\} \neq \{\emptyset\}. \tag{8.32}$$

So relations (8.29)–(8.32) form a **criterial base for recoverability of an affected vulnerable AIS** or, in other words, for the capability of an external industrial system to recover an AIS resource base reduced by an impact $\boldsymbol{\Delta v}$.

However, in the general case, the number of orders $\bar{q} \in \bar{Q}$ (ways of recovery) which may be completed by the *EIS* may be ***more than one***, so there arises the task of selection of ***one such order*** to be completed by the *EIS* for the recovery of the *RB* of the affected *AIS*.

This task may be solved rather obviously. First of all, it is reasonable to ***reduce*** the set of considered orders from $\overline{Q}$ to *min* $\overline{Q}$, thus extracting from $\overline{Q}$ all terminal multisets for which there are submultisets in $\overline{Q}$. It is evident that every excluded

TMS, as a potential order to the *EIS*, would require additional amounts of resources compared to its submultiset. If it is not necessary to reduce (or even minimize) amounts of resources spent by the *EIS* during recovery, this step may be omitted. And in any case, it is sufficient to define a filter $\overline{F}$ for selection from $\overline{Q}$ (or *min* $\overline{Q}$) of exactly one *TMS* $\bar{\boldsymbol{q}}$, which is an order to the *EIS*, so

$$\{\bar{q}\} = \overline{Q} \downarrow \overline{F}. \tag{8.33}$$

This one order $\bar{\boldsymbol{q}}$ would be completed by the *EIS* $\boldsymbol{I}_{\bar{q}} = \langle \bar{\boldsymbol{q}}, \boldsymbol{R}, \boldsymbol{v_0} \rangle$ assigned to this order, and the collection of resources $\bar{\boldsymbol{q}}$ produced as a result of its completion would allow ***recovery of the affected AIS I, after which completion of the initial order q would follow***.

Example 8.3 Let the manufacturing technological base of some *IS* be represented by the scheme *R* containing the following two unitary rules r_1 and r_2 in the technological interpretation and also two terminalizing *URs* r_3 and r_4:

$$r_1 : (auto) \to 1 \cdot (frame), 1 \cdot (engine), 4 \cdot (door), 4 \cdot (wheel), 400 \cdot (kwh),$$
$$50 \cdot (mnt : autos\ AL), 300 \cdot (eur);$$

$$r_2 : (engine) \to 1 \cdot (motor), 1 \cdot (transmission), 1 \cdot (fuel\ tank),$$
$$100 \cdot (kwh), \cdot 80 \cdot (mnt : engines\ AL), 900 \cdot (eur);$$

$$r_3 : (auto) \to 1 \cdot (\boldsymbol{auto});$$

$$r_4 : (engine) \to 1 \cdot (\boldsymbol{engine}).$$

Let the order q be the multiset $\{2 \cdot (auto)\}$, so $\bar{V}_S = \{v_1, v_2, v_3, v_4\}$, where

$$v_1 = \{2 \cdot (\boldsymbol{auto})\},$$

$$v_2 = \{1 \cdot (\boldsymbol{auto}), 1 \cdot (frame), 1 \cdot (\boldsymbol{engine}), 4 \cdot (door), 4 \cdot (wheel),$$
$$50 \cdot (mnt : autos\ AL), 400 \cdot (kwh), 300 \cdot (eur)\},$$

$$v_3 = \{1 \cdot (\boldsymbol{auto}), 1 \cdot (frame), 4 \cdot (door), 4 \cdot (wheel),$$
$$1 \cdot (motor), 1 \cdot (transmission), 1 \cdot (fuel\ tank),$$

$$50 \cdot (mnt : autos\ AL), 80 \cdot (mnt : engines\ AL),$$
$$500 \cdot (kwh), 1200 \cdot (eur)\},$$

$$v_4 = \{2 \cdot (\mathit{frame}), 8 \cdot (door), 8 \cdot (wheel), 2 \cdot (motor), 2 \cdot (transmission),$$
$$2 \cdot (\mathit{fuel\ tank}), 100 \cdot (mnt : autos\ AL), 160 \cdot (mnt : engines\ AL),$$
$$1000 \cdot (kwh), 2400 \cdot (eur)\}.$$

Let the resource base of the considered *IS* be

$$v_0 = \{1 \cdot (\boldsymbol{auto}),\ 2 \cdot (\boldsymbol{engine}),\ 1 \cdot (\mathit{frame}), 5 \cdot (door), 7 \cdot (wheel), 2 \cdot (motor),$$
$$3 \cdot (transmission), 2 \cdot (\mathit{fuel\ tank}),$$
$$120 \cdot (mnt : autos\ AL), 200 \cdot (mnt : engines\ AL),$$
$$1000 \cdot (kwh), 3000 \cdot (eur)\},$$

and the impact be the multiset

$$\Delta v = \{1 \cdot (\boldsymbol{auto}),\ 2 \cdot (door),\ 100 \cdot (kwh)\}.$$

As can be seen, the result of this impact will be the reduced *RB*

$$v_0 - \Delta v = \{\ 2 \cdot (\boldsymbol{engine}),\ 1 \cdot (\mathit{frame}), 3 \cdot (door), 7 \cdot (wheel), 2 \cdot (motor),$$
$$3 \cdot (transmission), 2 \cdot (\mathit{fuel\ tank}),$$
$$120 \cdot (mnt : autos\ AL), 200 \cdot (mnt : engines\ AL),$$
$$900 \cdot (kwh), 3000 \cdot (eur)\ \},$$

and because no $v \in \bar{V}_S$ is a submultiset of $v_0 - \Delta v$,the affected *AIS* is vulnerable to the impact Δv. The set of requirements to the external industrial system considered to recover this *AIS* will be $Q = \{q_1, q_2, q_3, q_4\}$, where

$$q_1 = v_1 - (v_0 - \Delta v) = \{2 \cdot (\boldsymbol{auto})\},$$
$$q_2 = v_2 - (v_0 - \Delta v) = \{1 \cdot (\boldsymbol{auto}),\ 1 \cdot (door)\},$$
$$q_3 = v_3 - (v_0 - \Delta v) = \{\ 1 \cdot (\boldsymbol{auto}),\ 1 \cdot (door)\},$$
$$q_4 = v_3 - (v_0 - \Delta v) = \{1 \cdot (\mathit{frame}), 5 \cdot (door), 1 \cdot (wheel), 100 \cdot (kwh)\},$$

and thus the set of possible orders to an *EIS* $\bar{Q} = \{\bar{q}_1, \bar{q}_2, \bar{q}_3\}$, where

$$\bar{q}_1 = \{2 \cdot (auto)\},$$
$$\bar{q}_2 = \{1 \cdot (auto),\ 1 \cdot (door)\},$$
$$\bar{q}_3 = \{1 \cdot (frame), 5 \cdot (door), 1 \cdot (wheel), 100 \cdot (kwh)\}.$$

As can be seen, $\overline{Q} = min\ \overline{Q}$, so there is no opportunity to reduce the set of possible orders to the *EIS*.

Now, let the manufacturing technological base of the *free external IS* considered to recover the affected *IS* be represented by the scheme $\boldsymbol{R}$ containing the following two unitary rules r_1 and r_2 in the technological interpretation and also two terminalizing *URs* r_3 and r_4:

$$r_1 : (auto) \to 1 \cdot (frame), 1 \cdot (engine), 4 \cdot (door), 4 \cdot (wheel),$$
$$70 \cdot (mnt : autos\ AL\ ES),$$
$$500 \cdot (kwh), 200 \cdot (eur);$$

$$r_2 : (engine) \to 1 \cdot (motor), 1 \cdot (transmission), 1 \cdot (fuel\ tank),$$
$$90 \cdot (mnt : engines\ AL\ ES),$$
$$200 \cdot (kwh), 400 \cdot (eur);$$

$$r_3 : (auto) \to 1 \cdot (\boldsymbol{auto});$$

$$r_4 : (engine) \to 1 \cdot (\boldsymbol{engine}).$$

(Note that durations of operation of the assembly lines of the external system have their own units of measurement different from the similar units of the *ALs* of the affected *IS*.)

Let the resource base of the free *EIS* be

$$\boldsymbol{v_0} = \{3 \cdot (\boldsymbol{auto}),\ 2 \cdot (frame), 9 \cdot (door), 5 \cdot (wheel), 3 \cdot (motor),$$
$$2 \cdot (transmission), 3 \cdot (fuel\ tank),$$
$$220 \cdot (mnt : autos\ AL\ ES), 300 \cdot (mnt : engines\ AL\ ES),$$
$$1200 \cdot (kwh), 5000 \cdot (eur)\}.$$

As can be seen, any of the three orders from the set $\overline{Q}$ may be completed by this *EIS*; each order is sufficient to recover the resource base of the affected *AIS* and to complete the initial order.

The above considerations concern the case when an impact is applied ***before the beginning of processing of an order*** q ***by an AIS*** I. However, applying the technique described in Sect. 8.1.1, it is quite simple to generalize it to the case when an impact is applied at a ***moment when a part*** $\Delta q \subset q$ ***of the initial order has already been completed***. In fact, it is sufficient to replace q by $q - \Delta q$ in the representation of the *AIS* I, which, hence, becomes $I = \langle q - \Delta q, R, v_0 \rangle$, where R and v_0 are, respectively, the *MTB* and the *RB* of this *AIS* after the moment of application of the impact Δv.

8.2.2 UMG-*Based Analysis of Supportability of Vulnerable* ISs

As was said above, support of an affected *AIS* by an external industrial system means an attempt to complete an order which was assigned to this *AIS* by applying the ***joined capabilities of both systems***, i.e., their technological and resource bases.

Let us consider a case where a destructive impact Δv is applied to an *AIS* $I = \langle q, R, v_0 \rangle$ and, as a result of this application, the ***affected AIS*** $I' = \langle q, R, v_0 - \Delta v \rangle$ ***is not capable of completing the order*** q. As above, we shall use a terminalized *UMG* $S = < q, R >$ for the ***generation of a set of resource bases, each enabling completion of the order*** q.

Joining the capabilities of an affected *AIS* and a free *EIS* $\boldsymbol{I} = \langle \{\varnothing\}, \boldsymbol{R}, \boldsymbol{v_0} \rangle$ means the creation of a ***new industrial system assigned to the order*** q, with manufacturing technological base $R \cup \boldsymbol{R}$, including all manufacturing devices of both systems, and resource base $v - \Delta v + \boldsymbol{v_0}$, including *ORs* occurring in the *RB* of either system. Let us note that in the case of the presence of identical manufacturing devices (unitary rules) in both R and $\boldsymbol{R}$, due to our use of set-theoretical join for the creation of the joint *MTB* of the new *IS*, the resulting *MTB* (the scheme of the constructed *UMG*) will include not two, but one manufacturing device (*UR*). However, the real capabilities of an *MTB* are represented, as everywhere above, by multiobjects $n \cdot (\Delta t : x)$, so due to summing of *RBs* $v - \Delta v$ and $\boldsymbol{v}$, the result will be $(n + \boldsymbol{n}) \cdot (\Delta t : x)$, where $n \cdot (\Delta t : x) \in R$ and $\boldsymbol{n} \cdot (\Delta t : x) \in \boldsymbol{R}$; thus, the ***real capabilities of a joint manufacturing technological base will be the sum of the capabilities of the joined ISs***, although the scheme of the resulting *UMG* will contain only one *UR* for every type of manufacturing device.

The criterion for an external *IS* to be able to support an affected *AIS* is rather evident.

Statement 8.4

Let $\boldsymbol{I}' = \langle q, R \cup \boldsymbol{R}, v_0 - \Delta v + \boldsymbol{v_0} \rangle$ be a joint *AIS* and $\boldsymbol{S}' = \langle q, R \cup \boldsymbol{R} \rangle$ be a terminalized *UMG* associated with $\boldsymbol{I}'$. Then the free *EIS* $\boldsymbol{I} = \langle \{\varnothing\}, \boldsymbol{R}, \boldsymbol{v_0} \rangle$ is capable of supporting the affected *AIS* if

$$\left(\exists \overline{v} \in \overline{V}_{S'}\right) \overline{v} \subseteq v_0 - \Delta v + \boldsymbol{v_0}. \tag{8.34}$$

There is one hidden nuance in this criterion: terminal objects $a \in \overline{A}_S$ may become non-terminal $a \in A_{S'} - \overline{A}_{S'}$, because there are *URs* belonging to the set $\boldsymbol{R}$ with header a (i.e., they represent *ORs* manufactured by an *EIS*, but occurring in the initial *RB* of the *AIS*). But because $\boldsymbol{R}$ also contains terminalizing rules for all non-terminal objects, this does not require any additional actions.

Example 8.4 Consider the manufacturing technological base of some *IS* represented by the same scheme R as in the previous Example 8.3. Let the order q, the *IS* resource base v_0, and the impact Δv also be the same as in this example, so the result of this impact will be the same reduced *RB*

$$v_0 - \Delta v = \{\, 2 \cdot (\boldsymbol{engine}),\ 1 \cdot (frame), 3 \cdot (door), 7 \cdot (wheel), 2 \cdot (motor), \\ 3 \cdot (transmission), 2 \cdot (fuel\ tank), \\ 120 \cdot (mnt : autos\ AL), 200 \cdot (mnt : engines\ AL), \\ 900 \cdot (kwh), 3000 \cdot (eur)\ \},$$

so, as above, the affected *AIS* is vulnerable to the impact Δv.

Now, let the manufacturing technological base of the free external *IS* considered to support the affected *IS* be represented also by the same scheme $\boldsymbol{R}$ as in the previous Example 8.3 and its resource base by $\boldsymbol{v_0}$.

According to Statement 8.4, the scheme $R \cup \boldsymbol{R}$ of the terminalized *UMG* associated with the joint *IS* will contain the following unitary rules:

$$r_1 : (auto) \to 1 \cdot (frame), 1 \cdot (engine), 4 \cdot (door), 4 \cdot (wheel), 400 \cdot (kwh), \\ 50 \cdot (mnt : autos\ AL), 300 \cdot (eur);$$

$$r_2 : (auto) \to 1 \cdot (frame), 1 \cdot (engine), 4 \cdot (door), 4 \cdot (wheel), 500 \cdot (kwh), \\ 70 \cdot (mnt : autos\ AL\ ES), 200 \cdot (eur);$$

$$r_3 : (auto) \to 1 \cdot (\boldsymbol{auto});$$

$$r_4 : (engine) \to 1 \cdot (motor), 1 \cdot (transmission), 1 \cdot (fuel\ tank), \\ 80 \cdot (mnt : engines\ AL), 100 \cdot (kwh), 900 \cdot (eur);$$

$$r_5 : (engine) \to 1 \cdot (motor), 1 \cdot (transmission), 1 \cdot (fuel\ tank),$$
$$90 \cdot (mnt : engines\ AL\ ES), 200 \cdot (kwh), 400 \cdot (eur);$$

$$r_6 : (engine) \to 1 \cdot (\boldsymbol{engine}).$$

(Note that the terminalizing *URs* are the same in the affected *IS* and the *EIS* due to the fact that they are present in the resulting scheme of the joint *IS* once, not twice.) According to the same statement, the resource base of the joint *IS* will be

$$v_0 - \Delta v + \boldsymbol{v_0} = \{3 \cdot (\boldsymbol{auto}), 2 \cdot (\boldsymbol{engine}), 3 \cdot (frame), 12 \cdot (door), 12 \cdot (wheel),$$
$$5 \cdot (motor), 5 \cdot (transmission), 5 \cdot (fuel\ tank),$$
$$120 \cdot (mnt : autos\ AL), 220 \cdot (mnt : autos\ AL\ ES),$$
$$200 \cdot (mnt : engines\ AL), 300 \cdot (mnt : engines\ AL\ ES),$$
$$2100 \cdot (kwh), 8000 \cdot (eur)\}.$$

As can be seen, the *MTB* and the *RB* of the joint *IS* allow order completion in many ways.

Let us note that an impact may be applied to an *AIS* not only before the beginning of processing of an order q, as considered above in this section, but also at a moment where some part of this order $\Delta q \subset q$ has already been completed. In the second case, similar to the technique described in the previous Sect. 8.2.1, the introduced criteria may be applied to the remaining parts of the order and of the resource base of the vulnerable *AIS*.

8.2.3 TMG-*Based Analysis of Supportability of Vulnerable* ISs

As in Sect. 8.1.3, let $I = \langle Q, R, v_0 \rangle$ be an ***IS MPR assigned to an order** $Q = \langle q, t \rangle$* with a resource base v_0 and a manufacturing technological base R and an impact on this *IS* be a triple $\boldsymbol{\delta} = \langle \boldsymbol{n}, \Delta R, \Delta v \rangle$, where $\boldsymbol{n} \in [n_0, \bar{n}]$ is the moment of application of the impact, eliminating the set of *MDs* ΔR from the *MTB* of this *IS* and the collection of resources Δv from its *RB*. The *AIS* I is vulnerable to the impact $\boldsymbol{\delta}$. Also let $\boldsymbol{I} = \langle \{\varnothing\}, \boldsymbol{R}, \boldsymbol{v} \rangle$ be a ***free external IS*** whose *MTB* $\boldsymbol{R}$ and *RB* $\boldsymbol{v}$ may be applied for support of the affected *IS* $\boldsymbol{I}$.

Consider an *FTMG* $S(\boldsymbol{\delta}, \boldsymbol{I}) = \langle v(\boldsymbol{n}) - \Delta v + \boldsymbol{v}, R - \Delta R \cup \boldsymbol{R}, F(\boldsymbol{n}) \rangle$ associated with the affected *AIS* $I = \langle \boldsymbol{Q}(\boldsymbol{n}), R - \Delta R, v(\boldsymbol{n}) - \Delta v \rangle$, which is supported by the *EIS* $\boldsymbol{I}$. As can be seen, the kernel of this multigrammar represents the ***joint resource base*** of I and $\boldsymbol{I}$, the scheme represents ***their joint manufacturing technological base***, and the filter represents the ***remaining part of the order*** which must be completed by the joint effort of both industrial systems. The ***criterion of supportability of a vulnerable AIS I, affected by an impact δ, by an EIS I*** is formulated as follows.

Statement 8.5 A free *EIS* $\boldsymbol{I} = \langle\{\varnothing\}, \boldsymbol{R}, \boldsymbol{v}\rangle$ is capable of supporting an *AIS* $I = <\boldsymbol{Q}, R, v_0>$ vulnerable to an impact $\boldsymbol{\delta} = <\boldsymbol{n}, \Delta R, \Delta v>$ if

$$V_{S(\delta,\, I)} \neq \{\varnothing\}. \tag{8.35}$$

Otherwise, this *AIS* is not supportable by this external *IS*.

There is one nuance, concerning the possibility of the presence of identical manufacturing devices in the joined manufacturing technological bases $R - \Delta R$ and $\boldsymbol{R}$. It is clear that, in the case such devices do exist, the result of joining these *MTBs* will contain ***only one not two*** identical devices. To avoid such incorrectness, we shall rename all *MDs* belonging to the supporting *EIS* by including in their names the prefix *, not used in the names of the *MDs* belonging to the affected *AIS*. By this technique, there will be no intersections between the sets of names of the *MDs* of the two industrial systems, and the result of the join of sets $R - \Delta R$ and $\boldsymbol{R}$ will be substantially correct. (The same technique was in fact applied in Sect. 4.4.3 to distinguish primary and secondary object-resources in the *RB* by concatenating the prefix * to names of secondary *ORs* and temporal rules allowing the extraction of such *ORs* from an *RB*.)

8.2.4 TMG-*Based Analysis of Recoverability of Vulnerable* ISs

To construct a criterion for the recoverability of a vulnerable *IS*, we shall apply the aforementioned technique of *OR* renaming by concatenating the prefix * to ***names of ORs belonging to the RB of the EIS and to temporal rules representing its MTB***.

Let us repeat that, as in Sects. 8.1.3 and 8.2.3, $I = <\boldsymbol{Q}, R, v_0>$ is an *IS MPR* ***assigned to an order*** $\boldsymbol{Q}=<\boldsymbol{q},\boldsymbol{t}>$, with resource base v_0 and manufacturing technological base *R*. An **impact eliminating a set of *MDs* ΔR from the *MTB* of this *IS* and the collection of resources Δv from its *RB*** is represented by a triple $\boldsymbol{\delta} = <\boldsymbol{n}, \Delta R, \Delta v>$, where $\boldsymbol{n} \in [n_0, \bar{n}]$ is the moment of its application. An *AIS I* is ***vulnerable to impact δ*** if it does not satisfy the condition of resilience to $\boldsymbol{\delta}$ (8.22).

Let $\boldsymbol{I} = \langle\{\varnothing\}, \boldsymbol{R}, \boldsymbol{v}\rangle$ be a ***free external IS*** whose *MTB* $\boldsymbol{R}$ and *RB* $\boldsymbol{v}$ may be applied to the recovery of an affected vulnerable *AIS I.*

Consider a filtering temporal multiset grammar

$$S'(\boldsymbol{\delta}, \boldsymbol{I}) = <v(\boldsymbol{n}) - \Delta v + {}^*\boldsymbol{v}, R - \Delta R \cup {}^*\boldsymbol{R} \cup {}^-\boldsymbol{R}, F(\boldsymbol{n})> \tag{8.36}$$

associated with the affected $AIS\ I = <\boldsymbol{Q}(\boldsymbol{n}), R - \Delta R, v(\boldsymbol{n}) - \Delta v>$ to be recovered by the *EIS* $\boldsymbol{I}$. Here,

$$^*(\{n_1 \cdot a_1, \ \ldots, \ n_m \cdot a_m\}) = \{n_1 \cdot (*a_1), \ \ldots, n_m \cdot (*a_m)\} \tag{8.37}$$

is an operation of renaming of objects belonging to a multiset by concatenating to their names the prefix *. Thus, $^* \ \Delta v$ is the collection of renamed object-resources to be produced by an *EIS* to recover the *RB* of the affected *AIS*. Hence, $^*\boldsymbol{v}$ is the "prefixed" *RB* of the external *IS*. A similar operation is defined regarding the set of temporal rules representing the *EIS MTB*:

$$^*\boldsymbol{R} = \{< *r; \ ^*v \rightarrow \ ^*v' + \{\Delta n \cdot \Delta t\} > | < r; v \rightarrow v' + \{\Delta n \cdot \Delta t\} > \ \in \boldsymbol{R}\}. \tag{8.38}$$

Recovery of any *OR* *a* is done by the application of a set of special "***deprefixing***" rules, similar to the terminalizing temporal rules introduced in Sect. 4.4.3:

$$^-\boldsymbol{R} = \{ \ \langle *a; \{1 \cdot (*a)\} \rightarrow \{1 \cdot a\} \rangle \mid a \in \beta(\Delta v) \ \}. \tag{8.39}$$

Now, it is clear that application of the temporal multiset grammar $S'(\boldsymbol{\delta}, \boldsymbol{I})$ generates ***all possible ways of recovery of ORs eliminated from the RB by the impact*** Δv. As a result of prefixing of the *EIS RB* and *MTB*, manufacturing processes of the *EIS* and the affected *IS* do not intersect, except for one type of action—***transfer of ORs produced by the EIS to the affected IS***. Such actions, which in fact ***play a substantial part in implementing recovery***, are modeled by the application of the "deprefixing" rules (8.39), and after any such rule is applied, the corresponding object-resource produced by the *EIS* is transferred from its resource base to the resource base of the affected *IS* and thus may be used in its manufacturing process. Of course, after this transfer, the *OR* is not available to the *MTB* of the *EIS*.

The introduced constructions allow the formulation of the following correct ***criterion of recoverability of a vulnerable IS***.

Statement 8.6 A free *EIS* $\boldsymbol{I} = \langle \{\varnothing\}, \boldsymbol{R}, \boldsymbol{v} \rangle$ is capable of recovering an *AIS* $I = \ < \boldsymbol{Q}, R, v_0 >$ vulnerable to an impact $\boldsymbol{\delta} = \ < \boldsymbol{n}, \Delta R, \Delta v >$ if

$$V_{S'(\boldsymbol{\delta}, \boldsymbol{I})} \neq \{\varnothing\}. \tag{8.40}$$

Otherwise, this *AIS* is not recoverable by this external *IS*.

Until now, we have considered *AISs* whose *MTBs* after an impact begin operation after ***full refreshment of their inputs***. Let us consider the more general case of *MTBs* with ***non-interrupted manufacturing devices***.

8.2.5 TMG-*Based Assessment of the Contribution of Non-interrupted Manufacturing Devices to the Supportability of Vulnerable* ISs

Let us recall that in the above two sections, we have assumed that all manufacturing devices which were not affected by an impact would begin their new operation cycles after an impact only ***after they were supplied afresh with all necessary ORs***. By this, we assume, in fact, that all *ORs* which were supplied to *MDs* at the beginning of a previous *OC*, which was interrupted by an impact, ***are lost***.

However, there may be ***non-interrupted MDs***, which, not being affected by an impact, continue their current *OCs* after the moment of impact and do not require refreshment of their input *ORs*. It is clear that such *MDs* may ***contribute to the recoverability and supportability of a vulnerable IS in three ways***:

1. By ***not losing*** collections of input *ORs*, already supplied and processed
2. By ***not expending*** their collection of input *ORs*, thus allowing their use by other *MDs*, which may begin operating in parallel
3. By ***producing*** a collection of output *ORs* earlier than might have been done if a new *OC* had begun after the moment of impact

Thus, it would be valuable to develop a technique of ***involvement of capabilities of non-interrupted MDs*** in the assessment of recoverability and supportability of vulnerable industrial systems and their rescheduling.

Let us define a set of ***auxiliary temporal rules*** R^+, which represent the application of non-interrupted *MDs* after the moment of an impact and addition to the *RB* of the affected *AIS* of those collections of *ORs* which were produced by these *MDs* during their current *OCs*, overlapping this moment. This definition is as follows:

$$R^+ = \bigcup_{\substack{<i,R(i)>\in\boldsymbol{\sigma} \\ i<\boldsymbol{n} \\ r\in R(i) \\ \langle r;\, v\rightarrow v'+\{\Delta n\cdot\Delta t\}\rangle\in R \\ \boldsymbol{n}<i+\Delta n}} \{\langle *r;\{1\cdot(*r)\}\rightarrow v'+\{\boldsymbol{\Delta n}\cdot\Delta t\}\rangle \,|\boldsymbol{\Delta n}=i + \Delta n-\boldsymbol{n}\}, \tag{8.41}$$

where, as in (6.41), $\boldsymbol{\sigma}=\{<n_0,R(n_0)>,\ldots,<n_p,R(n_p)>\}$ is an *AIS* schedule. Any *TR* belonging to this set may be applied if the *RB* at a moment following the moment of an impact contains a multiobject $1\cdot(*r)$, where $*r$ is an auxiliary name used as the name of a constructed *TR*, representing a fictive *MD* modeling the operation of a non-interrupted *MD* after the impact; evidently, $(*r)$ is the name of a fictive *OR* used for the application of the aforementioned *TR*. As can be seen, the result of an application of this *TR* to the *RB* containing this multiobject at the aforementioned moment would be the inclusion of the collection of *ORs* v', produced by the *MD r*, in the *RB* at $\boldsymbol{\Delta n}$ time units after the impact, where

$$\boldsymbol{\Delta n} = i + \Delta n - \boldsymbol{n}. \tag{8.42}$$

So an *MD r*, which started its current operation cycle at a moment *i*, would finish it at a moment

$$\boldsymbol{n} + \boldsymbol{\Delta n} = \boldsymbol{n} + (i + \Delta n - \boldsymbol{n}) = i + \Delta n, \tag{8.43}$$

i.e., ***when this OC would finish in the case that there was no impact***. Let us consider now a ***filtering temporal multiset grammar***

$$S^{+}(\boldsymbol{\delta}, \boldsymbol{I}) = < v(\boldsymbol{n}) - \Delta v + \boldsymbol{v}^{+}, R - \Delta R \cup R^{+}, F(\boldsymbol{n}) >, \tag{8.44}$$

where

$$\boldsymbol{v}^{+} = \bigcup_{\langle *r;\{1\cdot(*r)\}\to v'+\{\boldsymbol{\Delta n}\cdot\Delta t\}\rangle \in R^{+}} \{1\cdot(*r)\} \tag{8.45}$$

is a multiset containing the *ORs* to be added to the *AIS RB* in order to permit modeling of operation of non-interrupted *MDs*. This *FTMG* is very similar to that which was proposed to be used for the assessment of recoverability and supportability of affected *AISs*, but there is one nuance. Namely, the generation of *MSTs* by application of an *FTMG* $S^{+}(\boldsymbol{\delta}, \boldsymbol{I})$ may lead to a situation when, at the same time as a fictive *MD* $(*r)$ "operates" (i.e., in fact, a real non-interrupted *MD r* whose *OC* began before the impact operates), a parallel operation of another copy of the *MD r* may be implemented by a successful application of a *TR r*, which contradicts a basic principle of *IS* scheduling: ***any MD at any moment may process only one input collection of ORs***. So some additional corrections are needed to exclude this prohibited possibility.

Namely, we shall define

$$\boldsymbol{R}^{+} = R - \Delta R$$

$$\cup \left(\bigcup_{\langle *r;\{1\cdot(*r)\}\rightarrow v'+\{\boldsymbol{\Delta n}\cdot\Delta t\}\rangle \in R^{+}} \{\langle *r;\{1\cdot(*r)\}\rightarrow v' + \{1\cdot(\#r),\ \boldsymbol{\Delta n}\cdot\Delta t\}\rangle\} \right)$$

$$\cup \left(\bigcup_{\langle *r;\{1\cdot(*r)\}\rightarrow v'+\{\boldsymbol{\Delta n}\cdot\Delta t\}\rangle \in R^{+}} \{< r; v + \{1\cdot(\#r)\}\rightarrow v' + \{1\cdot(\#r),\ \Delta n\cdot\Delta t\} > \big| < r; v\rightarrow v' + \{\Delta n\cdot\Delta t\} > \in R\} \right)$$

$$- \left(\bigcup_{\langle *r;\{1\cdot(*r)\}\rightarrow v'+\{\boldsymbol{\Delta n}\cdot\Delta t\}\rangle \in R^{+}} \{\ < r; v\rightarrow v' + \{\Delta n\cdot\Delta t\} > \big| < r; v\rightarrow v' + \{\Delta n\cdot\Delta t\} > \in R\} \right). \tag{8.46}$$

This set of temporal rules contains the following subsets:Temporal rules $\langle *r;\{1\cdot(*r)\}\rightarrow v' + \{1\cdot(\#r), \boldsymbol{\Delta n}\cdot\Delta t\}\rangle$, each constructed by the addition of a multiobject $1\cdot(\#r)$ with the prefix #, different from *, to the right part of a *TR*

$$\langle *r;\{1\cdot(*r)\}\rightarrow v' + \{\boldsymbol{\Delta n}\cdot\Delta t\}\rangle \in R^{+} \tag{8.47}$$

Temporal rules

$$< r; v + \{1\cdot(\#r)\}\rightarrow v' + \{1\cdot(\#r), \Delta n\cdot\Delta t\} >, \tag{8.48}$$

each constructed by the addition of a multiobject $1\cdot(\#r)$ with the same prefix # to both the left and the right parts of a *TR*

$$< r; v\rightarrow v' + \{\Delta n\cdot\Delta t\} > \in R, \tag{8.49}$$

which is eliminated from the set *R*All other *TRs* belonging to $R - \Delta R$

Consider now a filtering temporal multiset grammar

$$\boldsymbol{S}^{+}(\boldsymbol{\delta}, \boldsymbol{I}) = < v(\boldsymbol{n}) - \Delta v + \boldsymbol{v}^{+}, \boldsymbol{R}^{+}, F(\boldsymbol{n}) > . \tag{8.50}$$

Let us consider an application of this *FTMG*. First of all, *TRs* (8.47), representing finalization of *OCs* of non-interrupted *MDs*, are applied along with other temporal rules. Any such application eliminates from the resource base a multiobject $1\cdot(*r)$ and replaces it with a multiobject $1\cdot(\#r)$. Until this replacement has been performed, no *TR* (8.48) can be applied, because its left part is not a submultiset of the current *MST*. So ***no non-interrupted MD is activated again until its OC overlapping the moment of impact is finished***. When this event occurs, a multiobject $1\cdot(*r)$ is replaced by an *MO* $1\cdot(\#r)$, present in the right part of the *TR* (8.48), and thus the appearance of $1\cdot(\#r)$ in the generated *MST* allows an opportunity to apply *TR* (8.48).

This temporal rule, in fact, is nothing but the initial *TR* $<r; v \rightarrow v' + \{\Delta n \cdot \Delta t\}>$ with its left and the right parts extended by a controlling marker $1 \cdot (\#r)$, which is "regenerated" by any application of (8.48) and thus does not change the logic of possible operation of the non-interrupted *MD r* after an impact.

Now, we may formulate ***the most general criterion of IS supportability and recoverability with respect to non-interrupted manufacturing devices***.

Statement 8.7 A free *EIS* $\boldsymbol{I} = \langle \{\varnothing\}, \boldsymbol{R}, \boldsymbol{v} \rangle$ is capable of supporting an *AIS* $I = \ < \boldsymbol{Q}, R, v_0 >$ which is vulnerable to an impact $\boldsymbol{\delta} = \ < \boldsymbol{n}, \Delta R, \Delta v >$, where the *MTB* remaining after the impact $R - \Delta R$ contains non-interrupted *MDs*, if

$$V_{S^{+}(\boldsymbol{\delta},\, \boldsymbol{I})} \neq \{\varnothing\}. \tag{8.51}$$

Otherwise, this *AIS* is not supportable by this external *IS*.

Similarly, we may obtain a ***criterion for the assessment of the recoverability of AISs whose MTBs contain non-interrupted manufacturing devices***.

Because we are considering only *ISs MPR*, their recoverability and supportability concern only their resource bases, but impacts may also destroy manufacturing devices belonging to the *MTBs* of affected *ISs*. So any such affected *MTB* remains reduced, and after completion of an order interrupted by an impact, the *IS* continues operating with this non-recovered *MTB*. The more general case, involving recovery of the *MTB* of the affected *IS MPR* by an external *IS MPAR* capable of producing manufacturing devices to replace those destroyed by impacts, will be considered in future publications, as will the most general case, where both the affected and external *STSs* are *ISs MPAR*.

8.3 Cascade Effects

Cascade (or chain) effects multiply the consequences of an initial impact by transmitting destruction or malfunctions through interconnections between subsystems (elements) of the affected IS (in the general case, STS) (Dobson et al. 2007). There are two kinds of cascade effects: passive and active.

Passive cascade effects *(PCEs)* result in a chain of stops of manufacturing devices, which are a consequence of a lack of the input resources necessary for their next operation cycles. The first to stop are those *MDs* whose input *ORs* were eliminated by an impact; the next to stop are *MDs* whose input *ORs* were expected to be produced by the aforementioned *MDs*, which were stopped directly by an impact and, hence, have not performed their operation cycles; and so on until all *MDs* of the affected *IS* stop. If in addition to the elimination of *ORs* also *MDs* are eliminated, then, naturally, the ***speed of degradation of the affected IS increases***. The dynamics of this process may be fully represented by the so-called stop schedule of the affected IS, whose formal description will be given in Sect. 8.3.1.

Active cascade effects *(ACEs)* result in waves of impacts, each eliminating resources available to an *IS* and, if possible, generating the next set of impacts applied to the remaining collection of resources. Such a process has its own internal logic (whether natural and/or deliberate), which can be represented by some set of temporal rules which are joined to the set of temporal rules representing the manufacturing technological base of the affected IS. In this case, the ***initial impact may be very limited, but its final results may be much more destructive than in the case of PCEs***. The dynamics of this process may be fully represented by a similar stop schedule as in the case of passive cascade effects. A formal description of active cascade effects will be given in Sect. 8.3.2.

8.3.1 Passive Cascade Effects

Let us define the above-introduced notion of the stop schedule of an affected *IS*.

Consider an *AIS* $I = \,<\boldsymbol{Q}, R, v_0>$ and its schedule

$$\boldsymbol{\sigma} = \{<n_0, R(n_0)>, \; \ldots, \; <n_l, R(n_l)>\}, \tag{8.52}$$

where $R(n_0) \subseteq R, \ldots, R(n_l) \subseteq R$ are sets of manufacturing devices activated, respectively, at moments $n_0, \ldots, n_l$. An impact on this *AIS*, as above, will be represented by a triple $\boldsymbol{\delta} = \,<\boldsymbol{n}, \Delta R, \Delta v>$, where $\boldsymbol{n} \in [n_0, \bar{n}]$ is the moment of application of this impact, resulting in the elimination of some subset ΔR of the *MTB* R of this *AIS*, and of the collection of resources Δv, which is a submultiset of its *RB* $v(\boldsymbol{n})$.

According to the above verbal description, an *AIS* stop schedule may be defined formally as follows:

$$\boldsymbol{\sigma}_{stop} = \{<\boldsymbol{n}, \Delta R>\} \cup \left(\bigcup_{n_i \in [\boldsymbol{n}, \; n_l]} \{<n_i, \; \Delta R_i>\} \right), \tag{8.53}$$

where

$$\begin{aligned} \Delta R_i = \{ \, r \mid r \in R(n_i) - \Delta R \& < r; v \rightarrow v' + \{\Delta n \cdot \Delta t\} > \\ \in \cdot R \& \neg (v \subseteq v(n_i)) \} \end{aligned} \tag{8.54}$$

and $v(n_i)$ is the resource base of the affected *AIS* at the moment n_i. As can be seen, every couple $<n_i, \Delta R_i>$ represents a set of names of *MDs* ΔR_i which stop at the moment n_i because there are object-resources in the collection $v - v(n_i)$ which are absent from the *RB* at this moment, thus preventing the start of the next operation cycle of this *MD*. The first element of an *AIS* stop schedule is, naturally, the couple $<\boldsymbol{n}, \Delta R>$, representing the set of *MDs* which were eliminated by the initial impact.

In the general case, a schedule including ***those MDs which do not stop but remain active (operating)*** would be of practical use. Such a schedule $\boldsymbol{\sigma}_{active}$ may be derived from (8.53) and (8.54) directly:

$$\boldsymbol{\sigma}_{active} = \{<\boldsymbol{n}, R - \Delta R>\} \cup \left(\bigcup_{n_i \in [\boldsymbol{n},\ n_l]} \{<n_i,\ R - \Delta R - \Delta R_i>\} \right). \tag{8.55}$$

Naturally, an integrated schedule, representing both stopping and remaining active *MDs*, may be used.

8.3.2 Active Cascade Effects

As was said above, active cascade effects have their internal logic, which must be represented in some adequate way. To avoid multiplying the number of such ways (recall Occam's razor), we shall use the same temporal multiset grammars for the representation of *ACEs*, taking into account that ***chain processes of destruction of elements of affected ISs are in fact resource-driven***.

Namely, we shall represent knowledge about *ACEs* in the form of a set of temporal rules whose only difference from *TRs* representing *MDs* of affected *ISs* is that multiobjects in the right parts of *TRs* may have ***negative multiplicities***, thus modeling the ***elimination (extraction) of amounts of ORs defined by these multiplicities from the resource bases of such ISs***.

Let us introduce a temporal rule

$$< r; v \to v'_+ - v'_- + \{\Delta n \cdot \Delta t\} >, \tag{8.56}$$

which means that an application of this fictive manufacturing device adds a collection of *ORs* v'_+ to the resource base of an affected *IS* and extracts a collection of *ORs* v'_- from this *RB,* both operations being executed in Δn time units since the moment when the resource base of the affected *IS* contains the collection of *ORs* v. In particular cases, we may have $v'_+ = \{\emptyset\}$, so no new *ORs* are added to the *RB*. However, $v'_+ \neq \{\emptyset\}$ allows addition to the *RB* of new *ORs*, which may activate the next, derived, impact, and so on. Thus, multiobjects belonging to multisets v'_+ may be used for the description of hidden internal interconnections, modeling propagation of impacts inside an affected *IS*, i.e., the real phenomenon of ***cascading of destructive effects***, while *MS* v'_- is used to represent the ***destructive power of cascaded impacts***.

The appearance of negative multiplicities in the right parts of temporal rules makes it necessary to change the mathematical semantics of temporal multiset grammars, but this correction is very local and boils down to a replacement of

(3.57) by the following definition based on the semantics of negative multiplicities introduced in Sect. 3.6.2:

$$L(v_R, n) = \left(\sum_{\substack{r/\tau \in v_R \\ \langle r; v \to v'_+ - v'_- + \{\Delta n \cdot \Delta t\} \rangle \in R \\ n - \Delta n \in \tau^*}} v'_+ \right) - \left(\sum_{\substack{r/\tau \in v_R \\ \langle r; v \to v'_+ - v'_- + \{\Delta n \cdot \Delta t\} \rangle \in R \\ n - \Delta n \in \tau^*}} v'_- \right). \tag{8.57}$$

As can be seen, at the same moment n when some *MDs* may finish their operation cycles and load into the resource base object-resources produced during these *OCs*, fictive *MDs*, representing some fragments of active cascade effects, may eliminate some parts of the *RB*.

Finally, a free industrial system with hidden active cascade effects may be represented as a triple $=\langle\{\varnothing\}, R \cup R_-, v_0\rangle$, where R_- is a set of temporal rules representing the set of the aforementioned *ACEs*. As may be seen, *ACEs* eliminate *ORs* from the *RB* not only as a result of their activation but also by utilizing *ORs* necessary for such activation. For example, biological viruses may destroy some parts of the human organism, but to do this damage, they need some resources present in this organism and necessary for their destructive action and replication. This obstacle is fully captured by the join of the *MTB R* and the set R_-, resulting in a joint *TMG*, as well as a unitary resource base utilized by both kinds of active resources—manufacturing devices and fictive *MDs*, representing *ACEs*. They may interact, at the first glance, in an unclear and even obscure way, so the proposed technique would find an **extremely effective application in predicting, understanding, and explaining such systemic effects, especially in such cases as pandemics like *COVID-19***.

Concerning implementation issues, i.e., finally, top-down generation of schedules of affected *ISs*, we may see that there are no difficulties in applying the techniques of dual *TMGs* in the case of temporal rules with negative multiplicities in the right parts of *TRs*. Speaking generally, the proposed application of negative multiplicities may be used to represent any process of interaction of two counterparts, each trying to eliminate resources of the other. Such processes have for a long time been studied in game theory. An application of the multigrammatical framework to a subject of this theory, called resource-based games, will be considered in Chap. 9.

8.4 Energy Infrastructures

Energy infrastructure, producing and delivering necessary amounts of electrical power and fuel to various stationary and mobile consumers, including industrial facilities, living houses, transportation vehicles, etc., is ***the basis of all modern STSs*** (Rinaldi et al. 2001; Katay 2010; Nepal and Jamasb 2013; Sheremet 2021b).

In the general case, amounts of electrical power to be delivered by an *EI* on demand to external customers in some predefined period of time are restricted by amounts of primary resources—crude oil, natural gas, and other possible energy carriers (*EnCs*)—available for *EP* generation, as well as by limited bandwidths of links forming electrical grids and fuel pipelines. The problem considered in this section is, given

A ***demand of consumers***, i.e., amounts of power and fuel to be consumed by them during a considered time period (this demand, as everywhere above in this book, will be called an ***order***)

An ***EI segment***, including ***fuel-producing and power-generating facilities***; ***links,*** providing power transmission and fuel transfer through distributed areas; as well as ***terminal units***, delivering fuel and power to their consumers

Primary resources available for power generation

A ***destructive impact***, eliminating some part of the considered *EI* segment and the aforementioned resources,

to ***assess whether the part of the considered segment and resources remaining after the impact would be capable of producing and delivering the amounts of power and fuel necessary to consumers*** (in the above-introduced terminology, ***to complete an order***).

If an affected *EI* is capable of completing an order, then it will be called ***resilient*** to this impact. Otherwise, the *EI* will be called ***vulnerable*** to it. Everywhere below in this section, we shall consider an *EI* as a ***closed*** system, which operates without direct application of any external resources or their availability for replenishment of the *EI's* own (internal) resources expended in order completion.

To develop a criterial base allowing the assessment of an *EI's* vulnerability to destructive impacts, let us begin with a detailed substantial description of energy infrastructure and its well-known graph representation.

8.4.1 Basic Graph Representation of Energy Infrastructures

An energy infrastructure is usually considered to consist of two strongly interconnected and mutually supplying segments producing fuel and electrical power (Brown 2008; Wang et al. 2019).

An ***electricity infrastructure*** (*ElcI*) in the most general case contains generation facilities (***power plants***, *PPs*), ***power transforming-distributing substations*** (*PTDSs*), and ***power terminal units*** (*PTUs*), delivering electrical power to its consumers. All these elements are connected by links, called ***power transmission lines*** (*PTLs*), each such line having its own technical parameters (voltage, length, power losses during transmission, etc.), and these lines are joined to form electric grids, which, in fact, in aggregate form an *ElcI*.

A ***fuel infrastructure*** (*FI*) (Ji et al. 2017; Jesse et al. 2019), similar to an *ElcI*, includes ***fuel-producing plants*** (*FPPs*), delivering fuel in the form of ***primary energy carriers*** (*PECs*) to ***fuel distribution stations*** (*FDSs*), as well as ***fuel terminal units*** (*FTUs*). All these elements are connected by pipes, which, in the general case, like *PTLs*, have individual technical parameters (diameter, length, pressure, fuel losses during transfer, consumed amounts of *EP*, etc.). Fuel produced by *FPPs* is used by power plants and other consumers. To limit the complexity under consideration here, we shall not expand an *FI* down to production of crude oil and natural gas from oil and gas fields and their transportation via oil and gas pipelines to *FPPs*; we shall assume that certain amounts of primary energy carriers, used for fuel production, are accumulated at ***fuel storages*** (*FSs*) colocated with *FPPs* and these amounts are part of the ***resource base*** of an *EI*.

An *ElcI* and an *FI* are joined to one another by ***terminal units***: any element of an *FI* consumes electrical power delivered to it by some *PTU*, while any *PP* depends for its operation on *FTUs* delivering fuels needed for power generation (in the general case, several energy carriers may be utilized by a single power plant). Also, there are *PTUs* and *FTUs* delivering power and fuels to external consumers. Regarding a considered time period (hour, day, etc.), any *FPP* may produce certain amounts of various fuels, and any *PP* may produce certain amounts of *EP* with various technical parameters.

Any output of any element of an *EI* is assumed to be ***consistent*** with a link transferring resources from it to another element, whose input, in turn, is assumed to be consistent with the aforementioned link, which is incoming to this other element and, thus, delivering to it the aforementioned resource. This overlapping of *EI* elements and boundary points of *EI* links is the background for modeling circulation of *EP* and fuel via an *EI*.

Any link has a ***limited bandwidth*** (or throughput capacity) as an integral technical parameter, determining the maximal amount of power (if it is a *PTL*) or fuel (if it is a pipe) which may be transmitted (transferred) via this link during a considered time period. Also, as was mentioned above, some power losses occur during power transmission via a *PTL*; similar losses of fuel are inherent to fuel transferring pipes.

So both electricity and fuel infrastructures have a ***tree-like concentric topology***, and, based on the above, an *EI* may be represented by a weighted oriented graph with nodes corresponding to *EI* elements and marked edges corresponding to *EI* links. This graph, in turn, in the algebraic representation is a ternary relation $G \subseteq A \times A \times N$, where A is the set of *EI* elements (*PPs*, *PTDSs*, *PTUs*, *FPPs*, *FDSs*, *FTUs*, *FSs*) and N is the set of positive rational numbers, representing bandwidths of *EI* links (*PTLs* and pipes). So $<a, a', n> \in G$ means that an element a is capable of transmitting (transferring) to an element a' by a link (*PTL* or pipe) $<a, a'>$ no more than n units (kilowatt-hours in the case of *EP* and barrels, cubic meters, kilograms, tons, etc. in the case of various fuels) during the considered time period. There may be only one triple <*a*, *a*′, *n*> ∈*G* for any link <*a*, *a*′>, i.e., a link has only one bandwidth (throughput capacity).

A destructive ***impact***, which in the general case is ***distributed***, may eliminate some elements and/or links of an *EI* as well as some amounts of resources stored at an *EI* resource base; naturally, an impact may be represented by some subset of nodes and edges eliminated from the initial graph *G*.

Let us illustrate the above by an example.

Example 8.5 Consider a small hypothetical segment of some *EI*, including a power plant, two power transformation-distribution stations, seven power terminal units, a fuel-producing plant, a fuel storage, two fuel distribution stations, and three fuel terminal units (*Fig.*8.1*a*). (Sequential numbers of *FDSs* and *FTUs* as well as names of fuel storage and the fuel-producing plant are denoted by bold symbols.)

There are also three external power customers. Generated power from a *PP* is delivered to both *PTDSs*, the first of which (numbered "1") delivers received power to four *PTUs* ("1," "2," "3," and "7") and the second ("2") delivers received power to five *PTUs* ("4," "5," "6," "8," and "9"). *PTUs* deliver power to the following elements of the considered *EI* segment: the *PTU* "1," to the *FDS* "**2**"; the *PTU* "2," to the *FDS* "**1**"; the *PTU* "3," to the ***FPP***; the *PTU* "4," to the ***FS***; the *PTU* "5," to the power customer "1"; the *PTU* "6," to the power customer "2"; and the *PTU* "7," to the power customer "3." In turn, elements of an *FI* by consumption of electrical power deliver fuel as follows: the ***FS***, to the ***FPP***; the ***FPP***, to the *FDS* "**1**"; the *FDS* "**1**," to the *FDS* "**2**"; and the *FDS* "**2**," to the *FTU* "**1**" colocated with the power plant. The *FTU*s "**2**" and "**3**," which both receive fuel from the *FDS* "**1**," deliver fuel to the fuel customers "**1**" and "**2**," respectively. These *FTUs* are provided with electrical power by the *PTUs* "8" and "9," which receive electrical power from the *PTDS* "2." An algebraic representation of the considered graph, including bandwidths (throughput capacities), is contained in Table 8.1.

The impact destroys the *PTDS* "1," the *PTUs* "5" and "7," as well as the *FDS* "**2**." Along with these destructions, the impact reduces the bandwidth of the link between the *PTDS* "2" and the *PTU* "6" from 300 to 100 kWh h. The resulting graph of the affected *EI* is represented in Fig. 8.1b.

Given this basic graph representation of *EIs*, we may move to their multigrammatical representation. Taking into account that a substantial formulation of *EI* resilience is based on the total amounts of *EP* and fuel produced and consumed during some considered time interval, we shall develop a *UMG*-based representation of energy infrastructures that would allow the possibility of application of the criterial base already developed for industrial *STSs*.

8.4.2 Basic Multigrammatical Representation of Energy Infrastructures

Let us begin with an ***electricity infrastructure***.

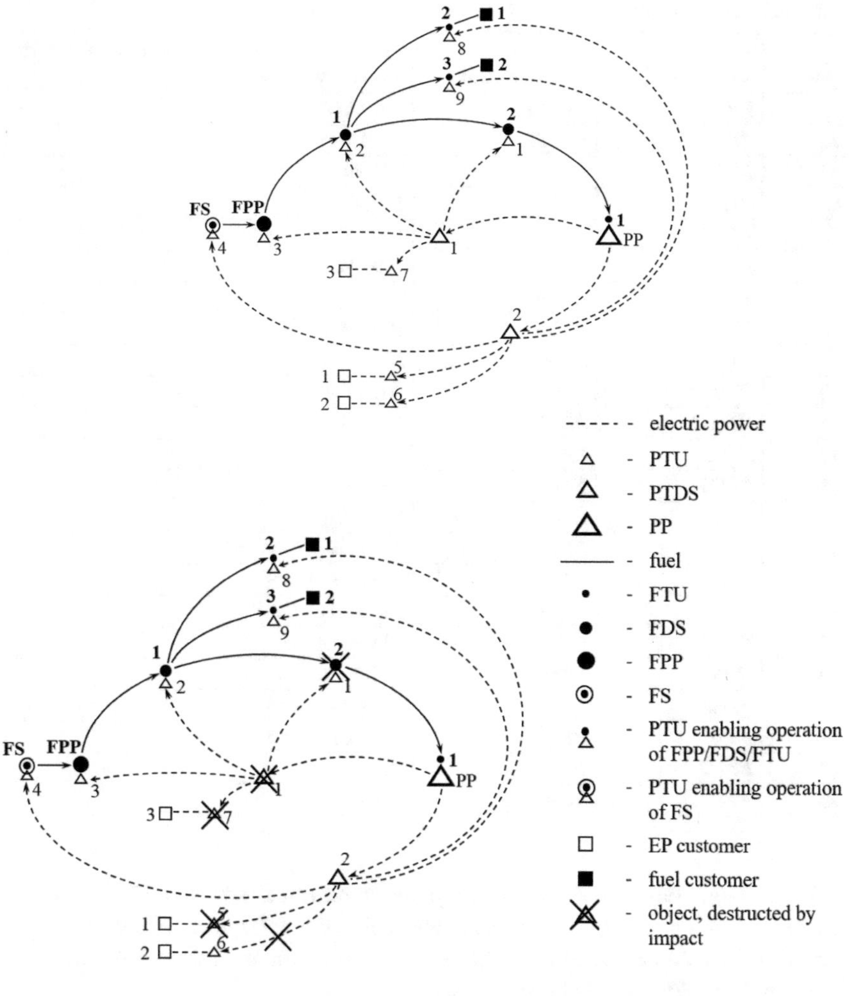

b

Fig. 8.1 Graph representation of a segment of an energy infrastructure: (**a**) initial state; (**b**) state after impact

In *URs* below, we shall use names of objects whose syntax will be (***kWh*** **:** *p*), where the string ***kWh*** denotes a measurement unit of *EP* transmitted via *PTLs* (kilowatt-hour) and *p* is a string representing the geographical point where an element of an *ElcI* is located (it may be designated by a unique symbolic name associated with specific geographic coordinates in a special database or directly by these coordinates). So a multiobject $n \cdot$ (***kWh*** **:** *p*) represents *n* kilowatt-hours generated or consumed at point (position, place) *p*. We assume that some basic alphabet *V* is used for the construction of strings representing geographical points.

Table 8.1 Algebraic representation of the graph of the segment of *EI*

No.	Source point	Receiver point	Channel upper threshold values of bandwidths (throughput capacities)
1.	*PP*	*PTDS1*	1000 kWh
2.	*PP*	*PTDS2*	1100 kWh
3.	*PTDS1*	*PTU1*	200 kWh
4.	*PTDS1*	*PTU2*	300 kWh
5.	*PTDS1*	*PTU3*	400 kWh
6.	*PTDS1*	*PTU7*	100 kWh
7.	*PTDS2*	*PTU4*	200 kWh
8.	*PTDS2*	*PTU5*	300 kWh
9.	*PTDS2*	*PTU6*	300 kWh
10.	*PTDS2*	*PTU8*	200 kWh
11.	*PTDS2*	*PTU9*	100 kWh
12.	*PTU5*	*EPC1*	300 kWh
13.	*PTU6*	*EPC2*	300 kWh
14.	*PTU7*	*EPC3*	100 kWh
15.	***FS***	***FPP***	200 tons of crude oil
16.	***FPP***	***FDS1***	200 tons of fuel
17.	***FDS1***	***FDS2***	100 tons of fuel
18.	***FDS1***	***FTU2***	50 tons of fuel
19.	***FDS1***	***FTU3***	50 tons of fuel
20.	***FDS2***	***FTU1***	100 tons of fuel
21.	***FTU1***	*PP*	100 tons of fuel
22.	***FTU2***	***FC1***	50 tons of fuel
23.	***FTU3***	***FC2***	50 tons of fuel

Let us begin our consideration with ***power terminal units***. Any *PTU* in order to deliver one unit of *EP* to a consumer connected to this *PTU* must receive it from the closest *PTDS* connected with it by a *PTL*. So a unitary rule representing this fragment of an *ElcI* would be as follows:

$$(\boldsymbol{kWh} : ptu) \rightarrow n \cdot (\boldsymbol{kWh} : ptds), n \cdot [ptds, ptu], \tag{8.58}$$

where *ptu* and *ptds* are strings representing locations of, respectively, a *PTU* and a *PTDS* supplying it, while $[ptds, ptu]$ is a string representing a *PTL* connecting them (here and everywhere below "[", "]" and "," are delimiters). In other words, [*ptds*, *ptu*] is an object representing a *PTL* whose start and final points are, respectively, *ptds* and *ptu*. The value $n \geq 1$ depends, finally, on the amounts of power losses occurring during its transmission via a *PTL* (in the case $n = 1$, there are no such losses); *n* is a rational number. So a multiobject $n \cdot [ptds, ptu]$ represents the fact that the considered *PTL* transmits 1 kWh h to a *PTU* located at a point *ptu* while receiving *n* kilowatt-hours from a *PTDS* located at a point *ptds*.

Let us note that the sense of (8.58) is fully similar to the sense of the ***technological interpretation of unitary rules*** regarding industrial systems, as it was introduced in Sect. 3.1.3 and illustrated by Example 3.11. Namely, to "create" 1 kWh h at a point *ptu*, it is necessary to have *n* kilowatt-hours at a point *ptds* and also a *PTL* connecting both points and able to transmit this amount of *EP* from *ptds* to *ptu*. Similar logic will be applied everywhere below to all components of *ElcI* and *FI*.

If a *PTDS* located at a point *ptds* is connected to power terminal units located at points $ptu_1, \ldots, ptu_m$, then this fragment of an *ElcI* is represented by the following *m* unitary rules:

$$(\boldsymbol{kWh} : ptu_1) \rightarrow n_1 \cdot (\boldsymbol{kWh} : ptds), n_1 \cdot [ptds, ptu_1],$$

$$\ldots$$

$$(\boldsymbol{kWh} : ptu_m) \rightarrow n_m \cdot (\boldsymbol{kWh} : ptds), n_m \cdot [ptds, ptu_m]. \tag{8.59}$$

Fragments of an *ElcI* consisting of connected *PTDSs* may be represented similarly. In this case, a string *ptds* is a representation of the location of a delivering power transforming-distributing substation, while $ptds_1, \ldots, ptds_l$ are locations of *PTDSs* which consume power transformed and transmitted by it:

$$(\boldsymbol{kWh} : ptds_1) \rightarrow n_1 \cdot (\boldsymbol{kWh} : ptds), n_1 \cdot [ptds, ptds_1],$$

$$\ldots$$

$$(\boldsymbol{kWh} : ptds_l) \rightarrow n_l \cdot (\boldsymbol{kWh} : ptds), n_l \cdot [ptds, ptds_l]. \tag{8.60}$$

In this way, ***all tree-like fragments of an ElcI are represented***, up to a power plant, producing electrical power. Any tree-like fragment of an *ElcI* containing some *PP* and connected with it *PTDSs* may be represented by the following *URs*:

$$(\boldsymbol{kWh} : ptds_1) \rightarrow n_1 \cdot (\boldsymbol{kWh} : pp), n_1 \cdot [pp, ptds_1],$$

$$\ldots$$

$$(\boldsymbol{kWh} : ptds_l) \rightarrow n_l \cdot (\boldsymbol{kWh} : pp), n_l \cdot [pp, ptds_l], \tag{8.61}$$

and if there are some power terminal units connected to a power plant directly, i.e., without any intermediate *PTDSs*, then also

$$(\boldsymbol{kWh} : ptu_1) \rightarrow n_1 \cdot (\boldsymbol{kWh} : pp), n_1 \cdot [pp, ptu_1],$$

$$\ldots$$

$$(\boldsymbol{kWh} : ptu_m) \rightarrow n_m \cdot (\boldsymbol{kWh} : pp), n_m \cdot [pp, ptu_m], \tag{8.62}$$

where *pp* is the location of a power plant.

A power plant, in turn, may be represented by a *UR*

$$(\boldsymbol{kWh} : pp) \rightarrow n_1 \cdot (res_1 : p_1), \ldots, n_k \cdot (res_k : p_k), \tag{8.63}$$

where $n_1, \ldots, n_k$ are amounts of resources $res_1, \ldots, res_k$ which must be delivered to locations $p_1, \ldots, p_k$, respectively, in order to generate 1 kWh h of electrical power at location *pp*, from which, in turn, it may be delivered by *PTLs* to the *PTDSs* (*PTUs*) closest to the *PP*.

Here, evidently, $p_1, \ldots, p_k$ are locations of fuel terminal units, which, in turn, deliver the aforementioned resources—most frequently, natural gas and various oil derivatives, which are transferred to power plants by pipelines.

So, evidently, it is reasonable to move on to consideration of a ***fuel infrastructure***.

Fuel terminal units delivering resources to consumers are represented as headers of unitary rules of the form

$$(res : ftu) \rightarrow n \cdot (res : fds), m \cdot (\boldsymbol{kWh} : ptu), n \cdot [fds, ftu], \tag{8.64}$$

where a multiobject $m \cdot (\boldsymbol{kWh} : ptu)$ represents a *PTU* of an electricity infrastructure located at a point *ptu* and supporting operation of an *FTU* located at a point *ftu* during delivery of one unit of a resource *res* from a point *fds* to point *ftu*. This amount of electrical power is consumed during a resource transfer via a pipe, whose start point is *fds* and final point is *ftu*. In the general case, due to losses of fuel during its transfer via a pipe, $n \geq 1$ units of fuel must be delivered to the pump at the start point of this pipe.

Distributing facilities (namely, *FDSs*) of fuel infrastructure may be represented similar to *PTDSs*:

$$(res : fds_1) \rightarrow n_1 \cdot (res : fds), m_1 \cdot (\boldsymbol{kWh} : ptu_1), n_1 \cdot [fds, fds_1],$$

$$\ldots$$

$$(res : fds_k) \rightarrow n_k \cdot (res : fds), m_k \cdot (\boldsymbol{kWh} : ptu_k), n_k \cdot [fds, fds_k]. \tag{8.65}$$

$$(res : ftu_1) \rightarrow n'_1 \cdot (res : fds), m'_1 \cdot (\boldsymbol{kWh} : ptu'_1), n'_1 \cdot [fds, ftu_1],$$

$$\ldots$$

$$(res : ftu_l) \rightarrow n'_l \cdot (res : fds), m'_l \cdot (\boldsymbol{kWh} : ptu'_l), n'_l \cdot [fds, ftu_l], \tag{8.66}$$

which means that the delivered resource incoming to any *FDS* is distributed to $k + l$ pipes by the application of the corresponding needed amounts of electrical power. The first *k* pipes perform fuel transfer to other *FDSs*, and the last *l* to *FTUs*. As above, $[fds, ftu_i]$, $i = 1, \ldots, l$, are pipes whose start point is *fds* and final points are ftu_i. Similarly, $[fds, fds_j]$, $j = 1, \ldots, k$, are pipes whose start point is *fds* and final points are fds_j. The presence of objects $(\boldsymbol{kWh} : ptu_i)$ in all unitary rules (8.65) and objects $(\boldsymbol{kWh} : ptu'_j)$ in all unitary rules (8.66) means that power terminal units belonging to an electricity infrastructure are installed and operate at some predefined points ptu_i and ptu'_j, respectively, to make possible physical contact with *FDSs* and *FTUs* and enable their power supply during transfer of resource *res*.

As was mentioned above, in the general case every pipe has its own technical parameters—finally, its own amounts of electrical power consumed, i.e., m_i and m'_j, as well as losses of fuel during its transfer via this pipe, i.e., n_i and n'_j.

As is clear, the described techniques may be applied up to the places of origination of energy carriers, i.e., ***fuel production plants***, which output pipeline gas and various oil derivatives used as a fuel by power plants. As was assumed above, ***PECs*** used for fuel production are accumulated at ***fuel storages*** colocated with *FPPs*. So operation of any such *FPP* may be represented as follows:

$$
\begin{aligned}
&(res : fpp)\\
&n \cdot (\boldsymbol{kWh} : ptu),\\
&m_1 \cdot (res_1 : fs_1), n_1 \cdot (\boldsymbol{kWh} : ptu_1),\\
&\ldots\\
&m_t \cdot (res_t : fs_t), n_t \cdot (\boldsymbol{kWh} : ptu_t),
\end{aligned}
\tag{8.67}
$$

where $fs_1, \ldots, fs_t$ are points where fuel storages with *PECs* $res_1, \ldots, res_t$ are located, so at these places, power terminal units are installed, thus allowing the relocation of amounts of these *PECs* necessary to an *FPP* for the production of one unit of fuel *res* at location *fpp*. The aforementioned relocation is possible if the needed amounts of electrical power, i.e., $n_1, \ldots, n_t$ kilowatt-hours, are available at points $ptu_1, \ldots, ptu_t$ where the respective *PTUs* are operating. In turn, to produce one unit of a fuel *res*, an *FPP* itself consumes n kilowatt-hours from a power terminal unit located at point *ptu*.

One more nuance connected with a multigrammatical representation of energy infrastructures and assessment of their resilience is the representation of ***active states of EI elements***. To represent the fact that any producing or transmitting (transferring) facility (*PP, FTP, PTDS, FTDS, PTU, FTU*) must be in an active state to carry out its functions, we shall apply techniques described above in the context of industrial systems and based on inclusion in the bodies of unitary rules of special multiobjects. So in the case of *URs* (8.58)–(8.63) concerning ***ElcIs***, any unitary rule

$$
(\boldsymbol{kWh} : x) \to X,
\tag{8.68}
$$

where X is the body of this *UR*, is transformed into

$$
(\boldsymbol{kWh} : x) \to X, 1 \cdot (+x),
\tag{8.69}
$$

where the symbol "+" means that ***facility*** x ***is in an active state and may produce*** **1 kWh h** ***of EP***. Similarly, unitary rules (8.64)–(8.67) concerning an *FI*

$$(res : x) \rightarrow X \tag{8.70}$$

are transformed into

$$(res : x) \rightarrow X, 1 \cdot (+x). \tag{8.71}$$

This means that ***facility** x **is in an active state and may produce one unit of resource** res*. As everywhere above, we shall use below the notion "***operation cycle of a facility** x*" (for short *OCF*), understanding it as an action performed by the facility to produce one unit of *EP*, fuel, or some other resource. A set (not obligatorily a sequence) of l such *OCFs* inside the considered time period of an *EI* operation has an obvious representation by a multiobject $l \cdot (+x)$.

We shall denote the set of unitary rules representing *ElcI*, *FI*, and their interconnections, as described above, by R_E. Let us illustrate techniques of construction of such a set given a graph representation of an *EI*, as described above in the previous Sect. 8.4.1.

Example 8.6 Consider the *EI* segment represented by the graph in Fig. 8.1a and Table 8.1. It may also be represented as the following set of unitary rules:

$$(\boldsymbol{kWh} : \boldsymbol{Ptds1}) \rightarrow 1 \cdot (\boldsymbol{kWh} : \boldsymbol{Pp}), 1 \cdot [\boldsymbol{Pp}, \boldsymbol{Ptds1}], 1 \cdot (+\boldsymbol{Ptds1}); \tag{8.72}$$

$$(\boldsymbol{kWh} : \boldsymbol{Ptds2}) \rightarrow 1 \cdot (\boldsymbol{kWh} : \boldsymbol{Pp}), 1 \cdot [\boldsymbol{Pp}, \boldsymbol{Ptds2}], 1 \cdot (+\boldsymbol{Ptds2}); \tag{8.73}$$

$$(\boldsymbol{kWh} : \boldsymbol{Ptu1}) \rightarrow 1 \cdot (\boldsymbol{kWh} : \boldsymbol{Ptds1}), 1 \cdot [\boldsymbol{Ptds1}, \boldsymbol{Ptu1}], 1 \cdot (+\boldsymbol{Ptu1}); \tag{8.74}$$

$$(\boldsymbol{kWh} : \boldsymbol{Ptu2}) \rightarrow 1 \cdot (\boldsymbol{kWh} : \boldsymbol{Ptds1}), 1 \cdot [\boldsymbol{Ptds1}, \boldsymbol{Ptu2}], 1 \cdot (+\boldsymbol{Ptu2}); \tag{8.75}$$

$$(\boldsymbol{kWh} : \boldsymbol{Ptu3}) \rightarrow 1 \cdot (\boldsymbol{kWh} : \boldsymbol{Ptds1}), 1 \cdot [\boldsymbol{Ptds1}, \boldsymbol{Ptu3}], 1 \cdot (+\boldsymbol{Ptu3}); \tag{8.76}$$

$$(\boldsymbol{kWh} : \boldsymbol{Ptu4}) \rightarrow 1 \cdot (\boldsymbol{kWh} : \boldsymbol{Ptds2}), 1 \cdot [\boldsymbol{Ptds2}, \boldsymbol{Ptu4}], 1 \cdot (+\boldsymbol{Ptu4}); \tag{8.77}$$

$$(\boldsymbol{kWh} : \boldsymbol{Ptu5}) \rightarrow 1 \cdot (\boldsymbol{kWh} : \boldsymbol{Ptds2}), 1 \cdot [\boldsymbol{Ptds2}, \boldsymbol{Ptu5}], 1 \cdot (+\boldsymbol{Ptu5}); \tag{8.78}$$

$$(\boldsymbol{kWh} : \boldsymbol{Ptu6}) \rightarrow 1 \cdot (\boldsymbol{kWh} : \boldsymbol{Ptds2}), 1 \cdot [\boldsymbol{Ptds2}, \boldsymbol{Ptu6}], 1 \cdot (+\boldsymbol{Ptu6}); \tag{8.79}$$

$$(\boldsymbol{kWh}:\boldsymbol{Ptu7}) \rightarrow 1\cdot(\boldsymbol{kWh}:\boldsymbol{Ptds1}),1\cdot[\boldsymbol{Ptds1},\boldsymbol{Ptu7}],1\cdot(+\boldsymbol{Ptu7}); \tag{8.80}$$

$$(\boldsymbol{kWh}:\boldsymbol{Ptu8}) \rightarrow 1\cdot(\boldsymbol{kWh}:\boldsymbol{Ptds2}),1\cdot[\boldsymbol{Ptds2},\boldsymbol{Ptu8}],1\cdot(+\boldsymbol{Ptu8}); \tag{8.81}$$

$$(\boldsymbol{kWh}:\boldsymbol{Ptu9}) \rightarrow 1\cdot(\boldsymbol{kWh}:\boldsymbol{Ptds2}),1\cdot[\boldsymbol{Ptds2},\boldsymbol{Ptu9}],1\cdot(+\boldsymbol{Ptu9}); \tag{8.82}$$

$$(\boldsymbol{kWh}:\boldsymbol{Pp}) \rightarrow 3\cdot(\boldsymbol{TonFuel}:\boldsymbol{Ftu1}),1\cdot[\boldsymbol{Ftu1},\boldsymbol{Pp}],1\cdot(+\boldsymbol{Pp}); \tag{8.83}$$

$$(\boldsymbol{TonFuel}:\boldsymbol{Ftu1})$$

$$1.05\cdot(\boldsymbol{TonFuel}:\boldsymbol{Fds2}),20\cdot(\boldsymbol{kWh}:\boldsymbol{Ptu1}),1\cdot[\boldsymbol{Fds2},\boldsymbol{Ftu1}],1\cdot(+\boldsymbol{Ftu1}); \tag{8.84}$$

$$(\boldsymbol{TonFuel}:\boldsymbol{Ftu2})$$

$$1.01\cdot(\boldsymbol{TonFuel}:\boldsymbol{Fds1}),\ 20\cdot(\boldsymbol{kWh}:\boldsymbol{Ptu2}),\ 1\cdot[\boldsymbol{Fds1},\boldsymbol{Ftu2}],1\cdot(+\boldsymbol{Ftu2}); \tag{8.85}$$

$$(\boldsymbol{TonFuel}:\boldsymbol{Ftu3})$$

$$1.02\cdot(\boldsymbol{TonFuel}:\boldsymbol{Fds1}),20\cdot(\boldsymbol{kWh}:\boldsymbol{Ptu2}),1\cdot[\boldsymbol{Fds1},\boldsymbol{Ftu3}],1\cdot(+\boldsymbol{Ftu3}); \tag{8.86}$$

$$(\boldsymbol{TonFuel}:\boldsymbol{Fds2})$$

$$1.01\cdot(\boldsymbol{TonFuel}:\boldsymbol{Fds1}),30\cdot(\boldsymbol{kWh}:\boldsymbol{Ptu2}),1\cdot[\boldsymbol{Fds1},\boldsymbol{Fds2}],1\cdot(+\boldsymbol{Fds2}); \tag{8.87}$$

$$(\boldsymbol{TonFuel}:\boldsymbol{Fds1})$$

$$1.01\cdot(\boldsymbol{TonFuel}:\boldsymbol{Fpp}),40\cdot(\boldsymbol{kWh}:\boldsymbol{Ptu3}),1\cdot[\boldsymbol{Fpp},\boldsymbol{Fds1}],1\cdot(+\boldsymbol{Fds1}); \tag{8.88}$$

$$(\boldsymbol{TonFuel}:\boldsymbol{Fpp})$$

$$2.9\cdot(\boldsymbol{TonCrudeOil}:\boldsymbol{Fs}),\ 50\cdot(\boldsymbol{kWh}:\boldsymbol{Ptu4}),\ 1\cdot[\boldsymbol{Fs},\boldsymbol{Fpp}],\ 1\cdot(+\boldsymbol{Fpp}). \tag{8.89}$$

As can be seen, *URs* (8.72) and (8.73) represent knowledge about the *PTDSs* "1" and "2," which are located, respectively, at the points ***Ptds*1** and ***Ptds*2,** and are connected by *PTLs*, represented by the multiobjects $1\cdot[\boldsymbol{Pp},\boldsymbol{Ptds1}]$ and $1\cdot[\boldsymbol{PP},\boldsymbol{Ptds2}]$, with the power plant located at the place ***Pp***; there are no significant losses of *EP* during its transmission from the *PP* to both *PTDSs*, so the same amount of *EP* which is put into either *PTL* by the *PP* is received by a *PTDS*; hence, the multiplicities of the object (***kWh : Pp***) in both URs (8.72) and (8.73) are equal to 1. The multiobjects $1 \cdot (+\boldsymbol{Ptds1})$ and $1 \cdot (+\boldsymbol{Ptds2})$ represent the fact that both *PTDSs* must be in active states to receive *EP* from the power plant producing it or to deliver *EP* to the power terminal units or *PTDSs* connected with them.

Knowledge about *PTUs* “1”–“9,” connected with the respective *PTDSs* in full accordance with the graph representation of the considered segment of the *EI*, is represented by *URs* (8.74)–(8.82).

The *UR* (8.83) represents that the power plant may produce 1 kWh h of *EP* by consuming for this objective 3 tons of fuel (represented by the multiobject $3 \cdot (\boldsymbol{TonFuel : Ftu1})$), receiving it via the pipe (represented by the *MO* $1 \cdot [\boldsymbol{Ftu1}, \boldsymbol{Pp}]$) from the fuel terminal unit “**1**” located at the point ***Ftu*1,** and being in the active state, which is represented by the *MO* $1 \cdot (+\boldsymbol{Pp})$.

The *URs* (8.84)–(8.86) represent knowledge about fuel terminal units “**1**”–“**3**.” The *UR* (8.84) represents knowledge about the resources necessary to the *FTU* “**1**” for receiving 1 ton of the fuel from the fuel-distributing station “**2**,” located at the place ***Fds*2**, via the pipe represented by the *MO* $1 \cdot [\boldsymbol{Fds2}, \boldsymbol{Ftu1}]$. Due to fuel losses during transfer, the *FDS* “**2**” delivering the fuel to the *FTU* “**1**” must put into the pipe represented by the *MO* $1 \cdot [\boldsymbol{Fds2}, \boldsymbol{Ftu1}]$ 1.05 tons of fuel, which is represented by the *MO* $1.05 \cdot (\boldsymbol{TonFuel : Fds2})$. The *FTU* “**1**” to receive 1 ton of the fuel consumes 20 kWh h of *EP* from the power terminal unit located at the point ***Ptu*1**, which is represented by the *MO* $20 \cdot (\boldsymbol{kWh : Ptu2})$. And, as usual, the *FTU* “**1**” must be in the active state, which is represented by the *MO* $1 \cdot (+\boldsymbol{Ftu1})$. The *URs* (8.85) and (8.86) in the same manner represent the knowledge about fuel terminal units “**2**” and “**3**,” which are supplied with *EP* from *PTU* “2,” and this *PTU* also consumes 20 kWh h for 1 ton of the received fuel.

The *URs* (8.87) and (8.88) represent similar knowledge about fuel-distributing stations “**1**” and “**2**,” provided with *EP* from the *PTUs* “3” and “2,” respectively; the *FDS* “**1**” consumes 40 kWh h of *EP* from *PTU* “3,” located at the point ***Ptu*3,** which is represented by the *MO* $40 \cdot (\boldsymbol{kWh : Ptu3})$, and *FDS* “**2**” consumes 30 kWh h of *EP* from *PTU* “2,” located at the point ***Ptu*2**, which is represented by the *MO* $30 \cdot (\boldsymbol{kWh : Ptu2})$. The *FDS* “**2**” receives fuel from the *FDS* “**1**” via the pipe represented by the *MO* $1 \cdot [\boldsymbol{Fds1}, \boldsymbol{Fds2}]$. The *FDS* “**1**,” in turn, receives fuel from the fuel-producing plant via the pipe represented by the *MO* $1 \cdot [\boldsymbol{Fpp}, \boldsymbol{Fds1}]$.

Finally, *UR* (8.89) represents the knowledge about the *FPP*, which is capable of producing 1 ton of fuel, receiving 2.9 tons of crude oil from the fuel storage via a pipe represented by the *MO* $1 \cdot [\boldsymbol{Fs}, \boldsymbol{Fpp}]$, and consuming 50 kWh h of *EP* from the *PTU* “4,” located at the point ***Ptu*4**, and this is represented by the *MO* $50 \cdot (\boldsymbol{kWh : Ptu4})$.

Finally, as can be seen, the considered segment of the *EI*, consuming crude oil from the fuel storage, provides external consumers with electrical power and fuel, respectively, via the *PTUs* “5,” “6,” and “7,” as well as via the *FTUs* “**2**” and “**3**.”

The ***resource base*** of any *EI* may be represented as a multiset v_E, including multiobjects of the following three types:

1. $m \cdot (res : p)$ for all fuel storages belonging to the considered *EI*, which means m units of material resource (*PEC* or produced fuel) *res* ***are available at some FS, located at a place*** p
2. $N \cdot [p, p']$ for all links, occurring in the considered *EI*, which means the value N is the bandwidth (throughput capacity) of the link $[p, p']$,i.e., the ***maximal amount of EP or material resource which may be transmitted (transferred)*** **via** ***this link during the considered time period*** (in the case when $[p, p']$ is a *PTL*, this amount is measured in kilowatt-hours, while in the case when $[p, p']$ is a pipe, this amount may be measured in barrels, cubic meters, kilograms, tons, etc.)
3. $L \cdot (+x)$ for all elements of the considered *EI*, thus establishing for any such element the ***maximal number of operation cycles which might be executed by it in the considered time period*** (in other words, L fixes the maximal productivity of the element x; a multiobject $L \cdot (+x)$ will be referred to below, as everywhere above, as the ***operation resource of the element*** x)

So, in fact, as in the case of industrial systems, the resource base of any ***energy infrastructure*** includes not only material resources (primary and produced energy carriers) but also the operation resources of its elements, as well as the throughput capacities of its links.

Example 8.7 The resource base of the segment of the *EI* considered in the previous Example 8.6 and corresponding to knowledge represented by Table 8.1 is as follows:

$$v_E =$$

$$\{100 \cdot (\boldsymbol{TonCrudeOil} : \boldsymbol{Fs}),$$

$$1000 \cdot [\boldsymbol{Pp}, \boldsymbol{Ptds1}], 1100 \cdot [\boldsymbol{Pp}, \boldsymbol{Ptds2}],$$

$$200 \cdot [\boldsymbol{Ptds1}, \boldsymbol{Ptu1}], 300 \cdot [\boldsymbol{Ptds1}, \boldsymbol{Ptu2}], 400 \cdot [\boldsymbol{Ptds1}, \boldsymbol{Ptu3}],$$

$$100 \cdot [\boldsymbol{Ptds1}, \boldsymbol{Ptu7}],$$

$$200 \cdot [\boldsymbol{Ptds2}, \boldsymbol{Ptu4}], 300 \cdot [\boldsymbol{Ptds2}, \boldsymbol{Ptu5}], 300 \cdot [\boldsymbol{Ptds2}, \boldsymbol{Ptu6}],$$

$$200 \cdot [\boldsymbol{Ptds2}, \boldsymbol{Ptu8}], 100 \cdot [\boldsymbol{Ptds2}, \boldsymbol{Ptu9}],$$

$$200 \cdot [\boldsymbol{Fs}, \boldsymbol{Fpp}],$$

$$200 \cdot [\boldsymbol{Fpp}, \boldsymbol{Fds1}],$$

$$100 \cdot [\boldsymbol{Fds1}, \boldsymbol{Fds2}], 50 \cdot [\boldsymbol{Fds1}, \boldsymbol{Ftu2}], 50 \cdot [\boldsymbol{Fds1}, \boldsymbol{Ftu3}],$$

$$100 \cdot [\boldsymbol{Fds2}, \boldsymbol{Ftu1}],$$

$$100 \cdot [\boldsymbol{Ftu1}, \boldsymbol{Pp}],$$

$$100 \cdot (+\boldsymbol{Pp}), 100 \cdot (+\boldsymbol{Ptds1}), 100 \cdot (+\boldsymbol{Ptds2}),$$

$$100 \cdot (+\boldsymbol{Ptu1}), 100 \cdot (+\boldsymbol{Ptu2}), 100 \cdot (+\boldsymbol{Ptu3}), 100 \cdot (+\boldsymbol{Ptu4}),$$

$$100\cdot(+\boldsymbol{Ptu5}),100\cdot(+\boldsymbol{Ptu6}),100\cdot(+\boldsymbol{Ptu7}),$$
$$10\cdot(+\boldsymbol{Ftu1}),10\cdot(+\boldsymbol{Ftu2}),10\cdot(+\boldsymbol{Ftu3}),10\cdot(+\boldsymbol{Fds2}),$$
$$10\cdot(+\boldsymbol{Fds1}),10\cdot(+\boldsymbol{Fpp})\}. \tag{8.90}$$

As can be seen, the fuel storage belonging to the considered segment of the *EI* contains 100 tons of crude oil.

The *PTL* connecting the power plant and the *PTDS* "1," during the considered time period, allows the transmission of no more than 1000 kWh h of *EP*, which is represented by the *MO* $1000\cdot[\boldsymbol{Pp},\boldsymbol{Ptds1}]$; the *PTL* connecting the *PP* and the *PTDS* "2" allows the transmission of no more than 1100 kWh h of *EP*, which is represented by the *MO* $1100\cdot[\boldsymbol{Pp},\boldsymbol{Ptds2}]$. The upper threshold values of the bandwidths of all other *PTLs* of the considered segment of the *EI* are represented similarly.

The pipe connecting the fuel storage and the fuel-producing plant allows the delivery of no more than 200 tons of crude oil, which is represented by the *MO* $200\cdot[\boldsymbol{Fs},\boldsymbol{Fpp}]$; the pipe connecting the fuel-producing plant and fuel-distributing station "**1**" allows the delivery of no more than 200 tons of fuel, which is represented by the *MO* $200\cdot[\boldsymbol{Fpp},\boldsymbol{Fds1}]$. The upper threshold values of the throughput capacities of all other pipes of the considered segment of the *EI* are represented similarly.

Any element of the *ElcI* belonging to this segment may execute 100 operation cycles during the considered time period, which is represented by the *MOs* $100\cdot(+\boldsymbol{Pp})$, …, $100\cdot(+\boldsymbol{Ptu9})$;any element of the *FI* belonging to this segment may execute 10 operation cycles during the considered time period, which is represented by the multiobjects $10\cdot(+\boldsymbol{Ftu1})$, …, $10\cdot(+\boldsymbol{Fpp})$.

After specifying a resource base, an *EI E* may be considered as a ***free industrial system*** $E = \ <\{\varnothing\}, R_E, v_E>$ in the sense of Chap. 6. Similarly, a demand on electrical power and fuel (***an order to be completed*** in the same sense) may be represented as a multiset q_E containing multiobjects like $n\cdot(\boldsymbol{kWh}:p)$, representing n kilowatt-hours to be delivered to a consumer located at a place p, where some *PTU* enabling this delivery is located, and multiobjects like $m\cdot(res:p)$, representing m units of fuel (or any other material resource) *res*, to be delivered to a consumer located at a place p, where an *FTU* enabling this delivery is located. As a result, an *EI* delivering amounts of power and fuel needed by customers may be considered as an ***industrial system*** $E_q = \ <q, R_E, v_E>$ ***assigned to an order*** q. This representation of an *IS* implies a ***filtering unitary multiset grammar*** $S_q = \ <q, R_E, F_E>$, where

$$F_E=\{\,a\le n\mid n\cdot a\in v_E\,\}\cup\{a=0\mid a\in\overline{A}_S\&a\notin v_E\}, \tag{8.91}$$

in such a way that this *FUMG* generates a set of terminal multisets, each representing some collection of resources sufficient for the completion of the order q by some particular collaboration of manufacturing devices (the second operand of the join is obligatory to eliminate ways of completing the order which satisfy the restrictions implied by the available resource base of the *EI*, but need some additional resources which are absent from the *RB* at all).

As may now be seen, a unitary multiset grammar $S_q = < q, R_E >$ defines a set $\overline{V}_{S_q}$ of terminal multisets each having the form

$$\Big\{M_1 \cdot (res_{i_1} : p_{i_1}), \ldots, M_s \cdot (res_{i_t} : p_{i_t}), N_1 \cdot \left[p_{j_1}, p'_{j_1}\right], \ldots, N_u \cdot \left[p_{j_u}, p'_{j_u}\right],$$

$$L_1 \cdot (+x_{k_1}), \ldots, L_z \cdot (+x_{k_z})\Big\}, \tag{8.92}$$

where $M_1, \ldots, M_s$ are ***amounts of***, respectively, ***resources*** $res_{i_1}, \ldots, res_{i_t}$ (*PECs* stored at *FSs*, fuels produced by *FPPs*, as well as *EP* produced by *PPs*) ***to be available at places*** $p_{i_1}, \ldots, p_{i_t}$ (via *PTUs* or *FTUs* or directly from fuel storages); N_1, …, N_u are ***amounts of energy carriers and electrical power to be transferred (transmitted)*** via, respectively, links $\left[p_{j_1}, p'_{j_1}\right], \ldots, \left[p_{j_u}, p'_{j_u}\right]$ (*PTLs* and pipes) during the considered time period; and L_1, …, L_z are ***numbers of operation cycles of***, respectively, ***facilities*** $x_{k_1}, \ldots, x_{k_z}$ involved in the completion of the order q.

So every *TMS* $v \in \overline{V}_{S_q}$ corresponds to some specific way of completing the order q (in the general case, there may be several ways, identical in resource consumption and facilities involvement).

Now, we may represent an *EI's* current resource base v_E as the sum of three multisets

$$v_E = v_E^{res} + v_E^{p} + v_E^{x}, \tag{8.93}$$

the first

$$v_E^{res} = \{\, \boldsymbol{M}_1 \cdot (res_{\boldsymbol{i}_1} : p_{\boldsymbol{i}_1}),\ \ldots,\ \boldsymbol{M}_s \cdot (res_{\boldsymbol{i}_t} : p_{\boldsymbol{i}_t}) \,\} \tag{8.94}$$

representing ***amounts of resources present at EI fuel storages***, the second

$$v_E^{p} = \Big\{\, \boldsymbol{N}_1 \cdot \left[p_{\boldsymbol{j}_1},\ p'_{\boldsymbol{j}_1}\right],\ \ldots,\ \boldsymbol{N}_u \cdot \left[p_{\boldsymbol{j}_u},\ p'_{\boldsymbol{j}_u}\right] \Big\} \tag{8.95}$$

representing ***current bandwidths and throughput capabilities of EI links***, and the third

$$v_E^{x} = \ \{\, \boldsymbol{L}_1 \cdot (+x_{\boldsymbol{k}_1}),\ \ldots,\ \boldsymbol{L}_z \cdot (+x_{\boldsymbol{k}_z}) \} \tag{8.96}$$

representing ***current operation resources of EI facilities***. (Bold indices $\boldsymbol{i}_1, \ldots, \boldsymbol{i}_t, \boldsymbol{j}_1, \ldots, \boldsymbol{j}_u, \boldsymbol{k}_1, \ldots, \boldsymbol{k}_z$, used in (8.94)–(8.96), differ from ordinary indices $i_1, \ldots, i_t, j_1, \ldots, j_u, k_1, \ldots, k_z$, used in (8.92).)

8.4.3 *Cyclicity of* FUMGs *Representing Energy Infrastructures and Their Finitarization*

Industrial systems are represented through the technological interpretation of unitary rules or, in other words, through their capability to manufacture (assemble) complex objects from their components beginning with atomic (non-splittable) elements (spare parts, microchips, etc.); thus, *FUMGs* representing *ISs* are ***essentially acyclic***, and, hence, sets of terminal multisets generated by the application of these multigrammars are finite.

Unlike industrial systems, an energy infrastructure operates in such a way that its fuel segment (namely, *FI*) consumes *EP* generated by its electricity segment (namely, *ElcI*), while the latter consumes fuel for *EP* production. Thus, *FUMGs* representing *EIs* are ***essentially cyclic***, and sets of multisets generated by their application are in the general case ***infinite***: for the core *UMG* $S'_q = < q, R_E >$ of an *FUMG* $S_q = < q, R_E, F_E >$, it holds that

$$\bar{V}_{S'_q} = \{\varnothing\} \tag{8.97}$$

and, simultaneously,

$$| V_{S'_q} | = \infty. \tag{8.98}$$

Hence, ***direct application of the multigrammatical representation and criterial base of ISs to EIs is in fact impossible***. So the task is to find a local correction of the aforementioned representation of *ISs* which ensures ***finitarity*** of *FUMGs* representing *EIs*. This correction will be called ***finitarization*** of *FUMGs*.

Here, we propose a simple solution of this problem based on the ***terminalization*** of non-terminal objects, introduced and described above in Sect. 8.1.1, as a tool for modeling *ISs* whose resource bases contain not only primary (non-splittable) components of *ORs* specified by an order to an *IS* but also components manufactured by an *IS*, beginning from the aforementioned primary components, at previous steps of its operation. Namely, we shall extend the set of *URs* R_E in the following way. Let R_E contains a unitary rule

$$(\boldsymbol{kWh} : x) \to X, \tag{8.99}$$

where X is a non-empty body. We shall join to R_E a unitary rule

$$(\boldsymbol{kWh} : x) \to k \cdot (res : x'), \tag{8.100}$$

which means that 1 kWh h appears at a location x as a result of consumption of k units of resource *res* located at a place x', and, most essentially, $(res : x')$ is a ***terminal*** object, which means that R_E contains no *UR* with a header $(res : x')$; the last, in turn, means that there is an alternative way for the terminal object to appear that does not involve the chain of mutual demands determined by the body X of the *UR* (8.100). In most cases, the resource *res* is ***power, accumulated at previous steps of operation of an EI*** or ***generated by some initiating action or operation*** (e.g., activation of a car ignition system).

In the first case, ***power storages*** (*PSs*), similar to fuel storages, are presumed, and, like *FSs*, they may be represented by multiobjects

$$n \cdot (\boldsymbol{kWh} : x) \in v_E, \tag{8.101}$$

which means that a *PS*, located at a place x, may provide on demand up to n kilowatt-hours.

The second case (***power generation by some initiating action***) is simply reduced to the first one by including in the resource base the same multiobject as in (8.101), which reflects the fact that the source of the aforementioned action, finally, is also some kind of power storage.

Thus, introducing by (8.100) and (8.101) the concept of a power storage, which, in fact, adequately reflects the essence of real processes of power supply, we have proposed the simplest way of ***finitarization of FUMGs representing EIs***. Now, evidently, ***although the set*** $V_{S'_q}$ ***remains infinite, the set*** $\bar{V}_{S'_q}$ in the general case is ***non-empty***, thus representing ***at least one way of completing an order*** q ***by the application of* a priori *accumulated power***. From the mathematical point of view, this means that the core *UMG* $S'_q = \ < q, R_E>$ of an *FUMG* $S_q = \ < q, R_E, F_E>$ generates at least one terminal *TMS*.

Now, we are ready to consider the main result of this section, which is a criterial base for the assessment of vulnerability of energy infrastructures to destructive impacts.

8.4.4 Criteria of Vulnerability of an Energy Infrastructure to a Destructive Impact

Let us begin with the ***initial task*** whose verbal formulation is as follows: ***given amounts of primary energy carriers at the fuel storages of an EI and demand for electrical power and fuels from its external consumers*** (an order to be completed by the *EI*), ***to assess whether the EI is or is not capable of completing the order*** (i.e., providing to these consumers the required amounts of *EP* and fuels).

Due to the introduced techniques of *EI* representation, it is sufficient to apply the criterion formulated by Statement 6.1 regarding industrial systems to energy infrastructures.

Statement 8.8 An energy infrastructure $E = \ < \{\varnothing\}, R_E, v_E>$ is not capable of completing an assigned order q if

$$\left(\forall v \in \bar{V}_{S_q}\right) \cdot v_E \subset v, \tag{8.102}$$

where $S_q = \ < q, R_E>$.

Speaking informally, an *EI E* is not capable of completing an assigned order q if there exists no way of generating (producing) and delivering the necessary amounts of *EP* and fuels, consuming for this objective amounts of primary energy carriers which are not greater than those available at the fuel storages of this *EI*, and using the capabilities of *EI* facilities and links for this generation (production) and delivery. This criterion may be represented by applying the respective filtering unitary multiset grammar $S_q = \ < q, R_E, F_E>$, where

$$F_E = \{\, a \leq n \mid n \cdot a \in v_E \,\} \cup \left\{a = 0 \mid a \in \overline{A}_S \& a \notin v_E\right\} \tag{8.103}$$

(the second operand of the join is obligatory to eliminate ways of order completion which satisfy restrictions implied by the available resource base of the *EI*, but need some additional resources which are absent from the *RB* at all).

Statement 8.9 An energy infrastructure $E = \ < \{\varnothing\}, R_E, v_E>$ is not capable of completing an assigned order q if

$$\bar{V}_{S_q} = \{\varnothing\}, \tag{8.104}$$

where $S_q = \ < q, R_E, F_E>$ and F_E is defined by (8.103).

All the above forms the basis for a formal consideration of the task of ***assessment of the vulnerability of EIs to destructive impacts***. Following the approach introduced in Sect. 8.1.1, we shall represent an impact as a multiset Δv which determines the capabilities of *EI* elements (facilities and links) and the amounts of primary energy carriers stored at *FSs* that are destroyed by this impact. After the impact application, the resource base v_E of the *EI* becomes $v_E - \Delta v$. This representation in the general case allows other possible variants of impact, which may destroy *EI* elements, reduce bandwidths (throughput capacities) of links (*PTLs* and pipes), as well as reduce amounts of *PECs* in *FSs*.

Namely,

$$m \cdot (res : p) \in \Delta v \tag{8.105}$$

means that an impact eliminates m units of a resource (fuel) from an *FS* located at a place p. Similarly,

$$n \cdot [p, p'] \in \Delta v \tag{8.106}$$

means that an impact reduces by n units the maximum amount of *EP* or fuel which may be transmitted (transferred) in the considered time period via a *PTL* (pipe) with start point p and final point p'. Finally, to represent destruction of any producing or transmitting (transferring) facility (*PP, FTP, PTDS, FTDS, PTU, FTU*), we may apply the same techniques, by including in an *MS* Δv multiobjects like $l \cdot (+x)$, representing that an element of an *EI* is affected, and a result of this action is a reduction of the operation resource of this element by l units. Obviously, the case of entire destruction of any component of an *EI* may be easily represented by inclusion in the multiset Δv of an object $\boldsymbol{N} \cdot (a)$, where $\boldsymbol{N}$ is a maximal number for the used implementation of *FUMG* algorithmics, so for any k

$$\{k \cdot (a)\} - \{\boldsymbol{N} \cdot (a)\} = \{\emptyset\}. \tag{8.107}$$

Example 8.8 Suppose the destructive impact destroys facilities *PTDS* "1" and *PTU* "7" of the segment of the *EI* considered in the previous Example 8.7, as well as reduces the amount of crude oil at the fuel storage by 20 tons, and also reduces the bandwidths (throughput capacities) of the *PTL* [***Ptds*2, *Ptu*4**] by 100 kWh h, the *PTL* [***Ptds*2, *Ptu*5**] by 200 kWh h, and the pipe [***Fds*2, *Ftu*1**] by 10 tons of fuel. So the result of this impact will be

$$\begin{aligned} & v_E - \Delta v = \\ & v_E - \{10{,}000 \cdot (+\boldsymbol{Ptds1}), 10{,}000 \cdot (+\boldsymbol{Ptu7}), 20 \cdot (\boldsymbol{TonCrudeOil} : \boldsymbol{Fs}), \\ & 100 \cdot [\boldsymbol{Ptds2},\ \boldsymbol{Ptu4}], 200 \cdot [\boldsymbol{Ptds2},\ \boldsymbol{Ptu5}], 10 \cdot [\boldsymbol{Fds2},\ \boldsymbol{Ftu1}]\}, \end{aligned} \tag{8.108}$$

where $\boldsymbol{N} = 10{,}000$.

Now, we may move directly to the proposed ***criterial base of EI vulnerability***.

Let us begin with the simplest case, when an impact is applied to an EI before the beginning of processing of an order $\boldsymbol{q}$.

Evidently, an impact Δv transforms an *EI* $E = \ <\{\emptyset\}, R_E, v_E>$ into an affected *EI* $E = \ <\{\emptyset\}, R_E, v_E - \Delta v>$, and to obtain the necessary criterion of vulnerability of energy infrastructures to destructive impacts, it is sufficient to apply to the affected *EI* Statements 8.8 and 8.9.

Statement 8.10 An energy infrastructure $E = \ <\{\emptyset\}, R_E, v_E>$ is vulnerable to an impact Δv applied before the beginning of processing of an assigned order q if

$$\left(\forall v \in \bar{V}_{S_q}\right) v_E - \Delta v \subset v \tag{8.109}$$

where $S_q = \ <q, R_E>$.

Verbally, if no way of completing order q is implementable (any way needs more resources than are available after the impact), then the energy infrastructure is vulnerable to the applied impact. Applying (8.103) to Statement 8.9, this criterion may be represented by the application of *FUMGs*.

Statement 8.11 An energy infrastructure $E = \ <\{\varnothing\}, R_E, v_E>$ is vulnerable to an impact Δv applied before the beginning of processing of an assigned order q if

$$\bar{V}_{S_q} = \{\varnothing\}, \tag{8.110}$$

where $S_q = \ <q, R_E, F_{E'}>$, and

$$F_{E'} = \{\, a \le n \mid n \cdot a \in v_E - \Delta v\} \cup \{a = 0 \mid a \in \overline{A}_S \& a \notin v_E - \Delta v\}. \tag{8.111}$$

Let us now consider a more general case, when an ***impact is applied to an EI at some time moment inside a time period of order processing***.

Up to this moment, some part Δq of an order q may be completed, and a respective part Δv_E of the *EI* resource base may already have been consumed. In this case, it is not difficult to formulate as corollaries of Statements 8.10 and 8.11 statements of the criteria of vulnerability of an *EI* affected by a destructive impact inside a time period of order processing.

Statement 8.12 An energy infrastructure $E = \ <\{\varnothing\}, R_E, v_E>$ is vulnerable to an impact Δv applied inside a time period of processing of an assigned order q, when a part Δq of this order is already completed and a part Δv_E of the *EI* resource base is already consumed, if

$$\left(\forall v \in \bar{V}_{S_{q-\Delta q}}\right) v_E - \Delta v_E - \Delta v \subset v. \tag{8.112}$$

Statement 8.13 An energy infrastructure $E = \ <\{\varnothing\}, R_E, v_E>$ is vulnerable to an impact Δv applied inside a time period of processing of an assigned order q, when a part Δq of this order is already completed and a part Δv_E of the *EI* resource base is already consumed, if

$$\bar{V}_{S_{q-\Delta q}} = \{\varnothing\}, \tag{8.113}$$

where $S_{q - \Delta q} = \ <q, R_E, F_{E''}>$, and

$$F_{E''} = \{\, a \le n \mid n \cdot a \in v_E - \Delta v_E - \Delta v\} \\ \cup \{a = 0 \mid a \in \overline{A}_S \& a \notin v_E - \Delta v_E - \Delta v\}. \tag{8.114}$$

Now, we may consider a more complicated case of *EIs* whose topology is designed and resource base is maintained in such a way that ***if some destructive impact is applied*** before or during order processing, and this impact renders an *EI* incapable of completing this order, then the ***EI recovers itself*** by activation of some amounts of operation resources prepared in advance as well as by application of some prestored amounts of material resources.

8.4.5 *Modeling Reservation and Recovery of Energy Infrastructures*

Taking into account the possibility of application of destructive impacts of various kinds to components of energy infrastructure, an *EI's* management usually ***prepares in advance some additional reserved facilities and primary or produced resources***, which are made available promptly after an impact is detected, and this measure in many cases provides as effective as possible mitigation of the consequences of the aforementioned impact (possibly including making an order feasible the impacted *EI* itself). The basis of such adaptability is some redundancy built into an *EI* before or during its operation (Brown 2008; Lavaei et al. 2014; Wang et al. 2019). The most usual measures implemented by *EIs'* designers and management are so-called backup power systems, providing *EP* generation for time periods when affected segments of *EIs* are recovering, and also ***bypasses***, providing electricity or fuel flows by some workarounds if the normal routes are broken by an impact.

It is not so difficult to apply *UMGs* to represent the described measures.

Namely, it is sufficient to join to an initial set of *URs*, representing the topology of an ***electricity infrastructure***, unitary rules reflecting alternative ways of *EP* transmission.

For any *UR* (8.58) representing a *PTU* located at a point *ptu*, it is sufficient to join to the set R_E one more *UR*

$$(\boldsymbol{kWh} : ptu) \rightarrow n' \cdot (\boldsymbol{kWh} : ptds'), n' \cdot [ptds', ptu], \tag{8.115}$$

representing the fact that ***this PTU may receive EP*** not only from the *PTDS* located at the point *ptds* but **also from a PTDS located at a point *ptds'***. (*It is assumed that there is a technological solution implementing this possibility.*) In the general case, $m \geq 1$ such alternative *PTDSs* may be capable of delivering *EP* to this *PTU*, and this is possible regarding any *PTU* belonging to the considered *EI*.

Similarly, the same technique may be applied to *PTDSs*. To any *UR* belonging to the set (8.60) and representing a *PTDS* located at a point $ptds_i$, it is sufficient to join to the set R_E one more *UR*

$$(\boldsymbol{kWh} : ptds_i) \rightarrow n'_i \cdot (\boldsymbol{kWh} : ptds'), n'_i \cdot [ptds', ptds_i], \tag{8.116}$$

representing the fact that ***this PTDS may receive EP*** not only from the *PTDS* located at the point *ptds* but ***also from a PTDS located at a point ptds'***. As in the case of *PTUs*, $m \geq 1$ such alternative *PTDSs* may be capable of delivering *EP* to this *PTDS*, and this is possible regarding any *PTDS* belonging to the considered *EI*.

The described technique without any changes may be applied also to *PTUs* and *PTDSs* connected directly with additional (reserve) power plants:

$$\left(\boldsymbol{kWh} : ptds_j\right) \rightarrow n_j \cdot \left(\boldsymbol{kWh} : pp'\right), n_j \cdot \left[pp', ptds_j\right], \tag{8.117}$$

$$\left(\boldsymbol{kWh} : ptu_k\right) \rightarrow n_k \cdot \left(\boldsymbol{kWh} : pp''\right), n_k \cdot \left[pp'', ptu_k\right]. \tag{8.118}$$

These *URs* represent the facts that a ***PTDS located at a point ptds$_j$ may receive EP*** not only from the *PP* located at the point *pp*, as defined by the respective *UR* from the set (8.61), but ***also from a PP located at a place pp$'$*** and similarly a *PTU* located at a point ptu_k and belonging to the set of *URs* (8.62) may receive *EP* from some *PP* located at a place pp'', which may differ from pp' or be the same. As in all cases considered above, there may be $m \geq 1$ such alternative power plants, capable of delivering *EP* to these *PTUs* and *PTDSs*, and this is possible regarding any *PTU* or *PTDS* connected with several power plants.

Any ***reserve power plant*** belonging to the considered *ElcI* and located at a point pp' may be represented by a *UR*

$$\left(\boldsymbol{kWh} : pp'\right) \rightarrow n'_1 \cdot \left(res'_1 : p'_1\right), \ldots, n'_{k'} \cdot \left(res'_{k'} : p'_{k'}\right), \tag{8.119}$$

where, as in (8.63), $n'_1, \ldots, n'_{k'}$ are amounts of resources $res'_1, \ldots, res'_{k'}$, which must be delivered to locations $p'_1, \ldots, p'_{k'}$ respectively, in order to generate 1 kWh h of electrical power at a location pp', from which, in turn, it may be delivered by *PTLs* to the *PTDSs* (*PTUs*) closest to this *PP*. Let us note that a reserve may be implemented not only by inclusion in an *ElcI* of some additional power plants but also by the implementation of alternative ways of *EP* generation and resources associated with them, by which a *PP* must be supplied. In this case, to a *UR* (8.63), a unitary rule

$$\left(\boldsymbol{kWh} : pp\right) \rightarrow n'_1 \cdot \left(res'_1 : p'_1\right), \ldots, n'_{k'} \cdot \left(res'_{k'} : p'_{k'}\right), \tag{8.120}$$

with the same header (***kWh*** *: pp*) and an alternative body representing a different supply set necessary for this way of *EP* generation, is joined to the set R_E.

As may be seen, with the application of ***alternating UMGs***, it is quite easy to represent electric grids of any complexity, not only of the tree-like structures that were considered above in the previous sections of this chapter.

Reservation of a ***fuel infrastructure*** may be represented in a similar way.

Namely, for any *UR* (8.64) representing an *FTU* located at a point *ftu*, it is sufficient to join to the set R_E one more *UR*

$$(res : ftu) \rightarrow n' \cdot (res : fds'), m' \cdot (\boldsymbol{kWh} : ftu), n \cdot [fds', ftu], \tag{8.121}$$

representing the fact that this *FTU* may receive the resource *res* not only from the *FDS* located at the point *fds* but also from an *FDS* located at a point fds'. (*As in the case of an ElcI, it is assumed that there is a technological solution implementing this possibility.*) In the general case, $m \geq 1$ such alternative *FDSs* may be capable of delivering the resource *res* to this *FTU*, and this is possible regarding any *FTU* belonging to the considered *FI*.

Reservation of distributing facilities (namely, *FDSs*) of fuel infrastructure may be represented similar to reservation of *PTDSs*:

$$(res : fds_i) \rightarrow n'_i \cdot (res : fds'), m'_1 \cdot (\mathbf{kWh} : ptu_i), n'_1 \cdot [fds', fds_i]. \tag{8.122}$$

Any such *UR* represents the fact that an *FDS* located at a point fds_i may receive the required amounts of resource *res* not only from the *FDS* located at the point *fds*, as defined by the respective *UR* from the set (8.65), but also from an *FDS* located at a place fds'. As in all cases considered above, $m \geq 1$ such alternative *FDSs* may be capable of delivering the resource *res* to this *FDS*, and this is possible regarding any *FDS* connected with several *FDSs* supplying it.

Finally, any ***reserve FPP*** producing fuel *res* used by power plants for *EP* generation and located at a point fpp' may be represented by a *UR*

$$\begin{gathered}(res : fpp') \\ n' \cdot (\mathbf{kWh} : ptu'), \\ m'_1 \cdot (res'_1 : fs'_1), n'_1 \cdot (\mathbf{kWh} : ptu'_1), \\ \ldots \\ m'_t \cdot (res'_t : fs'_{t'}), n'_{t'} \cdot (\mathbf{kWh} : ptu'_{t'}),\end{gathered} \tag{8.123}$$

where $fs'_1, \ldots, fs'_{t'}$ are points where fuel storages with *PECs* $res'_1, \ldots, res'_{t'}$ are located, so at these places, power terminal units are installed to allow the relocation of amounts of these *PECs* necessary to an *FPP* for the production of one unit of fuel *res* at a location fpp'. The aforementioned relocation is possible if needed amounts of electrical power, i.e., $n'_1, \ldots, n'_{t'}$ kilowatt-hours, are available at points $ptu'_1, \ldots, ptu'_{t'}$ where the respective *PTUs* are operating. In turn, to produce one unit of fuel *res*, the reserve FPP itself would consume n' kilowatt-hours from a power terminal unit located at a point ptu'.

Similar to power plants, any fuel-producing plant may be reserved not only by inclusion in an *FI* of some additional *FPPs* but also by the implementation of alternative ways of fuel production and the *PECs* associated with them with which an *FPP* must be supplied. In this case, for any *UR* (8.67), a unitary rule

$$\begin{gathered}(res : fpp) \\ n' \cdot (\mathbf{kWh} : ptu'), \\ m'_1 \cdot (res'_1 : fs'_1), n'_1 \cdot (\mathbf{kWh} : ptu'_1), \\ \ldots\end{gathered}$$

$$m'_t \cdot \left(res'_t : fs'_{t'}\right), n'_{t'} \cdot (\boldsymbol{kWh} : ptu'_{t'}), \tag{8.124}$$

with the same header (*res* : *fpp*) and an alternative body representing an alternative supply set necessary for this way of fuel production, is joined to the set R_E.

As may be seen, this generalization makes possible the application of the criteria (8.102)–(8.111) to the general case of *EIs*, without any changes. The main difference is that *UMGs* representing such *EIs* are alternating, and thus $\bar{V}_{S_q}$ is a ***multi-element set of terminal multisets***, i.e., $|\bar{V}_{S_q}| \geq 1$.

However, in practice, an *EI* operates using some subset of its components, and this subset as a whole has an ordinary concentric tree-like structure, while the remaining components stay in reserve until an impact, after which some or even all of the reserve components may be joined (switched) to the affected *EI*. To implement this approach, it is sufficient to represent a reserved *EI* as a ternary tuple (for short "quadruple") $E = \langle \{\emptyset\}, R_E, v_E, \boldsymbol{v_E} \rangle$, where $\boldsymbol{v_E}$ is the ***reserve resource base***, any part (submultiset) of which $\boldsymbol{\Delta v_E} \subseteq \boldsymbol{v_E}$ may be added to an *RB* reduced by an impact, transforming it from $v_E - \boldsymbol{\Delta v}$ to $v_E - \boldsymbol{\Delta v} + \boldsymbol{\Delta v_E}$. Following (8.93)–(8.96), the multiset $\Delta \boldsymbol{v_E}$ may be represented as the sum of three non-intersecting multisets similar to MSs v_E^{res}, v_E^p, v_E^x :

$$\boldsymbol{\Delta v_E} = \boldsymbol{\Delta v_E^{res}} + \boldsymbol{\Delta v_E^p} + \boldsymbol{\Delta v_E^x}, \tag{8.125}$$

the first summand

$$\boldsymbol{\Delta v_E^{res}} = \left\{ \boldsymbol{\Delta M}_1 \cdot \left(res_{i_1} : p_{i_1}\right), \ldots, \boldsymbol{\Delta M}_s \cdot \left(res_{i_s} : p_{i_s}\right) \right\} \tag{8.126}$$

representing amounts of resources (*PECs* and fuels) occurring at *EI* reserve fuel storages, as well as amounts of *EP* accumulated by backup power systems (in this case, res_{i_j} is nothing but $\boldsymbol{kWh}$), the second

$$\boldsymbol{\Delta v_E^p} = \left\{ \boldsymbol{\Delta N}_1 \cdot \left[p_{j_1}, p'_{j_1}\right], \ldots, \boldsymbol{\Delta N}_u \cdot \left[p_{j_u}, p'_{j_u}\right] \right\} \tag{8.127}$$

representing reserve throughput capabilities of an *EI* (*PTLs* and pipes kept out of operation until an impact), and the third

$$\boldsymbol{\Delta v_E^x} = \left\{ \boldsymbol{\Delta L_1} \cdot (+x_{k_1}), \ldots, \boldsymbol{\Delta L}_z \cdot \left(+x_{k_z}\right) \right\} \tag{8.128}$$

representing reserve operation resources of the *EI* (facilities kept ready to operate if necessary). Thus, addition to a reduced *RB* of an *MS* $\boldsymbol{\Delta v_E^{res}}$ adds to the affected *EI* some amounts of fuel, located at reserve *FSs*; an *MS* $\boldsymbol{\Delta v_E^p}$ adds to the *EI* transporting network some reserve links (*PTLs* and pipes); and, finally, an *MS* $\boldsymbol{\Delta v_E^x}$ adds to the *EI* some reserve producing, generating, and transmitting/transferring facilities. After this addition, criteria (8.102)–(8.104) may be applied to the *EI* $E = \langle \{\emptyset\}, R_E, v_E - \boldsymbol{\Delta v} + \boldsymbol{\Delta v_E} \rangle$ and an order *q*. If this *EI* is not vulnerable, then there exists an opportunity to recover it after impact, and thus there arises naturally the task of computation of "the best" of all possible multisets $\boldsymbol{\Delta v_E} \subseteq \boldsymbol{v_E}$ allowing the recovery of the affected *EI* to a state sufficient for the completion of the order. Otherwise, it is clear that the available reserve is insufficient for *EI* recovery and order completion.

8.4.6 Modeling Rechargeable Power Storages and Their Application

Until now, we have not determined either verbally or formally ***how power is accumulated in any specific PS*** nor how the last is ***recharged after or during order completion*** (e.g., in the case of a car accumulator, it is recharged ***during*** or, more correctly, ***by*** the car's motion). Let us consider the case of ***rechargeable power storages*** and their multigrammatical modeling.

We shall associate with ***any order*** q, whose objective is ***meeting a demand*** (request) by an external consumer for collections of resources located at predefined places, a so-called ***internal order*** q', whose objective is ***addition of a collection*** q' to the resource base, i.e., full or partial (or even redundant) ***replenishment of resources spent at the time of completion of order*** q. The simplest way of interpreting an internal order is as ***replacement of the RB*** v_E remaining after the completion of external order q ***by*** $v_E + q'$. However, this approach is not satisfactory, because in fact it applies the concept of an energy infrastructure as an ***open system***—the collection q' is ***not a part of the EI resource base*** and is obtained from systems which are ***external regarding EI***. The second reason for the rejection of this approach is that the collection q' is not at all correlated with the order q; it is assigned by the *EI* control system in some arbitrary way (the most natural one is $q' \in \bar{V}_{S_q}$, which means full replenishment of spent resources).

So it is necessary to develop techniques of representation of a logic of internal order q' construction and completion, which, from one side, allow ***compliance with an order*** q and, from the other, allow ***replenishment*** of ***power storages*** used during processing of order q, by consumption of any other resources occurring in the *EI* resource base. This approach fully fits the reality, where *PSs* are recharged on a regular basis by consumption of other resources belonging to the *EI RB*. Thus, firstly, the assumption that the *EI* is a ***closed*** system would be satisfied, and, secondly, ***all power storages*** applied during order q completion would be recharged by means of only the ***internal capabilities*** of the energy infrastructure.

The proposed multigrammatical representation of this important feature of *EIs* and their fragments is as follows.

Let $v \in \bar{V}_{S_q}$ be a collection of resources consumed for the completion of an order q, so $v \subseteq v_E$. We shall define a submultiset v' of the multiset v in such a way that multiobjects belonging to v' represent ***amounts of power delivered from PSs during the completion of the order*** q (all such multiobjects, obviously, have the form $n \cdot (\boldsymbol{kWh} : x)$), so

$$v' = \{\, n \cdot (\boldsymbol{kWh} : x) \mid n \cdot (\boldsymbol{kWh} : x) \in v \}. \tag{8.129}$$

Namely, these amounts of electrical power would be ***replenished in power storages*** before the next order arrives at the *EI*, so this multiset is nothing but a needed internal order, i.e., at first glance,

$$v' = q'. \tag{8.130}$$

However, the substantial difficulty, breaking (8.130), is that to be an order completed by some chain of energy transfers and transmissions, the *MS* v' in the general case would contain non-terminal objects, i.e., objects that are headers of unitary rules representing the *EI*. At the same time, the *MS* v', being a submultiset of the *MS* v_E, contains only terminal objects.

To avoid this deadlock, we propose the following solution. To define the logic of *PS* replenishment (recharge), the set R_E would contain *URs* of the form

$$(*\boldsymbol{kWh} : x) \to X, \tag{8.131}$$

where bold symbol " $*$ " means that to replenish 1 kWh h at a *PS* located at a place x, it is sufficient to complete an order which is a set containing all multiobjects belonging to the body X. So all *URs* like (8.131), in fact, define the logic of *PS* replenishment. Then, evidently,

$$q' = \{\, n \cdot (*\boldsymbol{kWh} : x) \mid n \cdot (\boldsymbol{kWh} : x) \in v' \}, \tag{8.132}$$

so after the completion of an initial order q, which results in the delivery of amounts of electrical power determined by q, an internal order q' is completed, resulting in the replenishment of the power storages used during the completion of q. As may be seen, power storages applied during the completion of an internal order q' are not replenished; otherwise, the process of replenishment may become ***recursive*** and too complicated in an implementation. From the practical point of view, this is quite natural. Let us note, as a conclusion of this section, that ***not all power storages belonging to a considered energy infrastructure are rechargeable*** (i.e., in a multigrammatical representation of an *EI*, not all *URs* with headers $(\boldsymbol{kWh} : x)$ are supplemented by *URs* (8.131) with headers $(*\boldsymbol{kWh} : x)$); all other *PSs* are presumed to be single-use, so they may be replaced after consumption of all initially accumulated power.

8.4.7 Further Tasks

Let us recall that the criterial base, introduced above in this section, permits an assessment of vulnerability of energy infrastructures to destructive impacts, on the following basis: if the *EI* and the impact satisfy the formulated conditions, then the *EI* is considered ***vulnerable***; but if the aforementioned conditions are not satisfied, then it ***does not mean*** that the affected *EI* is ***substantially resilient to the impact***. Let us underline that we do not affirm that in the case $\bar{V}_{S_q} \neq \{\varnothing\}$, the *EI* E is capable of completing the order q, because in the general case, there must be also assessed ***time delays*** associated with the production and delivery of material resources (though regarding electricity infrastructure and electrical power, circulating via its networks and grids, in the general case, such delays may be ignored). So, if an order includes a ***deadline*** for the delivery of all necessary resources to external consumers who are the source of this order, then, although the amounts of resources available for order completion may be sufficient for this objective, even an optimal schedule of order completion ***may not ensure timely delivery*** of all necessary resources to consumers. This is an implication of the ***non-additivity of time***; let us repeat once more that ***time is an additive resource regarding separate devices (facilities)***, but ***regarding an EI as a whole, it is non-additive***, because different devices may operate in parallel. So an *EI* not satisfying the above-introduced criteria and thus ***being not vulnerable in the above sense***, in the general case, may be ***not resilient to an impact***. For this reason, if in the general case an order includes restrictions on the duration of its completion, then to assess *EI* resilience, it would be necessary to apply more general mathematical tools than unitary multiset grammars—namely, ***temporal multiset grammars***.

There is also an ***inverse task*** to be solved—namely, given the remaining part of a considered *EI* segment and resources, to assess ***what maximal subset of the full set of consumers may be provided with power and fuel in accordance with their demand***. Another variation of this task is to assess whether ***some predefined subset of consumers may be provided with power and fuel according to their demand while all the remaining consumers may be provided with some part of their demand*** that is not less than some threshold values.

Also, the task of ***optimal recovery*** of an affected *EI* must be considered, and also the task of assessment of ***"the best" part of an order*** which may be completed given the part of an *EI* remaining after an impact (both with and without a reserve).

The next extremely important task to be considered in the future is a priori design of *EIs* that are ***maximally resilient to the most likely sets (sequences) of impacts*** and that consume for their recovery the minimal possible amounts of resources.

Let us note that just the same techniques as described above in this section regarding energy infrastructures may be applied also to ***heating systems, heating and cooling systems, combined heat and power systems, and also water supply systems*** (Werner 2017; Mazher et al. 2018). Application of the *MGF* to these systems as well as to ***sewer-mining systems*** (Makropoulos et al. 2018) was described briefly in Sheremet (2019c). All such special applications will be joined in the near future into an ***integrated application of the multigrammatical framework to the area of resilience and recovery of critical infrastructures***.

Chapter 9
Resource-Based Games

The foundation of any kind of conflict between sociotechnological systems is that both opposing sides possess ***resource bases*** and the objective of each is to eliminate (destroy) and/or capture resources of the counterpart, expending for this purpose some adequate amounts of his own resources. So the logic of the behavior of a competitor at any moment is defined by the current resources he possesses and the possible actions he may perform, expending some part of his own *RB* to eliminate/capture some part of the *RB* of the opponent. A ***conflict is implemented as a chain of actions*** (mutual impacts) of the opponents as long as at least one of the competitors has resources sufficient for the next action. The problem of ***planning the behavior of opponents during a conflict*** is to determine a sequence of actions ("***strategy***") leading to an end of the conflict (i.e., final resource bases of the competitors) which meets the objectives of both or at least one of the opponents, who is declared the winner. Such conflicts are called below *resource-based conflicts* (*RBCs*).

This chapter is dedicated to consideration of techniques applying the multigrammatical framework to representation of games and extraction of appropriate strategies of competitors. It is in some sense a logical continuation of the previous Chap. 8, dedicated to resilience of *STSs* affected by impacts or batches of impacts applied to an *STS* by some external actor.

We shall begin with the most frequently considered games, usually represented by trees and matrices, and demonstrate techniques for their multigrammatical representation. Afterward, we shall introduce a new class of games, named ***resource-based games*** and first announced in Sheremet (2020c), whose representation as a whole is based on the descriptive capabilities of the various classes of multiset grammars.

Let us pay attention to the principal difference between the multigrammatical modeling of games and their classic matrix and tree representations (Bierman and Fernandez 1997; Harrington 2014; Hillier and Lieberman 2014; Jiiang 2010; Naraynasamy et al. 2010; Papayoanou 2010; Roberts 2009). The classical approaches are based on an explicit description of moves and their sequences; this explication, as well as game results corresponding to all such sequences, must be

I. A. Sheremet, *Multigrammatical Framework for Knowledge-Based Digital Economy*, https://doi.org/10.1007/978-3-031-13858-4_9

done by an analyst. To extend this basic paradigm, the so-called repeated games (finite and infinite) were introduced (Benoit and Krishna 1985; Aumann and Maschler 1995; Waston 2013), but these, however, do not cover a lot of real conflicts well. The multigrammatical representation is much more convenient, simple, and, thus, applicable and adequate in a much greater number of practically interesting cases, because an analyst's job in this case is reduced to the creation of a moves base and game objectives, while the game explication according to the semantics of the applied class of multigrammars is done by an inference engine, which generates terminal multisets representing possible results of the game.

9.1 Multigrammatical Representation of Classic Games

As was said above, two basic forms of game representation are trees and matrices. The first are usually applied to a representation of games with ***sequential*** moves, and the second to a representation of games with ***simultaneous*** moves. Let us begin with games with sequential moves.

9.1.1 *Multigrammatical Representation of Classic Games with Sequential Moves*

Let us consider the decision tree of a game (such trees will also be named ***game decision trees*** and ***game trees***) with sequential moves (Fig. 9.1a), where 1 and 2 are moves of the player *A* from the initial state; similarly, possible moves of the player *B*, which may be made by him after a move of player *A*, are marked. Numbers m_{ij} and n_{ij} are ***payoffs*** of players *A* and *B* corresponding to the *i*-th move of *A* and *j*-th move of *B*.

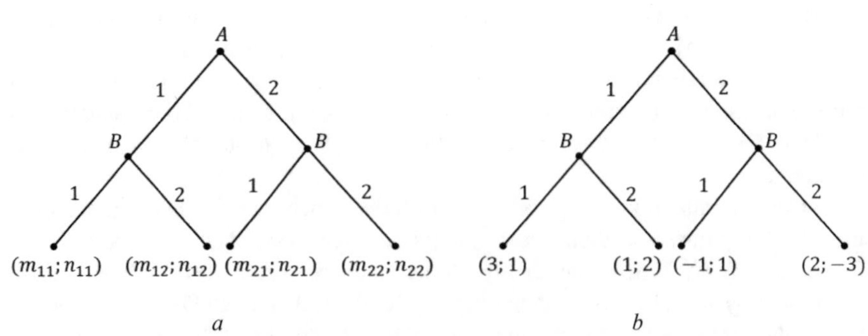

Fig. 9.1 Tree representations of games with sequential moves: (**a**) the general case of a game with two players, each able to make two different moves; (**b**) an example

The tree of the game depicted in Fig. 9.1b may be represented by the multiset grammar $S = \langle v_0, R \rangle$, where the kernel $v_0 = \{1 \cdot A\}$ represents the initial state of the game (the first move is made by the player A) and the scheme R contains the following rules:

$$\left(r_1^A\right) : \{1 \cdot A\} \to \{1 \cdot A_1\}, \tag{9.1}$$

$$\left(r_2^A\right) : \{1 \cdot A\} \to \{1 \cdot A_2\}, \tag{9.2}$$

$$\left(r_{11}^B\right) : \{1 \cdot A_1\} \to \{1 \cdot B_{11}\}, \tag{9.3}$$

$$\left(r_{12}^B\right) : \{1 \cdot A_1\} \to \{1 \cdot B_{12}\}, \tag{9.4}$$

$$\left(r_{21}^B\right) : \{1 \cdot A_2\} \to \{1 \cdot B_{21}\}, \tag{9.5}$$

$$\left(r_{22}^B\right) : \{1 \cdot A_2\} \to \{1 \cdot B_{22}\}, \tag{9.6}$$

$$\left(r_{11}^C\right) : \{1 \cdot B_{11}\} \to \{m_{11} \cdot a,\, n_{11} \cdot b\}, \tag{9.7}$$

$$\left(r_{12}^C\right) : \{1 \cdot B_{12}\} \to \{m_{12} \cdot a,\, n_{12} \cdot b\}, \tag{9.8}$$

$$\left(r_{21}^C\right) : \{1 \cdot B_{21}\} \to \{m_{21} \cdot a,\, n_{21} \cdot b\}, \tag{9.9}$$

$$\left(r_{22}^C\right) : \{1 \cdot B_{22}\} \to \{m_{22} \cdot a,\, n_{22} \cdot b\} \tag{9.10}$$

Here, non-terminal objects A_1 and A_2 represent possible moves of player A, while non-terminal objects B_{11}, B_{12}, B_{21}, and B_{22} are possible moves of player B. A multiobject $m_{ij} \cdot a$ represents the payoff of player A, while an MO $n_{ij} \cdot b$ is the payoff of player B, corresponding to the i-th move of A and the j-th move of B.

As can be seen, any possible course of a game may be represented by a generation chain

$$v_0 \overset{r_i^A}{\Rightarrow} v_1 \overset{r_{ij}^B}{\Rightarrow} v_{ij} \overset{r_{ij}^C}{\Rightarrow} \overline{v}_{ij}, \tag{9.11}$$

where $\overline{v}_{ij} = \{m_{ij} \cdot a,\ n_{ij} \cdot b\}$. ***Moves are made by applying the corresponding rules; every move is implemented by one generation step***.

The set of terminal multisets $\overline{V}_S$ represents the set of all possible results of a game.

Example 9.1

Consider the tree of the game in Fig. 9.1b. This tree may be represented by the *MG* $S = \langle v_0, R \rangle$, where the initial state of the game is represented by a kernel v_0 which is $\{1 \cdot A\}$ and R contains rules (9.1)–(9.6) as well as rules

$$\{1 \cdot B_{11}\} \to \{3 \cdot a,\ 1 \cdot b\},$$
$$\{1 \cdot B_{12}\} \to \{1 \cdot a,\ 2 \cdot b\},$$
$$\{1 \cdot B_{21}\} \to \{-1 \cdot a,\ 1 \cdot b\},$$
$$\{1 \cdot B_{22}\} \to \{2 \cdot a,\ -3 \cdot b\}.$$

Of course, instead of multiset grammars, ***unitary MGs*** might be applied, and then the *UMG* $S = \langle \{1 \cdot A\}, R \rangle$, representing the tree depicted in Fig. 9.1a, would contain the scheme R with the following unitary rules:

$$A \to 1 \cdot A_1\ , \tag{9.12}$$

$$A \to 1 \cdot A_2\ , \tag{9.13}$$

$$A_1 \to 1 \cdot B_{11}, \tag{9.14}$$

$$A_1 \to 1 \cdot B_{12}, \tag{9.15}$$

$$A_2 \to 1 \cdot B_{21}, \tag{9.16}$$

$$A_2 \to 1 \cdot B_{22}, \tag{9.17}$$

$$B_{11} \to m_{11} \cdot a, n_{11} \cdot b, \tag{9.18}$$

$$B_{12} \to m_{12} \cdot a, n_{12} \cdot b, \tag{9.19}$$

$$B_{21} \to m_{21} \cdot a, n_{21} \cdot b, \tag{9.20}$$

$$B_{22} \to m_{22} \cdot a, n_{22} \cdot b. \tag{9.21}$$

The proposed representation of games may be done even more conveniently by joining to the generated multisets information about moves made by the players. For an implementation of this useful feature, it is sufficient to write rules generating terminal multisets in the following form:

$$\{1 \cdot A_w\} \to \{1 \cdot [w], m_w \cdot a, n_w \cdot b\}, \tag{9.22}$$

$$\{1 \cdot B_w\} \to \{1 \cdot [w], m_w \cdot a, n_w \cdot b\}, \tag{9.23}$$

where $[w]$ is an object, "[" and "]" are auxiliary delimiters, and w is a sequence of moves leading to a terminal node (m_w, n_w) of the game decision tree.

As it is easy to see, the described techniques for the multigrammatical representation of game decision trees may be applied to trees with any number of nodes. It is sufficient to use objects with indices that are strings in the alphabet $\{1, 2\}$, so $A_w(B_w)$ where $w = s_1 s_2 \ldots s_k$ is the object representing a node in the game tree which is reachable from the root A of this tree by passing edges $S_i \in \{1, 2\}$, $i = 1, \ldots, k$. If this node is terminal, then the scheme R contains a rule

$$\{1 \cdot B_w\} \to \{m_w \cdot a, n_w \cdot b\}. \tag{9.24}$$

Otherwise, i.e., if the node is non-terminal, R contains two rules, corresponding to two possible moves of a player A (B):

$$\{1 \cdot A_w\} \to \{1 \cdot B_{w1}\}, \tag{9.25}$$

$$\{1 \cdot A_w\} \to \{1 \cdot B_{w2}\}, \tag{9.26}$$

$$\{1 \cdot B_w\} \to \{1 \cdot A_{w1}\}, \tag{9.27}$$

$$\{1 \cdot B_w\} \to \{1 \cdot A_{w2}\}. \tag{9.28}$$

The same techniques may be applied if there are $m > 2$ possible moves at every node and if the number of such moves may vary from node to node.

9.1.2 Multigrammatical Representation of Classic Games with Simultaneous Moves

Games with simultaneous moves are usually represented by matrices. A matrix G representing such a game contains information about possible moves of players A and B and the resulting payoff in such a way that the result of the i-th move of player A ($i = 1, \ldots, k$) and the j-th move of player B ($j = 1, \ldots, l$) is a couple $g_{ij} = \langle m_{ij}, n_{ij} \rangle$, where m_{ij} is the payoff of player A and n_{ij} is the payoff of player B (Fig. 9.2a).

A game matrix may be represented by a multigrammar $S = \langle v_0, R \rangle$, where $v_0 = \{1 \cdot A, 1 \cdot B\}$ and the scheme R contains the following rules:

$$r_1^A : \{1 \cdot A\} \rightarrow \{1 \cdot A_1\},$$
$$r_k^A : \{1 \cdot A\} \rightarrow \{1 \cdot A_k\}, \tag{9.29}$$

$$r_1^B : \{1 \cdot B\} \rightarrow \{1 \cdot B_1\},$$
$$r_l^B : \{1 \cdot B\} \rightarrow \{1 \cdot B_l\}, \tag{9.30}$$

$$r_{11}^{AB} : \{1 \cdot A_1, \, 1 \cdot B_1\} \rightarrow \{m_{11} \cdot a, \, n_{11} \cdot b\},$$
$$r_{ij}^{AB} : \{1 \cdot A_i, \, 1 \cdot B_j\} \rightarrow \{m_{ij} \cdot a, \, n_{ij} \cdot b\},$$
$$r_{kl}^{AB} : \{1 \cdot A_k, \, 1 \cdot B_l\} \rightarrow \{m_{kl} \cdot a, \, n_{kl} \cdot b\}. \tag{9.31}$$

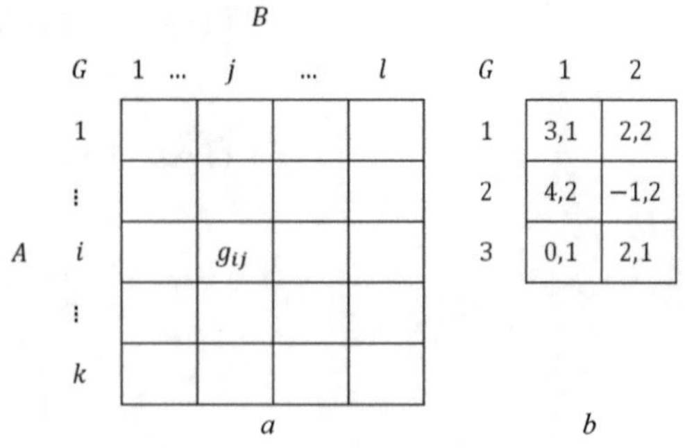

Fig. 9.2 A matrix representation of games with simultaneous moves: (**a**) the general case of a game with two players able to make k and l different moves; (**b**) an example with $k = 3$ and $l = 2$

G	1	2
1	3,1	2,2
2	4,2	−1,2
3	0,1	2,1

Here, non-terminal objects A_i ($i = 1, \ldots, k$) and B_j ($j = 1, \ldots, l$) represent moves of players, while the multiplicities m_{ij} and n_{ij} of terminal objects a and b are their payoffs corresponding to the i-th and the j-th moves.

Every possible course of a game may be represented by a generation chain

$$v_0 \overset{r_i^A}{\Rightarrow} v_i \overset{r_j^B}{\Rightarrow} v_{ij} \overset{r_{ij}^{AB}}{\Rightarrow} \boldsymbol{v}_{ij}, \tag{9.32}$$

where $\boldsymbol{v}_{ij} = \{m_{ij} \cdot a, n_{ij} \cdot b\}$. Because moves are made simultaneously, there may be an alternative generation chain:

$$v_0 \overset{r_j^B}{\Rightarrow} v_j \overset{r_i^A}{\Rightarrow} v_{ij} \overset{r_{ij}^{AB}}{\Rightarrow} \boldsymbol{v}_{ij}. \tag{9.33}$$

As can be seen, both generation chains create one and the same result $\boldsymbol{v}_{ij}$. The set of all possible results is, as above, $\overline{V}_S$.

All the above about games with simultaneous moves is illustrated by the following example.

Example 9.2

Let us consider the game matrix G from Fig. 9.2b. This matrix may be represented by the multiset grammar $S = \langle v_0, R \rangle$, whose kernel is $v_0 = \{1 \cdot A, 1 \cdot B\}$ and the scheme R contains the following rules:

$$\begin{gathered}
\{1 \cdot A\} \to \{1 \cdot A_1\}, \\
\{1 \cdot A\} \to \{1 \cdot A_2\}, \\
\{1 \cdot A\} \to \{1 \cdot A_3\}, \\
\{1 \cdot B\} \to \{1 \cdot B_1\}, \\
\{1 \cdot B\} \to \{1 \cdot B_2\}, \\
\{1 \cdot A_1, 1 \cdot B_1\} \to \{3 \cdot a,\ 1 \cdot b\}, \\
\{1 \cdot A_1, 1 \cdot B_2\} \to \{2 \cdot a,\ 2 \cdot b\}, \\
\{1 \cdot A_2, 1 \cdot B_1\} \to \{4 \cdot a,\ 2 \cdot b\}, \\
\{1 \cdot A_2, 1 \cdot B_2\} \to \{-1 \cdot a,\ 2 \cdot b\}, \\
\{1 \cdot A_3, 1 \cdot B_1\} \to \{1 \cdot b_1\}, \\
\{1 \cdot A_3, 1 \cdot B_2\} \to \{2 \cdot a,\ 1 \cdot b_1\}.
\end{gathered}$$

Every player participating in a game has his own objective—maximizing or minimizing his payoff. In classic games, this may be represented by a ***filtering multiset grammar*** $S = \langle v_0, R, F \rangle$, where v_0 and R are as above, while the ***filter*** F contains two optimizing conditions:

$$F = \{a = opt_A,\ b = opt_B\}, \tag{9.34}$$

where $\{opt_A, opt_B\} \subseteq \{max, min\}$. This technique may be applied to games with simultaneous moves as well as to games with sequential moves.

Example 9.3
Consider the game from the previous Example 9.2. Let the objective of each player be to maximize his own payoff. This may be represented by the *FMG* $S = \langle v_0, R, F \rangle$, where the kernel v_0 and the scheme R are as above, while the filter is

$$F = \{a = \ max\ ,\ b = \ max\ \}.$$

As can be seen,

$$\overline{V}_s = \{\{4 \cdot a,\ 2 \cdot b\}\},$$

which corresponds to the element g_{21} of the game matrix G, i.e., choice of the second move by player A and the first move by player B.

By analogy with games with sequential moves, we may extend the generated multisets by multiobjects representing the moves made by both players, in such a way that any terminal multiset contains information about the origin of the result of the game. For this purpose, we may use rules like

$$\{1 \cdot A_i,\ 1 \cdot B_i\} \to \{i \cdot \overline{A},\ j \cdot \overline{B},\ m_{ij} \cdot a,\ n_{ij} \cdot b\} \tag{9.35}$$

instead of (9.31); here, $i \cdot \overline{A}$ means player A makes the i-th move, while $j \cdot \overline{B}$ means player B makes the j-th move.

We shall not consider ***classic games in the multigrammatical representation*** any further, leaving this for future publications, and shall describe now a ***new MGF-based class of games*** whose modeling essentially involves new descriptional capabilities of multiset grammars compared to conventional tools like trees and matrices. These new features make such models much more useful for the solution of various practical problems.

9.2 Resource-Based Antagonistic Games

We shall begin with the simplest kind of resource-based games, where two participating sides (players A and B), each holding his resource base, make their moves sequentially, one after another.

Each move ("***impact***") results in the extraction (***elimination***) of some amounts of resources from the *RB* of the moving player as well as from the *RB* of his opponent, i.e., to reduce the opponent's resource base, a player must expend some part of his own *RB*. This situation is typical for conflicts of various kinds. Along with a resource base, each player has a ***moves base*** (***MB***), i.e., the set of his possible moves, and ***each move may be made only if his resource base contains all resources necessary for this move***. Also, each player may stop his activity, and after that, or when the next move cannot be made due to a lack of resources, the game is terminated. So a ***game continues as long as a next move is made***.

The ***game result is the players' resource bases at the moment when the game is stopped***. It is postulated that both players have full information about the *RB* and the *MB* of their opponents. Along with an *RB* and an *MB*, each player at the beginning of the game formulates his objective, i.e., what conditions should be satisfied by the resource bases of both players at the end of the game. Any player participating in a game wishes to achieve his objective. If there exists a final state of a game such that both players' objectives are achieved, the game is called ***bi-winner***; if only one player's goal is reachable, ***one-winner***; otherwise, ***no-winner***, or *empty*. According to the conceptual background of classical game theory, the described game may be classed as ***antagonistic*** and ***with sequential moves***.

Let us consider a multigrammatical representation of the above-described resource-based games.

9.2.1 Primary Definitions

First of all, we shall fix a structure of lexemes, denoting object-resources, following their syntax introduced in Chaps. 2, 3, and 7 and used for the representation of ***economical systems***; namely, the aforementioned lexemes are composite objects of the form $(a : x)$, where a is the name of an object-resource, $x \in \{A, B\}$ is the name of the holder (owner) of this *OR*, and "(", ")", and ":" are dividers. An *OR* holder participating a game is a player. So, similar to Chap. 7, a multiobject $n \cdot (a : x)$ describes the fact that n units of object-resources a belong to (are in ownership of) a player x. Thus, the initial resource base of a player A is a multiset

$$v_A = \left\{n_1^A \cdot (a_1 : A),\ \ldots,\ n_m^A \cdot (a_m : A)\right\}, \tag{9.36}$$

while the initial *RB* of a player B is a multiset

$$v_B = \left\{n_1^B \cdot (\,a'_1 : B),\ \ldots,\ n_l^B \cdot (\,a'_l : B)\right\}. \tag{9.37}$$

The multiset

$$v_{A,B} = v_A \cup v_B \tag{9.38}$$

will be called the ***initial resource base of the game***.

For a representation of moves, we shall use rules of the following form:

$$\{1 \cdot x, l_1 \cdot (a_{i_1} : x), \ldots, l_k \cdot (a_{i_k} : x)\}$$
$$\left\{1 \cdot y, -m_1 \cdot (a_{j_1} : y), \ldots, -m_g \cdot \left(a_{j_g} : y\right)\right\}, \tag{9.39}$$

where $x, y \in \{A, B\}$ and $x \neq y$. According to the semantics of *FMGs*, application of rule (9.39) is possible if the left part of this rule is a submultiset of the current multiset $\overline{v}$ generated at the previous steps.

A new and a principal thing in (9.39) is ***negative multiplicities*** $-m_1, \ldots, -m_g$, introduced in Sect. 3.5.2, which eliminate the corresponding amounts of object-resources from the *RB* of the opposing player y.

Example 9.4
Let the initial resource base of the game be

$$\overline{v} = \{1 \cdot A, 3 \cdot (eur : A), 4 \cdot (usd : A), 1 \cdot (eur : B), 3 \cdot (usd : B)\},$$

and the move represented by the rule r be

$$\{1 \cdot A, 2 \cdot (eur : A), 2 \cdot (usd : A)\} \rightarrow \{-3 \cdot (eur : B), 1 \cdot B\}.$$

According to (3.151)–(3.155),

$$v'_+ = \{1 \cdot B\},$$
$$v'_- = \{3 \cdot (eur : B)\},$$

and the result of the application of the rule r to the multiset $\overline{v}$ is the *MS*

$$\overline{v}' = \{1 \cdot (eur : A), 2 \cdot (usd : A), 1 \cdot (eur : B), 3 \cdot (usd : B)\} + \{1 \cdot B\} - \{3 \cdot (eur : B)\}$$
$$= \{1 \cdot B, 1 \cdot (eur : A), 2 \cdot (usd : A), 3 \cdot (usd : B)\}.$$

Now, it is clear that the result of an application of (9.39) is the appearance of a multiobject $1 \cdot y$ whose presence in the generated multiset enables the next move of player $y \neq x$. As can be seen, this form of rules exactly corresponds to the logic and dynamics of the considered games verbally described above.

9.2.2 *Multigrammatical Representation of Antagonistic RBGs*

Let us construct a filtering multigrammar representing a game in full accordance with its verbal description.

This *FMG* $S = \langle v_0, R, F\rangle$ is such that a multiset

$$v_0 = \{1 \cdot A\} + v_{A,B}, \tag{9.40}$$

is the kernel corresponding to the initial state of a game which starts with player's A move with the initial resource base of a game, defined by (9.36)–(9.38). The scheme R is the join of the moves bases of both players:

$$R = R_A \cup R_B, \tag{9.41}$$

where

$$R_A = \{r | r \in R\, has\, the\, form\, \{1 \cdot A\} + v \rightarrow v'\}, \tag{9.42}$$

$$R_B = \{r | r \in R\, has\, the\, form\, \{1 \cdot B\} + v \rightarrow v'\}, \tag{9.43}$$

and

$$F = F_A \cup F_B \tag{9.44}$$

is the filter which is the join of subfilters F_A and F_B, representing the objectives of the two players participating in the game. The filter F selects all terminal multisets which satisfy both players' objectives. These *TMSs* form a set $\overline{V}_S$, representing all possible results of the game.

Let $S_A = < v_0, R, F_A>$, $S_B = < v_0, R, F_B>$. Then, there are four possible results of the game:

1. If $\overline{V}_{S_A} = \{\varnothing\}$ and $\overline{V}_{S_B} = \{\varnothing\}$, then the game is ***no-winner*** (empty).
   If $\overline{V}_{S_A} \neq \{\varnothing\}$ and $\overline{V}_{S_B} = \{\varnothing\}$, then the winner is **player *A***.
   If $\overline{V}_{S_A} = \{\varnothing\}$ and $\overline{V}_{S_B} \neq \{\varnothing\}$, then the winner is **player *B***.
2. If $\overline{V}_{S_A} \neq \{\varnothing\}$ and $\overline{V}_{S_B} \neq \{\varnothing\}$, and, hence, $\overline{V}_S \neq \{\varnothing\}$, then the game is ***bi-winner***.

Every holder h may propose to another one several ways of performing an exchange, and there may be several holders $h'_1, \ldots, h'_k$ to which h may make his proposals. Every holder may (or may not) declare the objective of his participation in an exchange; this objective is represented by a set of boundary and optimization conditions $(a{:}h)\ \theta\ n$ and $a{:}h = opt$, respectively, where $\theta \in \{<, >, \leq, \geq, =\}$, $opt \in \{min, max\}$. Similar to the case of direct exchange economical systems, if all conditions declaring an objective of any player (holder) $h \in \{A, B\}$ have the form a: $h\ \theta\ n$ and/or $a{:}h = opt$ (i.e., concern amounts of his own resources), such a game (as a direct exchange *ES*) is called ***self-objective for a player (holder)*** h; if the game is self-objective for all participating players (holders), it is called ***self-objective;*** otherwise (i.e., if the objective of at least one player h includes some conditions on the amounts of resources of another player), such a game is called ***cross-objective***.

Table 9.1 Possible moves of the player *A*

No.	Spent by *A*	Eliminated from *B*
1	$2 \cdot a$, $3 \cdot b$	$1 \cdot a$, $1 \cdot d$
2	$1 \cdot a$, $3 \cdot b$	$2 \cdot b$
3	$1 \cdot a$, $2 \cdot b$, $3 \cdot c$	$1 \cdot b$, $4 \cdot d$

Table 9.2 Possible moves of the player *B*

No.	Spent by *B*	Eliminated from *A*
1	$3 \cdot a$, $2 \cdot b$	$2 \cdot a$, $1 \cdot b$
2	$1 \cdot b$, $1 \cdot c$, $4 \cdot d$	$4 \cdot c$

As can be seen, there are obvious common features between ***resource-based games*** and ***direct exchange economical systems***. In fact, moves made by players of an *RBG* implement some kind of exchange between them; the only difference is that exchange does not presume irrevocable elimination of *ORs* belonging to another holder—there is not elimination but circulation of *ORs* among holders. We shall refer to common and different features of ***economical systems of various classes and RBGs*** below more than once.

Let us illustrate the described technique of *RBG* multiset modeling by the following example.

Example 9.5
Let the initial resource base of the player *A* be $\{4 \cdot a, 8 \cdot b, 9 \cdot c\}$, while the *RB* of the player *B* is $\{6 \cdot a, 10 \cdot b, 3 \cdot c, 12 \cdot d\}$, and their capabilities (possible moves resulting in mutual impacts) are presented in the following tables.

Player *A's* objective is that, after the game stops, his resource base should contain no fewer than two objects *a*, no fewer than four objects *b*, and the maximal possible number of objects *c*, while the resource base of the player *B* should contain the minimal possible number of objects *d*, as well as no more than two objects *b*. Player *B's* goal is that, after the game stops, his resource base should contain no fewer than three objects *a* and no fewer than five objects *b,* while the resource base of player *A* should contain the minimal possible number of objects *c* and no more than three objects *b*.

The game, in which the first move is made by player *A*, may be represented by the filtering multigrammar $S = \langle v_0, R, F \rangle$, where

$$v_0 = \{4 \cdot (a : A), 8 \cdot (b : A), 9 \cdot (c : A), 6 \cdot (a : B), 10 \cdot (b : B),\ 3 \cdot (c : B),\ 12 \cdot (d : B),\ 1 \cdot A\}.$$

According to Tables 9.1 and 9.2, the scheme *R* will contain the following rules marked r_i^A and r_j^B:

$$r_1^A : \{1 \cdot A,\ 2 \cdot (a : A),\ 3 \cdot (b : A)\} \to \{-1 \cdot (a : B),\ -1 \cdot (d : B),\ 1 \cdot B\},$$

$$r_2^A : \{1 \cdot A,\, 1 \cdot (a:A),\, 3 \cdot (b:A)\} \to \{-2 \cdot (b:B),\, 1 \cdot B\},$$

$$r_3^A : \{1 \cdot A,\, 1 \cdot (a:A),\, 2 \cdot (b:A),\, 3 \cdot (c:A)\} \to \{-1 \cdot (b:B),\; -4 \cdot (d:B),\, 1 \cdot B\},$$

$$r_1^B : \{1 \cdot B,\, 3 \cdot (a:B),\, 2 \cdot (b:B)\} \to \{-2 \cdot (a:A),\; -1 \cdot (b:A),\, 1 \cdot A\},$$

$$r_2^B : \{1 \cdot B,\, 1 \cdot (b:B),\, 1 \cdot (c:B),\, 4 \cdot (d:B)\,\} \to \{-4 \cdot (c:A),\, 1 \cdot A\}.$$

The filter F corresponding to the above-described objectives of both players will contain the following conditions:

$$C_1^A : (a:A) \geq 2,$$

$$C_2^A : (b:A) \geq 4,$$

$$C_3^A : (c:A) = max\,,$$

$$C_4^A : (d:B) = min\,,$$

$$C_5^A : (b:B) \leq 2,$$

$$C_1^B : (a:B) \geq 3,$$

$$C_2^B : (b:B) \geq 5,$$

$$C_3^B : (c:A) = min\,,$$

$$C_4^B : (b:A) \leq 3.$$

As can be seen, the subfilter $F_A = \{C_1^A,\ \ldots,\ C_5^A\}$ defines the objective of player A, while the subfilter $F_B = \{C_1^B,\ \ldots,\ C_4^B\}$ defines the objective of player B.

Let us apply R to v_0. To simplify the representation of generation chains, we shall use a table whose columns correspond to objects occurring in generated multisets, each of which is represented by a row of this table. A number n appearing at the intersection of a column (x, X) and a row v_i is the multiplicity of object $(x : X)$ in the multiset v_i, i.e., $n \cdot (x : X) \in v_i$. Each row representing a multiset v_i is followed by a row representing a rule applied to v_i, in the form of the multiplicities of objects occurring in this rule; multiplicities which are in the left part of the rule are negative numbers, while multiplicities from the right part are represented by positive numbers. The table corresponds to one generation chain, and the two lowest rows represent the subfilters F_A and F_B in such a way that each boundary condition $(x : X)\theta n$ is represented by θn and each optimizing condition $(x : X) = opt$ is represented by opt, both of them appearing in the column (x, X).

The generation chain $v_0 \overset{r_1^A}{\Rightarrow} v_1 \overset{r_1^B}{\Rightarrow} v_2$ is represented by Table 9.3; as may be seen, the terminal multiset v_2 does not satisfy the filter F. We shall not continue the generation receiving it to the interested reader. As may be seen without any generation, this game is certainly not bi-winner because some of the conditions in the filter

Table 9.3 Generation chain

	A			B					
$\frac{v}{r}$	a	b	c	a	b	c	d	A	B
v_0	4	8	9	6	10	3	12	1	
r_1^A	−2	−3		−1			−1	−1	1
v_1	2	5	9	5	10	3	11		1
r_1^B	−2	−1		−3	−2			1	−1
v_2		4	9	2	8	3	11	1	
F_A	≥2	≥4	*max*		≤2		*min*		
F_B		≤3	*min*	≥3	≥5				

are contradictory (e.g., $(b : A) \geq 4 \in F_A$ and $(b : A) \leq 3 \in F_B$), i.e., no terminal multiset generated by the application of the scheme R to the kernel v_0 will satisfy both of them.

Let us consider some extensions of the proposed basic techniques for multigrammatical modeling of resource-based games.

9.2.3 *Multigrammatical Representation of Extensions of Antagonistic* RBGs

A more complicated case includes not only the elimination of resources of the opposing player by expending a player's own resources but also a change of their ownership ("their ***capturing***"). Such moves may be represented by rules having the following form:

$$\{1 \cdot x,\ l_1 \cdot (a_{i_1} : x),\ \ldots,\ l_k \cdot (a_{i_k} : x),$$
$$m_1 \cdot (a_{j_1} : y),\ \ldots, m_q \cdot \left(a_{j_q} : y\right)\}$$
$$\left\{1 \cdot y, m_1 \cdot (a_{j_1} : x),\ \ldots, m_q \cdot \left(a_{j_q} : x\right)\right\}. \tag{9.45}$$

As can be seen, a player x, expending l_1 object-resources $a_{i_1}, \ldots, l_k$ object-resources a_{i_k}, removes m_1 object-resources $a_{j_1}, \ldots, m_p$ object-resources a_{i_p} to his ownership from the opposing player. Unlike the previous case (9.39), where the result of every move was the elimination of object-resources from the resource bases of both players, here resources remain usable, but their holder becomes a different player. (Note that according to the mathematical semantics of *MGs*, rule (9.45) will be applied only if all removed resources in the necessary amounts are present in player's y ownership.) In this case, we have the same situation, in fact, ***as in direct exchange ESs***.

Example 9.6
Let us join to the moves base (i.e., the scheme R) of Example 9.5 two new moves (rules) allowing capturing of resources of the opposing player:

$$r_4^A : \{1 \cdot A, 3 \cdot (a : A), 1 \cdot (c : A), 1 \cdot (b : B), 1 \cdot (d : B)\}$$
$$\rightarrow \{1 \cdot B, 1 \cdot (b : A), 1 \cdot (d : A)\},$$

$$r_3^B : \{1 \cdot B, 1 \cdot (a : B), 3 \cdot (d : B), 1 \cdot (a : A), 2 \cdot (b : A)\}$$
$$\rightarrow \{1 \cdot A, 1 \cdot (a : B), 2 \cdot (b : B)\}.$$

As can be seen, an application of the rule r_4^A results in the capture of one object-resource b and one object-resource d, belonging to the player B, by the player A, who expends for this purpose three object-resources a and one object-resource c. Similarly, an application of the rule r_3^B causes the removal of one object-resource a and two object-resources b from A to B, the latter expending for this action one object-resource a and three object-resources d.

A more general case is a combination of the two previous cases, when one move causes both elimination (destruction) of some resources of the opposing player and capturing of other resources, which remain usable. Such "combined" moves have the following form:

$$\left\{1 \cdot x, l_1 \cdot (a_{i_1} : x), \ldots, l_k \cdot (a_{i_k} : x), m_1 \cdot (a_{j_1} : y), \ldots, m_p \cdot \left(a_{j_p} : y\right)\right\}$$
$$\left\{1 \cdot y, -l_1' \cdot \left(a_{h_1}' : y\right), \ldots, -l_q' \cdot \left(a_{h_q}' : y\right), m_1 \cdot (a_{j_1} : x), \ldots, m_p \cdot \left(a_{j_p} : x\right)\right\}. \tag{9.46}$$

As can be seen, this move causes the elimination of l'_1 object-resources $a'_{h_1}, \ldots,$ l'_q object-resources a'_{h_q}, belonging to a player y, and the removal of m_1 object-resources $a_{j_1}, \ldots, m_p$ object-resources a_{j_p} from his ownership to a player x, who expends for this purpose l_1 object-resources $a_{i_1}, \ldots, l_k$ object-resources a_{i_k}. Note that (9.46) is applicable only if the resource base of player y before the move of player x includes no fewer than l'_1 object-resources $a'_{h_1}, \ldots, l'_q$ object-resources a'_{h_q}. Otherwise, the operation cannot be implemented.

The next level of complexity presumes that some player is capable of producing new object-resources from those already in his ownership and afterward using the produced *ORs* during further steps of the game (conflict). As can be seen, in this case, players are nothing but ***interacting producing economical systems*** (however, with ***zero duration*** of *OR* manufacturing by *MDs*).

To implement this feature, we shall introduce so-called internal moves, which are defined by rules containing a "control" multiobject $1 \cdot x$ in both parts of a rule:

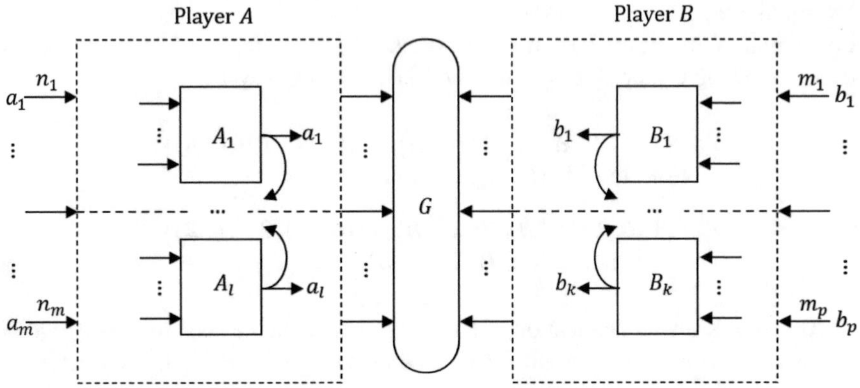

Fig. 9.3 A "black box" representation of a game in which players are producing economical systems

$$\{1 \cdot x, l_1 \cdot (a_{i_1} : x), \ \ldots, l_k \cdot (a_{i_k} : x)\}$$
$$\{1 \cdot x, l'_1 \cdot (a'_{i_1} : x), \ \ldots, l'_q \cdot (a'_{i_q} : x)\}. \tag{9.47}$$

As can be seen, after an application of this rule, player x remains active and will make the next move and so on until a rule with a multiobject $1 \cdot y$ instead of $1 \cdot x$ in the right part is applied. Moves described by rules like (9.47) will be called ***internal***.

Let us now describe how a ***player's producing capabilities may be defined by means of the technique of internal moves***. For this purpose, we shall apply an intermediate "black box" representation, similar to the one we used in Sect. 6.1, where we considered a *UMG*-based representation of industrial systems. Namely, we shall assume that a ***player x possesses a set of manufacturing devices***, each represented by a "black box" with $j \geq 1$ inputs and one output, as represented in Fig. 9.3, where the bifurcating output arrows meanthat produced *ORs* are loaded into the resource base of the player (as above, this *RB* is a part of the *RB* of the game) and afterward interaction between the players may be implemented via the *RB*. So the graphical representation shown in Fig. 9.3 defines that each object, produced by means of any manufacturing device, may be used in a "manufacturing (producing) chain," allowing the creation of new objects. So, the whole manufacturing technological base of a player x may be represented by a set of rules like

$$\{1 \cdot x, n_1 \cdot a_1, \ \ldots, n_m \cdot a_m\} \rightarrow \{1 \cdot x, 1 \cdot a\}, \tag{9.48}$$

each corresponding to an *MD*.

Some produced objects may be used in moves inflicting impacts on an opposing player, along with objects used in these moves directly. A graphical representation of such a game appears in Fig. 9.3, where the players' interaction is implemented through the depicted ellipse *G* (*"Game"*).

In the general case, all or some "black boxes" may have not one but $i > 1$ outputs, and, thus, the corresponding rule has the following form:

$$\{1 \cdot x, n_1 \cdot a_1, \ldots, n_m \cdot a_m\} \to \{1 \cdot x, n'_1 \cdot a'_1, .., n'_q \cdot a'_q\}. \tag{9.49}$$

Let us illustrate the introduced generalization by the following example.

Example 9.7
Let us join to the moves base (scheme R) obtained in the previous Example 9.6 two new moves (rules) manufacturing object-resources which may be used to inflict impacts:

$$r_5^A : \{1 \cdot A, 1 \cdot (a : A), 2 \cdot (c : A)\} \to \{1 \cdot A, 1 \cdot (b : A)\},$$

$$r_4^B : \{1 \cdot B, 1 \cdot (a : B), 3 \cdot (d : B)\} \to \{1 \cdot B, 1 \cdot (c : B)\}.$$

After these rules appear in the scheme R, generation chains (i.e., move sequences) will implement manufacturing operations allowing the inclusion in player $A's$ resource base of additional object-resources b, each produced by expending one *OR* a and two *ORs* c, as well as inclusion in player $B's$ resource base of additional *ORs* c, each produced by expending one *OR* a and three *ORs* d.

Let us recall that, as in the previous chapters, *ORs* expended to manufacture any new *OR* may be its spare parts or consumables (electrical energy, fuels, etc.) necessary for the operation of the manufacturing device. Among such object-resources, there must be an "***active state of the MD***," represented by a special multiobject $1 \cdot e$, where e is the name of this device. Such *ORs*, like all other *ORs*, may be present in a resource base, eliminated (destroyed) or captured as a result of some moves of the opposing side. This technique provides a simple representation of actions causing destructive impacts ***not only on the resource base but also on the manufacturing technological base of the opponent***.

All the above concerned ***antagonistic RBGs with sequential moves***.

Simultaneous moves, made by players synchronously, may be represented by rules of the following form:

$$\left\{\begin{array}{l} 1 \cdot x, 1 \cdot y, \\ l_1 \cdot (a_{i_1} : x), \ldots, l_k \cdot (a_{i_k} : x), \\ m_1 \cdot (b_{j_1} : y), \ldots, m_q \cdot \left(b_{j_q} : y\right) \end{array}\right\} \to \left\{\begin{array}{l} 1 \cdot x, 1 \cdot y, \\ l'_1 \cdot (a_{h_1} : x), \ldots, l'_{k'} \cdot \left(a_{h_{k'}} : x\right), \\ m'_1 \cdot (b_{g_1} : y), \ldots, m'_{q'} \cdot \left(b_{g_{q'}} : y\right) \end{array}\right\} \tag{9.50}$$

Taking into account (9.50), the kernel of the corresponding multigrammar in this case will be

$$v_0 = \{1 \cdot A,\ 1 \cdot B\} + v_A + v_B, \tag{9.51}$$

where v_A and v_B are the initial resource bases of both players. As above, the objectives of both players in this game will be represented by subfilters F_A and F_B, joined to make the filter $F = F_A \cup F_B$ of the corresponding filtering multigrammar.

Along with simultaneous moves made by both players, there may be internal moves of the same form as (9.48)–(9.49). By these individual moves, each player may prepare the next simultaneous move, made by both of them. All that was said above in this section about internal moves is applicable to the considered class of *RBGs*. As may be seen, a multigrammatical representation allows easy use of both sequential and simultaneous moves in one game.

As is known from game theory, there are also ***cooperative games***, where players interact without causing mutual damage, but, on the contrary, help one another in order to achieve their objectives. This class of *RBGs* is considered in the following section.

9.3 Cooperative and Coalitional Resource-Based Games

For simplicity, we shall begin with cooperative *RBGs*.

9.3.1 Cooperative RBGs

Cooperation means that there may be an exchange of resources between players or one of them may deliver his own object-resources to the other without any compensation. Also, of course, any player may produce new object-resources from his own *ORs* and also from *ORs* belonging to the second player. Such mutual actions are often performed in business and other areas of human activity whose internal logic presumes not only confrontation but also joining of efforts. Such games, in fact, are conceptually and formally very close to direct exchange economical systems.

A ***resource exchange*** may be represented by rules of the following form (in a simultaneous moves version):

$$\left\{\begin{matrix} 1\cdot x,\ 1\cdot y, \\ l_1\cdot(a_{i_1}:x),\ \ldots,\ l_k\cdot(a_{i_k}:x), \\ m_1\cdot(b_{j_1}:y),\ \ldots,\ m_q\cdot\left(b_{j_q}:y\right) \end{matrix}\right\}$$

$$\left\{\begin{matrix} 1\cdot x,\ 1\cdot y, \\ m_1\cdot(b_{j_1}:x),\ \ldots,\ m_q\cdot\left(b_{j_q}:x\right), \\ l_1\cdot(a_{i_1}:y),\ \ldots,\ l_k\cdot(a_{i_k}:y) \end{matrix}\right\}. \tag{9.52}$$

As may be seen, an application of this rule results in a transfer to a player's y property of l_1 object-resources a_{i_1},..., l_k object-resources a_{i_k}, all belonging to a player x. In return for the transferred resources, player x will get m_1 object-resources b_{j_1},..., m_q object-resources b_{j_q}, which until the exchange belonged to player y. In this case, players interact as in an exchange economical system.

In the "***altruistic***" case, when one player does not demand any compensation from his partner, a corresponding rule is as follows:

$$\{1\cdot x,\ 1\cdot y,\ l_1\cdot(a_{i_1}:x),\ \ldots,\ l_k\cdot(a_{i_k}:x)\}$$
$$\{1\cdot x,\ 1\cdot y,\ l_1\cdot(a_{i_1}:y),\ \ldots,\ l_k\cdot(a_{i_k}:y)\}. \tag{9.53}$$

Cooperation in a producing (manufacturing) process presumes the possible use of object-resources that belong to one player for the production of object-resources which, once manufactured, may belong to another player.

The structure of rules representing the producing (manufacturing) capabilities of players is the same as (9.48) and (9.49), but, for generality, ***multiobjects are extended by information about ownership***, which, as was said above, may be by either side. Such rules, implementing internal moves of players and in the simplest case corresponding to (9.48), may be as follows:

$$\{1\cdot x,\ n_1\cdot(a_{i_1}:x_1),\ \ldots,\ n_m\cdot(a_{i_m}:x_m)\}\to\{1\cdot x,\ 1\cdot(a:x_{m+1})\}. \tag{9.54}$$

This means that a player $x\in\{A,B\}$, making his move, produces one unit of an object-resource a, which belongs to a player $x_{m+1}\in\{A,B\}$, applying for this purpose n_1 object-resources a_{i_1} belonging to a player $x_1\in\{A,B\}$, ..., n_m object-resources a_{i_m} belonging to a player $x_m\in\{A,B\}$.

Of course, there may be a multiobject $1\cdot y$, where $y\neq x$, in the right part of rule (9.54). In this case, the move is ***not internal***, because the next move is made by another player.

So, finally, a filtering multiset grammar $S=\langle v_0,R,F\rangle$, whose kernel v_0 is the initial resource base joined with "starting" multiobject(s), whose scheme R contains rules like (9.52)–(9.54), and whose filter F contains conditions defining the objectives of both players, is a ***representation of a cooperative game with two players***.

9.3.2 *Multi-player Cooperative* RBGs *and Coalitional* RBGs

A further generalization of cooperative games with two players is ***cooperative games with an arbitrary number of players*** $m \geq 2$ (***multi-player cooperative games***).

The above-described technique for multigrammatical representation of games with two players allows a simple and natural description of ***games with any number of players***. It is sufficient to use multiobjects of the form $1 \cdot x$, where x is the name of a player, in any combinations, in the right and the left parts of rules. Such games may be easily represented by multigrammars whose schemes contain rules of the form

$$\{1 \cdot x_1, \ \ldots, \ 1 \cdot x_m, \ l_1 \cdot (a_{i_1} : x_{i_1}), \ \ldots, \ l_k \cdot (a_{i_k} : x_{i_k})\}$$
$$\left\{1 \cdot x_1, \ \ldots, \ 1 \cdot x_m, \ l'_1 \cdot \left(a_{j_1} : x_{j_1}\right), \ \ldots, \ l'_{k'} \cdot \left(a_{j_{k'}} : x_{j_{k'}}\right)\right\} \tag{9.55}$$

and the kernel is

$$\{1 \cdot x_1, \ \ldots, \ 1 \cdot x_m\} + v_1 + \ldots + v_m, \tag{9.56}$$

where x_i is the name of the i-th player and v_i is his initial resource base. As can be seen, rules like (9.55) represent cooperation of players in order to create a resource base which satisfies a filter

$$F = F_1 \cup \ldots \cup F_m, \tag{9.57}$$

where F_i is the subfilter, defining the objective of the i-th player. Of course, along with rules (9.55), there may be rules representing internal moves of any player, but also there may be moves made simultaneously by any subset of players $\left\{x_{i_1}, \ \ldots, x_{i_q}\right\} \subseteq \{x_1, \ \ldots, x_m\}$. This kind of cooperation may be implemented not only in the case when players help one another but also in the case when some of them join together in order to oppose another subset. Following the terminology of classical game theory, such stable sets (groups) of players, synchronizing their moves and matching their objectives, may be called ***coalitions***, and such games ***coalitional resource-based games***. As may be seen, these games are antagonistic, regarding opposing coalitions, but cooperative inside any coalition.

A multigrammatical representation of games makes it easy to describe the logic of behavior of any number of coalitions and the players belonging to them.

9.4 Distributed Resource-Based Games

A further generalization of the described multigrammatical technique for games modeling concerns the representation and application of players' locations.

Until now, we have assumed that a game takes place in one compact location, so the transportation of object-resources from one geographical point to another one is unnecessary. However, logistical issues are important when resources of both sides are distributed in space, and their elimination as well as capture is possible only when they are in some "predefined locations" relatively close or far from one another. Such games will be called ***distributed resource-based games*** (***DRBGs***).

9.4.1 *Antagonistic Distributed* RBGs *with Sequential Moves*

To cover the aforementioned logistical issues, we shall apply the same structure of lexemes in rules that was introduced above in Chaps. 6 and 7 for distributed industrial and economical systems. As above, we shall use a set of locations (positions) Z, where object-resources of any of the players may be present, and each such location will be denoted by $z \in Z$. (In the general case, z may be defined as a point in some coordinate system or as an area of any form—circle, quadrilateral, ellipse, or any shape defined by boundary points.) Below, we shall use a construction $(a : x, z)$ to denote an object-resource a existing at a location z and belonging to a player x. So a multiobject $n \cdot (a : x, z)$ denotes n *ORs* a located at z that are the property of x. Let us describe now how this new structure permits the representation of distributed resource-based games.

First of all, there is the possibility for every player to transport object-resources between locations, which may be represented by respective internal moves. Transportation of n object-resources that are the property of a player x, from a location z to a location z', expending for this purpose m of his object-resources b located at the same point (area) z, may be represented by the following rule:

$$\{1 \cdot x, n \cdot (a : x,\ z), m \cdot (b : x,\ z)\} \rightarrow \{1 \cdot x, n \cdot (a : x,\ z')\}. \tag{9.58}$$

There are many more complicated techniques for the representation of various ways of relocation of resources and the associated consumption. For example, if relocation may be implemented by another party x', who obtains payment for this work from player x of k object-resources c, the rule representing this possibility would have the following form:

$$\{1 \cdot x, n \cdot (a : x,\ z), k \cdot (c : x,\ z)\} \rightarrow \{1 \cdot x, n \cdot (a : x,\ z'), k \cdot (c : x',\ z)\}. \tag{9.59}$$

As can be seen, payment physically remains at the same location z of player x, but becomes the property of x'.

Let us note that this technique is quite general and can be applied for the representation of ***transportation infrastructure***, transportation vehicles, and their motion, allowing the relocation of material resources (Sheremet 2019c).

Table 9.4 Capabilities of the player A

No.	Consumed by A	Result
1	$1 \cdot (a : A; z_1)$, $1 \cdot (b : A; z_2)$	$1 \cdot (a : A; z_2)$ relocation
2	$2 \cdot (a : A; z_2)$, $2 \cdot (b : A; z_2)$	$1 \cdot (c : B; z_4)$ elimination from B
3	$2 \cdot (a : A; z_2)$, $1 \cdot (b : A; z_2)$	$1 \cdot (a : B; z_3)$ elimination from B

Secondly, the new structure allows a new form of representation of mutual impacts. Namely, impacts now may be represented in the following way:

$$\{1 \cdot x,\, l_1 \cdot (a_{i_1} : x,\ z_{i_1}),\ \ldots,\, l_k \cdot (a_{i_k} : x,\ z_{i_k})\}$$

$$\left\{1 \cdot y,\ -m_1 \cdot \left(a_{j_1} : y,\ z_{j_1}\right),\ \ldots,\ -m_g \cdot \left(a_{j_g} : y,\ z_{j_g}\right)\right\}. \tag{9.60}$$

This is an obvious modification of the initial definition (9.39). Note that there are no obligatory interconnections between locations $z_{i_1}, \ldots, z_{i_k}, \ldots, z_{j_1}, \ldots, z_{j_g}$; the real capabilities of a player x to eliminate object-resources of an opposing player y depend on what kinds of tools are applied for this purpose. (In warfare, with application of weapons operating in a physical space, these capabilities require the presence of the impacting tool within a distance that is no more than the weapon's firing range; in cyberwarfare, due to the global information infrastructure, the real distance between two points is immaterial.)

Rules representing ***"capturing" capabilities*** of players may be transformed in a similar way:

$$\{1 \cdot x,\, l_1 \cdot (a_{i_1} : x,\ z_{i_1}),\ \ldots,\, l_k \cdot (a_{i_k} : x,\ z_{i_k}),$$

$$m_1 \cdot \left(a_{j_1} : y,\ z_{j_1}\right),\ \ldots,\, m_g \cdot \left(a_{j_g} : y,\ z_{j_g}\right)\}$$

$$\left\{1 \cdot y,\, m_1 \cdot \left(a_{j_1} : x,\ z_{j_1}\right),\ \ldots,\, m_g \cdot \left(a_{j_g} : x,\ z_{j_g}\right)\right\}. \tag{9.61}$$

By combining various rules of the introduced form as well as applying various filters describing the set of possible objectives of the players, it is quite easy to describe any necessary logics of both players' behavior. Let us illustrate this by the following example.

Example 9.8
Let the initial resource base of the player A contain six object-resources a located at the point z_1 and five object-resources b located at the point z_2; the initial *RB* of the player B contains three object-resources a located at the point z_3 and two object-resources c located at the point z_4. The capabilities of both players are presented in the following Tables 9.4 and 9.5.

As can be seen, player A, by consuming one *OR* b located at the point z_2, may remove one *OR* a from the point z_1 to the point z_2. By consuming two *ORs* a and two *ORs* b—all located at the point z_2—player A may eliminate (destroy) one *OR* c

Table 9.5 Capabilities of the player B

No.	Consumed by B	Result
1	$1 \cdot (a : B; z_3)$, $1 \cdot (c : B; z_4)$	$1 \cdot (a : B; z_4)$ relocation
2	$1 \cdot (a : B; z_3)$, $1 \cdot (c : B; z_4)$	$1 \cdot (b : A; z_2)$ elimination from A
3	$2 \cdot (a : B; z_3)$, $1 \cdot (c : B; z_4)$	$2 \cdot (a : A; z_1)$ elimination from A

located at the point z_4 and belonging to player B. Finally, by consuming two *ORs* a and one *OR* b—all located at the point z_2—player A may eliminate one object-resource a located at the point z_3 and belonging to player B. The capabilities of player B are as follows. By consuming one *OR* c located at the point z_4, the player B may relocate one *OR* a from z_3 to z_4. By consuming one *OR* a located at the point z_3 and one *OR* c located at the point z_4, player B may eliminate one *OR* b located at the point z_2 and belonging to the player A. By consuming two *ORs* a located at the point z_3 and one *OR* c located at the point z_4, player B may eliminate two *ORs* a located at the point z_1 and belonging to player A.

The described capabilities may be represented by the following rules:

$$r_1^A : \{1 \cdot A, 1 \cdot (a : A,\ z_1), 1 \cdot (b : A,\ z_2)\} \rightarrow \{1 \cdot A, 1 \cdot (a : A; z_2)\},$$

$$r_2^A : \{1 \cdot A, 2 \cdot (a : A,\ z_2), 2 \cdot (b : A,\ z_2), 1 \cdot (c : B,\ z_4)\} \rightarrow \{1 \cdot B\},$$

$$r_3^A : \{1 \cdot A, 2 \cdot (a : A; z_2), 1 \cdot (b : A,\ z_2), 1 \cdot (a : B,\ z_3)\} \rightarrow \{1 \cdot B\},$$

$$r_1^B : \{1 \cdot B, 1 \cdot (a : B; z_3), 1 \cdot (c : B,\ z_4;)\} \rightarrow \{1 \cdot B, 1 \cdot (a : B,\ z_4)\},$$

$$r_2^B : \{1 \cdot B, 1 \cdot (a : B; z_3), 1 \cdot (c : B,\ z_4), 1 \cdot (b : A,\ z_2)\ \} \rightarrow \{1 \cdot A\},$$

$$r_3^B : \{1 \cdot B, 2 \cdot (a : B; z_3), 1 \cdot (c : B,\ z_4), 2 \cdot (a : A,\ z_1)\ \} \rightarrow \{1 \cdot A\},$$

$$r_4^A : \{1 \cdot A\} \rightarrow \{\varnothing\},$$

$$r_4^B : \{1 \cdot B\} \rightarrow \{\varnothing\}.$$

The application of these rules to the multiset

$$v_0 = \{1 \cdot A, 6 \cdot (a : A,\ z_1), 5 \cdot (b : A,\ z_2), 3 \cdot (a : B,\ z_3), 2 \cdot (c : B; z_4)\},$$

representing the initial resource base, allows the following possible generation chain:

$$v_0 \overset{r_1^A}{\Rightarrow} v_1 \overset{r_1^A}{\Rightarrow} v_2 \overset{r_1^A}{\Rightarrow} v_3 \overset{r_3^A}{\Rightarrow} v_4 \overset{r_1^B}{\Rightarrow} v_5 \overset{r_2^B}{\Rightarrow} v_6 \overset{r_4^A}{\Rightarrow} v_7.$$

It is represented in Table 9.6, similar to Table 9.3. The only difference is the additional string, containing locations. Zero multiplicities are represented, as above, by the empty positions in the table.

As can be seen, player A repeats the internal move r_1^A three times and then makes the move r_3^A, leading to the multiset v_4. Player B by his internal move r_1^B

Table 9.6 Generation chain

	A				*B*					
	z_1		z_2		z_3		z_4			
$\frac{v}{r}$	*a*	*b*	*a*	*b*	*a*	*c*	*a*	*c*	A	B
v_0	6			5	3			2	1	
r_1^A	−1		+1	−1					∓1	
v_1	5		1	4	3			2	1	
r_1^A	−1		+1	−1					∓1	
v_2	4		2	3	3			2	1	
r_1^A	−1		+1	−1					∓1	
v_3	3		3	2	3			2	1	
r_3^A			−2	−1	−1				−1	+1
v_4	3		1	1	2			2		1
r_1^B					−1		+1	−1		∓1
v_5	3		1	1	1		1	1		1
r_2^B				−1	−1			−1	+1	−1
v_6	3		1				1		1	
r_4^A									−1	
v_7	3		1				1			

transforms v_4 into v_5, which, in turn, is transformed by the move r_2^B into the multiset v_6. After this, player *A* makes the final move r_4^A, and the game is over. If the terminal multiset

$$\{3 \cdot (a : A,\ z_1;),\ 1 \cdot (a : A,\ z_2),\ 1 \cdot (a : B,\ z_4)\}$$

satisfies the filter established for the definition of the players' objectives, this generation chain is one possible chain leading to the desired objective.

All the above concerned antagonistic *DRBGs* with sequential moves.

9.4.2 *Distributed* RBGs *with Simultaneous Moves*

Simultaneous moves may be represented in the same way as when we considered classic games and local *RBGs*. Namely, it is sufficient to use rules of the following form (with locations in multiobjects):

$$\left\{ \begin{array}{l} 1 \cdot x,\ 1 \cdot y, \\ l_1 \cdot (a_{i_1} : x,\ z_{i_1}),\ \ldots,\ l_k \cdot (a_{i_k} : x,\ z_{i_k}), \\ m_1 \cdot \left(b_{j_1} : y,\ z_{j_1}\right),\ \ldots,\ m_q \cdot \left(b_{j_q} : y,\ z_{j_1}\right) \end{array} \right\}$$

$$\left\{\begin{array}{l} 1 \cdot x, 1 \cdot y, \\ l'_1 \cdot (a_{h_1} : x,\ z_{h_1}), \ \ldots, l'_{k'} \cdot (a_{h_{k'}} : x,\ z_{h_{k'}}), \\ m'_1 \cdot (b_{g_1} : y,\ z_{g_1},\ ), \ \ldots, m'_{q'} \cdot \left(b_{g_{q'}} : y,\ z_{g_{q'}}\right) \end{array}\right\}. \tag{9.62}$$

The kernel of the corresponding multigrammar will be, according to (9.51), nothing but $\{1 \cdot A, 1 \cdot B\} + v_A + v_B$, where v_A and v_B are the initial resource bases of both players. As everywhere above, the scheme, representing the moves base of this game, permits simultaneous moves if it is possible (i.e., as long as at least one rule may be applied).

The objectives of the game for both players, as above, are represented by subfilters F_A and F_B joined to make the filter $F = F_A \cup F_B$ of the relevant filtering multigrammar.

Along with simultaneous moves made by both players, there may be internal moves of the same form as (9.54) (extended, if necessary, by location components in the multiobjects). By such moves, each player may prepare simultaneous moves made by both of them. All that is said about such moves above in this section is applicable to this kind of antagonistic *DRBG*. As may be seen, the multigrammatical representation permits easy use of both sequential and parallel moves in one distributed resource-based game.

By use of simultaneous moves in *DRBGs*, we return in a natural way to cooperative and coalitional games, which, of course, may be not only local but also distributed.

9.4.3 Cooperative and Coalitional Distributed RBGs

Generalization of the rules (9.52) representing the basic logic of the local case of cooperative *RBGs* to the case of distributed *RBGs* is as simple as possible:

$$\left\{\begin{array}{l} 1 \cdot x, 1 \cdot y, \\ l_1 \cdot (a_{i_1} : x,\ z_{i_1}), \ \ldots, l_k \cdot (a_{i_k} : x,\ z_{i_k}), \\ m_1 \cdot (b_{j_1} : y,\ z_{j_1}), \ \ldots, m_g \cdot \left(b_{j_g} : y,\ z_{j_g}\right) \end{array}\right\}$$
$$\left\{\begin{array}{l} 1 \cdot x, 1 \cdot y, \\ m_1 \cdot (b_{j_1} : x,\ z_{j_1}), \ \ldots, m_g \cdot \left(b_{j_g} : x,\ z_{j_g}\right), \\ l_1 \cdot (a_{i_1} : y,\ z_{i_1}), \ \ldots, l_k \cdot (a_{i_k} : y,\ z_{i_k}) \end{array}\right\} \tag{9.63}$$

As may be seen, an application of this rule results in the transfer of l_1 object-resources a_{i_1} located at a position $z_{i_1}, \ldots, l_k$ object-resources a_{i_k} located at a position z_{i_k}, all belonging to a player *A*, to a player *B's* property. In return for the aforementioned transferred resources, player *A* will get m_1 *ORs* b_{j_1} located at a position

z_{j_1}, ...,m_g *ORs* b_{j_g} located at a position z_{j_g}, which until the exchange belonged to player *B*.

A rule, representing the "altruistic" case, is as follows:

$$\{1 \cdot x,\ 1 \cdot y,\ l_1 \cdot (a_{i_1} : x,\ z_{i_1}),\ \ldots,\ l_k \cdot (a_{i_k} : x,\ z_{i_k})\}$$
$$\{1 \cdot x,\ 1 \cdot y,\ l_1 \cdot (a_{i_1} : y,\ z_{i_1}),\ \ldots,\ l_k \cdot (a_{i_k} : y,\ z_{i_k})\}. \tag{9.64}$$

The structure of rules representing the producing (manufacturing) capabilities of players in the case of ***cooperative DRBGs*** is the same as (9.48) and (9.49), but, as everywhere above, ***multiobjects are extended by information about their location and ownership***. Such rules, corresponding to (9.48) and implementing internal moves of both players, may be as follows:

$$\{1 \cdot x,\ n_1 \cdot (a_{i_1} : x_1,\ z_{i_1}),\ \ldots,\ n_m \cdot (a_{i_m} : x_m,\ z_{i_m})\}$$
$$\{1 \cdot x,\ 1 \cdot (a : x_{m+1},\ z)\}. \tag{9.65}$$

This means that a player $x \in \{A, B\}$, making his move, produces one *OR* a located at a position z and belonging to a player $x_{m+1} \in \{A, B\}$, using for this purpose n_1 *ORs* a_{i_1} located at a position z_{i_1} and belonging to a player $x_1 \in \{A, B\}$, ..., n_m *ORs* a_{i_m} located at a position z_{i_m} and belonging to a player $x_m \in \{A, B\}$.

Representation of ***coalitional distributed RBGs*** is also as simple as possible due to application rules of the form

$$\{1 \cdot x_1,\ \ldots,\ 1 \cdot x_m,\ l_1 \cdot (a_{i_1} : x_{i_1},\ z_{i_1}),\ \ldots,\ l_k \cdot (a_{i_k} : x_{i_k},\ z_{i_k})\} \tag{9.66}$$

as well as rules representing internal moves of any player and moves made simultaneously by any non-empty subset of the set of players $\{x_{i_1},\ \ldots,\ x_{i_q}\} \subseteq \{x_1,\ \ldots,\ x_m\}$.

Let us note that until now, we have considered all classes of games on the assumption that ***all information about resources and moves at all steps of the game*** (implemented by rule applications) ***is available to all players***. However, this presumption is far from reality, where every player at every moment is aware only about some part of the resource bases and moves bases of other players. Moreover, this information in the general case may be incomplete, ambiguous, unreliable, contradictory, etc., in short, ***imperfect***. Such more sophisticated cases are studied in classical game theory under the name of *games with incomplete information*, or ***incomplete information games*** (Aumann and Maschler 1995; Sandomirsky 2014). ***Resource-based games with incomplete information***, as well as ***RBGs ongoing upon a time axis*** and called ***temporal RBGs*** (they are naturally represented by temporal multiset grammars), will be considered in future publications.

Chapter 10
Multigrammatical Representation of Classic Problems of Operations Research

The techniques described above for applying the multigrammatical framework to various classes of sociotechnological systems and their associated problems form a foundation for ***effective planning and scheduling in the rising digital economy***, as well as for ***dissemination of MG-represented knowledge*** about rational control and behavior of *STSs* ***in a real branch of modern science***. However, the real theoretical power and possible limitations of the proposed framework may be properly estimated and understood only by applying them to known problems which have for years been investigated from different points of view. This chapter is dedicated to such classic optimization problems from the operations research area. Their multigrammatical representation (formulation) aims at two objectives: ***first of all***, to describe one more collection of techniques for applying the *MGF* and ***secondly***, to establish a correspondence between various classes of such problems and classes of multigrammars.

There are two different approaches to the application of the *MGF* to classic *OpR* tasks. The first one is a ***"direct" application of the multigrammatical framework to classic optimization problems***, i.e., "***rewriting***" their well-known matrix/vector or graph/network representations into the form of the corresponding unitary multiset grammars or unitary multiset metagrammars.

But ***the MGF is much more flexible***, and this flexibility gives an opportunity to represent not only problems but also algorithms performing an effective search for a solution on the basis of the improved *UMMG* algorithmics described in Sect. 4.2.

We shall consider the *MGF* representations of classic *OpR* problems on networks (shortest path, traveling salesman, and maximal flow) in Sect. 10.1. Assignment, optimal pair matching, and transportation problems will be treated in Sect. 10.2. An integer linear programming (*ILP*) problem will be represented in Sect. 10.3.1, and, as an example to demonstrate the *MGF's* aforementioned flexibility, the well-known knapsack problem will be studied in Sect. 10.3.2. The relevant technique developed due to consideration of this example is described in Sect. 10.3.3 regarding multiobjective Boolean programming. Finally, Sect. 10.3.4 is dedicated to the

I. A. Sheremet, *Multigrammatical Framework for Knowledge-Based Digital Economy*, https://doi.org/10.1007/978-3-031-13858-4_10

multigrammatical representation of an *ILP* problem reduced to a problem of Boolean programming.

10.1 Shortest Path, Traveling Salesman, and Maximal Flow Problems

10.1.1 Shortest Path Problem (SPP)

Shortest path problem *(SPP)* (Bast et al. 2009; Roberts 2009; Abraham et al. 2010) is usually formulated as follows. Consider a weighted oriented graph $G \subseteq A \times A \times N$ such that $(a_i, a_j, l_{ij}) \in G$ is an edge (arc) of length l_{ij} connecting nodes a_i and a_j (N is the set of all non-negative integer numbers). A path p is a sequence of edges connected by common nodes

$$p = \langle (a_0,\ a_{i_1},\ l_{0i_1}),\ \ldots,\ (a_{i_j},\ a_{i_{j+1}},\ l_{i_j i_{j+1}}),\ \ldots,\ (a_{i_{k-1}},\ a_{i_k},\ l_{i_{k-1} i_k}) \rangle \qquad (10.1)$$

where a_0 is an initial node, a_{i_k} is a final node, and

$$L = l_{0i_1} + \ldots + l_{i_j i_{j+1}} + \ldots + l_{i_{k-1} i_k} \qquad (10.2)$$

is the length of this path. The problem is ***to find all paths from an initial to a final node with minimal length***.

This problem may be represented by a ***filtering unitary multiset grammar*** $S_G = \langle \{1 \cdot a_0\}, R_G, F_G \rangle$, where every edge $\langle a_i, a_j, l_{ij} \rangle \in G$ corresponds to the unitary rule

$$a_i \to 1 \cdot a_j, 1 \cdot e_{ij}, l_{ij} \cdot d, \qquad (10.3)$$

where a_i and a_j are objects corresponding to nodes of the graph G and e_{ij} is an object corresponding to an edge connecting a_i and a_j, while an object d is, in essence, a length measurement unit. (For simplicity, we shall use ***one-symbol names of objects*** in this chapter without brackets.) The filter F_G contains the optimizing condition

$$d = min\,. \qquad (10.4)$$

As it is easy to see, the set $\overline{V}_{S_G}$ generated by the application of the *FUMG* S_G includes terminal multisets of the form $\{1 \cdot a_{i_k}, 1 \cdot e_{0i_1}, \ldots, 1 \cdot e_{i_{k-1} i_k}, L \cdot d\}$, each corresponding to a path $\{(a_0,\ a_{i_1},\ l_{0i_1}),\ \ldots,\ (a_{i_{k-1}},\ a_{i_k},\ l_{i_{k-1} i_k})\}$ with length L, which is the sum of all l_{ij} occurring in unitary rules belonging to the generation chain beginning from the multiset $\{1 \cdot a_0\}$. A path as a sequence of edges is presented by multiobjects $1 \cdot e_{0i_1}, \ldots, 1 \cdot e_{i_{k-1} i_k}$. Condition (10.4) selects all paths which have minimal length. The set $\overline{V}_{S_G}$ includes not one but all such paths.

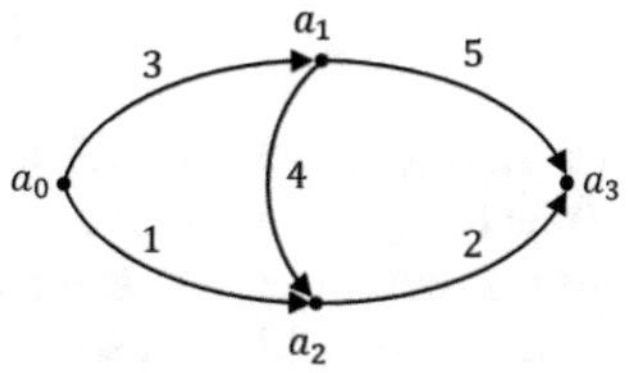

Fig. 10.1 The graph of a shortest path problem

Example 10.1

Let $G = \{(a_0, a_1, 3), (a_0, a_2, 1), (a_1, a_3, 5), (a_1, a_2, 4), (a_2, a_3, 2)\}$ (Fig. 10.1). Here, a_0 is the initial node and a_3 the final node.

According to (10.3), the scheme R_G contains the following unitary rules:

$$a_0 \rightarrow 1 \cdot a_1,\ 1 \cdot e_{01},\ 3 \cdot d,$$
$$a_0 \rightarrow 1 \cdot a_2,\ 1 \cdot e_{02},\ 1 \cdot d,$$
$$a_1 \rightarrow 1 \cdot a_3,\ 1 \cdot e_{13},\ 5 \cdot d,$$
$$a_1 \rightarrow 1 \cdot a_2,\ 1 \cdot e_{12},\ 4 \cdot d,$$
$$a_2 \rightarrow 1 \cdot a_3,\ 1 \cdot e_{23},\ 2 \cdot d.$$

As can be seen,

$$\overline{V}_{S_G} = \{\{1 \cdot a_3,\quad 1 \cdot e_{02},\quad 1 \cdot e_{23},\quad 3 \cdot d\}\},$$

i.e., there is one shortest path of length 3, consisting of two edges $\langle 0, 2\rangle$ and $\langle 2, 3\rangle$.

Unitary multiset grammars as a knowledge representation model allow easy and flexible formulation of ***various modifications of the classic shortest path problem***. For example, if every edge is marked not only by a length, which is directly associated with the time interval necessary for a transport vehicle (*TVh*) to pass this edge, but also with some necessary material resources spent for this passage, then the unitary rules belonging to the scheme R_G have the form

$$a_i \rightarrow 1 \cdot a_j,\ 1 \cdot e_{ij},\ l_{ij}^1 \cdot c_1, \ldots, l_{ij}^n \cdot c_n, \tag{10.5}$$

where $c_1, \ldots, c_n$ are the aforementioned resource measurement units, while $l_{ij}^1, \ldots, l_{ij}^n$ are their amounts necessary for passing the relevant edge. The filter may be

$$F_G = \{d = min\,,\ c_1 = min\,,\ \ldots,\ c_n = min\,\}, \tag{10.6}$$

when one wants to minimize all resource amounts spent while passing a path. If there is a softer formulation, presuming resources limits available to a traveler, then the filter would be of the form

$$F_G = \{d = min\,, c_1 \leq \boldsymbol{l}_1,\ \ldots,\ c_n \leq \boldsymbol{l}_n\}, \tag{10.7}$$

where $\boldsymbol{l}_1$, . . ., $\boldsymbol{l}_n$ are the aforementioned limits. Of course, there may be a threshold value for *d*, while some of the multiplicities of objects c_1, . . ., c_n in generated *TMSs* may be optimized. For example, if the unit *d* is "*mnt*" (*minute*), while the unit c_1 is "*usd*" or "*eur*," then one may solve the task of how to arrive at a final point in no more than $\boldsymbol{l}$ minutes, consuming for this action the minimal money. If so, then

$$F_G = \{d \leq \boldsymbol{l},\ c_1 = min\,\}. \tag{10.8}$$

As can be seen, there may be a lot of such combinations, providing a natural formulation of this kind of multiobjective optimization problems.

10.1.2 Traveling Salesman Problem (**TSP**)

Traveling salesman problem (*TSP*) (Gutin and Punnen 2006; MacGregor and Chu 2011; Taha 2016) is usually formulated as follows. There is a weighted oriented graph $G \subseteq A \times A \times \boldsymbol{N}$**,** where $\langle a_i, a_j, l_{ij}\rangle \in G$ has the same sense as in the previous Sect. 10.1.1. The difference is in consideration of only ***cyclic*** paths, starting and ending at the same node a_0, containing all nodes $a \in A$, and in such a way that every node is visited once (such a path is called a Hamiltonian cycle). The problem is to ***find all Hamiltonian cycles with the minimal length***. By application of the same techniques as in the case of the *SPP*, this problem may be represented by a ***filtering unitary multiset grammar*** $S_G = \langle\{1 \cdot a_0\}, R_G, F_G\rangle$, where every edge $\langle a_i, a_j, l_{ij}\rangle \in G$ corresponds to the unitary rule

$$a_i \rightarrow 1 \cdot a_j,\ \ 1 \cdot m_j,\ \ 1 \cdot e_{ij},\ \ l_{ij} \cdot d, \tag{10.9}$$

where a_j is a non-terminal object which serves for the continuation of generation (i.e., for passing nodes), m_j is used to count the number of visits to the *j*-th node, and e_{ij} represents an edge connecting the *i*-th and the *j*-th nodes, while *d*, as in the previous section, is a length measurement unit. All m_i, e_{ij}, and *d* are terminal objects. The filter F_G includes, as above, one optimizing condition $d = min$ and, besides, $|A|$ boundary conditions

$$m_i = 1, \tag{10.10}$$

for all $i = 0, \ldots, k$, where $k = |A| - 1$. The condition $d = min$ selects all paths with minimal length, while conditions (10.10) select those paths which contain all nodes, every one of which is visited exactly once. ***Due to*** (10.10)***, all paths which contain multiple node occurrences are eliminated***.

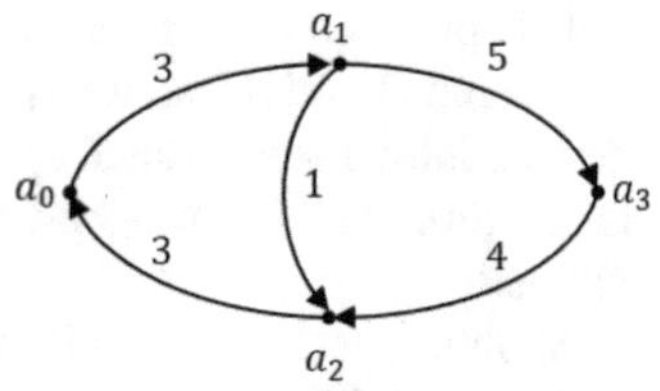

Fig. 10.2 The graph of a traveling salesman problem

As may be seen, every *TMS* generated by a *FUMG* S_G includes terminal multisets of the form

$$\{1\cdot m_0, 1\cdot m_1, \ldots, 1\cdot m_j, \ldots, 1\cdot m_k, 1\cdot e_{0i_1}, \ldots, 1\cdot e_{i_{k-1}0}, L\cdot d\}, \qquad (10.11)$$

each corresponding to its own closed path $a_0 \overset{+}{\Rightarrow} a_0$ with length L, where the last is the sum of all multiplicities l_{ij} occurring in bodies of *URs* applied during *TMS* generation beginning from the multiset $\{1 \cdot a_0\}$.

Example 10.2
Let $G = \{(a_0, a_1, 3), (a_1, a_3, 5), (a_1, a_2, 1), (a_2, a_0, 3), (a_3, a_2, 4)\}$ (Fig. 10.2). Then, according to (10.9) and (10.10), the scheme R_G will contain the following unitary rules:

$$a_0 \to 1\cdot a_1,\ 1\cdot m_1,\ 1\cdot e_{01},\ 3\cdot d,$$

$$a_1 \to 1\cdot a_3,\ 1\cdot m_3,\ 1\cdot e_{13},\ 5\cdot d,$$

$$a_1 \to 1\cdot a_2,\ 1\cdot m_2,\ 1\cdot e_{12},\ 1\cdot d,$$

$$a_2 \to 1\cdot a_0,\ 1\cdot m_0,\ 1\cdot e_{20},\ 3\cdot d,$$

$$a_3 \to 1\cdot a_2,\ 1\cdot m_2,\ 1\cdot e_{32},\ 4\cdot d.$$

The filter will be

$$F_G = \{m_0 = 1,\ m_1 = 1,\ m_2 = 1,\ m_3 = 1,\ m_4 = 1,\ d = min\}.$$

As can be seen,

$$\overline{V}_{S_G} = \{\{1\cdot m_0,\ 1\cdot m_1,\ 1\cdot m_2,\ 1\cdot m_3,\ 1\cdot m_4,\ 1\cdot e_{01},\ 1\cdot e_{13},\ 1\cdot e_{32},\ 1\cdot e_{20},\ 15\cdot d\}\}.$$

All that was said above during the consideration of the shortest path problem (particularly concerning ***multiobjective optimization***) is, of course, applicable to the traveling salesman problem. Let us, however, consider one more modification of this problem called usually "***the set TSP***" or "***the generalized TSP***" (Hillier and Lieberman 2014; Taha 2016).

This problem is formulated as follows. There are $k > 1$ traveling salesmen, who are located at nodes (points) $\boldsymbol{i}_1, \ldots, \boldsymbol{i}_p$. Starting from these points, they must visit all the remaining nodes in such a way that every node must be visited once, all salesmen must return to their starting points, and the total length of all salesman paths must be minimal.

Applying techniques introduced above for the *TSP*, we shall represent the generalized *TSP* by an ***FUMG*** $S_G = \langle \{1 \cdot a_0\}, R_G, F_G \rangle$, where, as in the classical case, every edge $\langle a_i, a_j, l_{ij} \rangle \in G$ corresponds to the unitary rule

$$a_i \rightarrow 1 \cdot a_j, 1 \cdot m_j, 1 \cdot e_{ij}, l_{ij} \cdot d. \tag{10.12}$$

The only difference is that a_0 is not a representation of a starting node, but the header of a *UR*

$$a_0 \rightarrow 1 \cdot a_{i_1}, \ldots, 1 \cdot a_{i_j}, \ldots, 1 \cdot a_{i_p}, \tag{10.13}$$

which allows generation to start beginning from objects corresponding to salesman starting points. All other components of the *UMG* S_G are just the same as in the classic *TSP*: the optimizing condition is $d = min$, and the boundary conditions are $m_i = 1$ for all $i = 0, 1, \ldots, |A| - 1$.

As may be seen, constructed *FUMGs* in all *TSP* modifications are ***cyclic***, and if the filter did not include conditions cutting off all redundant paths containing multiple visits to the same nodes, i.e., multisets with objects m_i and e_{ij} with multiplicities greater than 1, then these *FUMGs* would generate infinite sets of terminal multisets.

Evidently, it is sufficient to apply filtering unitary multiset grammars for the representation of the shortest path and the traveling salesman problems. However, the next problem considered, which, like the *SPP* and the *TSP*, is usually related to optimization on networks, requires the application of a more sophisticated toolkit, i.e., unitary multiset metagrammars.

10.1.3 Maximal Flow Problem (MFP)

Maximal flow problem (*MFP*) (Hillier and Lieberman 2014; Taha 2016) is usually formulated as follows. A transport network is a weighted oriented graph $\boldsymbol{G} \subseteq A \times A \times \boldsymbol{N}$ with an initial node $a_0 \in A$ (source) and a final node $a_k \in A$ (sink). Here, $(a_i, a_j, \boldsymbol{l}_{ij}) \in \boldsymbol{G}$ means that $\boldsymbol{l}_{ij}$ is the flow capacity of an edge connecting a_i and a_j. A flow in a network $\boldsymbol{G}$ is a weighted oriented graph $G \subseteq A \times A \times \boldsymbol{N}$ in which for every $(a_i, a_j, l_{ij}) \in G$, there exists $(a_i, a_j, \boldsymbol{l}_{ij}) \in \boldsymbol{G}$ such that $l_{ij} \leq \boldsymbol{l}_{ij}$.

A basic feature of any transport network is the so-called flow conservation property, written as follows:

$$\sum_{(a_i,\, a_j,\, l_{ij})\in G} l_{ij} = \sum_{(a_j,\, a_p,\, l_{jp})\in G} l_{jp}, \tag{10.14}$$

i.e., for every node a_j, ***the total input flow and the total output flow are equal***. It follows from the flow conservation property that the ***total flow from the source*** a_0 **is equal to the total flow entering the sink** a_k:

$$\sum_{(a_0,\, a_i,\, l_{0j})\in G} l_{0j} = \sum_{(a_j,\, a_k,\, l_{jk})\in G} l_{jk} = L. \tag{10.15}$$

The maximal flow problem is to ***find all flows with the maximal value L*** in the set of all possible flows.

Let us construct a ***unitary multiset metagrammar*** $S_G = \langle\{1 \cdot a_0\}, R_G, F_G\rangle$ representing this problem. We shall include in the scheme R_G one unitary metarule

$$a_0 \to x_{0i_1} \cdot e_{0i_1}, \ldots, x_{j_q k} \cdot e_{j_q k}, \tag{10.16}$$

whose multiplicity-variables x_{ij} correspond to the unknown variables l_{ij} in the classical formulation, while e_{ij} are non-terminal objects corresponding to the edges of the graph *G* (or, equivalently because of their isomorphism, of the graph **G**). Also, we shall include in the scheme R_G of this *UMMG* |*G*|, for every edge of the graph *G*, a unitary rule of the form

$$e_{ij} \to 1 \cdot e_i^-, 1 \cdot e_j^+, 1 \cdot \boldsymbol{e}_{ij}, \tag{10.17}$$

where e_i^- and e_j^+ are terminal objects whose multiplicities in generated terminal multisets are the total flows out of the *i*-th node into the *j*-th node, respectively, and $\boldsymbol{e}_{ij}$ is a terminal object whose multiplicity in any generated *TMS* is the flow from the *i*-th to the *j*-th node. The filter F_G contains $k - 1$ boundary conditions of the form

$$e_i^+ = e_i^-, \tag{10.18}$$

where $i = 1, \ldots, k - 1$, which reflect the equality of the input flow and output flow for every node except the source and the sink. The last are connected by one more boundary condition

$$e_0^- = e_k^+, \tag{10.19}$$

which defines the equality of the input flow and the output flow of the whole network. And, finally, F_G contains |*G*| obvious chain boundary conditions of the form

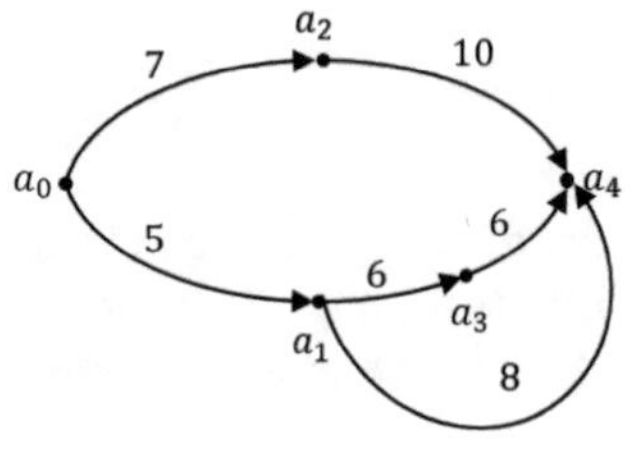

Fig. 10.3 The graph of a maximal flow problem

$$0 \leq x_{ij} \leq \boldsymbol{l}_{ij}, \tag{10.20}$$

for all $(a_i, a_j, \boldsymbol{l}_{ij}) \in \boldsymbol{G}$, as well as one optimizing condition

$$e_0^- = max. \tag{10.21}$$

As can be seen from the above description, all terminal multisets

$$v = \left\{ n \cdot e_0^-, \ldots, n_i^- \cdot e_i^-, n_{ij} \cdot \boldsymbol{e}_{ij}, n_j^+ \cdot e_j^+, \ldots, n \cdot e_k^+ \right\} \in \overline{V}_{SG} \tag{10.22}$$

which satisfy (10.18)–(10.21) are solutions of the problem, and the value n is nothing but the required ***maximal flow***.

Example 10.3
Let $\boldsymbol{G} = \{(a_0, a_1, 5), (a_0, a_2, 7), (a_1, a_3, 6), (a_1, a_4, 8), (a_3, a_4, 6), (a_2, a_4, 10)\}$ (Fig. 10.3). Then, according to (10.16)–(10.21), the scheme $R_{\boldsymbol{G}}$ of the *UMMG*
$S_{\boldsymbol{G}} = \langle \{1 \cdot a_0\}, R_{\boldsymbol{G}}, F_{\boldsymbol{G}} \rangle$ will contain the following unitary rules and metarules:

$$a_0 \to x_{01} \cdot e_{01},\ x_{13} \cdot e_{13},\ x_{14} \cdot e_{14},\ x_{34} \cdot e_{34},\ x_{02} \cdot e_{02},\ x_{24} \cdot e_{04},$$

$$e_{01} \to 1 \cdot e_0^-,\ 1 \cdot e_1^+,\ 1 \cdot \boldsymbol{e}_{01},$$

$$e_{13} \to 1 \cdot e_1^-,\ 1 \cdot e_3^+,\ 1 \cdot \boldsymbol{e}_{13},$$

$$e_{14} \to 1 \cdot e_1^-,\ 1 \cdot e_4^+,\ 1 \cdot \boldsymbol{e}_{14},$$

$$e_{34} \to 1 \cdot e_3^-,\ 1 \cdot e_4^+,\ 1 \cdot \boldsymbol{e}_{34},$$

$$e_{02} \to 1 \cdot e_0^-,\ 1 \cdot e_2^+,\ 1 \cdot \boldsymbol{e}_{02},$$

$$e_{24} \to 1 \cdot e_2^-,\ 1 \cdot e_4^+,\ 1 \cdot \boldsymbol{e}_{24}.$$

The filter $F_{\boldsymbol{G}}$ will contain the following boundary conditions:

$$e_1^+ = e_1^-,\ e_2^+ = e_2^-,\ e_3^+ = e_3^-,\ e_0^- = e_4^+,$$

$$0 \le x_{01} \le 5, \;\; 0 \le x_{02} \le 7, \;\; 0 \le x_{13} \le 6, \;\; 0 \le x_{14} \le 8,$$
$$0 \le x_{24} \le 10, \;\; 0 \le x_{34} \le 6.$$

The optimizing condition will be $e_0^- = max$.

As may be seen,

$$\overline{V}_{SG} = \{\{5 \cdot \boldsymbol{e}_{01}, \; 7 \cdot \boldsymbol{e}_{02}, \; 5 \cdot \boldsymbol{e}_{13}, \; 5 \cdot \boldsymbol{e}_{34}, \; 7 \cdot \boldsymbol{e}_{24}, \; 12 \cdot e_0^-, \; 5 \cdot e_1^+, \; 5 \cdot e_1^-,$$
$$7 \cdot e_2^+, \; 7 \cdot e_2^-, \; 5 \cdot e_3^+, \; 5 \cdot e_3^-, \; 12 \cdot e_4^+\},$$
$$\{5 \cdot \boldsymbol{e}_{01}, \; 7 \cdot \boldsymbol{e}_{02}, \; 5 \cdot \boldsymbol{e}_{14}, \; 7 \cdot \boldsymbol{e}_{24}, \; 12 \cdot e_0^-, \; 5 \cdot e_1^+, \; 5 \cdot e_1^-, \; 7 \cdot e_2^+, 7 \cdot e_2^-, 12 \cdot e_4^+\}\},$$

i.e., the maximal flow in this network is 12, and there are two different solutions of this problem.

By analogy with the shortest path problem, the maximal flow problem may be easily ***generalized*** if it is necessary to take into account the cost of transfer of one flow unit between two nodes. If so, then every unitary rule of the form (10.17) may be transformed into

$$e_{ij} \rightarrow 1 \cdot e_i^-, 1 \cdot e_i^+, 1 \cdot \boldsymbol{e}_{ij}, n_{ij} \cdot \boldsymbol{c}, \tag{10.23}$$

where n_{ij} is the aforementioned cost, while $\boldsymbol{c}$ is its measurement unit. After this transformation, it is sufficient to include in the filter F_G the additional boundary condition $\boldsymbol{c} \le C$, where C is an available resource (usually money). As can be seen, this extension selects those generated terminal multisets which ***fit the available cost of flow transfer from the source to the sink***. The selected *TMS* would have the form

$$v = \left\{N \cdot \boldsymbol{c}, \;\; n \cdot e_0^-, \; \ldots, n_i^- \cdot e_i^-, \;\; n_{ij} \cdot \boldsymbol{e}_{ij}, \;\; n_j^+ \cdot e_j^+, \; \ldots, n \cdot e_k^+\right\} \in \overline{V}_{SG}, \tag{10.24}$$

where N is the cost of this variant of flow transfer.

If it is necessary to select a variant with minimal cost, it is sufficient to include in the filter F_G the optimizing condition $\boldsymbol{c} = min$ instead of $\boldsymbol{c} \le C$. Of course, as in the cases of the generalized *SPP* and *TSP*, there may be more than one type of resources consumed during a flow transfer through a network, so the body of the unitary rule (10.23) may contain $k > 1$ multiobjects $N_{ij}^1 \cdot \boldsymbol{c}_1, \ldots, N_{ij}^k \cdot \boldsymbol{c}_k$. Similarly, there may be $k > 1$ optimizing conditions.

Let us note that there is multigrammatical representation of this task, which allows its solution by application of ***filtering unitary multiset grammars***, i.e., without unitary metarules containing multiplicity-variables in their bodies. In fact, the basis of this solution was for the first time considered in Sect. 8.4, concerning the much more complicated and practically oriented problem of assessment of resilience of energy infrastructures. We shall apply the aforementioned technique to a general case of the maximal flow problem.

Let us consider once more the weighted oriented graph $\boldsymbol{G} \subseteq A \times A \times \boldsymbol{N}$ of a transport network with initial node $a_0 \in A$ (source) and final node $a_k \in A$ (sink). As above, $(a_i, a_j, \boldsymbol{l_{ij}}) \in \boldsymbol{G}$ means that $\boldsymbol{l_{ij}}$ is the flow capacity of an edge (link) connecting a_i and a_j. A flow in the network $\boldsymbol{G}$ is a weighted oriented graph $G \subseteq A \times A \times \boldsymbol{N}$ in which for every $(a_i, a_j, l_{ij}) \in G$, there exists $(a_i, a_j, \boldsymbol{l_{ij}}) \in \boldsymbol{G}$ such that $l_{ij} \leq \boldsymbol{l_{ij}}$. The flow conservation property (10.14) ensures the delivery of all input flow from the source to the sink, while the limited flow capacities of links restrict the total flow which may be transported via the network. Now, we shall introduce another multigrammatical representation of the network, differing from (10.16) to (10.22).

We shall begin with a ***basic task***: ***to determine whether*** n ***units of some substance may be transported*** **via** ***the considered network***.

We shall construct a ***filtering unitary multiset grammar*** $\boldsymbol{S_G} = \langle\{n \cdot \boldsymbol{a}_0\}, \boldsymbol{R_G}, \boldsymbol{F_G}\rangle$ representing this task as follows (to distinguish the constructed *FUMG* from the *UMMG* considered above, we shall use bold symbols below). We shall include in the scheme $\boldsymbol{R_G}$ of this *UMMG* |*G*|, for every edge of the graph $\boldsymbol{G}$, a unitary rule of the form

$$\boldsymbol{a}_i \to 1 \cdot \boldsymbol{a}_j, 1 \cdot \boldsymbol{c}_{ij}, \tag{10.25}$$

where $\boldsymbol{a}_i$ and $\boldsymbol{a}_j$ are objects corresponding, respectively, to nodes a_i and a_j, while $\boldsymbol{c}_{ij}$ is a terminal object representing a link between these nodes. The filter $\boldsymbol{F_G}$ will contain boundary conditions for all edges of the graph $\boldsymbol{G}$, so there will be a *BC*

$$\boldsymbol{c}_{ij} \leq \boldsymbol{l}_{ij} \tag{10.26}$$

for every edge $(a_i, a_j, \boldsymbol{l}_{ij}) \in \boldsymbol{G}$. Also, $\boldsymbol{F_G}$ will contain one more boundary condition

$$\boldsymbol{a}_k = n, \tag{10.27}$$

formally ensuring the necessary identity of the input and output amounts of substance transported via the network.

As can be seen from the above description, any terminal multiset generated by the application of the core multigrammar $\boldsymbol{S'_G} = \langle\{n \cdot \boldsymbol{a}_0\}, \boldsymbol{R_G}\rangle$ of the *FUMG* $\boldsymbol{S_G} = \langle\{n \cdot \boldsymbol{a}_0\}, \boldsymbol{R_G}, \boldsymbol{F_G}\rangle$

$$v = \{m \cdot \boldsymbol{a}_k\} \cup \bigcup_{(a_i,\, a_j,\, \boldsymbol{l}_{ij}) \in \boldsymbol{G}} \{l_{ij} \cdot \boldsymbol{c}_{ij}\} \in \overline{V}_{S'G} \tag{10.28}$$

corresponds to some specific way of transportation of an input flow containing n units of the transported substance from the source a_0 to the sink a_k. In this multiset, there exists a multiobject $m \cdot a_k$ representing the number of units of the transported substance successfully delivered to the sink a_k, while l_{ij} represents a flow in the edge $(a_i, a_j, \boldsymbol{l}_{ij}) \in \boldsymbol{G}$ which occurs during the implementation of this way of delivery. Now, by application of the filter $\boldsymbol{F_G}$ to the set $\overline{V}_{s'_G}$, it is possible to obtain the needed set $\overline{V}_{s_G}$:

$$\overline{V}_{s_G} = \overline{V}_{s'_G} \downarrow \boldsymbol{F_G}. \tag{10.29}$$

As can be seen, if the set $\overline{V}_{s_G}$ is non-empty, then it is possible to transport the aforementioned amount of substance via the network. Otherwise, it is not possible.

Example 10.4

Let the graph

$\boldsymbol{G} = \{(a_0, a_1, 5), (a_0, a_2, 7), (a_1, a_3, 6), (a_1, a_4, 8), (a_3, a_4, 6), (a_2, a_4, 10)\}$ be the same as in the previous Example 10.3 (Fig. 10.3). It is necessary to assess whether it is possible to transport a flow of 3 units of substance via this network. To solve this task, we shall construct a suitable *UMG* $\boldsymbol{S_G} = \langle\{3 \cdot \boldsymbol{a}_0\}, \boldsymbol{R_G}, \boldsymbol{F_G}\rangle$. According to (10.25), the scheme $\boldsymbol{R_G}$ of this *UMG* will contain the following unitary rules:

$$\boldsymbol{a}_0 \to 1 \cdot \boldsymbol{a}_1, 1 \cdot \boldsymbol{c}_{01},$$

$$\boldsymbol{a}_0 \to 1 \cdot \boldsymbol{a}_2, 1 \cdot \boldsymbol{c}_{02},$$

$$\boldsymbol{a}_1 \to 1 \cdot \boldsymbol{a}_3, 1 \cdot \boldsymbol{c}_{13},$$

$$\boldsymbol{a}_1 \to 1 \cdot \boldsymbol{a}_4, 1 \cdot \boldsymbol{c}_{14},$$

$$\boldsymbol{a}_2 \to 1 \cdot \boldsymbol{a}_4, 1 \cdot \boldsymbol{c}_{24},$$

$$\boldsymbol{a}_3 \to 1 \cdot \boldsymbol{a}_4, 1 \cdot \boldsymbol{c}_{34}.$$

According to (10.26) and (10.27), the filter $\boldsymbol{F_G}$ will contain the following boundary conditions:

$$\boldsymbol{c}_{01} \leq 5,$$

$$\boldsymbol{c}_{02} \leq 7,$$

$$\boldsymbol{c}_{13} \leq 6,$$

$$\boldsymbol{c}_{14} \leq 8,$$

$$\boldsymbol{c}_{24} \leq 10,$$

$$\boldsymbol{c}_{34} \leq 6,$$

$$\boldsymbol{a}_4 = 3.$$

As may be seen, the set $\overline{V}_{s_G}$ is non-empty, so the input flow may be transported via this network.

Let us consider now the ***maximal flow problem,*** which we shall represent by applying the technique described above for the basic task.

To generate a flow of numbers which will be checked for the possibility of transportation of the corresponding flow of substance via a network, we shall join to the set $\boldsymbol{R_G}$ one more rule

$$\boldsymbol{a}_0 \to 2 \cdot \boldsymbol{a}_0, \tag{10.30}$$

thus generating a set of multisets

$$\{\{2 \cdot \boldsymbol{a}_0\}, \{3 \cdot \boldsymbol{a}_0\}, \{4 \cdot \boldsymbol{a}_0\}, \ldots\} \tag{10.31}$$

each such *MS*, in fact, representing a variant of the input flow whose viability of transportation via the network will be checked. To find the maximal number, we shall replace the boundary condition (10.27) in the filter $\boldsymbol{F_G}$ by an optimizing condition

$$\boldsymbol{a}_k = max\,. \tag{10.32}$$

Let us consider now the ***filtering unitary multiset grammar***

$$\boldsymbol{S}_{\boldsymbol{G}}^{max} = \langle\{1 \cdot \boldsymbol{a}_0\}, \boldsymbol{R'}_{\boldsymbol{G}}, \boldsymbol{F'}_{\boldsymbol{G}}\rangle, \tag{10.33}$$

where

$$\boldsymbol{R'}_{\boldsymbol{G}} = \boldsymbol{R_G} \cup \{ < \boldsymbol{a}_0 \to 2 \cdot \boldsymbol{a}_0 > \}, \tag{10.34}$$

$$\boldsymbol{F'}_{\boldsymbol{G}} = \boldsymbol{F_G} \cup \{\boldsymbol{a}_k = max\} - \{\boldsymbol{a}_k = n\}. \tag{10.35}$$

As is evident now, the core *UMG* of the *FUMG* $\boldsymbol{S}_{\boldsymbol{G}}^{max}$ generates an infinite set of terminal multisets $\overline{V}_{\boldsymbol{S}_{\boldsymbol{G}}^{max}}$, each *TMS* belonging to this set corresponding to some one variant of distribution of input flow in the network. Not all of these distributions will satisfy boundary conditions (10.26); moreover, since multiplicities of objects $\boldsymbol{c}_{ij}$ increase monotonically in every newly generated *TMS*, a finite subset of the set $\overline{V'}_{\boldsymbol{S}_{\boldsymbol{G}}^{max}}$ satisfies the filter $\boldsymbol{F'}_{\boldsymbol{G}}$, if indeed it is possible to transport at least one input flow via this network.

As can be seen, in fact, an *FUMG*-based representation is much more compact and natural due to the physical similarity of multiset generation and flow transportation. However, from a computational complexity point of view, a multimetagrammatical representation is much more effective, because the multigrammatical one allows a solution search by generation via single-object application. So it would be rational to represent a specific task with a multigrammar and, after construction of the equivalent unitary multimetagrammar (the general technique of such a construction was, in fact, described in Sect. 4.1), to apply this

UMMG for the generation of the set of terminal multisets representing the set of solutions of the initial task.

10.2 Assignment, Optimal Pair Matching, and Transportation Problems

10.2.1 Assignment Problem

Assignment problem (Hillier and Lieberman 2014; Taha 2016) is usually formulated as follows. There is an assignment matrix $X_{n \times n}$, where $x_{ij} = 1$ if the i-th job is assigned to the j-th person and $x_{ij} = 0$ otherwise. A cost matrix $C_{n \times n}$ contains information about costs of assignments, so c_{ij} is the cost of execution of the i-th job by the j-th person. The problem is to assign jobs to persons in such a way that the total cost, defined as

$$\sum_{i=1}^{n} \sum_{j=1}^{n} c_{ij} x_{ij}, \tag{10.36}$$

is minimal, under the restriction that one person may execute only one job and one job may be executed by one person only:

$$\sum_{i=1}^{n} x_{ij} = 1 \text{ for all } j = 1, \ldots, n, \tag{10.37}$$

$$\sum_{j=1}^{n} x_{ij} = 1 \text{ for all } i = 1, \ldots, n. \tag{10.38}$$

A ***unitary multiset metagrammar*** $S = < \{1 \cdot (problem)\}, R, F>$ represents this problem in such a way that the scheme R includes the following unitary metarules:

$$(problem) \to 1 \cdot (opt), 1 \cdot (rows), 1 \cdot (columns), 1 \cdot (solution), \tag{10.39}$$

$$(opt) \to c_{11} \cdot a_{11}, \ldots, c_{ij} \cdot a_{ij}, \ldots, c_{nn} \cdot a_{nn}, \tag{10.40}$$

$$a_{11} \to x_{11} \cdot \boldsymbol{e},$$

$$\ldots$$

$$a_{ij} \to x_{ij} \cdot \boldsymbol{e},$$

$$\ldots$$

$$a_{nn} \to x_{nn} \cdot \boldsymbol{e}, \tag{10.41}$$

$$(rows) \to x_{11} \cdot s_1, \ldots, x_{1n} \cdot s_1, \ldots, x_{ij} \cdot s_i, \ldots, x_{n1} \cdot s_n, \ldots, x_{nn} \cdot s_n, \tag{10.42}$$

$$(columns) \to x_{11} \cdot o_1, \ldots, x_{n1} \cdot o_1, \ldots, x_{ij} \cdot o_j, \ldots, x_{1n} \cdot o_n, \ldots, x_{nn} \cdot o_n, \tag{10.43}$$

$$(solution) \to x_{11} \cdot e_{11}, \ldots, x_{ij} \cdot e_{ij}, \ldots, x_{nn} \cdot e_{nn}. \tag{10.44}$$

Here, $\boldsymbol{e}$ is a terminal object whose multiplicity in generated terminal multisets is the ***value of the optimized function*** (10.36); s_i $(i = 1,\ldots, n)$ is a terminal object corresponding to the i-th row of the matrix X, and the multiplicity of this object in a generated *TMS* is the sum of values x_{ij} occurring in this i-th row; o_j $(j = 1,\ldots n)$ is a similar terminal object corresponding in the aforementioned sense to the j-th column of X. Multiplicity-variables x_{ij} correspond exactly to variables x_{ij} in the classical formulation (10.36)–(10.38). Terminal objects $e_{11}, \ldots, e_{nn}$ serve for a direct representation of a solution. It is quite evident that the filter F must contain the following boundary conditions, corresponding to the classical formulation of the assignment problem:

$$s_1 = 1, \ldots, s_n = 1, o_1 = 1, \ldots, o_n = 1, \tag{10.45}$$

$$0 \le x_{11} \le 1, \ldots, 0 \le x_{ij} \le 1, \ldots, 0 \le x_{nn} \le 1. \tag{10.46}$$

The optimizing condition in F is also obvious:

$$\boldsymbol{e} = min\,. \tag{10.47}$$

Now, one can see that terminal multisets generated by the unitary multiset metagrammar $S = \ < \{1 \cdot (problem)\}, R, F>$ have the form

$$\left\{\left(\sum_{i=1}^{n}\sum_{j=1}^{n} c_{ij}x_{ij}\right) \cdot \boldsymbol{e},\ 1 \cdot s_1,\ \ldots,\ 1 \cdot s_n,\ 1 \cdot o_1,\ \ldots,\ 1 \cdot o_n,\ x_{11} \cdot e_{11},\ \ldots,\ x_{nn} \cdot e_{nn}\right\}, \tag{10.48}$$

where multiobjects $x_{ij} \cdot e_{ij}$ with zero multiplicities are eliminated, so the number of multiobjects $1 \cdot e_{ij}$ in every such *TMS* is n.

Example 10.5

Consider the assignment problem whose objective is

$$2x_{11} + 3x_{12} + 4x_{21} + x_{22} \rightarrow min\,.$$

This problem is represented by the *UMMG* $S = \, < \{1 \cdot (problem)\}, R, F>$, where the scheme R along with the unitary rule (10.39) includes the following unitary metarules:

$$(opt) \rightarrow 2 \cdot a_{11},\ 3 \cdot a_{12},\ 4 \cdot a_{21},\ 1 \cdot a_{22},$$

$$a_{11} \rightarrow x_{11} \cdot e,$$

$$a_{12} \rightarrow x_{12} \cdot e,$$

$$a_{21} \rightarrow x_{21} \cdot e,$$

$$a_{22} \rightarrow x_{22} \cdot e,$$

$$(rows) \rightarrow x_{11} \cdot s_1, x_{12} \cdot s_1, x_{21} \cdot s_2, x_{21} \cdot s_2, x_{22} \cdot s_2,$$

$$(columns) \rightarrow x_{11} \cdot o_1, x_{21} \cdot o_1, x_{12} \cdot o_2, x_{22} \cdot o_2,$$

$$(solution) \rightarrow x_{11} \cdot e_{11}, x_{12} \cdot e_{12}, x_{21} \cdot e_{21}, x_{22} \cdot e_{22}.$$

According to (10.45) and (10.46),

$$F = \{s_1 = 1, s_2 = 1, o_1 = 1, o_2 = 1,$$

$$0 \leq x_{11} \leq 1, 0 \leq x_{12} \leq 1, 0 \leq x_{21} \leq 1, 0 \leq x_{22} \leq 1, \boldsymbol{e} = min\}.$$

Thus,

$$\overline{V_s} = \{\{3 \cdot \boldsymbol{e},\ 1 \cdot e_{11},\ 1 \cdot e_{22},\ 1 \cdot s_1,\ 1 \cdot s_2,\ 1 \cdot o_1,\ 1 \cdot o_2\}\}.$$

As everywhere above, it is quite easy to generalize this problem by additional optimality criteria as well as by additional boundary conditions.

10.2.2 Optimal Pair Matching Problem

Optimal pair matching problem (Hillier and Lieberman 2014; Taha 2016) is usually formulated as follows. Consider a weighted non-oriented graph

$$G = \{ < i_1, j_1, c_{i_1 j_1} >,\ \ldots,\ < i_m, j_m, c_{i_m j_m} > \}, \tag{10.49}$$

where $<r, q, c_{rq}> \ \in G$ means that c_{rq} is the weight of an edge connecting nodes r and q. A pair matching is a set

$$M' \subseteq M = \{<i_1, j_1>, \ \ldots, \ <i_m, j_m>\}, \tag{10.50}$$

such that no two edges have a common node:

$$(\forall <i,j> \in M')(\forall <k, l> \in M') \ \{i,j\} \cap \{k, l\} = \{\varnothing\}. \tag{10.51}$$

The pair matching weight is the sum of weights of all edges in M', and an optimal pair matching $\boldsymbol{M}$ is one which has a maximal weight:

$$C_{M'} = \sum_{<i,j> \in M'} c_{ij}, \tag{10.52}$$

$$(\forall M' \neq \boldsymbol{M}) \ C_{\boldsymbol{M}} \geq C_{M'}. \tag{10.53}$$

A ***unitary multiset metagrammar*** $S_G = \langle\{1 \cdot (problem)\}, R_G, F_G\rangle$ representing this problem is defined as follows. *URs* and *UMRs* belonging to the scheme R_G contain non-terminal objects, which are written as $[i,j]$, similar to $<i, j> \ \in M'$ in a usual matrix (the only difference is the different brackets). The scheme R_G includes one unitary rule

$$(problem) \rightarrow 1 \cdot [i_1, j_1], \ldots, 1 \cdot [i_m, j_m], \tag{10.54}$$

where every edge $[i,j] \in M$ is represented by a corresponding object $[i,j]$, and also R_G includes one *UMR* and one *UR* for each edge, as considered below. Namely, the unitary metarule corresponding to the k-th edge is

$$[i_k, j_k] \rightarrow x_k \cdot (cost_k), \ x_k \cdot [i_k * j_k], \ x_k \cdot [i_k], \ x_k \cdot [j_k], \tag{10.55}$$

where x_k is a multiplicity-variable whose value defines whether the k-th edge belongs to a pair matching (1 if yes; 0 otherwise); $[i_k]$ and $[j_k]$ are terminal objects corresponding to the nodes i_k and j_k which are incident to the k-th edge; $[i_k * j_k]$ is also a terminal object, corresponding to an edge $[i_k, j_k]$; and $(cost_k)$ is an auxiliary non-terminal object, serving for the representation of the k-th edge weight. The unitary rule representing the weight of an edge is

$$(cost_k) \rightarrow c_k \cdot \boldsymbol{e}, \tag{10.56}$$

i.e., the cost of the k-th edge is c_k units $\boldsymbol{e}$. The filter F_G includes boundary conditions

$$0 \leq x_k \leq 1, \tag{10.57}$$

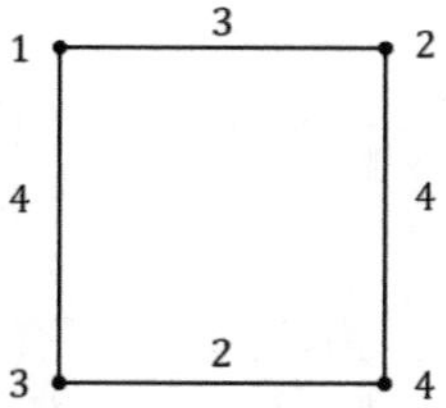

Fig. 10.4 The graph of an optimal pair matching problem

$$[i_k] = 1, \tag{10.58}$$

for all $k = 1, \ldots, m$, as well as one optimizing condition

$$\boldsymbol{e} = max\,. \tag{10.59}$$

From the above description, it is clear that terminal multisets generated by the *UMMG* S_G have the form

$$\left\{\left(\sum_{[i,\ \ j]\in M'} c_{ij}\right)\cdot \boldsymbol{e}\right\} \cup \left(\bigcup_{[i,\ \ j]\in M'} \{1\cdot[i*j],\ 1\cdot[i],\ 1\cdot[j]\}\right), \tag{10.60}$$

where the multiplicity of the terminal object $\boldsymbol{e}$ is the weight of the M' pair matching, while multiobjects of the form $1 \cdot [i * j]$ represent edges in this pair matching. Note that boundary condition (10.58) eliminates *TMSs* which contain terminal multiobjects like $l \cdot [i]$ where $l > 1$, as well as *TMSs* where some edges are absent, because of $l = 0$. So, (10.59) selects only those terminal multisets which correspond to the definition (10.50) and (10.51) of a pair matching.

Example 10.6
Consider the optimal pair matching problem with the graph $G = \; = \{[1,2,3],$ $[1,3,4],[2,4,4],[3,4,2]\}$ (Fig. 10.4).

According to (10.54)–(10.56), the scheme R_G of the unitary multiset metagrammar $S_G = \langle\{1 \cdot a_0\}, R_G, F_G\rangle$ representing this problem contains the following unitary rules and unitary metarules:

$$(problem) \to 1\cdot[1,\ 2], 1\cdot[1,\ 3], 1\cdot[2,\ 4], 1\cdot[3,\ 4],$$

$$[1,\ 2] \to x_1\cdot(cost_1), x_1\cdot[1], x_1\cdot[3], x_1\cdot[1*3]\,,$$

$$(cost_1) \to 3\cdot\boldsymbol{e},$$

$$[1,\ 3] \to x_2\cdot(cost_2), x_2\cdot[1], x_2\cdot[3], x_2\cdot[1*3],$$

$$(cost_2) \to 4\cdot\boldsymbol{e},$$

$$[2,\ 4] \to x_3\cdot(cost_3), x_3\cdot[2], x_3\cdot[4], x_3\cdot[2*4],$$

$$(cost_3) \to 4 \cdot \boldsymbol{e},$$

$$[3, 4] \to x_4 \cdot (cost_4), x_4 \cdot [3], x_4 \cdot [4], x_4 \cdot [3 * 4],$$

$$(cost_4) \to 2 \cdot \boldsymbol{e}.$$

According to (10.57)–(10.59), the filter F_G contains the boundary conditions $[1] = 1$, $[2] = 1$, $[3] = 1$, $[4] = 1$, $0 \le x_1 \le 1$, and $0 \le x_2 \le 1$, $0 \le x_3 \le 1$, $0 \le x_4 \le 1$ and the optimizing condition $\boldsymbol{e} = max$. Then

$$\overline{V}_{SG} = \{\{8 \cdot \boldsymbol{e},\ 1 \cdot [1 * 3],\ 1 \cdot [2 * 4],\ 1 \cdot [1],\ 1 \cdot [2],\ 1 \cdot [3],\ 1 \cdot [4]\}\}.$$

Thus, the optimal pair matching includes the edges [1,3] and [2,4] and has weight 8.

10.2.3 Transportation Problem

Transportation problem (Hillier and Lieberman 2014) is represented very similarly to the previous problems described above in this section. This problem is usually formulated as follows.

Find a matrix $X_{m \times n}$ from the condition

$$\sum_{i=1}^{m} \sum_{j=1}^{n} c_{ij} x_{ij} \to min, \tag{10.61}$$

under restrictions

$$\sum_{j=1}^{n} x_{ij} = a_i,\ i \in \{1, \ldots, m\}, \tag{10.62}$$

$$\sum_{i=1}^{m} x_{ij} = b_j,\ j \in \{1, \ldots, n\}, \tag{10.63}$$

$$x_{ij} \ge 0,\ i \in \{1, \ldots, m\},\ j \in \{1, \ldots, n\}. \tag{10.64}$$

Here, the value of the variable x_{ij} is the quantity in units of a load, transported from the i-th manufacturer (a source point) to the j-th consumer (a destination point), while c_{ij} is the cost of transportation of one aforementioned load unit, so the sum of products $c_{ij} \cdot x_{ij}$ over all possible values of i and j is the cost of transportation of all amounts x_{ij} from manufacturers to consumers. Condition (10.61) minimizes the total cost of this action, while a_i is the total load amount at the i-th source point, and b_j is the total load amount at the j-th destination point.

A ***unitary multiset metagrammar*** $S = \ < \{1 \cdot problem\}, R, F >$,representing (10.61)–(10.64), is as follows. The scheme R by analogy with (10.39)–(10.44) contains the following unitary rules and unitary metarules:

$$(problem) \to 1 \cdot (opt), 1 \cdot (sources), 1 \cdot (destination), \tag{10.65}$$

$$(opt) \to c_{11} \cdot (e_{11}), \ldots, c_{ij} \cdot (e_{ij}), \ldots, c_{mn} \cdot (e_{mn}), \tag{10.66}$$

$$(sources) \to x_{11} \cdot \boldsymbol{a}_1, \ldots, x_{1n} \cdot \boldsymbol{a}_1, \ldots, x_{m1} \cdot \boldsymbol{a}_m, \ldots, x_{mn} \cdot \boldsymbol{a}_m, \tag{10.67}$$

$$(destination) \to x_{11} \cdot \boldsymbol{b}_1, \ldots, x_{m1} \cdot \boldsymbol{b}_1, \ldots, x_{1n} \cdot \boldsymbol{b}_n, \ldots, x_{mn} \cdot \boldsymbol{b}_n, \tag{10.68}$$

where x_{ij} ($i \in \{1, \ldots, m\}, j \in \{1, \ldots, n\}$) are multiplicity-variables corresponding to the variables occurring in (10.39)–(10.44), while $\boldsymbol{a}_1, \ldots, \boldsymbol{a}_m, \boldsymbol{b}_1, \ldots, \boldsymbol{b}_n$ are terminal objects corresponding to the source ($\boldsymbol{a}_i$) and destination ($\boldsymbol{b}_j$) points. Also, R contains $m \times n$ unitary rules of the form

$$e_{ij} \to x_{ij} \cdot \boldsymbol{e}, x_{ij} \cdot (t_{ij}), \tag{10.69}$$

where $\boldsymbol{e}$ is a terminal object—a load measurement unit, x_{ij} has the same sense as above, and (t_{ij}) is a terminal object denoting transportation from the i-th source point to the j-th destination point, so that the presence of a multiobject $l \cdot (t_{ij})$ in a terminal multiset generated by the *UMMG S* means that l units of load will be transported from the i-th source point to the j-th destination point.

The filter F contains $m + n$ boundary conditions

$$\boldsymbol{a}_1 = a_1, \ldots, \boldsymbol{a}_m = a_m, \tag{10.70}$$

$$\boldsymbol{b}_1 = b_1, \ldots, \boldsymbol{b}_n = b_n, \tag{10.71}$$

and $m + n$ boundary conditions

$$0 \le x_{11} \le \boldsymbol{N}_{11}, \ldots, 0 \le x_{1n} \le \boldsymbol{N}_{1n}, \tag{10.72}$$

$$\ldots$$

$$0 \le x_{m1} \le \boldsymbol{N}_{m1}, \ldots, 0 \le x_{mn} \le \boldsymbol{N}_{mn}, \tag{10.73}$$

where

$$\boldsymbol{N}_{ij} = min\left\{a_i, b_j\right\}, \tag{10.74}$$

because the domain of the variable x_{ij} is the intersection of intervals $[0, a_i]$ and $[0, b_i]$, corresponding to maximal values of x_{ij} in (10.62) and (10.63), which, in turn, correspond to minimal (i.e., zero) values of all the remaining variables in the restriction. Also, F contains one optimizing condition

$$\boldsymbol{e} = min \quad . \tag{10.75}$$

As it is easy to see now, a solution of the problem in the multiset formulation is a set $\overline{V}_G$ containing terminal multisets of the form

$$\left\{\left[\sum_{i=1}^{m}\sum_{j=1}^{n}\left(c_{ij}\cdot\overline{x_{ij}}\right)\right]\cdot\boldsymbol{e}\right\}\cup\left(\bigcup_{i=1}^{m}\bigcup_{j=1}^{n}\left\{\overline{x_{ij}}\cdot\left(\boldsymbol{t}_{ij}\right)\right\}\right)$$
$$\cup\{a_1\cdot\boldsymbol{a}_1, \ldots, a_m\cdot\boldsymbol{a}_m, b_1\cdot\boldsymbol{b}_1, \ldots, b_n\cdot\boldsymbol{b}_n\}, \tag{10.76}$$

where the first one-element multiset of this join is $\left\{\overline{C}\cdot\boldsymbol{e}\right\}$, where $\overline{C}$ is the minimal total cost of transportation; the second multiset is a distribution of load amounts in such a way that $\overline{x_{ij}}$ units are transported from the i-th source point to the j-th destination point; and the third multiset contains terminal multiobjects whose multiplicities are the load amounts taken from the i-th source and received at the j-th destination for all $i \in \{1, \ldots, m\}$ and $j \in \{1, \ldots, n\}$. The multiplicities $\overline{x_{ij}}$ are the values of the corresponding variables x_{ij} from a classical formulation of the problem, so these values, in fact, represent a solution of this problem.

Example 10.7
Consider the matrix

$$C = \begin{Vmatrix} 2 & 4 & 1 \\ 3 & 7 & 5 \end{Vmatrix},$$

which defines the transport expenses of two manufacturers for the transportation of their production to three consumers. The manufacturers have 12 and 8 production units, while the consumers need 6, 9, and 5 units, respectively. The classical formulation of this problem is as follows:

$$2\cdot x_{11} + 4\cdot x_{12} + 1\cdot x_{13} + 3\cdot x_{21} + 7\cdot x_{22} + 5\cdot x_{23} \rightarrow min$$

under the restrictions

$$x_{11} + x_{12} + x_{13} = 12,$$

$$x_{21} + x_{22} + x_{23} = 8,$$

$$x_{11} + x_{21} = 6,$$

$$x_{12} + x_{22} = 9,$$

$$x_{13} + x_{23} = 5,$$

$$x_{11} \geq 0, x_{12} \geq 0, x_{13} \geq 0, x_{21} \geq 0, x_{22} \geq 0, x_{23} \geq 0.$$

The scheme *R* of the corresponding unitary multiset metagrammar $S = < \{1 \cdot (\textit{problem})\}, R, F >$ includes the following unitary rules and metarules:

$$(\textit{problem}) \rightarrow 1 \cdot (\textit{opt}), 1 \cdot (\textit{sources}), 1 \cdot (\textit{destination}),$$

$$(\textit{opt}) \rightarrow 2 \cdot e_{11}, 4 \cdot e_{12}, 1 \cdot e_{13}, 3 \cdot e_{21}, 7 \cdot e_{22}, 5 \cdot e_{23},$$

$$(\textit{sources}) \rightarrow x_{11} \cdot \boldsymbol{a}_1, x_{12} \cdot \boldsymbol{a}_1, x_{13} \cdot \mathbf{a}_1, x_{21} \cdot \boldsymbol{a}_2, x_{22} \cdot \boldsymbol{a}_2, x_{23} \cdot \boldsymbol{a}_2,$$

$$(\textit{destination}) \rightarrow x_{11} \cdot \boldsymbol{b}_1, x_{21} \cdot \boldsymbol{b}_1, x_{12} \cdot \boldsymbol{b}_2, x_{22} \cdot \boldsymbol{b}_2, x_{13} \cdot \boldsymbol{b}_3, x_{23} \cdot \boldsymbol{b}_3,$$

$$e_{11} \rightarrow x_{11} \cdot \boldsymbol{e}, x_{11} \cdot (t_{11}),$$

$$e_{12} \rightarrow x_{12} \cdot \boldsymbol{e}, x_{12} \cdot (t_{12}),$$

$$e_{13} \rightarrow x_{13} \cdot \boldsymbol{e}, x_{13} \cdot (t_{13}),$$

$$e_{21} \rightarrow x_{21} \cdot \boldsymbol{e}, x_{21} \cdot (t_{21}),$$

$$e_{22} \rightarrow x_{22} \cdot \boldsymbol{e}, x_{22} \cdot (t_{22}),$$

$$e_{23} \rightarrow x_{23} \cdot \boldsymbol{e}, x_{23} \cdot (t_{23}).$$

The filter *F* includes the boundary conditions

$$\boldsymbol{a}_1 = 12, \ \boldsymbol{a}_2 = 8, \ \boldsymbol{b}_1 = 6, \ \boldsymbol{b}_2 = 9, \ \boldsymbol{b}_3 = 5,$$

$$0 \leq x_{11} \leq 6, \ 0 \leq x_{12} \leq 9, \ 0 \leq x_{13} \leq 5,$$

$$0 \leq x_{21} \leq 6, \ 0 \leq x_{22} \leq 8, \ 0 \leq x_{23} \leq 5,$$

as well as the optimizing condition

$$\boldsymbol{e} = \textit{min} \, .$$

The solution of this problem is the one-element set

$$\overline{V_S} = \{\{79 \cdot \boldsymbol{e}, \ 6 \cdot (t_{11}), \ 3 \cdot (t_{12}), \ 3 \cdot (t_{13}), \ 6 \cdot (t_{22}), \ 2 \cdot (t_{23})\}\},$$

where the absence of the terminal multiobject t_{21} in the *TMS* $v \in \overline{V_S}$ means $x_{21} = 0$.

All three problems considered in this section may be modified and generalized by changes to the unitary multiset metagrammars representing them; some of the techniques for such changes were already described in this chapter.

Now, having some experience in *UMMG* representation of well-known and simply formulated optimization problems, we may consider one more general problem—integer linear programming (*ILP*) and some of its important particular cases.

10.3 Integer Linear Programming and Its Particular Cases

10.3.1 Integer Linear Programming

Integer linear programming problem is formulated as follows (Hillier and Lieberman 2014; Lasdon 2013; Taha 2016; Fiedler et al. 2006). Find a vector $\|x_1 \ldots x_n\|$ satisfying the condition

$$\sum_{i=1}^{n} c_i x_i \to max, \tag{10.77}$$

under constraints

$$\sum_{j=1}^{n} a_{1j} \cdot x_j \leq b_1,$$

$$\ldots$$

$$\sum_{j=1}^{n} a_{mj} \cdot x_j \leq b_m, \tag{10.78}$$

$$x_i \geq 0, i \in \{1, \ldots, m\}. \tag{10.79}$$

We shall consider only *ILP* with $a_{ij} \geq 0$, $i \in \{1, \ldots, m\}, j \in \{1, \ldots, n\}$.

The appropriate ***unitary multiset metagrammar*** $S = \ < \{1 \cdot (problem)\}, R, F>$ may be constructed by the application of a technique very close to that described in the previous section.

The scheme R contains the following elements:

$$(problem) \to 1 \cdot (opt), 1 \cdot (conditions), 1 \cdot (solution), \tag{10.80}$$

$$(opt) \rightarrow c_1 \cdot (y_1), \ldots, c_n \cdot (y_n), \tag{10.81}$$

$$(conditions) \rightarrow a_{11} \cdot (u_{11}), \ldots, a_{1n} \cdot (u_{1n}), \ldots, a_{m1} \cdot (u_{m1}), \ldots, a_{mn} \cdot (u_{mn}), \tag{10.82}$$

$$(solution) \rightarrow x_1 \cdot \boldsymbol{x}_1, \ldots, x_n \cdot \boldsymbol{x}_n, \tag{10.83}$$

as well as n unitary metarules

$$(y_j) \rightarrow x_j \cdot \boldsymbol{e}, \tag{10.84}$$

where $j \in \{1, \ldots, n\}$, and $m \times n$ unitary metarules

$$(u_{ij}) \rightarrow x_j \cdot \boldsymbol{u}_i, \tag{10.85}$$

where $i \in \{1, \ldots, m\}, j \in \{1, \ldots, n\}$.

Here in (10.81)–(10.85), $c_1, \ldots, c_n, a_{11}, \ldots, a_{1n}, \ldots, a_{m1}, \ldots, a_{mn}$ are multiplicity-constants identical to the constants occurring in the classical formulation (10.77)–(10.79); $x_1, \ldots, x_n$ are multiplicity-variables identical to the variables occurring in (10.77)–(10.79); the terminal object $\boldsymbol{e}$ is a measurement unit of values of the objective function; terminal objects $\boldsymbol{u}_1, \ldots, \boldsymbol{u}_n$ are measurement units of values of linear functions occurring in the left parts of inequalities (10.78); and $\boldsymbol{x}_1, \ldots, \boldsymbol{x}_n$ are terminal objects whose multiplicities in generated terminal multisets are the values of variables $x_1, \ldots, x_n$ from the classical formulation of the *ILP*, so $\boldsymbol{x}_1, \ldots, \boldsymbol{x}_n$ serve for a representation of its solutions. All the remaining objects appearing in (10.80)–(10.85) are intermediate non-terminal objects used for the purpose of generation of *TMSs*. As may be seen from (10.80) to (10.85), generated *TMSs* have the following form:

$$\{c \cdot \boldsymbol{e}, k_1 \cdot \boldsymbol{x}_1, \ldots, k_n \cdot \boldsymbol{x}_n, c_1 \cdot \boldsymbol{u}_1, \ldots, c_m \cdot \boldsymbol{u}_m\}, \tag{10.86}$$

where

$$c = \sum_{i=1}^{n} c_i \cdot k_i, \tag{10.87}$$

$$c_1 = \sum_{j=1}^{n} c_{1i} \cdot k_i,$$

$$\ldots$$

$$c_m = \sum_{j=1}^{n} c_{mi} \cdot k_i, \tag{10.88}$$

and $k_1, \ldots, k_n$ are nothing but the values of variables $x_1, \ldots, x_n$ from the classical formulation of the *ILP* problem.

From (10.86) to (10.88), it is obvious that the filter *F* of the constructed *UMMG* contains $m + n$ boundary conditions

$$\begin{aligned} &\boldsymbol{u}_1 \leq b_1, \\ &\ldots \\ &\boldsymbol{u}_m \leq b_m, \end{aligned} \tag{10.89}$$

$$\begin{aligned} &0 \leq x_1 \leq l_1, \\ &\ldots \\ &0 \leq x_n \leq l_n, \end{aligned} \tag{10.90}$$

as well as one optimizing condition

$$\boldsymbol{e} = max, \tag{10.91}$$

where

$$l_j = \mathbf{min}\left\{\boldsymbol{ent}\left(\frac{b_1}{a_{1j}}\right), \ \ldots, \boldsymbol{ent}\left(\frac{b_m}{a_{mj}}\right)\right\}, \tag{10.92}$$

is the maximal value of variable x_j; this value corresponds to the case where all other variable values are zero; because x_j occurs in all inequalities (10.78), its domain is the result of an intersection of domains, each corresponding to its own inequality, i.e.,

$$[0, l_j] = \bigcap_{i=1}^{m} \left[0, \boldsymbol{ent}\left(\frac{b_{ij}}{a_{ij}}\right)\right] = \left[0, \ min\left\{\boldsymbol{ent}\left(\frac{b_1}{a_{1j}}\right), \ \ldots, \ \boldsymbol{ent}\left(\frac{b_m}{a_{mj}}\right)\right\}\right], \tag{10.93}$$

where $\boldsymbol{ent}(a)$ is the maximal integer number which is not greater than the rational number a. Of course, there is simpler way to define l_i—for example, $l_i = MAX$, where *MAX* is the maximal integer number in the current implementation. However, (10.92) reduces the search space significantly (like (10.74) in the transportation problem), which is why it is important from a computational complexity point of view.

Let us illustrate the above by an example.

Example 10.8
Consider the following *ILP*:

$$2 \cdot x_1 + 3 \cdot x_2 + 6 \cdot x_3 \rightarrow max$$

under the constraints

$$3 \cdot x_1 + 5 \cdot x_2 + 7 \cdot x_3 \leq 15,$$

$$12 \cdot x_1 + 2 \cdot x_2 + x_3 \leq 20,$$

$$x_1 \geq 0, x_2 \geq 0, x_3 \geq 0.$$

According to (10.80)–(10.85), the unitary multiset metagrammar $S = < \{1 \cdot (problem)\}, R, F >$ representing this *ILP* contains the following *URs*, *UMRs*, and conditions:

$$(problem) \rightarrow 1 \cdot (opt), 1 \cdot (conditions), 1 \cdot (solution),$$

$$(opt) \rightarrow 2 \cdot (y_1), 3 \cdot (y_2), 6 \cdot (y_3), x_1 \cdot \boldsymbol{x}_1, x_2 \cdot \boldsymbol{x}_2, x_3 \cdot \boldsymbol{x}_3,$$

$$(conditions) \rightarrow 3 \cdot (u_{11}), 5 \cdot (u_{12}), 7 \cdot (u_{13}), 12 \cdot (u_{21}), 2 \cdot (u_{22}), 1 \cdot (u_{23}),$$

$$(y_1) \rightarrow x_1 \cdot \boldsymbol{e},$$

$$(y_2) \rightarrow x_2 \cdot \boldsymbol{e},$$

$$(y_3) \rightarrow x_3 \cdot \boldsymbol{e},$$

$$(u_{11}) \rightarrow x_1 \cdot \boldsymbol{u}_1,$$

$$(u_{12}) \rightarrow x_2 \cdot \boldsymbol{u}_1,$$

$$(u_{13}) \rightarrow x_3 \cdot \boldsymbol{u}_1,$$

$$(u_{21}) \rightarrow x_1 \cdot \boldsymbol{u}_2,$$

$$(u_{22}) \rightarrow x_2 \cdot \boldsymbol{u}_2,$$

$$(u_{23}) \rightarrow x_3 \cdot \boldsymbol{u}_2,$$

$$\boldsymbol{u}_1 \leq 15,$$

$$\boldsymbol{u}_2 \leq 20,$$

$$0 \leq \boldsymbol{x}_1 \leq 1,$$

$$0 \leq \boldsymbol{x}_2 \leq 3,$$

$$0 \leq \boldsymbol{x}_3 \leq 2,$$

$$\boldsymbol{e} = max.$$

The solution is the one-element set

$$\overline{V_s} = \{\{11 \cdot \boldsymbol{e},\ 1 \cdot \boldsymbol{x}_1,\ 1 \cdot \boldsymbol{x}_2,\ 1 \cdot \boldsymbol{x}_3,\ 15 \cdot \boldsymbol{u}_1,\ 15 \cdot \boldsymbol{u}_2\}\},$$

which corresponds to the solution $x_1 = 1$, $x_2 = 1$, $x_3 = 1$ in the classical formulation.

Evidently, it is quite simple to represent a multiobjective *ILP* problem, replacing (10.80) by

$$(problem) \rightarrow 1 \cdot (opt_1), \ldots, 1 \cdot (opt_k), 1 \cdot (conditions), 1 \cdot (solution), \quad (10.94)$$

and replacing (10.81) by $l > 1$ unitary metarules

$$(opt_i) \rightarrow c_1^i \cdot (y_1^i), \ldots, c_{n_i}^i \cdot \left(y_{n_i}^i\right), \quad (10.95)$$

where $i \in \{1, \ldots, k\}$.

As can be seen, this generalized result was obtained by a "direct" application of the multigrammatical framework to classic optimization problems, i.e., as was said in the introduction to this chapter, by "rewriting" their classical matrix/vector or graph/network representations into the form of the corresponding unitary multiset grammars or unitary multiset metagrammars.

But, as was mentioned in the same introduction, ***the MGF is much more flexible*** than is apparent from the considered cases. Below, we shall demonstrate this flexibility by its application to some well-known Boolean (or bivalent) programming problems. As will be clear, *UMGs/UMMGs* may represent not only a problem as it is but also an algorithm providing an effective search for its solution.

10.3.2 Knapsack Problem

Knapsack problem (Hillier and Lieberman 2014) is usually formulated as follows. There are n items, each having value c_i and weight a_i, $i = 1, \ldots, n$. Also, there is a knapsack with a maximum weight capacity b. The problem is to maximize the sum of the values of the items in the knapsack in such a way that their total weight is not greater than b.

In the usual *ILP* language, this problem is formulated as follows.

Find a vector $\|x_1 \ldots x_n\|$ satisfying the condition

$$\sum_{i=1}^{n} c_i x_i \rightarrow max, \quad (10.96)$$

under the restriction

$$\sum_{i=1}^{n} a_i x_i \le b, \tag{10.97}$$

where $x_i \in \{0, 1\}$, and $x_i = 1$ means that the i-th item is in the knapsack, while $x_i = 0$ corresponds to the opposite case.

We shall represent this problem as a ***filtering unitary multiset grammar*** (*not unitary metagrammar*) $S = \langle\{1 \cdot (q_0)\}, R, F\rangle$, where q_0 is a knapsack which is the start point of the generation process, the filter F contains two conditions

$$\boldsymbol{e} = max, \tag{10.98}$$

$$\boldsymbol{d} \le b, \tag{10.99}$$

where the object $\boldsymbol{e}$ is a measurement unit of items' values, while $\boldsymbol{d}$ is a similar measurement unit of items' weights, and b is the knapsack weight capacity.

The scheme R contains the following unitary rules:

$$(q_i) \rightarrow 1 \cdot (q_{i+1}), c_i \cdot \boldsymbol{e}, a_i \cdot \boldsymbol{d}, 1 \cdot \boldsymbol{q}_i, \tag{10.100}$$

$$(q_i) \rightarrow 1 \cdot (q_{i+1}), \tag{10.101}$$

for all $i = 0, 1, \ldots n$, and also one additional *UR* with an empty body (the syntax of unitary rules requires representation of the empty set $\{\varnothing\}$ as "nothing"):

$$(q_{n+1}) \rightarrow \ . \tag{10.102}$$

In these *URs*, every (q_i) and $\boldsymbol{q}_i$ such that $1 \le i \le n$ corresponds to the i-th item; (q_i) is a non-terminal object ensuring the continuation of *TMS* generation, while $\boldsymbol{q}_i$ is a terminal object whose occurrence in a terminal multiset means that the i-th item is in the knapsack; otherwise, i.e., if a *TMS* does not include $\boldsymbol{q}_i$, the i-th item is not in the knapsack.

As may be seen, generation, starting from the kernel $\{1 \cdot (q_0)\}$, proceeds in such a way that two multisets are generated from every already generated multiset

$$\left\{1 \cdot (q_i), C_i \cdot \boldsymbol{e}, \ D_i \cdot \boldsymbol{d}, \ 1 \cdot \boldsymbol{q}_{j_1}, \ \ldots, 1 \cdot \boldsymbol{q}_{j_i}\right\}, \tag{10.103}$$

where C_i and D_i are, respectively, the sums of values and weights of the j_1-th, ..., j_i -th items already included in the knapsack. The first generated *MS* corresponds to $x_i = 1$ in the classical formulation and adds the i-th item to the knapsack by an application of the *UR* (10.100), thus enabling the application of the next *UR*, whose head is (q_{i+1}). Also c_i and a_i are added to already accumulated multiplicities, and the multiobject $1 \cdot \boldsymbol{q}_i$ is included in the generated multiset. The second *MS*, corresponding to $x_i = 0$, is generated by an application of the *UR* (10.101), which transfers directly to a *UR* with the header (q_{i+1}), and the i-th item is not included in the knapsack; also c_i and a_i are not added to the accumulated value and weight, and the terminal object $\boldsymbol{q}_i$ is not included in the generated *MS*.

There is an additional *UR* (10.102) with an empty body, which allows the generation of multisets not including a terminal multiobject $1 \cdot \boldsymbol{q}_n$, which corresponds to the absence of the n-th item from the knapsack, i.e., $x_n = 0$ in the classical formulation.

Every terminal multiset generated as described above has the form (10.103), and after successful filtering, i.e., if $D_i \leq b$ and C_i is maximal among all such multiplicities occurring in a generated *TMS*, this *TMS* is included in $\overline{V_s}$, which represents the set of all solutions of the problem in question.

Let us illustrate the above by an example.

Example 10.9
Consider the knapsack problem

$$2x_1 + 4x_2 \rightarrow max,$$

where

$$3x_1 + x_2 \leq 5,$$

and $x_1, x_2 \in \{0, 1\}$. According to (10.100)–(10.102), the scheme R of the unitary multigrammar $S = \langle\{1 \cdot (q_0)\}, R, F\rangle$ representing this problem contains the following unitary rules:

$$r_1 : (q_0) \rightarrow 1 \cdot (q_1), 2 \cdot \boldsymbol{e}, 3 \cdot \boldsymbol{d}, 1 \cdot \boldsymbol{q}_1,$$
$$r_2 : (q_0) \rightarrow 1 \cdot (q_1),$$
$$r_3 : (q_1) \rightarrow 1 \cdot (q_2), 4 \cdot \boldsymbol{e}, 1 \cdot \boldsymbol{d}, 1 \cdot \boldsymbol{q}_2,$$
$$r_4 : (q_1) \rightarrow 1 \cdot (q_2),$$
$$r_5 : (q_2) \rightarrow .$$

As can be seen, the application of the *URs* $r_1 - r_5$ in the described manner generates four terminal multisets $v(x_1, x_2)$, where $x_i \in \{0, 1\}$:

$$v(1, 1) = \{6 \cdot \boldsymbol{e}, 4 \cdot \boldsymbol{d}, 1 \cdot \boldsymbol{q}_1, 1 \cdot \boldsymbol{q}_2\},$$
$$v(1, 0) = \{2 \cdot \boldsymbol{e}, 3 \cdot \boldsymbol{d}, 1 \cdot \boldsymbol{q}_1\},$$
$$v(0, 1) = \{4 \cdot \boldsymbol{e}, 1 \cdot \boldsymbol{d}, 1 \cdot \boldsymbol{q}_2\},$$

$$v(0,\ 0) = \{\varnothing\},$$

and

$$\{v(1,\ 1), v(1,\ 0), v(0,\ 1), v(0,\ 0)\} \downarrow \{\boldsymbol{d} \leq 5, \boldsymbol{e} = max\} = \{v(1,\ 1)\},$$

which corresponds to the solution $x_1 = x_2 = 1$ in the classical formulation.

The described techniques may be expanded to the general case of the multiobjective problem of Boolean (bivalent) programming, which is often referred to as the ***multi-dimensional knapsack problem***.

10.3.3 Multiobjective Boolean Programming

A generalization of the knapsack problem, the multiobjective problem of Boolean programming (Hillier and Lieberman 2014), may be formulated in the following way.

Find a vector $\|x_1 ... x_n\|$ satisfying the condition

$$\sum_{i=1}^{n} c_i^1 x_i \rightarrow max$$

$$\cdots$$

$$\sum_{i=1}^{n} c_i^k x_i \rightarrow max \tag{10.104}$$

under restrictions

$$\sum_{i=1}^{n} a_i^l x_i \leq b_1$$

$$\cdots$$

$$\sum_{i=1}^{n} d_i^l x_i \leq b_l \tag{10.105}$$

where $x_i \in \{0, 1\}$. Here, the $k > 1$ optimization criteria (10.104) correspond to k objectives of knapsack filling, while the $l > 1$ restrictions (10.105) correspond to $l > 1$ dimensions of knapsack weight capacity.

This problem may be represented by an ***FUMG*** $S = \langle\{1 \cdot (q_0)\}, R, F\rangle$, where (q_0) is the same as in the previous Sect. 10.3.2, and the filter F contains k optimizing conditions

$$\boldsymbol{e}_1 = max,$$
$$\ldots$$
$$\boldsymbol{e}_k = max, \tag{10.106}$$

as well as l boundary conditions

$$\boldsymbol{d}_1 \le b_1$$
$$\ldots$$
$$\boldsymbol{d}_l \le b_l, \tag{10.107}$$

where $\boldsymbol{e}_1, \ldots, \boldsymbol{e}_k$ are measurement units of items' value and $\boldsymbol{d}_1, \ldots, \boldsymbol{d}_l$ are measurement units of knapsack weight capacity.

The scheme R contains the following unitary rules:

$$(q_i) \to 1 \cdot (q_{i+1}), c_i^1 \cdot \boldsymbol{e}_1, \ldots, c_i^k \cdot \boldsymbol{e}_k, a_i^1 \cdot \boldsymbol{d}_1, \ldots, a_i^l \cdot \boldsymbol{d}_l, 1 \cdot \boldsymbol{q}_i \tag{10.108}$$

$$(q_i) \to 1 \cdot (q_{i+1}), \tag{10.109}$$

for all i=0,1,..., n, and also an additional unitary rule (10.102). Here, q_i and $\boldsymbol{q}_i$ have the same sense as above; $c_i^1, \ldots, c_i^k$ are multiplicities defining values of the i-th item in k various dimensions with measurement units $\boldsymbol{e}_1, \ldots, \boldsymbol{e}_k$; and $a_i^1, \ldots, a_i^l$ are multiplicities defining the weight of the i-th item in l various dimensions with measurement units $\boldsymbol{d}_1, \ldots, \boldsymbol{d}_l$. So every terminal multiset generated by an application of unitary rules (10.108), (10.109), and (10.102) is

$$\left\{ C_1 \cdot \boldsymbol{e}_1, \ldots, C_k \cdot \boldsymbol{e}_k, D_1 \cdot \boldsymbol{d}_1, \ldots, D_l \cdot \boldsymbol{d}_l, 1 \cdot \boldsymbol{q}_{i_1}, \ldots, 1 \cdot \boldsymbol{q}_{i_p} \right\}. \tag{10.110}$$

After successful filtering, i.e., if all $D_1 \le b_1, \ldots, D_l \le b_l$, and all $C_1, \ldots, C_k$ are maximal among all such multiplicities occurring in a generated *TMS*, this terminal multiset is included in the set $\bar{V}_S$, which represents all possible solutions of the problem.

Example 10.10

Consider the multiobjective problem of Boolean programming

$$x_1 + 3x_2 \to max$$
$$2x_1 + x_2 \to max,$$

where

$$2x_1 + 5x_2 \leq 6,$$

$$3x_1 + x_2 \leq 3,$$

and $x_1, x_2 \in \{0, 1\}$. According to (10.108), (10.109), and (10.102), the scheme R of the *UMG* $S = \langle\{1 \cdot q_0\}, R, F\rangle$ representing this problem contains the following unitary rules:

$$r_1 : (q_0) \rightarrow 1 \cdot (q_1), 1 \cdot \boldsymbol{e}_1, 2 \cdot \boldsymbol{e}_2, 2 \cdot \boldsymbol{d}_1, 3 \cdot \boldsymbol{d}_2, 1 \cdot \boldsymbol{q}_1,$$

$$r_2 : (q_0) \rightarrow 1 \cdot (q_1),$$

$$r_3 : (q_1) \rightarrow 1 \cdot (q_2), 3 \cdot \boldsymbol{e}_1, 1 \cdot \boldsymbol{e}_2, 5 \cdot \boldsymbol{d}_1, 1 \cdot \boldsymbol{d}_2, 1 \cdot \boldsymbol{q}_2,$$

$$r_4 : (q_1) \rightarrow 1 \cdot (q_2),$$

$$r_5 : (q_2).$$

The application of $r_1 - r_5$ generates four terminal multisets $v\ (x_1, x_2)$, where x_1, $x_2 \in \{0, 1\}$:

$$v(1,\ 1) = \{4 \cdot \boldsymbol{e}_1,\ 3 \cdot \boldsymbol{e}_2,\ 7 \cdot \boldsymbol{d}_1,\ 4 \cdot \boldsymbol{d}_2,\ 1 \cdot \boldsymbol{q}_1,\ 1 \cdot \boldsymbol{q}_2\},$$

$$v(1,\ 0) = \{1 \cdot \boldsymbol{e}_1,\ 2 \cdot \boldsymbol{e}_2,\ 2 \cdot \boldsymbol{d}_1,\ 3 \cdot \boldsymbol{d}_2,\ 1 \cdot \boldsymbol{q}_1\},$$

$$v(0,\ 1) = \{3 \cdot \boldsymbol{e}_1,\ 1 \cdot \boldsymbol{e}_2,\ 5 \cdot \boldsymbol{d}_1,\ 1 \cdot \boldsymbol{d}_2,\ 1 \cdot \boldsymbol{q}_2\},$$

$$v(0,\ 0) = \{\varnothing\},$$

and

$$(\{v(1,\ 1),\ v(1,\ 0),\ v(0,\ 1),\ v(0,\ 0)\} \downarrow \{\boldsymbol{d}_1 \leq 6,\ \boldsymbol{d}_2 \leq 3\}) \downarrow \{\boldsymbol{e}_1 = max,\ \boldsymbol{e}_2 = max\}$$

$$= \{\ v(1,\ 0), v(0,\ 1), v(0,\ 0)\} \downarrow \{\boldsymbol{e}_1 = max,\ \boldsymbol{e}_2 = max\} = \{\ v(0,\ 1)\} \cap \{\ v(1,\ 0)\} = \{\varnothing\},$$

so there are no solutions of this problem.

The proposed technique may be effectively applied not only to Boolean but, as a whole, to integer linear programming, as described in the following section.

10.3.4 Multigrammatical Representation of an ILP *Problem Reduced to a Problem of Boolean Programming*

As is known from Hillier and Lieberman (2014) and Taha (2016), a problem of integer linear programming in the case of finite domains of variables may be reduced to a problem of Boolean programming by the replacement of the set of integer-domain variables by an equivalent set of Boolean-domain variables.

Consider an *ILP* problem

$$c_1x_1 + \ldots + c_nx_n \rightarrow max \tag{10.111}$$

under restrictions

$$\begin{aligned} a_{11} \cdot x_1 + \ldots + a_{1n} \cdot x_n &\le b_1 \\ &\ldots \\ a_{m1} \cdot x_1 + \ldots + a_{mn} \cdot x_n &\le b_m, \end{aligned} \tag{10.112}$$

and also for all $i \in [1, n]$

$$x_i \in \left[0,\ 2^{M_i} - 1\right], \tag{10.113}$$

where $2^{M_i} - 1$ is the maximal value of variable x_i (in the simplest case, all M_i are equal to one and the same value M). By this assumption, every integer value l may be represented by a binary number $b_1b_2 \ldots b_{M_i} \in \{0,\ 1\}^{M_i}$, and, evidently, $l = 0$ corresponds to 00…0 (M_i times 0), while $l = 2^{M_i} - 1$ corresponds to 11…1 (M_i times 1). Let us replace every variable x_i by a sum

$$\begin{aligned} x_i &= x^i_{M_i} + 2 \cdot x^i_{M_{i-1}} + \ldots + 2^j \cdot x^i_{M_{i-j}} + \ldots + 2^{M_{i-1}} \cdot x^i_1 \\ &= \sum_{j=1}^{M_i} 2^{j-1} \cdot x^i_j\ , \end{aligned} \tag{10.114}$$

where x^i_j is a Boolean variable whose value is the j-th bit of the binary representation of x_i; as a result, the total number of variables will increase from n to

$$N = \sum_{i=1}^{n} M_i\ . \tag{10.115}$$

Example 10.11

Let the maximal value of the variable x_i be 7, i.e., $M_i = 3$. Then x_i may be replaced by

$$x_i = x^i_3 + 2x^i_2 + 4x^i_1.$$

If the value of the variable x_i is $5 = 101_2$, then $x^i_3 = 1$, $x^i_2 = 0$, and $x^i_1 = 1$.

In this way, an *ILP* problem (10.111)–(10.113) with n variables $x_1, \ldots, x_n$ may be replaced by a Boolean programming problem with variables $\boldsymbol{c_1, \ldots, c_N, a_{11}, \ldots, a_{mN}}$, recalculated in a full accordance with (10.114), and new Boolean variables $\boldsymbol{x_1}, \ldots, \boldsymbol{x_N}$:

$$\boldsymbol{c_1} \cdot \boldsymbol{x_1} + \ldots + \boldsymbol{c_N} \cdot \boldsymbol{x_N} \to max \tag{10.116}$$

under restrictions

$$\boldsymbol{a_{11}} \cdot \boldsymbol{x_1} + \ldots + \boldsymbol{a_{1N}} \cdot \boldsymbol{x_N} \leq \boldsymbol{b_1}$$

$$\ldots$$

$$\boldsymbol{a_{m1}} \cdot \boldsymbol{x_1} + \ldots + \boldsymbol{a_{mN}} \cdot \boldsymbol{x_N} \leq \boldsymbol{b_m}. \tag{10.117}$$

This problem, in turn, may be easily represented by a ***filtering unitary multiset grammar***, as was described above in Sects. 10.3.2–10.3.3.

Example 10.12
Consider the *ILP* problem

$$2x_1 + 3x_2 \to max$$

under the restrictions

$$x_1 + 2x_2 \leq 4,$$

$$2x_1 + x_2 \leq 5,$$

where $x_i \in [0, 3]$, $x_2 \in [0, 3]$. According to (10.111)–(10.117), this problem may be replaced by

$$2\left(x_2^1 + 2x_1^1\right) + 3\left(x_2^2 + 2x_1^2\right) \to max$$

under the restrictions

$$\left(x_2^1 + 2x_1^1\right) + 2\left(x_2^2 + 2x_1^2\right) \leq 4$$

$$2\left(x_2^1 + 2x_1^1\right) + \left(x_2^2 + 2x_1^2\right) \leq 5,$$

or, after simplification,

$$4x_1^1 + 2x_2^1 + 6x_1^2 + 3x_2^2 \to max$$

under the restrictions

$$2x_1^1 + x_2^1 + 4x_1^2 + 2x_2^2 \leq 4$$

$$4x_1^1 + 2x_2^1 + 2x_1^2 + x_2^2 \leq 5.$$

This problem is represented by the *FUMG* $S = \langle \{1 \cdot (q_0)\}, R, F \rangle$, where the filter F contains one optimizing and two boundary conditions:

$$\boldsymbol{e} = max,$$

$$\boldsymbol{d}_1 \leq 4,$$

$$\boldsymbol{d}_2 \leq 5.$$

The scheme R contains the following unitary rules:

$$r_1 : (q_0) \to 1 \cdot (q_1^1), 4 \cdot \boldsymbol{e}, 2 \cdot \boldsymbol{d}_1, 4 \cdot \boldsymbol{d}_2, 1 \cdot \boldsymbol{q}_1^1,$$

$$r_2 : (q_0) \to 1 \cdot (q_1^1),$$

$$r_3 : (q_1^1) \to 1 \cdot (q_2^1), 2 \cdot \boldsymbol{e}, 1 \cdot \boldsymbol{d}_1, 2 \cdot \boldsymbol{d}_2, 1 \cdot \boldsymbol{q}_2^1,$$

$$r_4 : (q_1^1) \to 1 \cdot (q_2^1),$$

$$r_5 : (q_2^1) \to 1 \cdot (q_1^2), 6 \cdot \boldsymbol{e}, 4 \cdot \boldsymbol{d}_1, 2 \cdot \boldsymbol{d}_2, 1 \cdot \boldsymbol{q}_1^2,$$

$$r_6 : (q_2^1) \to 1 \cdot (q_1^2),$$

$$r_7 : (q_1^2) \to 1 \cdot (q_2^2), 3 \cdot \boldsymbol{e}, 2 \cdot \boldsymbol{d}_1, 1 \cdot \boldsymbol{d}_2, 1 \cdot \boldsymbol{q}_2^2,$$

$$r_8 : (q_1^2) \to 1 \cdot (q_2^2),$$

$$r_9 : (q_2^2).$$

The application of the *URs* $r_1 - r_9$ generates the $2^4 = 16$ terminal multisets $v(x_1^1, x_2^1, x_1^2, x_2^2)$, where $x_j^i \in \{0, 1\}$:

$$v(0, 0, 0, 0) = \{\varnothing\},$$

$$v(0, 0, 0, 1) = \{3 \cdot \boldsymbol{e}, 2 \cdot \boldsymbol{d}_1, 1 \cdot \boldsymbol{d}_2, 1 \cdot \boldsymbol{q}_2^2\},$$

$$v(0, 0, 1, 0) = \{6 \cdot \boldsymbol{e}, 4 \cdot \boldsymbol{d}_1, 2 \cdot \boldsymbol{d}_2, 1 \cdot \boldsymbol{q}_1^2\},$$

$$v(0, 0, 1, 1) = \{9 \cdot \boldsymbol{e}, 6 \cdot \boldsymbol{d}_1, 3 \cdot \boldsymbol{d}_2, 1 \cdot \boldsymbol{q}_1^2, 1 \cdot \boldsymbol{q}_2^2\},$$

$$v(0, 1, 0, 0) = \{2 \cdot \boldsymbol{e}, 1 \cdot \boldsymbol{d}_1, 2 \cdot \boldsymbol{d}_2, 1 \cdot \boldsymbol{q}_2^1\},$$

$$v(0, 1, 0, 1) = \{5 \cdot \boldsymbol{e}, 3 \cdot \boldsymbol{d}_1, 3 \cdot \boldsymbol{d}_2, 1 \cdot \boldsymbol{q}_2^1, 1 \cdot \boldsymbol{q}_2^2\},$$

$$v(0,\ 1,\ 1,\ 0) = \{8 \cdot \boldsymbol{e},\ 5 \cdot \boldsymbol{d}_1,\ 4 \cdot \boldsymbol{d}_2,\ 1 \cdot \boldsymbol{q}_2^1,\ 1 \cdot \boldsymbol{q}_1^2\},$$

$$v(0,\ 1,\ 1,\ 1) = \{11 \cdot \boldsymbol{e},\ 7 \cdot \boldsymbol{d}_1,\ 5 \cdot \boldsymbol{d}_2,\ 1 \cdot \boldsymbol{q}_2^1,\ 1 \cdot \boldsymbol{q}_1^2,\ 1 \cdot \boldsymbol{q}_2^2\},$$

$$v(1,\ 0,\ 0,\ 0) = \{4 \cdot \boldsymbol{e},\ 2 \cdot \boldsymbol{d}_1,\ 4 \cdot \boldsymbol{d}_2,\ 1 \cdot \boldsymbol{q}_1^1,\ \},$$

$$v(1,\ 0,\ 0,\ 1) = \{7 \cdot \boldsymbol{e},\ 4 \cdot \boldsymbol{d}_1,\ 5 \cdot \boldsymbol{d}_2,\ 1 \cdot \boldsymbol{q}_1^1,\ 1 \cdot \boldsymbol{q}_2^2\},$$

$$v(1,\ 0,\ 1,\ 0) = \{10 \cdot \boldsymbol{e},\ 6 \cdot \boldsymbol{d}_1,\ 6 \cdot \boldsymbol{d}_2,\ 1 \cdot \boldsymbol{q}_1^1,\ 1 \cdot \boldsymbol{q}_1^2\},$$

$$v(1,\ 0,\ 1,\ 1) = \{13 \cdot \boldsymbol{e},\ 8 \cdot \boldsymbol{d}_1,\ 7 \cdot \boldsymbol{d}_2,\ 1 \cdot \boldsymbol{q}_1^1,\ 1 \cdot \boldsymbol{q}_1^2,\ 1 \cdot \boldsymbol{q}_2^2\},$$

$$v(1,\ 1,\ 0,\ 0) = \{6 \cdot \boldsymbol{e},\ 3 \cdot \boldsymbol{d}_1,\ 6 \cdot \boldsymbol{d}_2,\ 1 \cdot \boldsymbol{q}_1^1,\ 1 \cdot \boldsymbol{q}_2^1\},$$

$$v(1,\ 1,\ 0,\ 1) = \{9 \cdot \boldsymbol{e},\ 5 \cdot \boldsymbol{d}_1,\ 7 \cdot \boldsymbol{d}_2,\ 1 \cdot \boldsymbol{q}_1^1,\ 1 \cdot \boldsymbol{q}_2^1,\ 1 \cdot \boldsymbol{q}_2^2\},$$

$$v(1,\ 1,\ 1,\ 0) = \{12 \cdot \boldsymbol{e},\ 7 \cdot \boldsymbol{d}_1,\ 8 \cdot \boldsymbol{d}_2,\ 1 \cdot \boldsymbol{q}_1^1,\ 1 \cdot \boldsymbol{q}_2^1,\ 1 \cdot \boldsymbol{q}_1^2\},$$

$$v(1,\ 1,\ 1,\ 1) = \{15 \cdot \boldsymbol{e},\ 9 \cdot \boldsymbol{d}_1,\ 9 \cdot \boldsymbol{d}_2,\ 1 \cdot \boldsymbol{q}_1^1,\ 1 \cdot \boldsymbol{q}_2^1,\ 1 \cdot \boldsymbol{q}_1^2,\ 1 \cdot \boldsymbol{q}_2^2\}.$$

After filtering by the boundary conditions $\boldsymbol{d}_1 \leq 4$ and $\boldsymbol{d}_2 \leq 5$, we obtain the set of terminal multisets $\{v(0,0,0,0), v(0,0,0,1), v(0,0,1,0), v(0,1,0,0), v(0,1,0,1), v(1,0,0,0), v(1,0,0,1)\}$, which, being filtered by the optimizing condition $\boldsymbol{e} = max$, is reduced to

$$\bar{V}_S = \{v(1,\ 0,\ 0,\ 1)\} = \{\{7 \cdot \boldsymbol{e},\ 4 \cdot \boldsymbol{d}_1,\ 5 \cdot \boldsymbol{d}_2,\ 1 \cdot \boldsymbol{q}_1^1,\ 1 \cdot \boldsymbol{q}_2^2\}\}.$$

This means that the initial *ILP* problem has the solution $x_1 = 10_2 = 2$ and $x_2 = 01_2 = 1$ and the objective function value $2x_1 + 3x_2 = 7$.

The proposed and illustrated logic of the construction of the *FUMG* representing the *ILP* problem with finite domains of variables is the basis for a somewhat non-evident statement.

Statement 10.1 If the domains of variables of an *ILP* problem are finite, it may be represented by a filtering unitary multiset grammar.

Proof is made, in fact, by the above description of the logic of construction of an *FUMG*.

Because all classic optimization problems considered in this chapter are particular cases of the *ILP* problem, it is clear that, although some of them were represented by unitary multiset metagrammars, in the case of finite domains of variables, all these problems may be represented by filtering unitary multiset grammars. ***Due to the descriptional capabilities of the MGF, no additional effort needs to be applied if, instead of the one-objective problem considered in this section, we proceed to a multiobjective one***.

Until now, we have considered interconnections between operations research and the *MGF* by multigrammatical representation of some well-known problems of *OpR*. Not less interesting and important is an investigation of the aforementioned interconnections in the reverse direction, namely, by the reduction of the task of generation of the set of terminal multisets defined by some multigrammar to a corresponding problem of operations research. Such a reduction may be very useful from the point of view of assessment of the computational complexity of the aforementioned generation. We shall follow this approach regarding unitary multiset metagrammars, which are the most practically applicable class of non-temporal *MGs*.

10.4 Correspondence of Unitary Multiset Metagrammars and Multiobjective Problems of Discrete Polynomial Programming

Following Sheremet (2010, 2011a), we shall begin with the simplest case of the announced task—acyclic non-variative *UMMGs*.

Let us recall the basic constructions defined in Sect. 3.3.2 regarding unitary multiset metagrammars.

Let $S = \langle \{1 \cdot \boldsymbol{a}_0\}, R, F \rangle$ be a non-variative *UMMG* whose filter F is the join of two subfilters: $F_{\leq}$, comprising boundary conditions, and F_{opt}, comprising optimizing conditions.

Let us recall that boundary conditions may be of two types:

$$n \leq a \leq n', \tag{10.118}$$

$$n \leq \gamma \leq n', \tag{10.119}$$

the first one (10.118) determining a set of values (domain) of multiplicities of a terminal object a in terminal multisets belonging to the set $\overline{V_s}$ and the second (10.108) determining the domain of values of a multiplicity-variable γ, occurring in bodies of unitary metarules appearing in the scheme R.

Optimizing conditions, in turn, may be of the following four types:

$$a = min, \tag{10.120}$$

$$a = max, \tag{10.121}$$

$$\gamma = min, \tag{10.122}$$

$$\gamma = max, \tag{10.123}$$

where (10.120) and (10.121) determine that the set $\overline{V_s}$ includes *TMSs* in which the multiplicity of a terminal object a is minimal (maximal) from all *TMSs* belonging to the set $\overline{V_{s'}}$, where $S' = \langle\{1 \cdot \boldsymbol{a}_0\}, R\rangle$ is the core *MG* of the *UMMG S*; (10.122) and (10.123), in turn, determine that the set $\overline{V_s}$ includes *TMSs* created by generation chains in which the minimal (maximal) value of a variable-multiplicity γ over all generation chains is used.

Because a *UMMG S* is non-variative, there exists a single unitary metarule (in the case of the absence of *MVs* in its body a unitary rule)

$$\boldsymbol{a} \rightarrow \mu_1 \cdot a_1, \ldots, \mu_m \cdot a_m, \tag{10.124}$$

where any μ_i may be an integer number or a variable-multiplicity, for every non-terminal object $\boldsymbol{a}$.

Let us determine in a general form the multiplicity of a terminal object a in a terminal multiset generated by a *UMMG S*. The summand of this value M_a, corresponding to some generation chain from the kernel $\{1 \cdot \boldsymbol{a}_0\}$ to the terminal object a, is nothing but the product of all multiplicities belonging to this chain:

$$\begin{gathered}
\{1 \cdot \boldsymbol{a}_0\} \overset{r_{l1}}{\Rightarrow} \{\ldots, \mu_{i1} \cdot \boldsymbol{a}_{j1}, \ldots\}, \\
\{1 \cdot \boldsymbol{a}_{j1}\} \overset{r_{l2}}{\Rightarrow} \{\ldots, \mu_{i2} \cdot \boldsymbol{a}_{j2}, \ldots\}, \\
\ldots \\
\{1 \cdot \boldsymbol{a}_{jk}\} \overset{r_{lk+1}}{\Rightarrow} \{\ldots, \mu_{ik+1} \cdot \boldsymbol{a}_{jk+1}, \ldots\}, \\
\ldots \\
\{1 \cdot \boldsymbol{a}_{jm-1}\} \overset{r_{lm}}{\Rightarrow} \{\ldots, \mu_{im} \cdot a, \ldots\},
\end{gathered} \tag{10.125}$$

so

$$M_a = \prod_{l=1}^{m} \mu_{il}, \tag{10.126}$$

or, after multiplication,

$$M_a = h_a \cdot \left(\gamma_{p_1}\right)^{q_1} \cdot \ldots \cdot \left(\gamma_{p_t}\right)^{q_t}, \tag{10.127}$$

where h_a is the result of multiplication of all multiplicity-constants belonging to the chain (10.125), while $\left(\gamma_{p_1}\right)^{q_1} \cdot \ldots \cdot \left(\gamma_{p_t}\right)^{q_t}$ is the product of all multiplicity-variables belonging to this chain. The fact that any *MV* γ_{pj} may occur in this chain some $q_j > 1$ times is reflected by the power q_j of this *MV*. In the general case, a chain may include only *MCs*, so

$$M_a = h_a. \tag{10.128}$$

Let us note that in the most general case, a terminal object a may occur in the bodies of more than one *UMR*, so its multiplicity in a terminal multiset generated by the application of the *UMMG S* is

$$\boldsymbol{M}_a\left(\gamma_1^a, \ldots, \gamma_{m_a}^a\right) =$$

$$h_1^a \cdot \left(\gamma_1^1\right)^{q_{1,1}} \cdot \ldots \cdot \left(\gamma_{p_1}^1\right)^{q_{1,p_1}} + \ldots + h_u^a \cdot \left(\gamma_1^u\right)^{q_{u,1}} \cdot \ldots \cdot \left(\gamma_{p_u}^u\right)^{q_{u,p_u}}, \tag{10.129}$$

where u is the number of generation chains whose results (i.e., *TMSs* generated by them) include the terminal object a, and

$$\left\{\gamma_1^a, \ldots, \gamma_{m_a}^a\right\} = \bigcup_{i=1}^{u} \bigcup_{j=1}^{p_i} \left\{\gamma_j^i\right\} \tag{10.130}$$

is the set of *MVs* occurring in this sum. As can be seen, $\boldsymbol{M}_a\left(\gamma_1^a, \ldots, \gamma_{m_a}^a\right)$ is a polynomial in the variables $\gamma_1^a, \ldots, \gamma_{m_a}^a$.

Now, if the filter *F* includes a boundary condition

$$n \le a \le n', \tag{10.131}$$

it corresponds to a restriction

$$n \le \boldsymbol{M}_a\left(\gamma_1^a, \ldots, \gamma_{m_a}^a\right) \le n' \tag{10.132}$$

in the sense of the formulation of a conventional problem of mathematical programming, and if the filter *F* includes an optimizing condition $a = min$ or $a = max$, it corresponds to an objective

$$\boldsymbol{M}_a\left(\gamma_1^a, \ldots, \gamma_{m_a}^a\right) \to min \tag{10.133}$$

or

$$\boldsymbol{M}_a\left(\gamma_1^a, \ldots, \gamma_{m_a}^a\right) \to max, \tag{10.134}$$

respectively, in the same sense.

After the execution of the described steps regarding all terminal objects occurring in the filter F, we obtain the following optimization problem:find a vector $\|\gamma_1 ... \gamma_n\|$ from conditions

$$\boldsymbol{M}_{a_{i1}}\left(\gamma_1^{a_{i1}}, \ldots, \gamma_{m_{a_{i1}}}^{a_{i1}}\right) \rightarrow \boldsymbol{opt}_1,$$

$$\ldots$$

$$\boldsymbol{M}_{a_{ik}}\left(\gamma_1^{a_{ik}}, \ldots, \gamma_{m_{a_{ik}}}^{a_{ik}}\right) \rightarrow \boldsymbol{opt}_k \tag{10.135}$$

under restrictions

$$n_1 \leq \boldsymbol{M}_{a_{j1}}\left(\gamma_1^{a_{j1}}, \ldots, \gamma_{m_{a_{j1}}}^{a_{j1}}\right) \leq n'_1,$$

$$\ldots$$

$$n_l \leq \boldsymbol{M}_{a_{jl}}\left(\gamma_1^{a_{jl}}, \ldots, \gamma_{m_{a_{jl}}}^{a_{jl}}\right) \leq n'_l, \tag{10.136}$$

where

$$\boldsymbol{opt}_j = \begin{cases} max, & \text{if } a_{ij} = max \in F, \\ min & \text{otherwise.} \end{cases} \tag{10.137}$$

As can be seen, because the objective functions and restrictions are polynomials in variables $\gamma_1, \ldots, \gamma_n$, this is the standard formulation of a ***multiobjective problem of discrete polynomial programming.***

Let us illustrate the above by the following example.

Example 10.13
Let $S = \langle\{1 \cdot \boldsymbol{a}_0\}, R, F\rangle$ be the acyclic non-variative *UMMG* whose scheme R includes the following unitary rules and metarules:

$$\boldsymbol{a}_0 \rightarrow \gamma_1 \cdot \boldsymbol{a}_1, 5 \cdot \boldsymbol{a}_2,$$

$$\boldsymbol{a}_1 \rightarrow 4 \cdot \boldsymbol{a}_2, 2 \cdot \boldsymbol{a}_4,$$

$$\boldsymbol{a}_2 \rightarrow \gamma_2 \cdot \boldsymbol{a}_3, 5 \cdot \boldsymbol{a}_4,$$

$$\boldsymbol{a}_3 \rightarrow 3 \cdot \boldsymbol{a}_4, 5 \cdot \boldsymbol{a}_5,$$

while the filter F includes the following conditions:

$$0 \leq \gamma_1 \leq 10\,,$$

$$0 \leq \gamma_2 \leq 3,$$

$$0 \leq a_5 \leq 6,$$

$$a_4 = min\,,$$

$$a_5 = max\,.$$

According to (10.124)–(10.137), this *UMMG* is reduced to the following multiobjective problem of discrete polynomial programming:find the vector $\|\gamma_1\ \gamma_2\|$ from the condition
$12\gamma_1\ \gamma_2 + 22\gamma_1 + 15\gamma_2 + 25 \rightarrow \boldsymbol{min}$

$$20\gamma_1\ \gamma_2 + 25\gamma_2 \rightarrow \boldsymbol{max}$$

under restrictions

$$0 \leq 20\gamma_1\ \gamma_2 + 25\gamma_2 \leq 6,$$

$$0 \leq \gamma_1 \leq 10\,,$$

$$0 \leq \gamma_2 \leq 3.$$

We have considered the basic case of non-variative acyclic *UMMGs*. However, as shown in Sect. 4.2, any variative acyclic *UMMG* may be replaced by an equivalent non-variative acyclic *UMMG* whose application enables the generation of the same set of terminal multisets. Hence, the presented result may be expanded correctly on a general case of acyclic *UMMGs*.

As one can see, the multigrammatical framework provides a relatively simple and understandable representation of classic optimization problems of operations research, although we have demonstrated the capabilities of *FUMGs/UMMGs* without any of the various refinements of their basic syntax and semantics, which would make a multigrammatical representation of the aforementioned problems even more compact and natural. Having in mind this conclusion, as well as the evident opportunities for a highly parallel hardware implementation of the *MGF*, let us consider its interconnections with some closely related models of computation.

Chapter 11
Interconnections Between Multiset Grammars and Models of Computation

A comparative study of the multigrammatical framework and various well-known models of computation is valuable from the point of view of theoretical computer science, because it provides one more estimation of the real descriptional power of the *MGF* compared to other formal systems developed earlier. There is a wide range of models of computation whose semantics may be directly represented by various classes of multiset grammars and, in some cases, vice versa. This chapter concerns the following models which also may be applied to the formulation and solution of some of the above-considered problems from the *STS* area—***vector addition and substitution systems*** (Sect. 11.1); ***production functions*** (Sect. 11.2); ***Petri nets*** (*PNs*), whose nexus with the *MGF* is studied in Sect. 11.3; and ***string-operating grammars*** (Sect. 11.4). Interconnections between *MGs* and one of the most advanced *AI* models—multi-agent systems—were, in fact, considered in Sect. 4.3, where a *MAT*-based technique of implementation of filtering multiset grammars was described.

11.1 Vector Addition and Substitution Systems

As is known, a finite multiset

$$v = \{n_1 \cdot a_1, \ \ldots, n_m \cdot a_m\} \tag{11.1}$$

may be presented as an integer-valued vector

$$x = \|n_1 \ldots n_m\|. \tag{11.2}$$

This mapping from a set of finite multisets to a set of integer-valued vectors is called ***Parikh's mapping*** (Parikh 1966). Both classes of systems considered below, operating on such vectors, are based on this mapping.

I. A. Sheremet, *Multigrammatical Framework for Knowledge-Based Digital Economy*, https://doi.org/10.1007/978-3-031-13858-4_11

11.1.1 Vector Addition Systems

These systems, introduced in Karp and Miller (1969), are a direct analog of multigrammars regarding vector representation of multisets.

A ***vector addition system (VAS)*** X is a couple $<x_0, X_\Sigma>$, where x_0 is an initial vector and X_Σ is a set of ***basis vectors*** $\{x_1, \ldots, x_k\}$; all of $x_0, x_1, \ldots, x_k$ are integer-valued vectors, and all components of the vector x_0 are non-negative.

The ***reachability set of the vector addition system*** X, designated below X^*, is determined in the following way:

$$\overline{X}_{(0)} = \{x_0\}, \tag{11.3}$$

$$\overline{X}_{(i+1)} = \overline{X}_{(i)} \cup \left[\bigcup_{x \in \overline{X}_{(i)}} \bigcup_{x_j \in X_\Sigma} \{x + x_j \mid x + x_j \geq 0\} \right], \tag{11.4}$$

$$X^* = \overline{X}_{(\infty)}. \tag{11.5}$$

Example 11.1

Assume we have the vector addition system $X = <x_0, X_\Sigma>$, where $x_0 = \|3\ 4\ 1\|$ and X_Σ includes three basis vectors

$$x_1 = \| -1 \quad 0 \quad 2\|,$$

$$x_2 = \| \ 0 -2 \quad 1\|,$$

$$x_3 = \| \ 1 \quad 1 -3\|.$$

Let us consider some two steps of the process of creation of the reachability set of this system, as represented by Table 11.1.

Note that the vectors $x_0 + x_3$ and $x_5 + x_3$ are not included in the sets $X_{(1)}$ and $X_{(2)}$ due to the negativity of their third components; nor is the vector $x_5 + x_1$—due to its identity with the vector $x_4 + x_2$. An interested reader may write down the further steps independently, if desired.

As may easily be seen, any vector addition system $X = <x_0, X_\Sigma>$ may be put in compliance with some multiset grammar $S = <v_0, R>$ in such a way that:

Table 11.1 Creation of the reachability set

x_0	$x_0 + x_1$	$x_0 + x_2$	$x_0 + x_3$	$X_{(1)} = X_{(0)} \cup$
3 4 1	2 4 3	3 2 2	4 5 − 2	$\cup\{\|3\ 4\ 1\|, \|2\ 4\ 3\|, \|3\ 2\ 2\|\}$
x_4	$x_4 + x_1$	$x_4 + x_2$	$x_4 + x_3$	$X_{(2)} = X_{(1)} \cup$
2 4 3	1 4 5	2 2 4	3 5 0	$\cup\{\|1\ 4\ 5\|, \|2\ 2\ 4\|, \|3\ 5\ 0\|, \|3\ 0\ 3\|\}$
x_5	$x_5 + x_1$	$x_5 + x_2$	$x_5 + x_3$	
3 2 2	2 2 4	3 0 3	4 3 − 1	

1. The dimensionality of vectors $x \in \{x_0\} \cup X_\Sigma$ is equal to the number of objects in the *MG S*, i.e., $|A_S|$.
2. The vector x_0 corresponds to the multiset v_0, so that the i-th component of this vector x_0^i is determined as follows:

$$x_0^i = \begin{cases} n_i, & \text{if } \ n_i \cdot a_i \in v_0, \\ 0 & \text{otherwise;} \end{cases} \tag{11.6}$$

3. The rule $r_j \in R$

$$\{n_1^j \cdot a_{i_1}^j, \ \ldots, n_m^j \cdot a_{i_m}^j\} \to \left\{n_{m+1}^j \cdot a_{i_{m+1}}^j, \ \ldots, n_{m+k}^j \cdot a_{i_{m+k}}^j\right\} \tag{11.7}$$

corresponds to the basis vector $x_j \in X_\Sigma$ whose i_l-th component is determined as follows:

$$x_{i_l}^j = \begin{cases} -n_l^j & \text{if } l = 1, \ \ldots, m, \\ n_l^j & \text{if } l = m+1, \ \ldots, m+k, \\ 0 & \text{otherwise.} \end{cases} \tag{11.8}$$

In other words, if $k \in \{1,\ldots,m\}$, then the i_k-th component of a vector x_j is assumed equal to minus the multiplicity n_k^j of the multiobject $n_k^j \cdot a_{i_k}^j$ from the left part of the rule r_j. On the other hand, if $k \in \{m + 1,\ldots, m + k\}$, the i_k-th component of a vector x_j is assumed equal to the multiplicity n_k^j of the multiobject $n_k^j \cdot a_{i_k}^j$ from the right part of the rule r_j. All other components of a vector x_j are zero.

Due to (11.6)–(11.8), the ***process of generation of multisets, implemented by application of rules, is modeled naturally and adequately by the process of addition of vectors corresponding to these rules***. In this case, each such addition of any negative component of a vector corresponds to a multiobject from the left part of a rule, causing extraction of the appropriate number of objects from a multiset created at the previous steps, and any addition of a positive component of this vector corresponds to multiobjects from the right part of a rule, causing inclusion of the appropriate number of objects in this multiset. The logic of modeling multigrammars by vector addition systems as a whole is illustrated by the following example.

Example 11.2
Consider the multiset grammar $S = \ <v_0, R>$, where the kernel

$$v_0 = \{1 \cdot a_1, 3 \cdot a_2\},$$

and the scheme R includes two rules

$$r_1: \ \{1\cdot a_1\}\rightarrow\{1\cdot a_2\},$$
$$r_2: \ \{5\cdot a_2\}\rightarrow\{1\cdot a_1\}.$$

As may be seen,

$$V_S=\{\{1\cdot a_1,\ 3\cdot a_2\},\{4\cdot a_2\}\}.$$

According to (11.6)–(11.8), the vector addition system $X = \ < x_0,\ X_\Sigma>$ corresponding to this multigrammar is as follows:

$$x_0=\|\ \ 1 \quad 3\ \|,$$
$$x_1=\|-1 \quad 1\ \|,$$
$$x_2=\|\ \ 1-5\ \|.$$

The reachability set of this *VAS* is computed in the following way:

$$X_{(0)}=\{\|\ 1\ 3\ \|\},$$
$$x_0+x_1=\|\ 0 \quad 4\ \|,$$
$$x_0+x_2=\|\ 2\ \ -2\ \|,$$

Therefore,

$$X_{(1)}=\{\|\ 1\ 3\ \|,\|\ 0\ 4\ \|\},$$

and $X_{(2)}=X_{(1)}$. Thus, $X^*=X_{(1)}$, which is equivalent to V_S.

Vector addition systems may be classified by analogy with the corresponding multiset grammars as general, context-sensitive, context-free, or linear. No additional restrictions are put on the form of basis vectors in ***general*** *VASs*. In ***context-free*** *VASs*, each basis vector contains a single negative component whose value is –1. In ***linear*** *VASs*, each basis vector along with a single -1 component has a single positive component.

Let us pay attention to the fact that in attempting to determine the notion of a ***context-sensitive*** *VAS*, we face ***conceptual restrictions common to these systems***. Let us recall that any context-sensitive rule has the form

$$v_c+(v-v_c)\rightarrow v_c+(v'-v_c), \tag{11.9}$$

where v_c is a context occurring in a multiset $\overline{v}$ generated at the previous steps. If such a context is present in a multiset $\overline{v}$, then replacement of $v - v_c$ by $v' - v_c$ is done. So, as can be seen, in this case, the components of the basis vector corresponding to the objects belonging to v_c have ***both negative and positive values***. This fact illustrates that ***for each VAS, a multigrammar equivalent to it may be created***, but an

equivalent VAS cannot be created for each MG. Thus, if we replace the rule $\{1 \cdot a_1\} \rightarrow \{1 \cdot a_2\}$ from Example 6.2 by the rule $\{1 \cdot a_1\} \rightarrow \{2 \cdot a_1, 1 \cdot a_2\}$, we fail to create for this rule a basis vector.

11.1.2 Vector Substitution Systems

A couple $\overline{X} = < x_0, X_{\rightarrow} >$ is called a ***vector substitution system*** (*VSS*), where the set of substitutions

$$X_{\rightarrow} = \{ < x_1, x_1' >, \ \ldots, \ < x_k, x_k' > \} \tag{11.10}$$

defines the reachability set of the *VSS* $\overline{X}$, designated below by $\overline{X}^*$, in the following way:

$$\overline{X}_{(0)} = \{x_0\}, \tag{11.11}$$

$$\overline{X}_{(i+1)} = \overline{X}_{(i)} \cup \left[\bigcup_{x \in \overline{X}_{(i)}} \ \bigcup_{< x_j, \, x_j' > \in X_{\rightarrow}} \varphi\left(x, \ x_j, \ x_j'\right) \right], \tag{11.12}$$

$$\overline{X}^* = \overline{X}_{(\infty)}, \tag{11.13}$$

where

$$\varphi\left(x, x_j, x_j'\right) = \begin{cases} x - x_j + x_j', & \text{if } x - x_j \geq 0 \\ x & \text{otherwise.} \end{cases} \tag{11.14}$$

Example 11.3
Let us consider the vector substitution system $\overline{X} = < x_0, X_{\rightarrow} >$, where $x_0 = \| \ 2 \ 3 \ 4 \ \|$ and $X_{\rightarrow} = \{< \| \ 0 \ 1 \ 0 \ \|, \| \ 3 \ 0 \ 0 \ \| >, < \| \ 2 \ 0 \ 3 \ \|, \| \ 1 \ 2 \ 0 \ \| >\}$.

According to (11.11)–(11.14),

$$\overline{X}_{(0)} = \{\| \ 2 \ 3 \ 4 \ \|\},$$

$$\overline{X}_{(1)} = \overline{X}_{(0)} \cup \{\| \ 5 \ 2 \ 4 \ \|, \| \ 1 \ 5 \ 1 \ \|\},$$

$$\overline{X}_{(2)} = \overline{X}_{(1)} \cup \{\| \ 8 \ 1 \ 4 \ \|, \| \ 6 \ 4 \ 1 \ \|, \| \ 4 \ 4 \ 1 \ \|\},$$

etc.

As may be seen, a vector pair $<x_j, x'_j>$ is a direct analog of a rule of a multiset grammar, and a set $X_{\rightarrow}$ is the same analog of its scheme.

As in the case of vector addition systems, each multigrammar $S = <v_0, R>$ may be associated with a *VSS* $\overline{X} = <x_0, X_{\rightarrow}>$, where:

1. The dimensionality of vectors $x \in \{x_0\} \cup \left(\bigcup_{j=1}^{k} \{x_i, x'_i\} \right)$ is equal to the number of objects in the *MG S*, i.e., $|A_S|$
2. The vector x_0 corresponds to the multiset v_0 in such a way that its *i*-th component x_0^i is defined by the expression (10.6)
3. A rule $r_j \in R$

$$\{n_1^j \cdot a_{i_1}^j, \ldots, n_m^j \cdot a_{i_m}^j\} \rightarrow \left\{n_{m+1}^j \cdot a_{i_{m+1}}^j, \ldots, n_{m+k}^j \cdot a_{i_{m+k}}^j\right\} \quad (11.15)$$

corresponds to a pair of vectors $<x_j, x'_j> \in X_{\rightarrow}$, where $x_{i_l}^j = n_l^j$, $l = 1, \ldots, m$, and $(x')_{i_l}^j = n_{m+l}^j$, $l = 1, \ldots, k$, and all other components of vectors x_j and x'_j are zero.

The logic of modeling multigrammars by vector substitution systems is illustrated by the following example.

Example 11.4
Consider the multigrammar $S = <v_0, R>$, where the kernel

$$v_0 = \{1 \cdot a_1, 2 \cdot a_2\}$$

and the scheme R includes two rules

$$r_1 : \{1 \cdot a_1\} \rightarrow \{2 \cdot a_1, 1 \cdot a_2\},$$
$$r_2 : \{3 \cdot a_2\} \rightarrow \{1 \cdot a_1, 2 \cdot a_2\}.$$

This *MG* may be represented by the vector substitution system $\overline{X} = <x_0, X_{\rightarrow}>$, where

$$x_0 = \|\ 1\ \ 2\ \|$$
$$<x_1, x'_1> = <\|\ 1\ \ 0\ \|, \|\ 2\ \ 1\ \|>$$

$<x_2, x'_2> = <\|\ 0\ \ 3\ \|, \|\ 1\ \ 2\ \|>$.

Due to the structural and conceptual similarity of *MGs* and *VSSs*, the latter, similar to *VASs*, may be classified without any additional conditions as general, context-sensitive, context-free, or linear.

11.2 Production Functions

Production functions are a convenient tool for the representation and analysis of stochastic context-free grammars and the stochastic context-free languages generated by their application (Chi 1999) and for a long time have been successfully used in structural pattern recognition. This experience forms the conceptual background for their similar application regarding multiset grammars.

To use the production functions toolkit, it is sufficient to represent a multiset as a ***monomial***, in which ***variables correspond to objects*** and ***degrees of variables*** to ***objects' multiplicities***. In other words, by setting up a ***bijective mapping*** between a set of objects and a set of variables, we determine a ***similar mapping between multisets and monomials***. By generalizing this approach to sets of multisets, ***each such set may be associated with a polynomial***, i.e., a ***sum of monomials***, ***each corresponding to its multiset***. Formally, by defining a bijective function

$$f : A \leftrightarrow Y, \tag{11.16}$$

where $A = \{a_1, \ldots, a_m\}$ is a set of objects and $Y = \{y_1, \ldots, y_m\}$ is a set of variables, we can associate each multiset

$$v = \{n_1 \cdot a_1, \ \ldots, n_m \cdot a_m\} \tag{11.17}$$

with a monomial

$$z = y_1^{n_1} \cdot \ \ldots \ \cdot y_m^{n_m} \tag{11.18}$$

and a set of multisets

$$V = \left\{ \left\{ n_1^1 \cdot a_1^1, \ \ldots, \ n_{m_1}^1 \cdot a_{m_1}^1 \right\}, \ldots, \left\{ n_1^k \cdot a_1^k, \ \ldots, \ n_{m_k}^k \cdot a_{m_k}^k \right\} \right\} \tag{11.19}$$

with a polynomial

$$z = \left(y_1^1\right)^{n_1^1} \cdot \ \ldots \ \cdot \left(y_{m_1}^1\right)^{n_{m_1}^1} + \ldots + \left(y_1^k\right)^{n_1^k} \cdot \ \ldots \ \cdot \left(y_{m_k}^k\right)^{n_{m_k}^k}. \tag{11.20}$$

Thus, by associating each multiobject $n \cdot a$ with a monomial y^n, we directly represent any multiset or set of multisets with basis A in the form of a polynomial with variables $y \in Y$.

Example 11.5
The multiset

$$v = \{3 \cdot a_1, 2 \cdot a_2, 7 \cdot a_3\}$$

may be represented by the monomial

$$z = y_1^3 \cdot y_2^2 \cdot y_3^7,$$

and the set of multisets

$$v = \{\{5 \cdot a_1,\ 1 \cdot a_3\}, \{3 \cdot a_2,\ 2 \cdot a_4\}, \{1 \cdot a_1,\ 2 \cdot a_2,\ 6 \cdot a_3,\ 2 \cdot a_4\}\}$$

may be represented by the polynomial

$$z = y_1^5 \cdot y_3 + y_2^3 \cdot y_4^2 + y_1 \cdot y_2^2 \cdot y_3^6 \cdot y_4^2.$$

Let us consider a 1-multigrammar $S = \ < a_0, R>$, and, following (11.17)–(11.18), we shall represent each unitary rule

$$a \to n_1 \cdot a_1, \ldots, n_m \cdot a_m, \tag{11.21}$$

by an equality

$$y = y_1^{n_1} \cdot \ \ldots \ \cdot y_m^{n_m}, \tag{11.22}$$

where $y, y_1, \ldots, y_m$ are variables corresponding, respectively, to objects $a, a_1, \ldots, a_m$. In this way, a generation step

$$v \underset{\Rightarrow}{r} v', \tag{11.23}$$

being implemented by means of a rule $r \in R$ (11.21), is naturally modeled by substitution of the monomial $y_1^{n_1} \cdot \ \ldots \ \cdot y_m^{n_m}$ for the variable y in the monomial z corresponding to the multiset v. If in this case $n \cdot a \in v$, i.e., the variable y_n occurs in the monomial z, then the result of the substitution will be the monomial

$$z' = z'' \cdot \left(y_1^{n_1} \cdot \ \ldots \ \cdot y_m^{n_m}\right)^n, \tag{11.24}$$

or, after removing parentheses,

$$z' = z'' \cdot y_1^{n_1 \cdot n} \cdot \ \ldots \ \cdot y_m^{n_m \cdot n}, \tag{11.25}$$

and, after collecting monomials,

$$z' = y_{i_1}^{n_{i_1}} \cdot \ldots \cdot y_{i_k}^{n_{i_k}}, \tag{11.26}$$

where

$$v' = \{n_{i_1} \cdot a_{i_1}, \ldots, n_{i_m} \cdot a_{i_m}\}, \tag{11.27}$$

and

$$\beta(v') = \beta(v) \cup \{a_1, \ldots, a_m\}, \tag{11.28}$$

where, recall, $\beta(v)$ is the basis of the multiset v, i.e., the set of objects occurring in this *MS*.

Example 11.6
Let us consider the 1-multigrammar $S = \ < a_0, R>$, where the unitary rules have the following form:

$$a_0 \to 1 \cdot a_1, 2 \cdot a_3;$$
$$a_1 \to 3 \cdot a_2, 1 \cdot a_3;$$
$$a_2 \to 1 \cdot a_3.$$

The following equations may be associated with these unitary rules:

$$y_0 = y_1 \cdot y_3^2,$$
$$y_1 = y_2^3 \cdot y_3,$$
$$y_2 = y_3.$$

As can be seen, after all substitutions,

$$y_0 = y_3^6,$$

which means $\bar{V}_S = \{6 \cdot a_3\}$.

In the case of k-multigrammars, each group of alternating unitary rules

$$a \to n_1^1 \cdot a_1^1, \ldots, n_{m_1}^1 \cdot a_{m_1}^1,$$
$$\ldots$$
$$a \to n_1^l \cdot a_1^l, \ldots, n_{m_l}^l \cdot a_{m_l}^l, \tag{11.29}$$

at first glance may be associated with the following equation:

$$y = \sum_{i=1}^{l} \left(\prod_{j=1}^{m_i} \left(y_j^i \right)^{n_j^i} \right). \tag{11.30}$$

However, in this case, substitution of the right part of (11.30) into a monomial

$$z = z' \cdot y^n \tag{11.31}$$

with parentheses removed leads to a completely different result regarding the semantics of k-multigrammars. So an ***application of production functions is correct only in the case of 1-multigrammars***.

11.3 Petri Nets

A ***Petri net*** (Murata 1989; Reisig 1991; David and Alla 2005) represents a complex of two elements—a structure and a state. The ***structure*** of a Petri net is a tuple $C = \ < P, T, I, O>$, where P is a finite collection of positions ("***places***") and T is a finite collection of ***transitions*** ($P \cap T = \{\varnothing\}$), while I and O are an input function and an output function, respectively:

$$\begin{aligned} I &: P \times \mathbf{N_1} \to T, \\ O &: T \to P \times \mathbf{N_1}, \end{aligned} \tag{11.32}$$

where $\mathbf{N_1} = [1, \infty]$ is a set of positive integer numbers. A position p is an ***input position*** of a transition t if

$$t = I(p, n) \tag{11.33}$$

and an ***output position*** of a transition t' if

$$t' = O(p, n'). \tag{11.34}$$

A ***state***, or a *marking*, of a Petri net is a mapping $\mu : P \to \mathbf{N_1}$, so that the value $\mu(p)$ is the number of so-called tokens located on the specific step of net operation in the position p. A *PN* operation proceeds via single-step ***transition firings***. A transition firing t with a set of input positions $p_1^t, \ldots, p_{k_t}^t$, such that

$$I(p_1^t, n_1^t) = t,$$

$$\ldots$$

$$I\left(p_{k_t}^t, n_{k_t}^t\right) = t, \tag{11.35}$$

and a set of output positions $\left\{\overline{p}_1^t, \ldots, \overline{p}_{l_t}^t\right\}$, such that

$$O\left(\overline{p}_1^t, \overline{n}_1^t\right) = t,$$

$$\ldots$$

$$O\left(\overline{p}_{l_t}^t, \overline{n}_{l_t}^t\right) = t, \tag{11.36}$$

takes place when the current state μ of the net C is such that

$$n_1^t \leq \mu\left(p_1^t\right),$$

$$\ldots$$

$$n_{k_t}^t \leq \mu\left(p_{k_t}^t\right). \tag{11.37}$$

A transition firing t takes a net C into a ***new state*** μ' associated with the state μ by the following relations:

$$\mu'\left(p_1^t\right) = \mu\left(p_1^t\right) - n_1^t,$$

$$\ldots$$

$$\mu'\left(p_{k_t}^t\right) = \mu\left(p_{k_t}^t\right) - n_{k_t}^t,$$

$$\mu'\left(\overline{p}_1^t\right) = \mu\left(\overline{p}_1^t\right) - \overline{n}_1^t,$$

$$\ldots$$

$$\mu'\left(\overline{p}_{l_t}^t\right) = \mu\left(\overline{p}_{l_t}^t\right) - \overline{n}_{l_t}^t. \tag{11.38}$$

In other words, a transition firing results in the ***removal from input positions of a number of tokens defined by the input function*** I and, after this, an ***addition to the output positions of a number of tokens defined by the output function*** O. A Petri net operates in discrete time in such a way that at each moment, ***all transitions whose activation conditions satisfy*** (11.37) ***may be activated***. Petri nets are generally used as a convenient model of parallel computation that provides a formalization and solution of tasks related to parallelizing and synchronization of computation processes in the context of limited hardware resources.

Example 11.7
Let us consider the Petri net represented in Fig. 11.1, where the number of tokens in the positions (initial marking) and the number of arcs connecting the nodes are

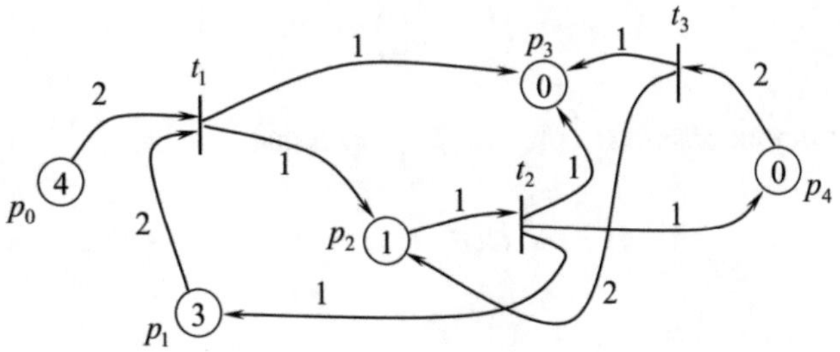

Fig. 11.1 A Petri net

Table 11.2 Operation of the Petri net

	Marking				
	p_0	p_1	p_2	p_3	p_4
Transition firing	4	3	1	0	0
t_1	2	1	2	1	0
t_2	2	2	1	2	1
t_1	0	0	2	3	1
t_2	0	1	1	4	2
t_3	0	1	3	5	0

represented by integer numbers. The probable initial steps of operation of this net are given in the following Table 11.2.

Further steps may be performed by an interested reader if desired.

Each Petri net $C = < P, T, I, O>$ may be associated with a set of rules R of a certain multigrammar S. Namely, ***each position*** $p \in P$ ***is associated with an object*** $a \in |A_S|$, i.e., $|A_S| = |P|$, while ***each transition*** t ***is associated with a rule*** $r \in R$

$$\{n_1 \cdot a_1, \ldots, n_m \cdot a_m\} \to \{n'_1 \cdot a'_1, \ldots, n'_m \cdot a'_m\}, \tag{11.39}$$

such that

$$I(p_i, n_i) = t,$$
$$O(p'_j, n'_j) = t, \tag{11.40}$$

where $i = 1, \ldots, k$; $j = 1, \ldots, l$; and objects $a_1, \ldots, a_k, a'_1, \ldots, a'_l$ correspond in the abovementioned sense to positions $p_1, \ldots, p_k, p'_1, \ldots, p'_l$. Each state μ of a net C such that $\mu(p) = n$ is associated with a multiset

$$v = \{n_1 \cdot a_1, \ldots, n_m \cdot a_m\}, \tag{11.41}$$

where

$$n_i = \mu(p_i) \tag{11.42}$$

and a_i corresponds to p_i. An ***initial state of this Petri net*** μ_0, thus, corresponds to a ***multiset*** v_0, while a ***net*** C in this state corresponds to a ***multigrammar*** $S = < v_0, R>$. In this case, operation of a ***Petri net allows the modeling of the generation of multisets by a corresponding multigrammar***, so that ***any transition firing implements an application of the corresponding rule***.

Example 11.8
The kernel of the multiset grammar $S = < v_0, R>$ corresponding to the Petri net considered in the previous Example 11.7 is

$$v_0 = \{4 \cdot a_0, 3 \cdot a_1, 1 \cdot a_2\},$$

while the scheme R includes the following rules:

$$r_1 : \{2 \cdot a_0, 2 \cdot a_1\} \rightarrow \{1 \cdot a_2, 1 \cdot a_3\},$$
$$r_2 : \{1 \cdot a_2\} \rightarrow \{1 \cdot a_1, 1 \cdot a_3, 1 \cdot a_4\},$$
$$r_3 : \{2 \cdot a_4\} \rightarrow \{1 \cdot a_3, 2 \cdot a_2\}.$$

Thus, there is a ***bijective correspondence between Petri nets and multiset grammars***, allowing the application of various *PNs* to the study of properties of *MGs* (firstly, algorithmic solvability of various problems concerning different classes of *MGs* and their subclasses), and vice versa, permitting the application of the results obtained for multigrammars to the analysis of various classes of Petri nets.

It is appropriate to note here that in mathematical logic and its applications, two classes of objects are usually studied—the ***calculi*** and the ***devices*** (automata), so that ***each calculus is associated with a certain class of devices,*** allowing the application (interpretation) of sentences of this calculus, i.e., in fact, giving an algorithmic representation of its semantics. This correlation is the most natural and usual for the ***associative calculi operating on symbol strings*** (in particular, ***grammars*** (Chomsky 2005)) and ***different types of Turing machines*** (in particular, ***push-down automata***) (Hopcroft et al. 2001; Meduna 2014). In this sense, the ***Petri nets may be considered as a class of devices directly associated with multigrammars as a class of calculi***. As was mentioned in Gvishiani et al. (2018), there exists a correspondence between ***timed colored Petri nets*** and ***temporal multiset grammars***.

However, ***filtering MGs and TMGs, self-generated TMGs and FTMGs, and the metagrammatical extensions of all classes of MGs cannot be represented by Petri nets***. Moreover, all classes of multiset grammars are associated with a ***universal not existential semantics***, i.e., they allow the representation and search of ***all solutions*** of the problem in question, not ***only one***. Petri nets have no embedded tools for the definition of the ***set of solutions*** of a problem and for the generation of this whole set of solutions.

11.4 String-Operating Grammars and "Set-of-Strings" Databases

The ***"set-of-strings" framework*** (*SSF*) was proposed and developed by the author in Sheremet (1994) as a generalization of earlier models of data (first of all, the relational model (Darwen 2012; Date 2012)) and knowledge representation models (Akerkar and Sajja 2010; Apt 2003; Bratko 2012; Cross 2017; Frunkwirth and Abdennadher 2003; Kendal and Green 2007; Marriott and Stucky 2003; Pannu 2015; Siekmann 2014; Wallace 2002), which were universally based on ***predicate-based information representation***. Nowadays, the *SSF* is known and understood as a comprehensive theoretical toolkit for Big Data and modeling and implementation of *AI* technologies associated with it (Sheremet 2011b, 2012, 2013, 2016b, 2017, 2020a, b). In fact, ***the SSF is a historical predecessor of the MGF, and the latter inherits many features peculiar to the SSF***.

As the *MGF* is founded on a generalized representation of a processed data item as a multiset (*a **set of multiobjects***), similarly, the *SSF* was founded on a generalized representation of a database (*DB*) as a ***set of strings***, and this framework was practically the first where ***classical string-operating grammars were applied not for the description of the syntax of programming languages but for the definition of the structure of elements of a DB*** (**i.e.,** ***the structure of the aforementioned strings)***. Let us consider the main features of the *SSF* data model, which are the foundation of this framework, and its interconnections with the *MGF*.

The axiomatics of the *SSF* is based on a representation of the operational logic of a database management system (*DBMS*), which is the operational component of a data storage (*DS*) , as a mapping

$$\boldsymbol{\varphi} : \boldsymbol{\Sigma} \times \boldsymbol{I} \rightarrow \boldsymbol{\Sigma} \times \boldsymbol{A}, \tag{11.43}$$

where $\boldsymbol{\Sigma}$ is a set of states of a data storage; $\boldsymbol{I}$ a set of possible access messages (*AMs*) to the *DS*, e.g., an input language; and $\boldsymbol{A}$ a set of possible replies of the *DBMS* to *AMs*. Access messages arrive at the *DS* at discrete time moments $t = 0, 1, \ldots$, and a reply to an *AM*, incoming at a moment t, is given at a moment $t + 1$, while the information component of the *DS* at the same time transfers from a state $\sigma_t \in \boldsymbol{\Sigma}$ to a state $\sigma_{t+1} \in \boldsymbol{\Sigma}$. For simplicity, we assume that no more than one *AM* is incoming at each time moment.

The ***state of the DS*** σ_t is a couple $<W_t, D_t>$, where $W_t = \{w_1, \ldots, w_m\} \subset V^*$is a ***database*** and $D_t = \{d_1, \ldots, d_m\} \subset V^*$is a ***metadatabase*** (*MDB*). A database element $w_i \in W_t$ is a string in an alphabet V (V^* is the set of all strings in this alphabet) and is called a ***fact***. At the initial point of time, $t_0 = 0$, $W_0 = D_0 = \{\varnothing\}$.

An ***access message*** to a *DS* is a triple $<o, c, x>$, where o means an operation performed (inclusion, deletion, selection), c means the *DS* component which is the object of the operation (database, metadatabase), and x is the informative part of the *AM*, i.e., a query (a more or less complex condition for the selection of facts from a

DB) or a data update (inclusion in/deletion from a database). A reply to an *AM* represents a finite set

$$A_{t+1} \subset V^*. \tag{11.44}$$

Unlike predicate-based models of data, there are no* a priori *restrictions on the structures of database elements (*i.e.*, facts) in the SSF, and a specific structure of facts may be user-defined (or administrator-defined) by means of a metadatabase.

From the point of view of theoretical computer science, a function **φ** is a formal definition of the ***semantics of a language I***. There are four types of such semantics in the *SSF*: ***set-theoretical*** (*ST*-), ***mathematical*** (*M*-), ***operational*** (*Op*-), and ***implementational*** (*I*-).

ST-semantics is a formal set-theoretical definition of *DBMS* functions.

M-semantics corresponds to *ST*-semantics, but operates on concrete specifications of a metadatabase and an input language, which together form a model of a real *DBMS* language.

Op-semantics corresponds to *M*-semantics, but, unlike the latter, is defined algorithmically, so that an algorithm of interpretation of an access message (i.e., a sentence of an input language) creates a result (reply) in full compliance with the expressions of the *M*-semantics by execution of a finite number of steps or reveals the impossibility of such creation as a result of errors detected in the *AM*.

I-semantics is based on an *Op*-semantics and corresponds to the latter in the sense of the results of interpretation of any access message. However, if the definition of an *Op*-semantics is based on algorithmic solvability, the objective of the definition of an *I*-semantics is an effective implementation of the input language, i.e., minimization of the average time to process access messages (finally, the computation complexity of the corresponding algorithms) with regard to specific features of the *DBMS* hardware. An *I*-semantics is a basis for the design of real software and hardware of data storages.

The ***different types of semantics of a DBMS input language correspond to the different levels of abstraction of the notion "data storage."***

Let us consider the ***set-theoretical semantics*** of a *DS* input language, which, in turn, includes sublanguages providing independent access to a database and to a metadatabase.

The *ST-semantics of a database sublanguage*, usually referred to as a ***data manipulation language (DML)***, is based on the assumption that the informative part of an *AM* x is a finite description of some set $I_t \subseteq V^*$, which corresponds to the degree of the user's information awareness about a certain aspect of a problem area. Expressions of a *DML ST*-semantics connect W_{t+1} (the database after operation completion) and A_{t+1} (the reply to an access message) with the set I_t and the set W_t (the database at the moment of access):

$$W_{t+1} = W_t \cup I_t,$$

$$A_{t+1} = W_{t+1} - W_t \tag{11.45}$$

for ***inclusion*** (A_{t+1} contains new facts which were absent in the database before the moment of access), where I_t is ***finite*** by definition;

$$W_{t+1} = W_t - I_t,$$
$$A_{t+1} = W_t - W_{t+1} \tag{11.46}$$

for ***deletion*** (A_{t+1} contains facts which are removed from the database), where I_t may be ***infinite*** in the general case;

$$W_{t+1} = W_t,$$
$$A_{t+1} = W_t \cap I_t, \tag{11.47}$$

for ***data selection*** (***query***), i.e., A_{t+1} contains facts that occur in the database W_t and at the same time satisfy the conditions specified by the set I_t, which also may be ***infinite*** in the general case.

Example 11.11
Consider the database W_t, whose elements (facts) contain information about car types, state registration numbers, and owners:

$$W_t = \{NissanXTrail\ C930XK\ Jones,$$
$$AudiA6\ E311SA\ Smith,$$
$$ToyotaCamry\ B151VA\ Bale\}.$$

Inclusion of the fact about a Kia Rio with the state registration number P615EF, whose owner is James, is implemented in accordance with (11.45) by the set-theoretical join of the set W_t and the set

$$I_t = \{KiaRio\ P615EF\ James\}.$$

Deletion from the obtained *DB* of all facts regarding Toyota cars is implemented according to (11.46) by means of the set-theoretical subtraction of the set

$$I_t = \{\ Toyota\quad A000AA,$$
$$\dots$$
$$ToyotaZZZZZZZZ\ Z000ZZ\ ZZZZZZZZZZZZZZZ\ \}$$

from the set W_t (on the assumption that the maximal length of a Toyota car's brand is 8 symbols and the maximal length of an owner's surname is 15 symbols). This operation will result in two sets W_{t+1} and A_{t+1}:

$$W_{t+1} = \{NissanXTrail\ C930XK\ Jones,$$
$$AudiA6\ E311SA\ Smith,$$
$$KiaRio\ P615EF\ James\},$$
$$A_{t+1} = W_t - W_{t+1} = \{ToyotaCamry\ B151VA\ Bale\}.$$

Selection from the database of facts relating to a car owned by Smith may be implemented according to (11.47) by the set-theoretical intersection of the set W_t and the set

$$I_t = \{A\ A000AA\ Smith,$$
$$\ldots$$
$$ZZZZZZZZZZ\ Z999ZZ\ Smith\}$$

(on the assumption that the name of a car brand is 1–10 symbols).

The result of this operation is the set

$A_{t+1} = \{AudiA6\ \ E311SA\ Smith\}$.

A ***correct database*** W_t is one whose elements meet certain syntactic (structural) restrictions fixed by a metadatabase, i.e., have a structure determined by an *MDB*:

$$W_t \subseteq \overline{W}_t, \tag{11.48}$$

where $\overline{W}_t \subseteq V^*$ is the set of permissible elements of the database.

A *DML M-semantics* is based on a representation of an *MDB* as the ***set of rules of a context-free grammar***. Each such rule has the form $\alpha \rightarrow \beta$, where α is a non-terminal symbol (non-terminal) and β is a string consisting of symbols of the terminal alphabet *V* and non-terminals which represent the names of facts' structural elements (substrings). An axiom of a context-free grammar G_t, whose scheme (set of rules) is D_t, is designated α_0 and has the sense "***fact***." As a result of this primary axiomatization,

$$\overline{W}_t = L(G_t), \tag{11.49}$$

where $L(G_t)$ is the language, described by the context-free grammar $G_t = < V, V_t^N, \alpha_0, D_t >$, and

$$V_t^N = \{\ \alpha \mid \alpha \rightarrow \beta \in D_t\ \} \tag{11.50}$$

is the non-terminal alphabet of this grammar. Thus, a ***metadatabase determines the structure of facts and allows their naming as well as the naming of their structural fragments***. Finally, the set of permissible elements of a correct database $\overline{W}_t$ is represented as the context-free language defined by a CF grammar G_t:

$$\overline{W}_t = \{w | \alpha_0 \Rightarrow^* w \& w \in V^*\}, \tag{11.51}$$

where $\Rightarrow^*$ is a denotation of the ***relation of mutual derivability of sentential forms*** (*SFs*) of a CF grammar, similar to the relation $\overset{*}{\Rightarrow}$ defined in Sect. 3.1.3 regarding multisets. The sentential forms of a CF grammar G_t may be used as the simplest tool for finitely representing a set I_t, so that the informative part of an access message is a string $c \in SF(G_t)$ (here, $SF(G_t)$ is the set of sentential forms of the grammar G_t), which determines the set I_t as a subset of the language $L(G_t)$, including all its words generated from c:

$$I_t = \{ w | w \in L(G_t) \& c \Rightarrow^* w \}, \tag{11.52}$$

and formally

$$SF(G_t) = \{ x | x \in \left(V \cup V_t^N\right)^* \& \alpha_0 \Rightarrow^* x \}. \tag{11.53}$$

Example 11.12
The metadatabase defining the structure of the database elements from the previous Example 6.11 may be as follows:

$$< fact > \to < car >,$$

$$< car > \to < brand > < number > < owner >,$$

$$< brand > \to < text >,$$

$$< number > \to < letter > < digit > < digit > < digit > < letter > < letter >,$$

$$< owner > \to < text >,$$

$$< text > \to < symbols >,$$

$$< symbols > \to < symbol >,$$

$$< symbols > \to < symbol > < symbols >,$$

$$< symbol > \to < letter >,$$

$$< symbol > \to < digit >,$$

$$< symbol > \to \quad ,$$

$$< letter > \to A,$$

$$\ldots$$

$$< letter > \to Z,$$

$$< digit > \to 0,$$

...

$$<digit> \to 9.$$

In this case, the informative part of a query for the selection of the database elements containing information about owners of Nissan cars of all kinds will be as follows:

$$Nissan <text> \quad <number> \quad <owner>,$$

while the similar query for the selection of a car with the state registration number S333AT will be as follows:

$$<brand> S333AT <owner>,$$

while a query for the selection of a car whose owner is Lewis will be as follows:

$$<brand> \quad <number> Lewis.$$

A ***DML Op-semantics*** corresponding to the described *DML* M-semantics is elementary: it is based on the algorithmics of recognition of derivability of a string $c \in \left(V \cup V_t^H\right)^*$ in a CF grammar G_t (i.e., recognition of whether a string c belongs to the set of sentential forms of this grammar) and on a total search of facts $w \in W_t$ in order to determine their derivability from c. In the case of an inclusion, it is determined whether the set I_t is finite. All the aforementioned problems are algorithmically solvable, and algorithms for their solution are well known (Hopcroft et al. 2001; Meduna 2014).

A ***DML I-semantics*** corresponding to the described *DML Op*-semantics is nothing but a set of algorithms ***minimizing redundant search in a DB*** during the processing of access messages. The basis of these algorithms is the application of a partial order on a set of sentential forms of the CF grammar G_t to exclude from further processing subsets of the set W_t which do not contain the necessary elements (facts), without searching for such elements individually.

If G_t is an unambiguous context-free grammar, the set $SF(G_t)$ is partially ordered by the relation of derivability (Sheremet 1994, 2013, 2016b), so for any two *SFs* $x, x' \in SF(G_t)$, there is a set of *SFs* from which x and x' are derived (generated)—upper bounds of the *SF* set $\{x, x'\}$—and a least upper bound $\mathbf{sup}\{x, x'\}$. For some $x, x' \in SF(G_t)$, there may be a set of lower bounds and a greatest lower bound $\mathbf{inf}\{x, x'\}$.

Statement 11.1 (Sheremet 1994, 2013). If a CF grammar G_t is unambiguous, x and x' are its sentential forms, and $\mathbf{inf}\{x, x'\}$ does not exist, then for each *SF* x'' derived from x' in the grammar G_t, $\mathbf{inf}\{x, x''\}$ does not exist as well.

A general tool of minimization of redundant search of *DB* elements during the processing of access messages (queries or commands for inclusion/deletion) is the so-called SF-trees.

An ***SF-tree*** corresponding to a *DB* W (index t in W_t is omitted for simplicity), denoted $\tau(W)$, possesses the following properties:

1. The root of the tree is α_0—an axiom of the grammar G (index t in G_t is also omitted).
2. Leaves (terminal nodes) of the tree are facts $w \in W$.
3. Internal (non-terminal) tree nodes are sentential forms of the grammar G, and the condition $x \Rightarrow^* x'$ is true for each successor x' of a node x.
4. The branching factor of the tree is variable, but it is limited by some constant, which represents the maximal number of alternatives for each non-terminal in the scheme D of the grammar G, increased by one:

$$m = 1 + \frac{1}{max \{ l\alpha \in V_H \mid \{\alpha \to \beta_1, \ldots, \alpha \to \beta_l\} \subseteq D \}} \tag{11.54}$$

(we can fix $m = 3$ by transforming the grammar G into a bi-alternative form).

Statement 11.1 provides exclusion from consideration during the processing of *AMs* with informative part $x \in SF(G)$ all subtrees of the tree $\tau(W)$ with root x' in the case that $\mathbf{inf}\{x, x'\}$ does not exist, since from this statement and property 3 of a tree $\tau(W)$, it follows that there does not exist $\mathbf{inf}\{x, w\}$ for all leaves $w \in W$ of the aforementioned subtrees as well.

In general, ***there is a direct conceptual interconnection between the "set-of-strings" framework and the multigrammatical framework*** presented in this monograph.

Namely, a ***context-free grammar*** G_t defining the set of correct database elements $\overline{W}_t$ according to (11.51) corresponds to a ***unitary multiset grammar***, and a request to a *DB* with informative part c corresponds to a filter, so that a couple $<G_t, c>$produces the following set:

$$\overline{A}_t = \overline{W}_t \cap I_t, \tag{11.55}$$

where I_t is determined by the expression (11.52). As can be seen, ***a set of facts*** $\overline{A}_t$ ***is a string analog of a set of multisets generated by a filtering unitary multiset grammar***. In this case, the ***scheme*** R **of an FUMG** $S = \ < v_0, R, F>$ is an ***analog of a metadatabase*** D_t; the ***kernel*** v_0 is an ***analog of an axiom*** a_0 of a **CF** ***grammar*** G_t; a ***rule***

$$\alpha \to \beta \in D_t \tag{11.56}$$

is an ***analog of a unitary rule***

$$a \to n_1 \cdot a_1, \ldots, n_m \cdot a_m \in R; \tag{11.57}$$

a ***symbol of an alphabet*** $V \cup V_t^N$ is an ***analog of an object*** $a \in A_S$; a **terminal symbol belonging to a terminal alphabet** V is an ***analog of a terminal object of an FUMG***; and a ***string***

$$w \in \overline{A}_t \tag{11.58}$$

is an ***analog of a terminal multiset***

$$v \in \overline{V}_S. \tag{11.59}$$

The aforementioned analogies are not accidental, since the author relied naturally on the experience of application of the context-free grammars in the development of the *SSF* during the development of the *MGF*.

Various computation models may be studied regarding the multigrammatical framework, and their consideration will be the scope of future publications.

Chapter 12
Implementation Issues and Future Developments

The components of the multigrammatical framework developed and considered in the previous chapters form the foundation for its effective implementation, which requires the solution of the following three basic problems:

Development of a detailed ***general algorithm of operation*** of an *MGF*-centered smart sociotechnological system, which will be considered in Sect. 12.1

Creation of a ***basic toolkit allowing maintenance of multigrammatical knowledge bases*** appropriate to the current state of an *STS* technological base, which will be considered in Sect. 12.2

Development of ***software/hardware solutions*** for the implementation of *STS* controllers and schedulers for various classes of *STSs* and multigrammars

Concerning software/hardware solutions, it would be reasonable to suppose that the first developed research software releases of *STSCs* and *STSSs* might be installed and made available as shareware in the *Software-as-a-Service* mode, which would allow their broad application in testing regimes. By acquisition of implementation experience as well as by the appearance of effective hardware solutions, effective *STS* controllers and schedulers as *Systems-on-a-Chip* (*SoCs*) may be created. Due to the heterogeneity of the *MGF* algorithmics, it would, apparently, be not such a technologically complicated task to develop and produce such *SoCs* in sufficient quantities. However, there may be various approaches to further development of the *MGF*, including its implementation issues; these topics will be considered in Sect. 12.3.

12.1 Operation of *MGF*-Centered Smart Industrial *STSs*

We have introduced a generalized scheme of a smart (knowledge-based) industrial sociotechnological system in Sect. 2.3 (Figs. 2.3 and 2.4). However, this scheme was presented without any details regarding knowledge representation or the data circulating between key components of an *IS*. Now, we have all the necessary details as a

I. A. Sheremet, *Multigrammatical Framework for Knowledge-Based Digital Economy*, https://doi.org/10.1007/978-3-031-13858-4_12

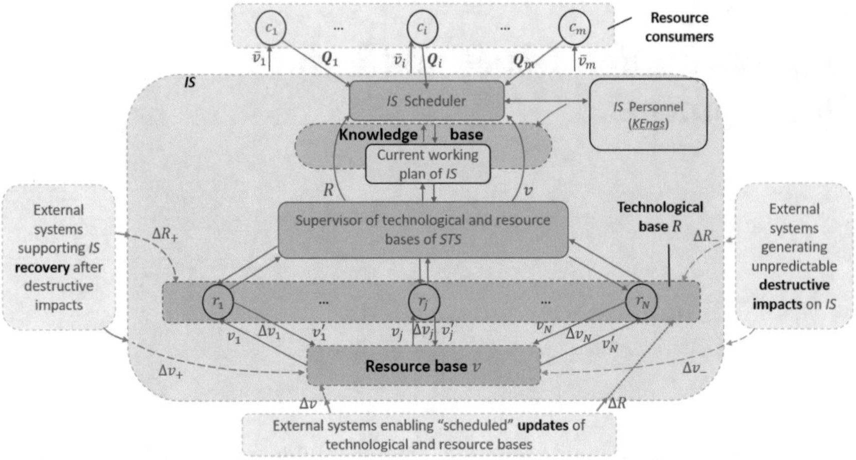

Fig. 12.1 A general scheme of operation of an *MGF*-based industrial *STS*

result of the development of the multigrammatical framework, so let us describe the aforementioned scheme in the context of the *MGF* (Fig. 12.1).

All interactions with *MDs* and storages allowing the accumulation of resources and their utilization, as well as interactions with external systems, are implemented by an ***IS controller***, which is the ***top-level management entity for all other subjects of the IS*** participating in the process of *IS* operation. ***The IS controller, in fact, applies all schedules created and corrected by the IS scheduler.***

Consumers $c_1, \ldots, c_m$ at a priori unpredictable moments make requests to an industrial system, and each such request represents an order $\boldsymbol{Q}=<\boldsymbol{q},\boldsymbol{t}>$, where $\boldsymbol{q}$ is the substantial part of this order, while $\boldsymbol{t}$ is its temporal part. A multiset

$$\boldsymbol{q} = \{\boldsymbol{n_1} \cdot \boldsymbol{a_1}, \ \ldots, \boldsymbol{n_m} \cdot \boldsymbol{a_m}\} \tag{12.1}$$

defines a collection of object-resources to be manufactured by this *IS* and delivered to a consumer at a time moment satisfying the filter

$$\boldsymbol{t} = \{n \leq t \leq n', t = min\}, \tag{12.2}$$

determining the earliest possible time moment in the interval $[n, n']$.

Any request representing an order $\boldsymbol{Q}$ and arriving at an *IS* at a moment $\boldsymbol{n}$, when some $k \geq 0$ previous orders are already being processed, is processed by an *IS* scheduler operating in a multi-flow mode and creating an *IS* schedule allowing the completion of the order $\boldsymbol{Q}$ without affecting the *IS* schedules concerning previous orders. The *ISS* operates on knowledge and data about the manufacturing technological base and the resource base of this *IS*, and their current states $v(\boldsymbol{n})$ and $R(\boldsymbol{n})$, respectively, are available to the *ISS* via the *IS* controller. Both $v(\boldsymbol{n})$ and $R(\boldsymbol{n})$ are formed by the *SIS* according to impacts generated by a hostile external *STS* and/or a hazardous environment and according to more or less regular recovery and refreshment actions performed by a friendly external *STS*, so

$$v(\boldsymbol{n}) = v(\boldsymbol{n}-1) + \Delta v^{+}(\boldsymbol{n}) - \Delta v^{-}(\boldsymbol{n}), \tag{12.3}$$

$$R(\boldsymbol{n}) = R(\boldsymbol{n}-1) \cup \Delta R^{+}(\boldsymbol{n}) - \Delta R^{-}(\boldsymbol{n}), \tag{12.4}$$

where $\Delta v^{+}(\boldsymbol{n})$ and $\Delta R^{+}(\boldsymbol{n})$ are, respectively, increments of the *RB* and the *MTB* provided by a friendly external *STS*, while $\Delta v^{-}(\boldsymbol{n})$ and $\Delta R^{-}(\boldsymbol{n})$ are, on the contrary, their decrements, caused by impacts of a hostile external *STS* and/or the environment. $R(\boldsymbol{n})$, $R(\boldsymbol{n}-1)$, $\Delta R^{+}(\boldsymbol{n})$, $\Delta R^{-}(\boldsymbol{n})$ are, respectively, sets of temporal rules representing manufacturing devices, and $v(\boldsymbol{n})$, $v(\boldsymbol{n}-1)$, $\Delta v^{+}(\boldsymbol{n})$, $\Delta v^{-}(\boldsymbol{n})$ are multisets representing collections of object-resources.

The current *IS* schedule at a moment $\boldsymbol{n}$ in the considered general multi-order case is as follows:

$$\boldsymbol{\sigma}(\boldsymbol{n}) = \{ <\boldsymbol{n}, \boldsymbol{R}(\boldsymbol{n})>, \ \ldots, \ <n_p, \boldsymbol{R}(n_p)> \}, \tag{12.5}$$

where for any $n \in \{\boldsymbol{n}, \ldots, n_p\}$ (n_p is the maximal moment when at least one *MD* will be activated)

$$\boldsymbol{R}(n) = \{ r_1^n/\boldsymbol{i}_1^n, \ \ldots, \ r_m^n/\boldsymbol{i}_m^n \}, \tag{12.6}$$

and $\boldsymbol{i}_j^k$ is the unique identifier of a request representing a corresponding order.

An IS controller sequentially activates MDs at moments $n \in \{\boldsymbol{n}, \ldots, n_p\}$***, thus enabling coherent operation of the manufacturing technological base of the industrial system in full correspondence with the set of schedules created and, if necessary, permanently corrected by the IS scheduler and being completed by the IS to satisfy requests of consumers.***

The simplest case is when no impact is applied to the considered *IS* or when an impact causing decrements $\Delta R^{-}(\boldsymbol{n})$ and $\Delta v^{-}(\boldsymbol{n})$ does not eliminate any of the object-resources and/or manufacturing devices to be used in the current schedule.

If an order $\boldsymbol{Q}$ arriving at a moment $\boldsymbol{n}$ may be completed by an *IS*, then its subsequent actions depend on how many possible variants of *IS* schedules allowing the completion of this order are generated by the *ISS*.

If there is the only one such variant (which is thus the actual schedule), then the *ISS* transfers to the consumer the output schedule $\boldsymbol{\sigma}_{out}$, which represents a timed sequence of operations delivering to the customer the produced *ORs*, and this schedule is used for monitoring the current process of order completion and for the customer's acceptance of the manufactured and delivered *ORs*.

If there are $l \geq 2$ possible ways of completing the order, there may be two options: a free choice by the *ISS* of the variant to be executed or an offer to the consumer to choose the preferable variant by a manual search of the set of possible schedules or to formulate an additional filter to be applied by the *ISS* for the selection of the variant to be executed. After one variant of the schedule is selected in one or another way, it is processed as described above.

A more complicated case is when an order ***Q*** arriving at a moment ***n*** is not feasible for the considered industrial system with its current *MTB* and *RB*. The following steps depend on whether this system is open or closed.

If it is ***open***, then the *ISS* forms a set of variants of input schedules which would permit the completion of this order and displays it to the consumer for the selection of one or none of them. If one of these schedules is selected, then the *ISS* forms a batch of requests to a free external *IS*, which, in turn, may reply positively or negatively. If all orders represented by requests in this batch are feasible for this *EIS*, then their processing begins, and during this prolonged action, the resulting *ORs* are delivered to the *IS* which is completing the initial order ***Q***. Otherwise, i.e., if the free *EIS* is not capable of completing the necessary batch of orders, then the considered *IS* operates as if it were closed.

If the *IS* is ***closed***, then the *ISS* forms a set of variants of partial order completion and displays it to the consumer for the selection of the preferable one or for rejection. If some variant is selected, the corresponding order is completed in the standard way.

The most complicated case is when an impact causing decrements $\Delta R^{-}(\boldsymbol{n})$ and $\Delta v^{-}(\boldsymbol{n})$ eliminates some object-resources and/or manufacturing devices that are scheduled for use. In this extreme situation, the *ISS* selects a batch of orders whose completion would involve the eliminated *ORs* and *MDs* and attempts to create *IS* schedules which would allow the completion of these orders with the current *RB* $v(\boldsymbol{n})$ and *MTB* $R(\boldsymbol{n})$. If this attempt is successful, then further operation is as usual. Otherwise, the *ISS* operates as described above in the case of a single order, but, if necessary, interacting with all consumers whose orders, as a consequence of the impact, are in a jeopardy.

So ***due to the developed multigrammatical knowledge representation and associated ISS algorithmics, the management system of an industrial STS operates as an adaptive control system, neutralizing the destructive consequences of impacts of hostile external systems and/or a hazardous environment and allowing rational behavior of the affected STS. The aforementioned representation and algorithmics are the foundation of the smartness and resilience of an MGF-based STS.***

12.2 Creation and Maintenance of Multigrammatical Knowledge Bases

Let us once more underline that our general approach to the development and application of tools, the background of which is associated with the area of artificial intelligence, is radically different from the approaches consolidated as a result of long-standing scientific activity in this area.

The aforementioned approaches are driven by the objective of the transfer to computers of human ability to solve more or less complicated intellectual

problems. So knowledge-based decision support systems (*DSSs*), originating from the historically earlier expert systems, are based on just this presumption of the possibility and feasibility of replication of human experience in the solution of some kinds of problem. An intermediate step is the representation of this experience as a knowledge base, containing a set of relations and interconnections applied by humans operating in the given problem area. Being loaded into a knowledge base by ***knowledge engineers*** (*KEngs*), who serve as an ***interface between knowledge carriers*** ("experts") ***and computers***, the aforementioned relations and interconnections may be applied by the ***ExS inference engine***, which executes mental actions appropriate to users' queries, as it might be done by the experts whose knowledge was acquired and loaded by the *KEngs* to the *KB*. So it is presumed that for a non-expert user, the ***result of access to an ExS would be the same as access to the experts themselves***.

This concept has at least four ***fundamental drawbacks***:

1. It is ***very difficult to extract knowledge*** from its carriers, especially in ill-formalized problem areas.
2. ***Knowledge extracted may be very subjective***, depending on the background and current experience of the experts involved, and may ***vary not only from one person to another***, but, as well, ***when one and the same person is involved in a knowledge acquisition process at different periods of time***.
3. As a consequence of the previous reason, ***knowledge accumulated in the KB*** very often may be ***inconsistent, incomplete, and non-verifiable***.
4. The volumes of *KBs* for practically applicable *DSSs* would be so large that ***their creation, verification, and maintenance would be of great difficulty and would require a lot of time and a great number of highly skilled KEngs***.

The listed drawbacks are considered to be bypassed by the development and application of techniques based on ***machine learning***, and there are two main approaches in current focus: non-classical logic and input-output learning.

The ***first*** one, collecting together ***non-monotonic reasoning*** (Reinfrank M. et al. 1989; Alferes et al. 1995; Aliseda 2017; Bochman 2018; Schurz 2005), ***default reasoning*** (Delgrande et al. 2000; Lukaszewicz 1988; Nute 1994; Pelletier and Elio 1997), ***abductive reasoning*** (Magnani 2015; Queiroz and Merrell 2006), etc., forms an actively developed and interesting *AI* area, but ***its effective practical application so far is not evident***.

The ***second***, on the contrary, due to its implementation by ***neural networks*** (Goodfellow et al. 2016; Piccinini and Bahar 2012) is on the rise, and many applications are covered by this approach. Application of a neural network presumes two steps: ***learning*** and ***work***.

The first step is implemented via ***teaching input***, which ***trains the network by operating it in the reverse mode.***

After the network is trained, it is put to work ***computing outputs from inputs*** in accordance with its initial topology and the weights of connections between neurons determined while training.

However, this approach is also not free of ***some evident drawbacks***:

1. It may be efficient, but only in cases when the ***problem*** considered may be represented and ***decomposed to the point where it is possible to solve it by some neural algorithm*** implementing massive parallel computation (as, e.g., in the area of recognition of video images).
2. It may be implementable if there exists the capability to ***obtain*** somewhere ***teaching input*** which ***fully corresponds to the expected application area and problems to be solved***.
3. It may be valid if there is confidence that the ***teaching input covers all or at least close to all situations to appear during the whole future period of network application***.

The last drawback is especially crucial, because the result of training of a network fully depends on the teaching input, and if there is no certainty that this input is representative, sufficient, and fully relevant to the current and all future situations, so there is ***no trust in the outputs received from such a network***. However, universal techniques for the assessment of the quality of teaching inputs and their relevance to the expected situations are not known.

In developing the multigrammatical framework, we initially intended to design an approach to the application of *AI* technology to *STS* planning and scheduling that would be ***free in its interpretation for problem-solving of any subjective and heuristic knowledge as well as from probabilistic algorithms***. Due to these intensions, we have obtained the resulting knowledge representation and, as a whole, a concept of smart (knowledge-based) sociotechnological system, based on ***another set of background principles***.

The multigrammatical knowledge base of any of the above-considered sociotechnological systems may contain two types of rules and metarules:

1. ***Unitary***, allowing operation by ***STS management staff*** in an ***offline*** mode by sets of integral parameters (most often, total amounts of resources consumed and/or produced by the system) necessary for ***planning*** *STS* behavior over the considered time intervals (primarily, ***system goal setting***)
2. ***Temporal***, applied, mainly, by the ***STS scheduler***, allowing in an ***online*** mode ***rational hard real-time STS control***, ensuring coherent operation of its technological base for processing of incoming orders, ***but also*** applied by ***STS management staff*** in an ***offline "what-if"*** mode for ***planning*** *STS* operations regarding expected flows of orders, destructive impacts, recovery actions, and the relevant detailed *TB* dynamics during order completion (also, primarily, ***goal setting***)

URs representing *STSs* are extracted unambiguously from the regulations establishing organizational structures and the supply arrangements associated with them. *URs* representing technological systems (devices and complexes) are extracted from technological documents and/or *CAD/CALS* systems defining the design and life cycle of these devices.

TRs representing manufacturing devices, similarly, are extracted from technological documents which define input and output collections as well as durations of operation cycles of *MDs*. All this knowledge is ***not empirical and/or probabilistic***, and maintenance (update) is simply and naturally synchronized with ongoing changes of an *MTB* (inclusion of new *MDs* in the *MTB*, elimination of already operating *MDs* from an *MTB*, as well as replacement of one of them by another).

Not less natural and initially valid is our ***representation of a resource base***, which is a collection of object-resources, ***as a multiset***, and ***representation of an impact as a pair of decrements of the RB and MTB*** related to a specific moment on the time axis. None of these constructions is the result of a subjective expert estimate; all numbers of *ORs* are ***measurable*** or ***countable***; hence, the same is true of multiplicities of objects representing these *ORs*.

On the other hand, the multigrammatical framework provides consumers with a simple and natural language for representation of orders, including their substantial part (***what to get and where to deliver***) along with time restrictions (***when to get***).

And, finally, the *MGF* algorithmics is a collection of mathematically correct algorithms with provable logics that are adequate to the mathematical semantics of the corresponding actions of consumers and *STS* schedulers. ***No heuristics or heuristics-containing transformations are used in these algorithms***; by these features, ***the MGF inherits the ideology of operations research***, and, as is discussed in detail in Chap. 10, the nexus between them is maximally close and natural.

So the ***role of knowledge engineers in the operation of MGF-based STSs*** is ***rather different*** than in the case of ***conventional knowledge-based decision support systems and expert systems***. Namely, they ***do not extract knowledge from experts***. They work as an interface between the aforementioned sources of knowledge about *STS* and the *STS* controller, applying all available tools ***for the creation of the STS KB by knowledge extraction from its heterogeneous electronic carriers*** and ***for the maintenance of this KB during the STS life cycle***. So the ***functions of STS KEngs are close to the functions of Data Stewards***, who, as a newly originated segment of the *IT* community, perform creation and maintenance of Big Data storages and ensure its availability to the end users (Wilkinson et al. 2016). Taking into account the number of *URs* and *TRs* to be accumulated in an *STS KB*, i.e., the ***volume of such knowledge bases, which for practically interesting cases are estimated in hundreds of thousands*** (Karasev and Sheremet 2008; Sheremet and Karasev 2013; Sheremet 2018), we may suppose that the *MGF* implementation leads us to the concept of ***Big Knowledge,*** which is a natural extension and generalization of ***Big Data***. So it would be reasonable to introduce a concept of ***FAIR*** (*Findable, Accessible, Interoperable, Reusable*) ***Knowledge***, assigning to it all the features inherent so far to ***FAIR Data*** (Wilkinson et al. 2016). Implementation of this concept would require a specific toolkit allowing ***all necessary activities for Big Knowledge stewardship***. Key features of the first software-implemented version of such a toolkit, allowing the creation, maintenance, and application of *UMG* knowledge bases, are described in Karasev and Sheremet (2008), Sheremet and Karasev (2013), and Sheremet (2018).

12.3 Future Developments of the Multigrammatical Framework

The multigrammatical framework in its current state is at the early beginning of its development and wide application. As we may suppose, to become widely applicable and profitable, the *MGF* must be developed in three main directions:

1. Applications
2. Mathematical background
3. Implementation

Let us consider the listed directions.

12.3.1 Applications

The area of possible *MGF* applications is rather wide and covers, in turn, also three directions:

1. Industry
2. Economy
3. Critical infrastructures

The most evident, ideologically closest, and most useful *MGF* application is the ***development of Industry 4.0***. Due to knowledge-based smart (optimal or at least rational) planning and scheduling, ***MGF-centered industrial systems may really become the "brain" of Industry 4.0 and the smart basis of the future advanced technosphere***. The main advantage of such an approach would be the natural unification of *IS* scheduling and, thus, *ISs'* flexibility, interoperability, and scalability. All these features would simplify and make easy the integration of different *ISs*, as well as increasing their capacity. From the other side, due to the initial *MGF*-based consideration of resilience, recovery, and support of industrial systems affected by destructive impacts, and its unified solution of the appropriate tasks, ***MGF-based IS schedulers would ensure maximally rational behavior of the affected IS in a hazardous environment and possible extreme situations***. This as a whole is the background for the implementation of the concept of ***smart and resilient ISs***.

Regarding the ***economy*** area, application of *MGFs* provides an opportunity for the solution of complicated combinatorial tasks inherent in concrete economical and financial activities and the behavior of subjects of the local, national, and global economy. A research effort would be applied to develop and accumulate techniques of multigrammatical representation of such tasks and their implementation for *STS* online scheduling. Along with such tasks, the *MGF* would be used for modeling economic systems in an offline regime in order to obtain reasonable values of some basic economical parameters (finally, prices/costs of various produced and consumed assets as well as of persons' activities), allowing sustainable operation of any *ES*.

Application of the *MGF* to the area of ***critical infrastructures*** would provide opportunities for an ***assessment of their real resilience*** to possible destructive impacts, as well as for a ***search for their weak components***. Multigrammatical representation of *CIs'* terrestrial and/or functional segments and their interconnections, as well as of ***hidden active cascade effects***, would make *MGF*-based ***CI simulators*** an effective tool for such assessment. Of course, ***CI schedulers*** might be created and installed at *CIs'* control centers to ensure rational operation of *CI* segments in extreme situations. To achieve the full adequacy of multigrammatical models of the aforementioned segments as well as quality of control decisions, it would be valuable to ***combine the MGF technology*** with ***already known and software-implemented technologies used for modeling and simulation of the aforementioned segments*** (power, transportation, industry, commerce, banking and finance, information and communications, emergency, public works, etc.). The value of such an *MGF* application is that ***MGF-based schedulers may serve as systems integrators, joining heterogeneous models of different local functional assets of CIs to their terrestrial multifunctional segments***.

12.3.2 Mathematical Background

A lot of interesting and complicated problems to be considered may be declared here:

1. Further improvements of generation algorithms, minimizing the number of redundant generation steps by early cut-off of unpromising branches, for all classes of filtering multiset grammars and metagrammars (general, unitary, temporal)
2. Development of parallel generation algorithms for various distributed computing environments and all classes of multiset grammars and metagrammars
3. Development of top-down generation algorithms for filtering self-generating *MGs*, *MMGs*, *TMGs*, and *TMMGs*
4. Development of an improved algorithmics of multi-order scheduling
5. Further study of the mathematical properties of all classes of *MGs* and *MMGs*
6. A wider consideration of the interconnections between all classes of *MGs* and *MMGs* and known problems of operations research
7. Further study of the interconnections between all classes of *MGs* and *MMGs* and known models of computation
8. Integration of all classes of *MGs* and *MMGs* and augmented Post systems, being the background of the *SSF*, to obtain additional capabilities for problem-solving via interconnected operations on multisets and sets of strings

All research in this area, along with such evident objectives as reduction of the computational complexity of generation of sets of multisets and hence efficiency of

STS schedulers, aims also at further positioning of the MGF among known mathematical toolkits and theoretical constructions.

A separate direction would be ***resource-based games***. *RBGs* are an intermediate area between *MGF* applications and its mathematical background, allowing relatively easy representation of the capabilities and goals of players and thus a compact and adequate description of resource-based conflicts of any nature. It would be of both theoretical and practical profit to further study the interconnections between *RBGs* and various classes of games considered in modern game theory. From the other side, the *RBG* toolkit itself might be used for the representation and investigation of classes of games never considered before by reason of the descriptional limitations of available conventional mathematical toolkits.

12.3.3 Implementation

The area of *MGF* implementation includes directions of research associated with future development of:

1. *MGF*-centered software regarding conventional von Neumann architectures as well as highly parallel homogeneous computing environments based on multicore processors and/or on various multiprocessor solutions
2. Techniques of hardware *MGF* implementation via the creation of heterogeneous computing environments based on specialized *RISC* processors
3. Hardware based on non-conventional models of computation for multiset generation

An effective intermediate tool for all listed directions would be a ***multi-agent implementation of basic algorithms for generating sets of multisets defined by filtering MGs, UMGs, and TMGs, as well as by relevant self-generating multigrammars and multiset metagrammars***. The first steps in this direction were described in Sect. 4.4 of this book. So all parallelization issues may be considered from this valuable background.

As was several times repeated throughout the whole book, there may be ***two main regimes of application of the MGF***: ***online*** (direct *STS* scheduling) and ***offline*** (predictive "what-if" modeling of *STS* operation under some kinds of flows of orders and destructive impacts).

To implement effective *STS* schedulers operating on very large knowledge bases, it would be reasonable to design ***non-conventional macro-pipeline architectures compatible with the aforementioned RISC processors***. There are evident reasons for the development of such hardware: in fact, ***multiset generation is based on only two operations—subtraction and addition of multisets*** (in particular cases, like non-variative *UMGs/UMMGs*, multiple addition may be replaced by multiplication by a constant). So there may be designed ***the simplest and hence the cheapest basic processor unit***, close in its function to the artificial neuron, for ***direct implementation of massive parallelism in the area of multiset generation***. Macro-pipeline architectures, implemented as fields of such neuron-like *RISC* processors, would be an effective tool, firstly, for ***multi-order scheduling of STSs operating in hard real time***.

Let us consider now the possibility of development of multiset generation hardware based on ***non-conventional models of computation***.

The closest to a real implementation is, in our opinion, an approach based on the representation of some class of mathematical programming problems by ***electrical circuits*** (*ECs*) (Dennis 1959). Namely, it has been known for a long time that ***the problem of quadratic mathematical programming may be represented by an electrical circuit***, which is constructed from the simplest elements (resistors, inductors, and capacitors). The main advantage of this approach is that the time of solution of the problem, i.e., of computation of the variable values, is defined by the time of transient process in the *EC* representing this problem. In turn, as is known from Sheremet (2010, 2011a), the ***problem of synthesis of a hierarchical resource-consuming STS, the solution of which may be obtained*** **via** ***its representation by an appropriate UMMG with specific boundary and optimizing conditions, implies a multiobjective problem of polynomial programming***. Taking into account that the algorithm described in Sect. 4.2 for the generation of the set of terminal multisets defined by a unitary multiset metagrammar is fundamental for practically all other *MGF* algorithmics, we may be sure that an ***attempt to apply EC-based technology to the implementation of a significant subset of the MGF would be productive***.

Regarding a ***quantum implementation of the MGF***, a starting point in this direction may be a multigrammatical representation of the problem of Boolean programming described in Sect. 10.3.4, which is known to be solved by an appropriate quantum-like algorithm ("***adiabatic optimization***") (Finnila et al. 1994; Rønnow et al. 2014; Tosatti 2008; Boixo et al. 2014). If so, there may be some creative feedback allowing the application of such algorithms for effective implementation of various classes of multiset grammars. However, there is some uncertainty regarding the appearance of really reliable and widely available quantum computers due to the number of technical challenges in building a large-scale quantum computer with a sufficient number of qubits.

Concerning the *DNA* approach, or ***molecular computation*** (Calude et al. 2001; Jonoska and Saito 2002), it seems promising that there is a natural and rather simple representation of some chemical reactions (*ChRs*) by multiset grammars, thus raising the possibility of a feedback—namely, application of some of the aforementioned *ChRs* for the simulation of multiset generation. It is not difficult to represent rules by some primary chemical reactions and, beginning from some initial chemical substance representing the initial multiset, to implement a generation chain by a sequence of the aforementioned primary reactions leading to the substance representing the terminal multiset. To apply the *MGF* to *ChR* modeling, however, it would be necessary to move on from multisets to their generalization—***macrosets***—which in their simplest form (sets of multisets) were addressed here for the first time in Sect. 4.2.3, dedicated to the *UMMG* algorithmics. A macroset in the most general form is defined recursively as a multiset of objects and macrosets, and this provides a flexible and universal representation of any chemical substance and any kind of *ChR*. However, a formal definition of macrosets, as well as of macroset grammars and metagrammars, would be the subject of further research.

One more non-conventional model of computation which may be considered as a possible way to implement the *MGF* is ***membrane computing*** (Ciobanu et al. 2007). This model may be considered as a biological version of molecular computation, and all that has been said above is true also in this case.

As a whole, there are many interesting and complicated problems originating from the multigrammatical framework, and a strong research effort should be applied to its development and wide dissemination.

Chapter 13
Conclusion

To finish our consideration of the main issues of the multigrammatical framework, developed to enable the creation and efficient operation of smart and resilient sociotechnological systems forming together a future global digital economy, let us make some general remarks.

The *MGF* includes three classes of multiset grammars: general, which are the background of all the rest, unitary, and temporal. Any such grammars may have a filter, thus being filtering, and also, as in the case of temporal multigrammars, may have self-generating rules, thus allowing self-generation, which is a unique fundamental property for modeling and study of industrial systems capable of producing manufacturing devices (active resources), not only passive (consumed) resources (Fig. 13.1).

Unitary multigrammars are intended as a tool for the solution of tasks associated with ***planning*** *STS* behavior over the considered time intervals, i.e., finally, with ***system goal setting***. Knowledge bases containing *URs* are used by *STS* schedulers, providing ***STS management staff*** in an ***offline*** mode with sets of integral parameters of systems operation—most often, the total amounts of resources consumed and/or produced by a system.

Temporal multigrammars, in turn, are intended to be applied, mainly, by *STS* schedulers, providing in an ***online*** mode ***rational hard real-time STS control***, ensuring coherent operation of their technological bases for the completion of incoming orders.

From the other side, *TMG* knowledge bases applied by *STS* schedulers in an ***offline "what-if"*** mode also allow ***STS planning*** and, finally, ***goal setting***, but in a much more sophisticated detailed dynamic formulation, including expected flows of orders, hostile destructive impacts, friendly recovery actions, and relevant detailed (in the form of sets of partial schedules of devices) dynamics of *STS* technological bases during the processing of orders.

To enable solution of the most complicated problems considered in operations research and systems analysis, metagrammatical extensions were proposed for each of the three introduced classes of *MGs*. Finally, the most advanced and powerful

I. A. Sheremet, *Multigrammatical Framework for Knowledge-Based Digital Economy*, https://doi.org/10.1007/978-3-031-13858-4_13

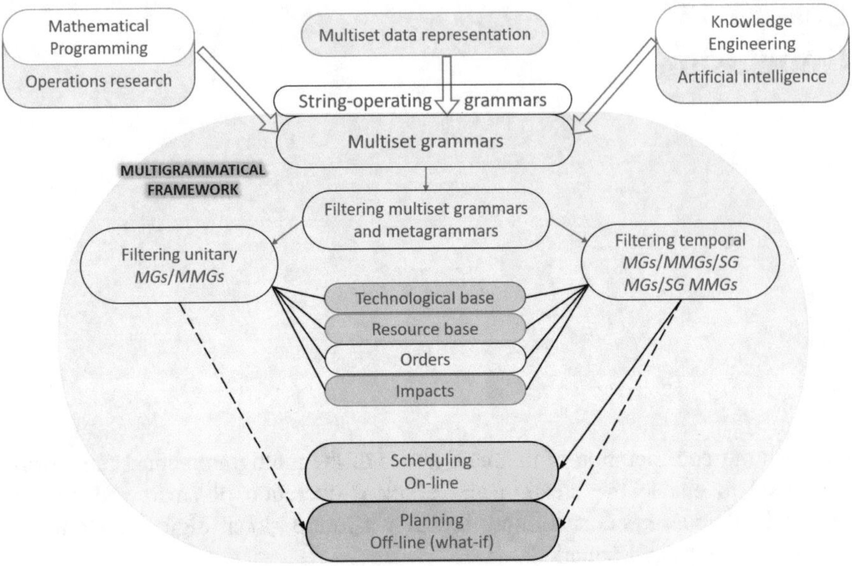

Fig. 13.1 Integral representation of the multigrammatical framework

class of multigrammars are self-generating multiset metagrammars, which allow the most general and detailed modeling of a modern cyberphysical industry.

There are many problems to be considered for all mentioned classes of multigrammars and their subclasses. Here in this book, we have only outlined basic *MGF* techniques and approaches. Two key groups of tasks are relevant for any of the *MG* classes and subclasses: ***algorithmics*** and ***applications***.

The objective of the solution of tasks of the ***first*** group is, finally, ***reduction of the computational complexity of generation of sets of multisets*** in the context of very large knowledge bases representing real *STSs* and their resource bases.

Tasks of the ***second*** group concern the application of specific *MGs* to the representation of relevant *STSs* and problems associated with these systems. As may be seen from results presented in this book on ***multigrammatical analysis of STS capabilities***, as well as on ***synthesis of STSs possessing predefined properties***, the *MGF* creates a flexible and general mathematical toolkit for the formal mathematical formulation and solution of such complicated tasks. In a closest perspective, the *MGF*-based software engines may be considered as a natural and effective extension of modern Product Lifecycle Management (*PLM*) and Enterprise Resource Planning (*ERP*) systems (Bendoly and Jacobs 2005; Malakooti 2013; Menon 2019; Saaksvuori 2008).

Let us make some concluding remarks on the assessment of ***new opportunities for the multigrammatical framework regarding various areas of modern science***.

Let us begin with the ***artificial intelligence*** area and underline once more that, in introducing the concept of *MGF*-based smart scheduling, we ***have not followed the conventional for an AI understanding of a KB as a store of heuristic***

experts-originated implications which may be applied by a knowledge interpretation engine to solve the problem in question. The efficiency of decision support systems implementing such an approach is crucially dependent on the experts' skills, experience, and subjectively understood ways of problem-solving. So, as this varies from one to another expert or even expert community, knowledge engineers responsible for *KB* maintenance may load into a *KB* different logics of scheduling, which would result, finally, in ***ambiguous*** and, hence, ***non-applicable*** results in practical situations.

The ***multigrammatical framework is based on a fundamentally different approach to extraction of the knowledge to be accumulated in a knowledge base and applied by STS schedulers***—whether in ***online*** or ***offline*** mode.

The conceptual background of the *MGF* regarding industry is understanding a knowledge base as a set of "black boxes" implicitly interconnected via a resource base, each black box representing some manufacturing device consuming some collection of input resources and producing a collection of output resources during some adequate time interval. There is ***no place*** for any subjectivity or ***ambiguity*** in such a *KB*, and any schedule created by an *STSS* (or, in the case of an industrial system, an *ISS*) allows processing of a consumer's order, which strictly defines a collection of resources to be produced as well as terms of production and limits of resources available to the *STS* (*IS*) manufacturing technological base for the completion of this order. Similarly, there is ***no place for any heuristics in the process of schedule creation*** by an *STSS* (*ISS*), and all reduction of redundant search during this process is done by ***provable cutting-off of deadlock-generating branches not leading to a solution***. (The earlier a cut-off is done, the greater the reduction of the redundant search.) In fact, explication of the aforementioned interconnections between *MDs* in the form of an *STS* (*IS*) schedule allowing the completion of an order is a ***core intellectual function of an STSS (ISS)***. By this interpretation of a knowledge base, an order, and solution space pruning, ***the MGF is very close to the ideology of operations research***—if speaking about industrial systems modeled by temporal multiset grammars and then to the ideology of such segments of *OpR* as scheduling theory. This is not by chance, because the *MGF* was initially developed as a basic knowledge representation model for the formulation and solution of just such problems. From the other side, one may consider known ***OpR models as specific problem-oriented KRMs***, with inference engines performing reduced search for solutions strictly defined by conventional non-procedural (declarative) mathematical tools based mainly on the ***matrix-vector calculus*** or ***graph-network models***.

From the other side, the *MGF*, due to temporal multisets, temporal multigrammars, metagrammatical extensions of multigrammars, and, especially, self-generated *TMGs*, provides some ***fundamentally new features compared to known KRMs***. These features, illustrated in this book by the application of the *MGF* to the solution of various tasks concerning sociotechnological systems of different classes, very possibly may be useful in wider problem areas.

Let us pay attention to the real simplicity of the creation and maintenance of multigrammatical knowledge bases: when including in a *KB* the digital twin of a new device being added to the technological base of an *STS*, it is not obligatory to pay attention to all the rest of the *TB* and the *KB* representing it. This is possible due to the absence of explicit connections between devices, which is a consequence of the conceptual background of the *MGF*—namely, generation of new multisets by the independent application of rules to the current *MS*—thus allowing interaction between devices via a resource base. Such a simple technique for updating multigrammatical *KBs* (in fact, it may be implemented with a very limited participation of knowledge engineers by application of *CALS*-based tools capable of extracting digital twins of devices directly from their digital passports or application manuals) makes possible the maintenance and application of extra-large volumes of knowledge bases, which, in fact, represents a ***next natural step of global digitalization*** which is expected to follow Big Data—namely, ***Big Knowledge***.

From the theoretical ***computer science*** point of view, the *MGF* is a ***non-conventional model of computation***, enabling "goal- and knowledge-driven computation by ubiquitous multiset generation," which combines the ideology of string-operating grammars and constraint programming on such a flexible and general data structure as a multiset, in a natural way integrating in one data item symbolic and numeric information. A multiset representation of data is much more convenient than their conventional vector (matrix) representation. To operate on vector-represented data in practically oriented systems, it is necessary to store and process a large amount of extra-long vectors, and inclusion of one additional data item, equivalent to a multiobject with a new object, is a hardest case, demanding an increase of vector dimensionality and, thus, correction of all stored vectors containing the relevant component; the same is true in the case of exclusion of some one component, demanding a decrease of vector dimensionality.

From the ***mathematical economy*** point of view, the *MGF* would be a convenient tool for the representation and solution of multiple problems, first of all, for ***concrete economical systems, business intelligence,*** and ***economical combinatorics***. However, macroeconomical applications are not less natural for the *MGF*, which operates on basic categories inherent to this area of research and activity (as was illustrated by the *UMG*-based modeling of *ISs* and *Ess* with the brief reference to the Leontief model).

From the ***systems analysis*** point of view, the *MGF* may be considered as a convenient mathematical background for the adequate representation of various large-scale hierarchical systems and any other systems of any nature (industrial, economical, sociotechnological, political, ecological, etc.) and their investigation in a "what-if" mode, simply implementing, if necessary, the now well-advanced techniques of agent-based systems modeling and simulation (Niazi and Hussain 2009; Wilensky and Rand 2015). In this area, the *MGF* may be considered as a new approach, strongly supporting the establishment of systems analysis as ***mathematics of the twenty-first century***.

So the development and application of the *MGF* has created new opportunities for the consideration and solution of various problems which until now were hardly amenable to representation by conventional mathematical tools. From the other side, the unified algorithmics associated with the *MGF* is a vital precondition for its ***effective hardware implementation***.

The state of the modern technosphere may permit fast and wide dissemination of *MGF*-based software and hardware and, finally, creation of ***MGF-based ecosystems*** as a unified background for an implementation of the aforementioned ***GaaS concept***.

In any case, the ***multigrammatical framework represents a new way of thinking about sociotechnological systems, about the global digital economy, and, as a whole, about the modern and future technosphere.***

Working on this book, the author has believed that the proposed framework would be interesting and useful to systems analysts, economists of different levels, practical decision-makers, mathematicians, computer scientists, as well as scholars and engineers solving various complicated problems associated with a global digital economy.

Acronyms

ACE	Active cascade effect
AE	Application engine
AES OPR	Assigned to order economical system operating on passive resources
AI	Artificial intelligence
AIS	Assigned to order industrial system
AIS MPR	Assigned to order industrial system manufacturing passive resources
AL	Assembly line
AM	Access message
Am	"Alert" message
B	Borrower
BC	Boundary condition
CAD	Computer-aided design
CALS	Computer-aided logistics and support
CBC	Chain boundary condition
CF	Context-free
CI	Critical infrastructure
ChR	Chemical reaction
CODATA	Committee on Data
CSA	Committee for Systems Analysis
DB	Data base
DBMS	Data base management system
DE	Digital economy
DiT	Digital twin
DML	Data manipulation language
DMS	Dynamic multiset
DRBG	Distributed resource-based game
DS	Data storage
DSS	Decision support system
DSTS	Distributed sociotechnological system
EBC	Elementary boundary condition

I. A. Sheremet, *Multigrammatical Framework for Knowledge-Based Digital Economy*, https://doi.org/10.1007/978-3-031-13858-4

EC	Electrical circuit
EES	Exchange economical system
EESS	Exchange economical system scheduler
EI	Energy infrastructure
EIS	External industrial system
ElcI	Electricity infrastructure
EnC	Energy carrier
EP	Electrical power
ERP	Enterprise resource planning
EP	Electrical power
EnC	Energy carrier
ES	Economical system
ES OPAR	Economical system operating on passive and active resources
ES OPR	Economical system operating on passive resources
ExS	Expert system
FAIR	Findable, accessible, interoperable, reusable
FDS	Fuel distribution station
FES OPR	Free economical system operating on passive resources
FI	Fuel infrastructure
FIS	Free industrial system
FIS MPR	Free industrial system manufacturing passive resources
FMG	Filtering multiset grammar
Fm	"Finish" message
FPP	Fuel producing plant
FS	Fuel storage
FSG	Filtering self-generating
FSMS	Finitarized set of multisets
FTMG	Filtering temporal multiset grammar
FTU	Fuel terminal unit
FUMG	Filtering unitary multiset grammar
GaaS	Government as a service
GII	Global information infrastructure
GTI	Global transportation infrastructure
GMG	General multiset grammar
HaaD	Human as a device
I	Implementational (semantics)
IAB	Immediate-activation-based
IC	Input collection
IIASA	International Institute of Applied Systems Analysis
IIoT	Industrial internet of things
ILP	Integer linear programming
IM	Interval multiplicity
IoT	Internet of things
IPS	Individual pseudoschedule

IS	Industrial system
ISC	IS controller
ISh	Individual schedule
ISM	IS manager
IS MPAR	Industrial system manufacturing passive and active resources
IS MPR	Industrial system manufacturing passive resources
ISS	IS scheduler
IT	Information technology
KB	Knowledge base
KEng	Knowledge engineer
KIE	Knowledge interpretation engine
k-MG	k-multigrammar
KRM	Knowledge representation model
L	Lender
LES	Lending economical system
LESS	Lending economical system scheduler
LSTS	Local sociotechnological system
M	Mathematical (semantics)
MAT	Multi-agent technology
MATS	Multiple access technological system
MAS	Multi-agent system
MaaS	Mobility as a service
MB	Moves base
MC	Multiplicity-constant
MD	Manufacturing device
MdD	Manufactured device
MDB	Metadata base
MFMD	Multifunctional manufacturing device
MFP	Maximal flow problem
MG	Multiset grammar (multigrammar)
MGF	Multigrammatical framework
MMG	Multiset metagrammar (multimetagrammar)
MO	Multiobject
MR	Metarule
MS	Multiset
MSIM	Multiset with interval multiplicities
MST	Temporal multiset
MTR	Mirror temporal rule
MV	Multiplicity-variable
MTB	Manufacturing technological base
NTO	Non-terminal object
O	Organizational
OC	Operation cycle
OCF	Operation cycle of facility

Op	Operational (semantics)
OpC	Optimizing condition
OpR	Operations research
OR	Object-resource
OT	Organizational-technological
OU	Organizational unit
PCE	Passive cascade effect
PEC	Primary energy carrier
PES	Producing economical system
PLM	Product lifecycle management
PP	Power plant
PPBS	Planning, programming and budgeting system
PS	Power storage
PTDS	Power transforming-distributing substation
PTL	Power transmission line
PTU	Power terminal unit
RAS	Russian Academy of Sciences
RB	Resource base
RBC	Resource-based conflict
RBG	Resource-based game
RCPSP	Resource-constrained project scheduling problem
RCS	Resource-consuming system
RISC	Reduced instruction set computer
ROT	Resource-consuming organizational-technological
RSK	Resource-safekeeping
SDG	Sustainable development goals
SF	Sentential form
SG	Self-generating
SAE	Substance affecting the environment
SIS	Supervisor of manufacturing technological base and resource base of IS
Sm	“Start” message
SMS	Set of multisets
SoC	System-on-a-crystal
SPP	Shortest path problem
SQL	Structured query language
SSF	“Set of strings” framework
ST	Set-theoretical (semantics)
STMS	Set of terminal multisets
STS	Sociotechnological system
STSC	STS controller
STSS	STS scheduler
TB	Technological base
TE	Technological equipment

TM	Time marker
TMG	Temporal multiset grammar
TMMG	Temporal multiset metagrammar
TMR	Temporal metarule
TMS	Terminal multiset
TMST	Terminal temporal multiset
TO	Terminal object
ToM	Tools of manufacturing
TR	Temporal rule
TSP	Traveling salesman problem
TVh	Transport vehicle
UMG	Unitary multiset grammar
UMMG	Unitary multiset metagrammar
UMR	Unitary metarule
UR	Unitary rule
VCM	Variable-containing multiplicity
VMS	V-multiset
WAN	Wide area network
1-MG	1-Multigrammar

References

Abdolshah M (2014) A review of resource-constrained project scheduling problems (RCPSP) approaches and solutions. Int Trans J Eng Manag Appl Sci Technol 5(4):253–286

Abraham I, Fiat A, Goldberg AV, Werneck RF (2010) Highway dimension, shortest paths, and provably efficient algorithms. In: Proceedings of ACM-SIAM symposium on discrete algorithms (SODA10), January 2010. SIAM, New York

Ackoff RL (1981) Creating the corporate future. Plan or be planned for. Wiley, Hoboken

Akama S (2015) Elements of quantum computing: history, theories and engineering applications. Springer, New York

Akerkar R, Sajja P (2010) Knowledge-based systems. Jones and Bartlett Publishers, Sudbury

Alcaraz C, Zeadally S (2015) Critical infrastructure protection: requirements and challenges for the twenty-first century. Int J Crit Infrastruct Prot 8:53–66. https://doi.org/10.1016/j.ijcip.2014.12.002

Alferes JJ, Damasio CV, Pereira LM (1995) A logic programming system for non-monotonic reasoning. J Autom Reason 14(1):93–147

Aliseda A (2017) The logic of abduction: an introduction. In: Magnani L, Bertlotti T (eds) Springer handbook of model-based science. Springer, pp 219–230

Apt K (2003) Principles of constraint programming. Cambridge University Press, Cambridge, p 420. https://doi.org/10.1017/CBO9780511615320

Aumann RJ, Maschler RB (1995) Repeated games with incomplete information. MIT Press, Cambridge

Baller S, Dutta S, Lanvin B (eds) (2016) The global information technology report 2016. Innovating in the digital economy. WEF and INSEAD, Geneva

Banâtre J-P, Le Métayer D (1993) Programming by multiset transformation. Commun ACM 36(1): 98–111. https://doi.org/10.1145/151233.151242

Bast H, Funke S, Matijevic D (2009) Ultrafast shortest-path queries via transit nodes. In: Demetrescu C, Goldberg AV, Johnson DS (eds) The shortest path problem: ninth DIMACS implementation challenge. AMS, New York, pp 175–192

Baziari T (2021) Urban mobility in digital era the shift from vehicle ownership to usership. Risalant Consultants. https://www.researchgate.net/publication/350767756

Beer S (1959) Cybernetics and management. English Universities Press

Beer S (1988) The heart of enterprise. Wiley, London

Belardinelli F, Argento F (eds) 2017 Multi-agent systems and agreement technologies 15th European conference, EUMAS, and fifth international conference, AT 2017: Evry, France, December 14–15, Revised Selected Papers

Bendoly E, Jacobs FR (2005) Strategic erp extension and use. Stanford University Press, Stanford

Benoit JP, Krishna V (1985) Finitely repeated games. Econometrica 53(4):905–922. https://doi.org/10.2307/1912660

I. A. Sheremet, *Multigrammatical Framework for Knowledge-Based Digital Economy*, https://doi.org/10.1007/978-3-031-13858-4

Bentley LD, Whitten JL (2007) Systems analysis and design for the global enterprise. McGraw-Hill/Irwin, New York

Bierman HS, Fernandez IF (1997) Game theory for applied economists, 2nd edn. Addison-Wesley, Boston

Bjerkholt O, Kuzz HD (2006) Introduction: the history of input-output analysis, Leontief's path and alternative tracks. Econ Syst Res 18(4):331–333. https://doi.org/10.1080/09535310601030850

Bjørner D, Ershov AP, Jones ND (eds) (1988) Partial evaluation and mixed computation. North Holland, Amsterdam

Blanchard BS, Fabrycky WJ (2010) Systems engineering and analysis. Prentice-Hall, Englewood Cliffs

Blizard WD (1989) Multiset theory. Notre Dame J Form Log 30:36–66

Bochman A (2018) On laws and counterfactuals in causal reasoning. In: Proceedings of the sixteenth international conference on principles of knowledge representation and reasoning (KR 2018). Association for the Advancement of Artificial Intelligence. pp 494–503

Boixo S, Rønnow TF, Isakov SV, Wang Z, Wecker D, Lidar DA, Martinis JM, Troyer M (2014) Evidence for quantum annealing with more than one hundred qubits. Nat Phys 10:218–224

Bonds E, Downey L (2012) "Green" technology and ecologically unequal exchange: the environmental and social consequences of ecological modernization in the world-system. J World-Syst Res 18(2):167–186. https://doi.org/10.5195/jwsr.2012.482

Bratko I (2012) Prolog programming for artificial intelligence. Addison-Wesley, New York

Brown RE (2008) Electric power distribution reliability. CRC Press, Boca Raton

Burkart O (1997) Automatic verification of sequential infinite-state processes. Lect Notes Comput Sci 1354:163. https://doi.org/10.1007/3-540-69678-4

Calude CS, Paun G, Rozenberg G, Salomaa A (2001) Multiset processing: mathematical, computer science and molecular computing points of view, Lecture notes in computer science, vol 2235. Springer, New York, p 359. https://doi.org/10.1007/3-540-45523-X

Carreras B, Lynch V, Dobson J, Newman D (2005) Critical points and transitions in an electric power transmission model for cascading failure blackouts. Chaos 12(4):985–994. https://doi.org/10.1063/1.1505810

Cavdaregla B, Hammel E, Mitchell JE, Sharkey TC, Wallace WA (2013) Integrating restoration and scheduling decisions for disrupted interdependent infrastructure systems. Ann Oper Res 203:279–294. https://doi.org/10.1007/S10479-011-0959-3

Chi Z (1999) Statistical properties of probabilistic context-free grammars. Comput Linguist 25(1): 131–160

Chomsky N (2005) Syntactic structures. Mouton de Gruyter, The Hague

Ciobanu G, Pérez-Jiménez MJ, Păun G (eds) (2007) Applications of membrane computing. Springer

Colombo AW, Bangemann T, Karnouskos T, Delsing J, Stluka P, Harrison R, Jammes F, Lastra JL (eds) (2014) Industrial cloud-based cyber-physical systems. Springer, Cham. https://doi.org/10.1007/978-3-319-05624-1_9

Conway RW, Maxwell WL, Miller LW (2003) Theory of scheduling. Dover Publications, Mineola

Cross TB (2017) The uses of artificial intelligence in business. Prentice Hall, New York. TECHtionary.com

D'Andrea R, Dullerud G (2003) Distributed control design for spatially interconnected systems. IEEE Trans Autom Control 48(9):1478–1495. https://doi.org/10.1109/TAC.2003.816954

Dabrowski C, Hunt F (2011) Using Markov chain and graph theory concepts to analyze behaviour in complex distributed systems. In: Proceedings of European Modelling and Simulation Conference (EMSS). Rome, Italy, pp 658–669

Dabrowski C, Hunt F, Morrison K (2011) Improving the efficiency of Markov chain analysis of complex distributed systems. National Institute of Standards and Technology Interagency Report 7744

Dannen C (2017) Introducing Ethereum and solidity—foundations of cryptocurrency and blockchain programming for beginners. Apress, New York

Darwen H (2012) An introduction to relational database theory. bookboon.com, London, UK
Date CI (2012) An introduction to database systems, 8th edn. Pearson, London
David R, Alla H (2005) Discrete, continuous, and hybrid petri nets. Springer, Berlin
Delgrande JP, Schaub T, Tompits H (2000) Logic programs with compiled preferences. In: Baral C, Truszczynski M (eds) Proceedings of the eight international workshop on nonmonotonic reasoning, NMR'2000, p ArXiv
Dennis JB (1959) Mathematical programming and electrical networks. Wiley, New York
Dobson I, Carreras BA, Lynch VE, Newman DE (2007) Complex systems analysis of series of blackouts: cascading failure, critical points, self-organization. Chaos 17:026103. https://doi.org/10.1063/1.2737822
Don Vito PA (1969) The essentials of a planning-programming-budgeting system. RAND Corporation, Santa Monica
Dullerud G, Paganini F (2005) A course in robust control theory: a convex approach. Springer, New York
Ershov AP (1977) On the partial computation principle. Inf Process Lett 2(2):38–41
Eshra A, Shah S, Song T, Reif J (2019) Renewable DNA hairpin-based logic circuits. IEEE Trans Nanotechnol 18:252–259. https://doi.org/10.1109/tnano.2019.2896189
Eusqeld I, Nan C, Dietz S (2011) "System–of–Systems" approach for interdependent critical infrastructures. Reliab Eng Syst Saf 96:679–686. https://doi.org/10.1016/j.ress.2010.12.010
Factory (2013) Factory-in-a-day. Eur Comm 2013. http://www.factory-in-a-day.eu
Farhi E, Goldstone J, Gutmann S (2015) A quantum approximate optimization algorithm applied to a bonded occurrence constraint problem. In: Quantum physics. Report MIT-CTP/4628. Cornell University, New York. arXiv: 1412.6062
Fiedler M, Nedoma J, Ramik J, Rohn J, Zimmerman K (2006) Linear optimization problems with inexact data. Springer, New York
von Bertalanffy L (1988) General system theory: foundations, development, application. George Braziller, New York
Finnila AB, Gomez MA, Sebenik C, Stenson C, Doll JD (1994) Quantum annealing: a new method for minimizing multidimensional functions. Chem Phys Lett 219(5–6):343–348
Franks B (2014) The analytics revolution: how to improve your business by making analytics operational in the big data era. Wiley, New York
Frunkwirth T, Abdennadher S (2003) Essentials of constraint programming. Springer, Berlin, p 398. https://doi.org/10.1007/978-3-662-05138-2
Gass SI, Assad AA (2005) An annotated timeline of operations research: an informal history. Kluwer Academic Publishers, New York
Geraldi J, Lechter T (2012) Gantt charts revisited. Int J Manag Proj Bus 5(4):578–594. https://doi.org/10.1108/17538371211268889
Giusti L (2009) A review of waste management practices and their impact on human health. Waste Manag 29(8):2227–2239. https://doi.org/10.1016/j.wasman.2009.03.028
Goh CJ, Yang XQ (2002) Duality in optimization and variational inequalities. Taylor & Francis, New York, p 330. https://doi.org/10.1201/9781420018868
Gollakota ARK, Gautam S, Shu C-M (2020) Inconsistencies of e-waste management in developing nations—facts and plausible solutions. J Environ Manage 261:110234. https://doi.org/10.1016/j.jenvman.2020.11023
Gondran M, Minoux M (2008) Graphs, dioids and semirings: new models and algorithms. Springer, New York
Goodfellow I, Bengio Y, Courville A (2016) Deep learning. MIT Press
Gutin G, Punnen AP (eds) (2006) The traveling salesman problem and its variations. Kluwer Academic Publishers, New York
Gvishiani AD, Roberts FS, Sheremet IA (2018) On the assessment of sustainability of distributed sociotechnical systems to natural disasters. Russ J Earth Sci 18:ES4004. https://doi.org/10.2205/2018ES000627

Haimes YY, Jiang P (2001) Leontief-based model of risk in complex interconnected infrastructures. J Infrastruct Syst 7:1–12. https://doi.org/10.1061/(ASCE)1076-0342(2001)7
Hansen E (1979) Global optimization using interval analysis—the one-dimensional case. J Optim Theory Appl 29:331–334
Hansen E (1992) Global optimization using interval analysis. Marcel Dekker, New York
Hansen E, Walster GW (2004) Global optimization using interval analysis. Marcel Dekker, New York
Harrington JE (2014) Games, strategies, and decision making, 2nd edn. Worth Publishers, New York
He X, Cha EJ (2018) Modeling the damage and recovery of interdependent critical infrastructure systems from natural hazards. Reliab Eng Syst Saf 177:162–175. https://doi.org/10.1016/j.ress.2018.04.029
Hemmecke R, Koppe M, Lee J, Weismantel R (2009) Nonlinear integer programming. In: 50 Years of integer programming 1958–2008: the early years and state-of-the-art surveys. Springer, New York, pp 1–57. https://doi.org/10.1007/978-3-540-68279-0-15
Hermann JW (2006) Handbook of production scheduling. Springer, New York, p 318. https://doi.org/10.1007/0-387-33117-4
Hespanha J, Naghstabrizi P, Xu Y (2007) A survey of recent results in networked control systems. Proc IEEE 95(1):138–162. https://doi.org/10.1109/JPROC.2006.887288
Hickman JL (1980) A note on the concept of multiset. Bull Aust Math Soc 22:211–217. https://doi.org/10.1017/5000497270000650X
Hillier SF, Lieberman GJ (2014) Introduction to operations research. McGraw Hill, Boston
Hopcroft JE, Motwani R, Ullman J (2001) Introduction to automata theory, languages, and computation. Addison-Wesley, Reading
Horn RA, Johnson CR (1991) Matrix analysis. Cambridge University Press, New York, p 287
Houben R (2015) Bitcoin: there are two sides to every coin. ICCLR 26(5):193–208
Huws A (2014) Labor in the global digital economy: the Cybertariat comes of age. NYU Press, New York
Ignatova Z, Martinez-Perez I, Zimmermann K-H (2008) DNA computing models. Springer, New York
Itkin VE (1991) An algebra of mixed computation. Theor Comput Sci 90(1):81–93
Jadbabiae A, Lin J, Morse A (2003) Coordination of groups of mobile autonomous agents using nearest neighbor rules. IEEE Trans Autom Control 48(6):988–1001
Jesse B-J, Heinrichs HU, Kuckshinrichs E (2019) Adapting the theory of resilience to energy systems: a review and outlook. Energy Sustain Soc 9(27):1–19
Ji C, Wei Y, Poor HV (2017) Resilience of energy infrastructure and services: modeling, data analytics, and metrics. Proc IEEE 105(7):1354–1366
Jiiang DY (2010) Situation analysis of double action games with entropy. Science Press, New York
Jonoska N, Saito M (2002) Boundary components of thickened graphs. In: Jonoska N, Seeman NC (eds) Proceedings of the seventh international meeting on DNA based computers, Lecture notes in computer science, vol 2340. Springer, pp 70–81
Kaper H, Roberts F, Sheremet I (2022) Preparing for a crisis: improving the resilience of digitized complex systems. In: Dhersin J-S, Kaper H, Ndifon W, Roberts F, Rousseau C, Ziegler GM (eds) Mathematics for action. Supporting science-based decision-making. UNESCO, Paris. ISBN 978-92-3-100517-6. https://unesdoc.unesco.org/ark:/48223/pf0000380883.locale=en
Karasev RS, Sheremet IA (2008) Assessment of the duration of the objective-driven program with the shifted terms of its contracts. Def Technol Issues 1:31–40. In Russian
Karlot JK (2005) Integer programming: theory and practice. CRC Press, Boca Raton
Karp RM, Miller RE (1969) Parallel program schemata. J Comput Syst Sci 3(2):147–195. https://doi.org/10.1016/S0022-0000(69)80011-5
Katay ME (2010) Electric power industry as critical infrastructure. Network World. https://www.networkworld.com/article/2217677/datacenter/electric-power-industry-as-critical-infrastructure.html

Kendal SL, Green M (2007) An introduction to knowledge engineering. Springer, London

Klavins E, Christ R, Lipsley D (2006) A grammatical approach to self-organizing robotic systems. IEEE Trans Autom Control 51(6):949–962. https://doi.org/10.1109/TAC.2006.876950

Klein R (1999) Scheduling of resource-constrained projects. Kluwer Academic Publishers, Boston

Kovalchuk MV, Naraikin OS, Yatsishina EB (2019) Nature-like technologies: new opportunities and new challenges. Bull Russ Acad Sci 89(5):455–465. https://doi.org/10.31857/S0869-5873895455-465. In Russian

Kurian M, McCarney P (eds) (2010) Peri-urban water and sanitation services. Policy, planning and method. Springer, New York, p 300. https://doi.org/10.1107/978-90-481-9425-4_11

Kuznetsov I (2021) Iceland's four-day work week experiment billed "Enormous Success". Sputnik. https://sptnkne.ws/GGYv

Lade SJ, Gross T (2012) Early warning signals for critical transitions: a generalized modeling approach. Comput Biol 8(2):e1002360. https://doi.org/10.1371/journal.pcbi.1002360

Lake J (1976) Sets, fuzzy sets, multisets and functions. J Lond Math Soc 12:323–326

Larsen ER, Osorio S, van Ackere A (2017) A framework to evaluate security of supply in the electricity sector. Renew Sustain Energy Rev 79:646–655. https://doi.org/10.1016/j.rser.2017.05.085

Lasdon SL (2013) Optimization theory for large systems. Dover Publications, New York

Lavaei J, Tse D, Zhang B (2014) Geometry of power flows and optimization in distributed networks. IEEE Trans Power Syst 29(2):572–583

Lee J, Bagheri B, Hung-An K (2015) A cyber-physical systems architecture for industry 4.0-based manufacturing systems. Manuf Lett 3:18–23. https://doi.org/10.1016/j.mfglet.2014.12.001

Leung JY-T (ed) (2004) Handbook of scheduling: algorithms, models, and performance analysis. Chapman & Hall/CRC, New York

Levin DA, Peres Y, Wilmer EL (2009) Markov chains and mixing times. Providence. American Mathematical Society

Li H, Rosenwald GW, Jung J, Liu C-C (2005) Strategic power infrastructure defense. Proc IEEE 93(5):918–933. https://doi.org/10.1109/JPROC.2005.847260

Liu K, Wang M, Zhu W, Wu J, Yan X (2018) Vulnerability analysis of an Urban gas pipeline network considering pipeline-road dependency. Int J Crit Infrastruct Prot 22:125–138. https://doi.org/10.1016/j.ijcip.2018.08.008

Lloyd JW, Shepherdson JC (1991) Partial evaluation in logic programming. J Log Program 11(3–4):217–242. https://doi.org/10.1016/0743-1066(91)90027-M

Lukaszewicz W (1988) Considerations on default logic: an alternative approach. Comput Intell 4(1):1–16

Lund H, Werner S, Wiltshire R, Svendsen S, Thorsen J-E, Hvelplund F, Vad Mathiesen B (2014) Fourth generation district heating (4GDH): integrating smart thermal grids into future sustainable energy systems. Energy 68:1–11. https://doi.org/10.1016/j.energy.2014.02.019

MacGregor JN, Chu Y (2011) Human performance on the traveling salesman and related problems: a review. J Probl Solving 3, No. 2. https://doi.org/10.7771/1932-6246.1090

Magnani L (2015) Naturalizing logic. J Appl Log 13(1):13–36

Makropoulos C, Rozos E, Tsoukalas I, Plevri A, Karakatsanis L, Karagiannidis L (2018) Sewer-mining: a water reuse option supporting circular economy, public service provision and entrepreneurship. J Environ Manage 216:285–298

Malakooti B (2013) Operations and production systems with multiple objectives. Wiley, New York

Marriott K (1994) Constraint multiset grammars. In: Proceedings of IEEE symposium on visual languages. IEEE Computer Society Press, pp 118–125. https://doi.org/10.1109/VL.1994.363633

Marriott K (1996) Parsing visual languages with constraint multiset grammars. In: Programming languages: implementation, logic and programs, Lecture notes in computer science, vol 1292. Springer, New York, pp 24–25

Marriott K, Meyer B (1997) On the classification of visual languages by grammar hierarchies. J Vis Lang Comput 8:375–402. https://doi.org/10.1006/jvlc.1997.0053

Marriott K, Stucky PG (2003) Programming with constraints: an introduction. MIT Press, Cambridge

Marx K (2018) Capital. A critique of political economy. 1. www.marxists.org

Mazher AR, Liu S, Shukla A (2018) A state of art review on the district heating systems. Renew Sustain Energy Rev 96:420–439. https://doi.org/10.1016/j.rser.2018.08.005

McCreary D, Kelly A (2013) Making sense of NoSQL: a guide for managers and the rest of us. Manning Publications, Shelter Island

Meduna A (2014) Formal languages and computation: models and their application. CRC Press, New York

Menon S (2019) Benefits and process improvements for ERP implementation: results from an exploratory case study. Int Bus Res 12(8):124–132. https://doi.org/10.5539/ibr.v12n8p124

Mesbahi M, Egerstedt M (2010) Graph-theoretic methods in multiagent networks. Princeton University Press, Princeton. https://doi.org/10.1515/9781400835355

Meyer RK, McRobbie MA (1982) Multisets and relevant implication. I, II. Aust J Philos 60:107–139. https://doi.org/10.1080/00048408212340551

Mills K, Dabrowski C (2006) Investigating global behaviour in computing grids, Lecture notes in computer science, vol 4124. Springer, Berlin, pp 120–136

Mills K, Dabrowski C (2008) Can economics-based resource allocation prove effective in a computation marketplace? J Grid Comput 6(3):291–311. https://doi.org/10.1007/s10723-007-9094-4

Mills K, Filliben J, Cho D-Y, Schwartz E (2011) Predicting macroscopic dynamics in large distributed systems. Proceedings of ASME 2011 conference on pressure vessels & piping. Baltimore, MD, July 17–22, 2011

Mills K, Filliben J, Dabrowski C (2012) Predicting global failure regimes in complex information systems. DoE COMBINE Workshop

Montanaro A (2015) Quantum algorithms: an overview. Npj Quantum Inf 2:15023. https://doi.org/10.1038/npjqi.2015.23

Munoz JM (2017) Global business intelligence. Routledge, Abingdon

Murata T (1989) Petri nets: properties, analysis and applications. Proc IEEE 77(4):541–558. https://doi.org/10.1109/5.24143

Naraynasamy V, Wong KW, Rai S, Chiou A (2010) Complex game design modeling. In: Cultural computing: second IFIP TC 14, entertainment computing symposium, ECS 2010. Brisbane, Australia, September 20–23, pp 65–74. https://doi.org/10.1007/978-3-642-15214-6_7

Nepal R, Jamasb T (2013) Security of European electricity systems: conceptualizing the assessment criteria and core indicators. Int J Crit Infrastruct Prot 6(3–4):182–196. https://doi.org/10.1016/j.i.cip.2013.07.001

Niazi M, Hussain A (2009) Agent-based tools for modeling and simulation of self-organization in Peer-to-Peer, ad-hoc and other complex networks. IEEE Commun Mag 47(3):163–173. https://doi.org/10.1109/MCOM.2009.4804403

Nute D (1994) Defeasible logic. In: Gabbay DM, Hogger CJ, Robinson JA (eds) Handbook of logic in artificial intelligence and logic programming, vol 3. Oxford University Press, pp 355–395

Olfati-Saber R, Fax JA, Murray RM (2007) Consensus and cooperation in networked multiagent systems. Proc IEEE 95(1):215–233. https://doi.org/10.1109/JPROC.2006.887293

Pannu A (2015) Artificial intelligence and its application in different areas. Int J Eng Innov Technol 4(4):79–84

Papayoanou P (2010) Game theory for business: a primer in strategic gaming. Probabilistic Publishing, Sugar Land

Parikh RJ (1966) On context-free languages. J ACM 13(4):570–581

Pederson P, Dudenhoeffer D, Hartley S, Permann M (2006) Critical infrastructure interdependency modeling: a survey of U.S. and International Research. Idaho National Laboratory, Idaho Falls, Idaho

Pelletier FJ, Elio R (1997) What should default reasoning be by default? Comput Intell 17:165–187

Pena D, Tschernykh A, Nesmachnow S, Massobrio R, Feoktistov A, Bychkov I, Radchenko G, Drozdov AY, Garichev SN (2019) Operating cost and quality of service optimization for multi-vehicle-type timetabling for urban bus systems. J Parallel Distrib Comput 133:272–285. https://doi.org/10.1016/j.jpdc.2018.001.009

Petrovskiy AB (2002) Main notions of the multisets theory. Editorial URSS, Moscow. (In Russian)

Petrovskiy AB (2003) Spaces of Sets and Multisets. Editorial URSS, Moscow. (In Russian)

Petrovskiy AB (2018) Theory of measured sets and multisets. Nauka, Moscow. (In Russian)

Piccinini G, Bahar S (2012) Neural computation and the computational theory of cognition. Cogn Sci A Multidiscip J 37(3). https://doi.org/10.1111/cogs.12012

Pospelov GS, Irikov VA (1976) Program-objective planning and control. The Soviet Radio, Moscow. (In Russian)

Queiroz J, Merrell F (2006) Semiosis and pragmatism: toward a dynamic concept of meaning. Sign Syst Stud 34(1):37–64

Raizberg BA, Lobko AG (2002) Program-objective planning and control. INFRA-M, Moscow. (In Russian)

Rausch P, Sheta A, Ayesh A (eds) (2013) Business intelligence and performance management: theory, systems, and industrial applications. Springer, London

Rehak D, Senovsky P, Hromada M, Lovecek T (2019) Complex approach to assessing resilience of criterial infrastructure elements. Int J Crit Infrastruct Prot 25:125–138. https://doi.org/10.1016/i.ijcip.2019.03.003

Red'ko VN, Bui DB, Grishko YA (2015) Modern state of multisets theory from the entity point of view. Cybern Syst Anal 51:171–178

Rehak D, Senovsky P, Hromada M, Lovecek T, Novotny P (2018) Cascading impact assessment in a critical infrastructure system. Int J Crit Infrastruct Prot 22:125–138. https://doi.org/10.1016/j.ijcip.2018.06.004

Reisig W (1991) Petri nets and algebraic specifications. Theor Comput Sci 80(1):1–34. https://doi.org/10.1016/0304-3975(91)90203-e

Reinfrank M, de Kleer J, Ginsberg ML, Sandewall E (eds) (1989) Non-monotonic reasoning. Second international workshop Grassau, FRG, June 13–15, proceedings 1988, Lecture notes in artificial intelligence, vol 346. Springer, Berlin

Reyers B, Selig ER (2020) Global targets that reveal the social–ecological interdependencies of sustainable development. Nat Ecol Evol 4(8):1011–1019. https://doi.org/10.1038/s41559-020-1230-6

Rinaldi S, Peerenboom JP, Kelly TK (2001) Identifying, understanding, and analyzing critical infrastructure interdependencies. IEEE Control Syst Mag 21:11–25. https://doi.org/10.1109/37.960131

Roberts FS (2009) Applied combinatorics. Chapman & Hall/CRC, London

Roberts F (2016) What is big data and how has it changed. Invited talk at international conference on data intensive systems analysis for geohazard studies. Sochi, Russia, July 18–21, 2016

Robinson M (ed) (2007) Performance budgeting: linking funding and results. Palgrave Macmillan, London

Rodrik D (2014) Green industrial policy. Oxf Rev Econ Pol 30(3):470–471. https://doi.org/10.1093/oxrep/gru025

Rønnow TF, Wang Z, Job J, Boixo S, Isakov SV, Wecker D, Martinis JM, Lidar DA, Troyer M (2014) Defining and detecting quantum speedup. arXiv:1401.2910

Rzevski G, Skobelev P (2014) Managing complexity. WIT Press, London

Saaksvuori A (2008) Product lifecycle management. Springer, Berlin

Sainter P, Oldham K, Larkin A, Murton A, Brimble R (2000) Product knowledge management within knowledge-based engineering systems. In: Proceedings of ASME 2000 design engineering technical conference. ASME, Baltimore, pp 1–8

Salamon T (2011) Design of agent-based models. Bruckner Publishing, Repin-Zivolin

Sandomirsky F (2014) Repeated games of incomplete information with large sets of states. Int J Game Theory 43(4):767–789. https://doi.org/10.1007/s00182-013-0404-8

Sharkey TC, Pinkley SGN (2019) Quantitative models for infrastructure restoration after extreme events: network optimization meets scheduling. In: Kaper HG, Roberts FS (eds) Mathematics of planet earth: protecting our planet, learning from the past, safeguarding for the future. Springer, Cham, pp 313–336

Shchukin P (2021) Russia eradicates fresh food worth 1.7 Bln. RUR. Lenta.ru, 21 July, 2021. newsland.com/7446080

Schatten M, Tomicic I, Duric BO (2015) Multi-agent modeling methods for massively multi-player on-line role-playing games. In: 2015 38th international convention on information and communication technology, electronics and microelectronics (MIPRO)

Scheffer M (2009) Critical transitions in nature and society. Wiley, Hoboken

Scheffer M, Bascompte J, Brock WA, Brovkin V, Carenter SR, Dakos V, Held V, van Nes EH, Rietkerk M, Sugihara G (2009) Early warning signals for critical transitions. Nature 461(3): 53–59. https://doi.org/10.1038/nature08227

Schmidt G (2010) Relational mathematics. Encyclopedia of mathematics and its applications, vol 132. Cambridge University Press, Cambridge. https://doi.org/10.1017/CBO9780511778810

Schurz G (2005) Non-monotonic reasoning from an evolutionary viewpoint: ontic, logical and cognitive foundations. Synthese 146:37–51

Schwab K (2015) The fourth industrial revolution. What it means and how to respond -Foreign Affairs December 12, available from: https://foreignaffairs.com/articles/2015-12-12/fourth-industrial-revolution

Sheremet IA (1994) Intelligent software environments for computerized information processing systems. Nauka, Moscow. (In Russian)

Sheremet IA (2010) Recursive multisets and their applications. Nauka, Moscow. (In Russian)

Sheremet IA (2011a) Recursive multisets and their applications. NG Verlag, Berlin

Sheremet IA (2011b) Word equations on context-free languages. EANS, Hannover

Sheremet IA (2012) Grammatical codings. EANS, Hannover

Sheremet IA (2013) Augmented post systems: the mathematical framework for knowledge and data engineering in network-centric environment. EANS, Berlin

Sheremet IA, Karasev RS (2013) On the approach to optimization of algorithmics of recursive multisets. Inform Technol Design Manuf 3:3–10. (In Russian)

Sheremet IA (2016a) Multiset approach to the estimation of consequences of natural disaster impacts on industrial systems. Geoinformatics Res Papers 4:BS4002. https://doi.org/10.2205/2016BS01Sochi

Sheremet IA (2016b) Data and knowledge bases with incomplete information in a "Set of Strings" framework. Int J Eng Appl Sci 3:8–21

Sheremet I, Zhukov I (2016) Optimizing multiset metagrammars. Formal Definitions Int J Comput Eng Inform Technol 8(7):106–114

Sheremet IA (2017) Augmented post systems: string-operating knowledge representation for big data and internet of things applications. Geoinformatics Res Papers 5:1

Sheremet IA (2018) Multiset analysis of consequences of natural disasters impacts on large-scale industrial systems. Data Sci J 17(4):1–17. https://doi.org/10.5334/dsj-2018-004

Sheremet IA, Karasev RS (2018) A pilot software package for multiset optimization, and its application to program-objective planning. Inform Technol Design Manuf 2:3–12. (In Russian)

Sheremet I (2019a) Multiset-based knowledge representation for the assessment and optimization of large-scale sociotechnical systems. In: Vizureanu P (ed) Enhanced expert systems. IntechOpen, London. https://doi.org/10.5772/intechopen.81698. https://www.intechopen.com/books/enhanced-expert-systems/multiset-based-knowledge-representation-for-the-assessment-and-optimization-of-large-scale-sociotech

Sheremet I (2019b) Unitary multiset grammars and metagrammars algorithmics and applications. In: Vizureanu P (ed) Enhanced expert systems. IntechOpen, London. https://doi.org/10.5772/intechopen.82713. https://www.intechopen.com/books/enhanced-expert-systems/unitary-multiset-grammars-an-metagrammars-algorithmics-and-application

Sheremet I (2019c) Multiset-based assessment of resilience of sociotechnological systems to natural hazards. In: Tiefenbacher J (ed) Natural hazards—risks, exposure, response, and resilience. IntechOpen, London. https://doi.org/10.5772/intechopen.83508. https://www.intechopen.com/online-first/multiset-based-assessment-of-resilience-of-sociotechnological-systems-to-natural-hazards

Sheremet I (2020a) "Set of Strings" framework for big data modeling. In: Sud K, Erdogmus P, Kadry S (eds) Introduction to data science and machine learning. IntechOpen, London. https://doi.org/10.5772/intechopen.85602. https://www.intechopen.com/books/introduction-to-data-science-and-machine-learning/-set-of-strings-framework-for-big-data-modeling

Sheremet I (2020b) Augmented post systems: syntax, semantics, and applications. In: Sud K, Erdogmus P, Kadry S (eds) Introduction to data science and machine learning. IntechOpen, London. https://doi.org/10.5772/intechopen.86207. https://www.intechopen.com/online-first/augmented-post-systems-syntax-semantics-and-applications

Sheremet I (2020c) Resource-based games. In: Bychkov I, Tchernykh A, Feoktistov A (eds) ICCS-DE 2020. Proceedings of second international workshop on information, computation, and control systems for distributed environments. Irkutsk, Russia, July 6–7, pp 234–251. https://doi.org/10.47350/ICCS-DE.2020.22. http://CEUR-WS.org/Vol-2638

Sheremet I (2020d) Multi-agent implementation of filtering multiset grammars. In: Sarfraz M, Abdul Karim SA (eds) Computational optimization techniques and appplications. IntechOpen, London. https://doi.org/10.5772/intechopen.93303. https://www.intechopen.com/online-first/multi-agent-implementation-of-filtering-multiset-grammars

Sheremet I (2020e) Multigrammatical approach to the assessment of sustainability of intelligent transport systems. In: 2020 international conference on information technology and nanotechnology (ITNT), Samara, Russia, pp 1–6. https://doi.org/10.1109/ITNT49337.2020.9253334. https://ieeexplore.ieee.org/document/9253334

Sheremet I (2021a) Application of the multigrammatical framework to the assessment of resilience and recoverability of large-scale industrial systems. In: Roberts FS, Sheremet IA (eds) Resilience in the digital age, Lecture notes in computer science, vol 12660. Springer, Berlin, pp 16–34. https://doi.org/10.1007/978-3-030-70370-7_2

Sheremet I (2021b) Multiset-based assessment of vulnerability of energy infrastructures to destructive impacts. In: Proceedings of the third international workshop on information, computation, and control systems for distributed environments, vol 2913. CEUR-WS Proceedings, pp 139–163. https://doi.org/10.47350/ICCS-DE.2021.12. http://ceur-ws.org/Vol-2913/paper12.pdf

Siekmann JH (ed) (2014) Handbook of the history of logic, Computational logic, vol 9. Elsevier, Amstedam

Simovici DA, Djeraba C (2008) Mathematical tools for data mining: set theory, partial orders, combinatorics. Springer Science & Business Media, London

Singh D, Ibrahim AM, Yohanna T, Singh JN (2007) An overview of applications of multisets. Novi Sad J Math 37:37–92

Shoham Y, Leyton-Brown K (2009) Multiagent systems. Algorithmic, game-theoretic, and logical foundations. Cambridge University Press, Cambridge

Shokin YI (1996) On interval problems, interval algorithms and their complexity. Comput Technol 1(1):3. (In Russian)

Stergiopoulos G, Kotzanikolaou P, Theocharidou M, Lykou G, Gritzalis D (2016) Time-based critical infrastructure dependency analysis for large-scale and cross-sectoral failures. Int J Crit Infrastruct Prot 12:46–60. https://doi.org/10.1016/j.ijcip.2015.12.002

Stewart W (1994) Introduction to the numerical solutions of Markov chains. Princeton University Press, Princeton

Sycara KP (1998) Multiagent Systems. AI Mag 19(2):79–92

Sustainable Development (2019) Sustainable development in the European Union Monitoring report on progress towards the SDGs in an EU context. Eurostat. https://doi.org/10.2785/44964

Taha HA (2016) Operations research: an introduction. Pearson, London

Tilly S, Rosenblatt HJ (2016) System analysis and design. Cengage Learning, Boston. Ebook-dl.com

Tosatti E (2008) Optimization using quantum mechanics: quantum annealing through adiabatic evolution. J Phys A Math Theor 41:209801. https://doi.org/10.1088/1751-8113/41/20/209801

Trojet M, H'Mida F, Lopez P (2011) Project scheduling under resource constraints: application of the cumulative global constraint in a decision support framework. Comput Ind Eng 61:357–363

Vaish G (2013) Getting started with NoSQL. Packt Publishing, Birmingham

Vandezande N (2018) Virtual currencies: a legal framework. Intersentia, Antwerp

van der Hoog S (2018) Deep learning in (and of) agent-based models: a prospectus. ArXiv. 1706.06302

Vernadsky V (1998) The biosphere: complete. Annotated edition. Copernicus, New York

Vespignani A (2010) Complex networks: the fragility of interdependency. Nature 464:984–985. https://doi.org/10.1038/464984a

Wallace M (2002) Constraint logic programming. In: Computational logic: logic programming and beyond, Lecture notes in computer science, vol 2407. Springer, New York, pp 512–556

Waldrop MM (2018) Free agents: monumentally complex models are gaming out disaster scenarios with millions of simulated people. Science 360(6385):144–147

Wang L, Wang G (2016) Big data in CyberPhysical systems, digital manufacturing and industry 4.0. Int J Eng Manuf 2016(4):1–8. https://doi.org/10.5815/ijem.2016.04.01

Wang J, Zuo W, Rhode-Barbagidos L, Lu X, Wang J, Lin Y (2019) Literature review on modeling and simulation of energy infrastructures from a resilience perspective. Reliab Eng Syst Saf 183: 360–373

Waston J (2013) Strategy: an introduction to game theory. W.W. Norton and Company, New York

Werner S (2017) International review of district heating and cooling. Energy 137:617–631. https://doi.org/10.1016/j.energy.2017.04.045

Whitten JL, Bentley LD (2006) Introduction to systems analysis and design. McGraw-Hill/Irwin, New York

Wilensky U, Rand W (2015) An introduction to agent-based modeling: modeling natural, social, and engineered complex systems with NetLogo. MIT Press, Cambridge

Wilkinson MD et al (2016) The FAIR guiding principles for scientific data management and stewardship. Scientific Data 3:160018. https://doi.org/10.1038/sdata.2016.18

Wooldridge M (2009) An introduction to multi-agent systems. Wiley, Chichester

Yeoh W, Yokoo M (2012) Distributed problem solving. AI Mag 33(3):53–65

Zheng Z, Xie S, Dai H-N, Chen X, Wang H (2018) Blockchain challenges and opportunities: a survey. Int J Web Grid Serv 14(4):352–376. https://doi.org/10.1504/IJWGS.2018.095647

Index

I. A. Sheremet, *Multigrammatical Framework for Knowledge-Based Digital Economy*, https://doi.org/10.1007/978-3-031-13858-4

Printed in the United States
by Baker & Taylor Publisher Services